Elementary Particle Physics

This modern introduction to particle physics equips students with the skills needed to develop a deep and intuitive understanding of the physical theory underpinning contemporary experimental results.

The fundamental tools of particle physics are introduced and accompanied by historical profiles charting the development of the field. Theory and experiment are closely linked, with descriptions of experimental techniques used at CERN accompanied by detail on the physics of the Large Hadron Collider and the strong and weak forces that dominate proton collisions. Recent experimental results are featured, including the discovery of the Higgs boson. Equations are supported by physical interpretations, and end-of-chapter problems are based on data sets from a range of particle physics experiments including dark matter, neutrino, and collider experiments. A solutions manual for instructors is available online. Additional features include worked examples throughout, a detailed glossary of key terms, appendices covering essential background material, and extensive references and further reading to aid self-study, making this an invaluable resource for advanced undergraduates in physics.

Andrew J. Larkoski is a Visiting Assistant Professor at Reed College. He earned his Ph.D. from Stanford University working at SLAC National Accelerator Laboratory and has held postdoctoral research appointments at Massachusetts Institute of Technology and Harvard University. Andrew is a leading expert on the theory of quantum chromodynamics (QCD) and has won the LHC Theory Initiative Fellowship and the Wu-Ki Tung Award for Early-Career Research on QCD.

Elementary Particle Physics

An Intuitive Introduction

ANDREW J. LARKOSKI

Reed College

Shaftesbury Road, Cambridge CB2 8EA, United Kingdom

One Liberty Plaza, 20th Floor, New York, NY 10006, USA

477 Williamstown Road, Port Melbourne, VIC 3207, Australia

314–321, 3rd Floor, Plot 3, Splendor Forum, Jasola District Centre, New Delhi – 110025, India

103 Penang Road, #05–06/07, Visioncrest Commercial, Singapore 238467

Cambridge University Press is part of Cambridge University Press & Assessment,
a department of the University of Cambridge.

We share the University's mission to contribute to society through the pursuit of
education, learning and research at the highest international levels of excellence.

www.cambridge.org
Information on this title: www.cambridge.org/9781108496988
DOI: 10.1017/9781108633758

First published 2019 (version 2, August 2022)

Printed in the United Kingdom by TJ Books Ltd, Padstow Cornwall

A catalogue record for this publication is available from the British Library.

Library of Congress Cataloging-in-Publication Data
Names: Larkoski, Andrew J., 1985– author.
Title: Elementary particle physics : an intuitive introduction /
Andrew J. Larkoski.
Description: Cambridge, United Kingdom ; New York, NY : Cambridge University
Press, 2019. | Includes bibliographical references and index.
Identifiers: LCCN 2019003347 | ISBN 9781108496988 (hardback ; alk. paper) |
ISBN 1108496989 (hardback ; alk. paper)
Subjects: LCSH: Particles (Nuclear physics)
Classification: LCC QC793.2 .L37 2019 | DDC 539.7/2–dc23
LC record available at https://lccn.loc.gov/2019003347

ISBN 978-1-108-49698-8 Hardback

Additional resources for this publication at www.cambridge.org/larkoski

Cambridge University Press & Assessment has no responsibility for the persistence
or accuracy of URLs for external or third-party internet websites referred to in this
publication and does not guarantee that any content on such websites is, or will
remain, accurate or appropriate.

For Patricia and Henry

Contents

Preface

Not only God knows, I know, and by the end of the semester, you will know.

Sidney Coleman

Particle physics is a subject that strikes both awe and fear into students of physics. Awe because particle physics is extremely far-reaching: its realm ranges from the inner workings of atoms to the mechanisms for fusion in the center of stars to the earliest moments of the universe. From a small number of fundamental principles, all of these phenomena can be consistently described and understood. On the other hand, fear because particle physics is notorious for being a mathematically dense and abstract topic, and one for which its experimental validation is often reduced to interpreting obscure plots. Fancy mathematics can be mistaken for physical rigor, and a mathematics-heavy approach to particle physics often hides a much simpler structure.

Textbooks on particle physics for undergraduates are often organized historically, which can add to confusion. Throughout the twentieth century, more and more was learned about the subatomic world, but the way it progressed was never linear. For example, hundreds of particles that we now call hadrons were discovered in the mid-twentieth century, with no clear organizing principle at the time. It wasn't until the development of the strong force, quantum chromodynamics (QCD), in the 1970s that an explanation of all of these hadrons as combinations of only five fundamental particles, the quarks, was firmly established. Only after introducing this zoo of particles would a textbook that proceeds historically identify the simple principles underlying this structure. The **why** should take precedence over **how** physical phenomena manifest themselves, and a book that builds from the ground up can't proceed historically.

Additionally, particle physics is very much an active field of physics with new data and discoveries. A modern book on particle physics needs to include discussions of recent results, the most prominent of which is the discovery of the Higgs boson at the Large Hadron Collider (LHC) in 2012. However, to describe and motivate why the discovery of the Higgs was so important requires significant background, covering topics ranging from electroweak symmetry breaking to the dynamics of proton scattering, quantum loops in Feynman diagrams, particle detector experiments, and statistical analyses, among others. Therefore, in some sense, there simply isn't space in a modern particle physics textbook to describe every major result since the 1920s. By the end of the course a student should be able to understand almost any plot produced by the experiments at the LHC.

This book was born out of the particle physics class at Reed College, which I taught during the spring semester of 2017. The twin goals of this textbook are to be up-to-date and to build concepts from the ground up, based firmly on physical intuition. This book provides an intuitive explanation for the physics being introduced. This is necessarily an ahistorical approach, which has consequences for how topics are introduced and motivated as compared to other textbooks. With a modern viewpoint, we can identify past results and predictions that had an outsized impact on the field as a whole. For example, interpreting results at the LHC requires use of proton collision simulations, referred to as parton shower programs. The physical basis for the parton shower is the DGLAP splitting functions, which were developed in the 1970s as a consequence of QCD. Thus it is vital for interpretation of results from the LHC to understand and appreciate the DGLAP splitting functions. Chapter 9 covers this topic.

A potential drawback of this approach is that it is not encyclopedic. Any undergraduate textbook on particle physics suffers from this, however. A full mathematical treatment of particle physics requires quantum field theory, which is (at least) a year-long graduate-level course. So, there will be some things for which the motivation is less than ideal. The most prominent of these is the construction and calculation of Feynman diagrams, which are motivated in this book in analogy to circuit diagrams, but their mathematical justification lies well beyond such a course. Similarly, to understand all of the intricacies of experimental measurements requires years of actually working on the experiments. Only then can you understand where the systematic uncertainties come from, the limitations of your detector, and all of the blood, sweat, and tears that went into a measurement, which is sometimes just a single number.

Overview of This Book

This textbook is organized into three broad themes:

- The Tools of Particle Physics
- The Strong Force
- The Weak Force.

I don't claim to have invented this organization; at least two other modern particle physics textbooks use a similar organizational scheme. However, I do think that this is the correct approach for such a course. Of the four fundamental forces, three are relevant for particle physics (strong, weak, electromagnetism), and of those three, the strong and weak forces have no long-distance classical counterpart. So, it is natural, then, to focus on them, especially because their phenomena dominate the description of the physics probed at the LHC.

The Tools of Particle Physics

The first five chapters cover the tools of particle physics and are material that I think is required in such a course. Chapter 1 sets the stage, introducing the Standard Model of Particle Physics and the LHC to frame the content of the rest of the book. Additionally, just like at the beginning of an introductory physics course, appropriate units to describe particle physics phenomena are introduced. Chapter 2 is a review of special relativity and relativistic wave equations from a Lagrangian viewpoint. The Lorentz invariance of the wave equations is verified mathematically, as well as understood physically, by demonstrating that total angular momentum of the Klein–Gordon and Dirac Lagrangians is 0. Perhaps the fundamental guiding principle of particle physics is Noether's theorem, which provides the connection of group symmetries and conservation laws. Chapter 3 introduces groups and their importance in particle physics, starting from identification of the symmetries of an equilateral triangle. This chapter also motivates Hermitian operators in quantum mechanics from probability conservation and the way this framework enables a concrete definition of what a "particle" is. Fermi's Golden Rule and Feynman diagrams are introduced in Chapter 4. While this chapter will provide enough detail for students to perform calculations of Feynman diagrams and construct cross sections, my goal here is to de-emphasize Feynman diagrams somewhat, as compared to some other textbooks. Feynman diagrams are particle physics, but particle physics is much more than just Feynman diagrams. Chapter 5 introduces the LHC and its two largest experiments, ATLAS and CMS. Detailed discussions of proton acceleration, proton collision, detector components, and statistics are provided to present students with the tools to understand experimental results. Similar topics are not often covered in other books.

The Strong Force

Chapters 6 through 9 cover the phenomena of the strong force, QCD. Chapter 6 is the introduction to QCD, where electron–positron collisions are studied in detail. This chapter begins with a detailed study of $e^+e^- \to \mu^+\mu^-$ scattering within quantum electrodynamics (QED). This provides a framework for discussion of the importance of inclusive cross sections and evidence for both the three colors of QCD as well as the spin-1/2 nature of quarks. With evidence for quarks established, Chapter 7 introduces partons and Bjorken scaling as evidence for point-like, nearly free constituents of the proton. A detailed interpretation of Bjorken scaling is provided by Fourier transforming to position space, where its consequences become clear. This chapter also discusses evidence for the gluon from three-jet events, in analogy to photon emission in QED. These pieces then set the stage for Chapter 8 in which the three colors of QCD, the spin-1 gluon, and the spin-1/2 quarks are put together in a consistent theoretical framework. Physical arguments are provided to augment the geometrical construction of QCD and non-Abelian gauge theories in general. This chapter also surveys some of the more non-trivial consequences of QCD, of which the most profound is the property of asymptotic freedom. The discussion of QCD ends in Chapter 9 with its most shocking prediction: the formation of high-energy, collimated streams of particles called jets. The prediction of jets is guided by the observation that QCD at high energies is approximately scale invariant, which has consequences for parton evolution manifested in the DGLAP equations. Very uniquely, this chapter also has a simple, explicit, all-orders prediction of jet structure for an observable in electron–positron collisions called thrust.

The Weak Force

The final third of the book, Chapters 10 through 14, is devoted to the weak force. Chapter 10 invites the reader to study this force with the observation of parity violation in nuclear decays, from detailed discussions of the Wu experiment and its motivation. The $V - A$ theory is introduced as a phenomenological model of parity violation and the decay rate of the muon is calculated. Numerous idiosyncrasies of this parity-violating interaction are mentioned in Chapter 11. These motivate spontaneous symmetry breaking and the Higgs mechanism, which is introduced by analogy with similar situations in quantum mechanics. By connecting electromagnetism with charged and neutral currents observed in electron–positron scattering, we are able to construct the electroweak theory and its pattern of symmetry breaking. Consequences of the weak force for properties of the fermions of the Standard Model is the topic of Chapter 12. The mechanism of flavor mixing and CP violation in the quark sector is provided in detail and motivated by non-commutation of mass and flavor operators. Neutrino oscillation is also introduced, but no apology is made for imprecision of the calculation. When and why neutrinos oscillate was only relatively recently clearly elucidated and involves ideas of entanglement, interference, and decoherence. Chapter 13 is the culmination of the book with the discovery of the Higgs boson. This is also one of the few places in this book where a historical organization is presented, with the method of discovery of the Higgs motivated from searches at the Large Electron–Positron Collider (LEP), to early searches at Tevatron and LHC, and finally to its discovery in 2012. A review of the current established properties of the Higgs closes the chapter. As with any book on particle physics, the final chapter, Chapter 14, looks forward to the open questions and where the field will go in the future.

Key Features

Worked Examples and Supplementary Appendices

Along with the intuitive discussion of topics, each chapter contains worked examples focused around understanding a relevant measurement. There is really nothing as satisfying in physics as seeing a prediction which started from some very simple assumptions validated by concrete data. The goal of the worked examples is both to show the student the application of these ideas and to share the excitement of working in the field of particle physics, where experiment and theory are so closely connected. Additionally, appendices provide background or summary information as a quick and easy reference for students. The appendices cover a background of quantum mechanics (likely from a perspective students haven't seen), details about Dirac δ-functions, Fourier transforms, a collection of results from the main body of the text, and a bibliography of suggested reading for delving further. Key particle physics terms are emphasized in bold throughout the text, and a glossary of a substantial number of the terms is also provided, as significant jargon is used in particle physics.

Exercises on Recent Results

The exercises at the end of each chapter cover a broad range of applications to test the student's understanding of the topic of the particular chapter. Most of the exercises are relatively standard calculations for the student, but two or three of the exercises are in much more depth and involve studying data from experiment in the context of the material of the chapter. This broadens and deepens the topics covered in the worked examples, and exposes students to relevant experiments and results that couldn't be covered within the main text. Examples of these exercises include analyzing dark matter mass and interaction rate bounds, estimating the mass of the top quark from its decay products, studying LHC event displays, a simple extraction of quark parton distribution functions from data for the Z boson rapidity, lowest-order predictions for jet masses at the LHC, predicting neutrino scattering rates in the IceCube experiment, validating the left-handed nature of the top quark decay, and estimating backgrounds in searches for the Higgs boson decay. Additionally, the final exercise in each chapter is the statement of an open problem in particle physics, intended to expose the student to some of the big questions of the field.

Historical Profiles

It is sometimes easy to forget that physics is a human endeavor done by people. I have included historical profiles throughout the text to provide context and a bit of humanity to the topic. I have highlighted scientists who contributed significantly to the subject at hand, but have attempted to focus on those people who haven't been overly deified (e.g., not Fermi or Feynman). Historical profiles include mini-biographies of Emmy Noether (p. 16), Paul Dirac (p. 34), Fabiola Gianotti (p. 127), Mary Gaillard and Sau Lan Wu (p. 188), Gerardus 't Hooft (p. 235), Guido Altarelli (p. 258), Chien-Shiung Wu (p. 291), Helen Quinn (p. 367), and Benjamin Lee (p. 407). A few "legendary" particle physics stories are presented, including the etymology of the barn unit of cross section (p. 81), the origin of penguin diagrams (p. 337), and the Higgs boson discovery announcement (p. 417).

Extensive In-Text Referencing

I have also worked to provide extensive (and where possible, exhaustive) references to the original literature for every topic covered in this book. References are provided as footnotes, so that one can immediately identify the paper without flipping back and forth to the end of the chapter or end of the book. I have also collected all in-text references in the bibliography for ease of searching. The only way that the referencing could be as thorough as it is is through innumerable searches of my own on InSpire (http://inspirehep.net) and arXiv (http://arxiv.org). InSpire is an online database of essentially every publication relevant to particle physics in history. The reference format used in this book is that provided by InSpire, which is ubiquitous in technical papers on particle physics. arXiv is the preprint archive for particle physics (and now many more fields), where scientists post their completed papers before journal publication. It enables the rapid transmission of ideas, and every paper on particle physics written in the past 25 years is available there for free.

How to Use this Book

My class at Reed College had about 24 students in it with roughly an equal mix of juniors and seniors (third- and fourth-year undergraduates). This was an interesting challenge for a subject like particle physics: the seniors had completed classes on electromagnetism and quantum mechanics, while the juniors were taking quantum mechanics concurrently. This required a shift in the presentation of the material focusing on analogies and physical intuition, hence the motivation for this book. The level of the course seemed to strike a happy medium in which both sets of students were satisfied with the level of the lectures. That said, I do feel that a course on particle physics requires students to have completed at least a first semester of electromagnetism and a sophomore-level (second-year) modern physics course. Not having previous exposure to quantum mechanics or classical field theory severely restricts the breadth of topics that can be covered.

In the semester-long course of 26 80-minute lectures, I succeeded in introducing most of the topics covered in this book. However, that isn't to say that much time was spent on them. For example, the treatment of neutrino oscillations in class consisted of a single lecture, and most of that time was used to perform the standard two-state interference calculation. To completely and honestly motivate the reason for neutrino oscillation would require at least one more lecture on the topic, which may not be possible depending on time constraints and interests of the instructor. Nevertheless, I do think that topics covered in every chapter of this book could fill a course, regardless of the time available.

That said, some topics are more important than others. As mentioned earlier, I see the first five chapters of this book as required. Units, special relativity, group theory, Feynman diagrams, and experimental techniques are fundamental to being able to speak the language of particle physics. A substantial number of experimental measurements are provided in these first chapters so that, even if the course does not cover much more, students would see modern results. For a course with limited time, a number of topics in the strong and weak force sections could be skipped. For the strong force, I view Chapter 6, Chapter 7 through the beginning of Section 7.3, and the consequences of QCD discussed in Section 8.3 as required. If there's a bit more time in the course, then covering one of the parton evolution or jets topics in Chapter 9 would add significant content. For the weak force, I view Chapter 10, the first half of Chapter 11, and Section 13.2 as required. With a bit more time, a course could cover one of the topics of Chapter 12 (quark mixing or neutrino oscillations), or add more details about the Higgs boson discovery in Chapter 13.

Finally, I have attempted to keep the prose light and the enthusiasm high because, after all, this is physics and it should be fun. I hope students can enjoy reading this book and gain an appreciation for this beautiful subject.

Acknowledgements

First, I want to thank the 2017 Physics 366 class at Reed College for their enthusiasm for particle physics and especially their feedback on the original exercises for that course.

Much has improved in going from hand-written lectures to typeset chapters, and these students were the original motivation.

The existence of this textbook and the exposition contained in it are due to a large number of people who have influenced the way I think about particle physics. I have had the exceptional fortune to have excellent mentors and advisors throughout my career who were patient enough to answer my questions and provide detailed explanations. I thank Brian Batell, David Ellis, Stephen Ellis, Matthew Schwartz, Iain Stewart, Matthew Strassler, Jesse Thaler, and especially my Ph.D. advisor Michael Peskin for their guidance. As a graduate student I benefitted from amazing fellow students at Stanford with whom I had numerous illuminating discussions, reading groups, and seminar series. I wish to especially thank Camille Boucher-Veronneau, Kassahun Betre, Randall Cotta, Martin Jankowiak, Jeffrey Pennington, and Tomas Rube.

The intuitive approach of the explanations in this book is due to several influences. As an undergraduate, I worked as a teaching assistant for the Physics Education Group at the University of Washington, where physics tutorials were developed. I thank Mila Kryjevskaia, Peter Shaffer, and MacKenzie Stetzer for their guidance, helping me focus on the fundamental issues and work to a qualitative explanation of physical phenomena. In graduate school, I benefitted from excellent courses on particle physics taught by Savas Dimopoulos and Michael Peskin. While their teaching styles differed widely, I have attempted to balance Savas's heuristic approach with Michael's "shut up and calculate" approach in this book. I owe a huge debt of gratitude to Patricia Burchat, for whom I was a teaching assistant in her undergraduate courses on quantum mechanics and particle physics. Patricia's excellent courses were the first time that I taught particle physics to undergraduates and were some of the first physics classes at Stanford to employ physics by inquiry techniques. I also thank Lauren Tompkins and Paul Simeon for discussions and sharing of materials from the particle physics class that Lauren taught at Stanford in 2017.

Getting this book to completion was due to a number of people. I thank my colleague Joel Franklin at Reed College for suggesting that I turn my lecture notes into a book and publish it. The writing benefitted from my participation in the Stanford Writers' Group, and especially from comments from Priscilla Burgess. Bruce Van Buskirk at Reed College was very helpful in addressing questions that I had about copyrights. I thank Fred Olness for inviting me to lecture at the 2017 CTEQ Summer School, where I developed the explanation for the resummation of thrust presented in Chapter 9. I thank my editors at Cambridge University Press, Nicholas Gibbons and Heather Brolly, for their comments and criticisms on the manuscript, and especially to Heather for her assistance in making the final product so polished.

My education in particle physics and particularly in QCD has benefitted from outstanding long-time collaborators. I thank Christopher Frye, Simone Marzani, Gavin Salam, Matthew Schwartz, Peter Skands, Gregory Soyez, Iain Stewart, Kai Yan and especially Ian Moult, Duff Neill, and Jesse Thaler. In a field like QCD where data are rich, my collaborators taught me that precision is of the utmost importance because no one can fool Nature.

There are a number of people to thank for results reprinted throughout this book. I thank David Kaplan for permission to reprint the representation of the Standard Model in Fig. 1.1; I thank Scott Hertel for permission to reprint the LUX dark matter bound plot in Fig. 2.3; I thank Melissa Franklin for permission to reprint the CDF top quark discovery plot in Fig. 5.11; I thank Leslie Rosenberg for permission to reprint the thrust angle distribution in Fig. 6.5; I thank Dmitri Denisov for permission to reprint the Z boson rapidity spectrum from D$\emptyset$ in Fig. 7.2; I thank Raymond Frey for permission to reprint the three-jet energy fractions from SLD in Fig. 7.4; I thank Stanley Brodsky for permission to reprint an excerpt about jets from a paper in the footnote on page 263; I thank Ignacio Taboada for permission to reprint results from IceCube neutrino measurements in Figs. 10.4 and 10.5; and I thank Karsten Heeger for permission to reprint neutrino oscillation results from Daya Bay in Fig. 12.3.

I also thank several people for clarification about history or context presented in this book. I thank Joel Walker for a careful reading, finding some typos, and suggestions for improving explanation throughout the book; I thank Benjamin Nachman for suggesting the reorganization of the sections on particle tracking and calorimetry in Chapter 5 to emphasize ionization in the tracker; I thank James Bjorken for a brief history of the strong force and searches for jets before asymptotic freedom and the "November Revolution"; I thank Deepak Kar for urging me to clarify what is actually measured in studies of the underlying event in Exercise 9.7; and I thank Mary James for the relationship of the physicists mentioned in the Historical Profile in Box 11.1.

In writing and describing the physics presented in this book, I relied on a number of well-worn references. Michael Peskin and Daniel Schroeder's *An Introduction to Quantum Field Theory* and Matthew Schwartz's *Quantum Field Theory and the Standard Model* were particularly used, especially Schwartz's description of the spin group (in my Chapter 3) and Peskin and Schroeder's description of the β-function and asymptotic freedom (in my Chapter 8).

I thank my parents, Tim and Colleen, who always supported my education and have wholeheartedly attempted to understand my world of physics academia.

Finally, I thank my wife, Patricia, for detailed criticism, support, and discussions about this project. We have been together since both of us started as physicists, through graduate school and beyond, and I see this book as a culmination of that part of our lives.

It goes without saying that any successes of this book are due to these people, and any failures are all my own.

Andrew James Larkoski
Portland, Oregon

1 Introduction

Particle physics is the study of the fundamental principles of Nature. Within the purview of particle physics are some of the deepest questions we can ask, like "What is responsible for mass?" or "Why are there three spatial and one time dimensions?" These are such big questions that no individual or even individual country can hope to answer them alone. Contemporary particle physics is truly an international endeavor, with scientists from nearly every country on Earth involved in the major experiments. Today's particle physicist may regularly travel to conferences in Argentina, visit collaborators in Japan, watch a live news conference about a major discovery from Switzerland, or even collect data at the South Pole. It is also a dynamic field, with numerous new results in particle physics published every week testing those theories that we have or suggesting new ones. The liveliness and brisk rate at which ideas are transferred in this field is largely due to particle physics having one of the largest and most widely used preprint article servers in all of science. These reasons also make taking a course on particle physics attractive to many physics students.

All of the machinery, formalism, insight, and tools that you have gained as a physics student is essential for studying particle physics. This involves the whole range of advanced physics courses:

- **Classical Mechanics.** Lagrangians and Hamiltonians are the principle way in which we express a system in particle physics.
- **Special Relativity.** The particles we explore are traveling at or near the speed of light, c.
- **Quantum Mechanics.** The particles and physical systems we investigate are extremely small, so the fundamental quanta of action, $\hbar$, is necessary in our analysis.
- **Statistical Mechanics.** Particles are classified by their intrinsic spin, which defines them as fermions or bosons.
- **Electromagnetism.** Likely electromagnetism, through Maxwell's equations, is the first field theory that you encounter in physics courses.

The language of particle physics is mathematics. From complex analysis to Fourier transforms, group theory and representation theory, linear algebra, distribution theory, and statistics myriad fields of mathematics are vital to articulate the principles, theories, and data of particle physics. As we will see in this book, the physics is extremely helpful in guiding the mathematical expressions. The goal of this book is to use the intuition gained through other physics courses and apply it to particle physics, which gets us a long way toward understanding, without just blindly following the mathematics.

The particle physics introduced in this book is also the gateway to quantum field theory, the result of the harmonious marriage of quantum mechanics and special relativity.

A complete treatment of quantum field theory is beyond the scope of this book, but we will see glimpses of a richer underlying structure as the book progresses. In particular, quantum field theory is the framework in which three of the four fundamental forces of Nature are formulated. The three forces are electromagnetism, the strong force, and the weak force. The strong and weak forces are the focus of most of this book, with aspects of electromagnetism studied throughout. Quantum field theory enables a formalism which produces predictions that can be compared to data, and it is often (and rightfully!) stated that quantum field theory is the most wide-reaching and precise theory of Nature that exists.

This chapter serves as the overview that invites you to study this rich field. Our goal is to frame the rest of the book, which necessitates a review of the forces of Nature, a preview of the Standard Model of Particle Physics, and a glimpse of the Large Hadron Collider, the currently running and most superlative particle physics experiment ever. We also need to introduce natural units to describe particle physics phenomena, and we find that familiar SI units are woefully inadequate.

1.1 A Brief History of Forces

Interactions between particles can be expressed through the four fundamental forces. Gravity is the force that was first understood at some analytical level. Gravity is a universally attractive force that couples to energy and momentum. By "universally attractive" we mean that two particles are always attracted to one another through gravity. By "couples" we mean that the strength of the gravitational force is proportional to the energy of the particle. For particles with slow velocities with respect to the speed of light, the energy to which gravity couples is just the mass of the particle. The strength of the force of gravity, defined by either Newton's universal law of gravitation or general relativity, is quantified by Newton's constant, G_N. For example, in Newton's theory, the force of gravity between two masses m_1 and m_2 separated by distance $\vec{r}$ is

$$\vec{F}_g = -\frac{G_N m_1 m_2}{|\vec{r}|^2}\hat{r},\qquad(1.1)$$

where $\hat{r}$ is a unit vector in the direction of $\vec{r}$. We say that G_N is the "strength of coupling" of gravity, or "coupling constant" for short. If G_N is larger, the force is larger; if G_N is smaller, the force is smaller. In SI units, the value of G_N is

$$G_N = 6.67 \times 10^{-11}\ \mathrm{m^3\,kg^{-1}\,s^{-2}}.\qquad(1.2)$$

It turns out that, in appropriate units that we will discuss further later in this chapter, G_N is incredibly tiny. Gravitational forces are completely ignorable for any microscopic experiment involving individual particles, like electrons or protons.

The next force that was understood is electromagnetism. Unlike gravity, which is universally attractive because mass is always positive, electromagnetism can be either attractive or repulsive (or neutral). Particles or other objects can have positive, negative, or no charge and the relative sign of charges determines whether the force is attractive or repulsive. The electric force between two charges q_1 and q_2 separated by distance $\vec{r}$ is

$$\vec{F}_e = \frac{1}{4\pi\epsilon_0} \frac{q_1 q_2}{|\vec{r}|^2} \hat{r}.$$ (1.3)

Here, the factor of $(4\pi\epsilon_0)^{-1}$ is the coupling constant of electromagnetism. The value of ϵ_0 in SI units is

$$\epsilon_0 = 8.85 \times 10^{-12} \text{ F} \cdot \text{m}^{-1},$$ (1.4)

where F is the SI unit of the farad. In appropriate units to enable comparison, this is billions and billions of times larger than the coupling of gravity, G_N. Electricity and magnetism are intimately related as an electric field in one reference frame produces a magnetic field in another reference frame. This is also the starting point for special relativity, which we'll review in Chapter 2.

This was the story at the end of the nineteenth century. Knowing the mass and charge of an object is sufficient to determine how it will interact with any other object, assuming that the only forces are gravity and electromagnetism. This is also the point where this book begins, at the beginning of the twentieth century. At this time, physics was undergoing huge revolutions: in addition to the formulation of the modern pillars of relativity and quantum mechanics, the electron was recently discovered, as was the nuclear structure of the atom, and even odder things like superconductivity. A nineteenth century physicist was completely powerless to address these phenomena and understand them. They are not described strictly within the paradigm of Newtonian gravity and Maxwellian electromagnetism.

Throughout the twentieth century, more and more particles and interactions were discovered: the positron, the anti-particle of the electron; neutrinos, very light cousins to the electron that are electrically neutral and seem to pass through nearly everything; the muon, similar to the electron but more massive; and so on. Near the end of the 1960s, hundreds of new particles had been discovered and their properties (like mass, charge, and intrinsic spin) measured. It was looking like quite a mess, with no clear organizing principle. However, in the late 1960s through the late 1970s, heroic efforts from theoretical and experimental physicists around the world yielded a simple underlying framework that could explain all experimental results. It became known as the Standard Model of Particle Physics.

1.2 The Standard Model of Particle Physics

The **Standard Model** consists of all but one of the fundamental particles and forces that are important in our experiments. It provides an organizing principle for how to construct more complicated objects from these basic building blocks. A **fundamental particle** is one which we believe is truly elementary: it has no spatial extent (it is a point) and is not made up of any more fundamental parts. For example, hydrogen is not fundamental because it

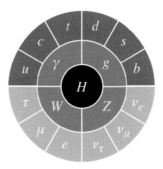

Fig. 1.1   Artistic representation of the 17 Standard Model particles. The top outer ring are the six quarks, the bottom outer ring are the six leptons, the middle ring are the four force-carrying bosons, and the center is the Higgs boson. Courtesy of Particle Fever, LLC.

consists of a proton and an electron, while it is believed that the electron is fundamental. In this book, we will study the theoretical predictions and experimental justification of the Standard Model.

The particles of the Standard Model can be artistically arranged and represented as a series of concentric rings displayed in Fig. 1.1. These 17 particles and their interactions are responsible for almost all observed phenomena. Their relationship to one another in this figure is indicative of their intrinsic properties and interactions. Gravity and its force-carrying particle, the graviton, is the one conspicuously absent force. There are four major areas of the Standard Model represented by the different regions in Fig. 1.1:

- the **quarks** (top outer ring)
- the **leptons** (bottom outer ring)
- the **force carriers** (middle ring)
- and the **Higgs boson** (center).

You have likely heard of many of these particles and their properties before. Here, we will just briefly introduce them, and we will get to know them all intimately throughout this book.

The rings in this representation are indicative of the spin of the particle. **Spin** is the intrinsic angular momentum of a particle which we will introduce in detail in Chapter 3. The quarks and leptons all have half-integer spin and so are **fermions**, while the force carriers and Higgs have integer spin and so are **bosons**. The Higgs boson H has spin 0 and was predicted in the 1960s but was only discovered in 2012 at CERN in Switzerland. Of the force carriers, one is very familiar: the photon γ is the force carrier of electromagnetism. In addition to electromagnetism, the Standard Model has two other forces: the **strong force** (called quantum chromodynamics or QCD) and the **weak force**. Unlike electromagnetism or gravity, the strong and weak forces exist only at very short distances; they have no classical mechanics counterparts. The force carrier of the strong force is called the **gluon** g and is responsible for binding atomic nuclei together. The force carriers of the weak force are the W and Z **bosons**. The weak force mediates radioactive decay of unstable elements, such as uranium.

On the top of the Standard Model rings are the six quarks: up u, down d, charm c, strange s, top t, and bottom b. They couple to all four forces and form bound states called **hadrons**. The two hadrons most relevant to everyday life are protons and neutrons, which are composed of up and down quarks. The four other quarks are only produced in high-energy collisions of particles. The leptons, the bottom of the Standard Model ring, consist of the electrically charged leptons (electron e, muon μ, and tau τ) and their electrically neutral cousins, the **neutrinos** ν. The only lepton you would encounter during your regular day is the electron, the least massive electrically charged particle of the Standard Model. You can only produce the other charged leptons in high-energy collisions, and you'd never know it, but about ten quadrillion (10^{16}) neutrinos passed right through you while you read this.

Our focus in this book will be on the strong and weak forces, as they have no counterpart in classical mechanics. Because of this, they exhibit extremely weird phenomena that will challenge our abilities to describe them theoretically. The properties of all of the particles of the Standard Model (like mass, charge, or spin) and the experimental results that measured them are collected in the **Particle Data Group's (PDG)** *Review of Particle Physics*. You can find it online at `http://pdg.lbl.gov` or you can order the book yourself from the website (it's free!).

1.3 The Large Hadron Collider

The largest scientific experiment ever is located outside of Geneva, Switzerland, accelerates protons to near the speed of light, is a Sagittarius, and loves international travel. It is the **Large Hadron Collider**, or LHC. Figure 1.2 is a bird's eye view from high over Geneva, looking to the south. To orient you, Lake Geneva is the slice coming from the left of the photo, and downtown Geneva is located at its tip. This photo is taken from the Jura Mountains, and far off in the distance you can see Mont Blanc. The ring of the LHC is denoted with the oval. This is for illustration; the ring is located 100 meters underground. Also, note the size of the ring. The Geneva airport is located just to the south and the runway is about 2 miles (3.2 kilometers) long. This LHC ring is about 18 miles (27 kilometers) in circumference. In it, two counterrotating beams of protons are accelerated to enormous energies. Each proton at the LHC has the kinetic energy of a flying mosquito, and a mosquito has about 10^{20} protons!

To study elementary particles, we use a sophisticated technique that could be called the "Neanderthal method." To look inside the accelerated protons, we smash them together, exploding them apart into a huge number of particles. Just like a detective at the scene of a car crash, a particle physicist must reconstruct the moment of proton collision using only the remnants and debris from the collision. This is indeed a tall order, but also like a detective, there are guiding principles that can be used to infer what happened. For example, energy and momentum must be conserved in a collision, and this greatly restricts how all those particles might have been produced.

Fig. 1.2 Aerial view of the region near Geneva, Switzerland, with the illustration of the Large Hadron Collider ring as the large oval. Lake Geneva is the slash from the left, downtown Geneva is located at the end of Lake Geneva, and Mt. Blanc is visible off in the distance. The main CERN site at Meyrin and the experiments along the ring are denoted. Credit: CERN © CERN.

As particle detectives, we must collect as much evidence as possible. This is accomplished with enormous detector experiments that measure nearly all the particles produced in proton collisions. Figure 1.3 shows a picture of one of the experiments at the LHC, called **ATLAS** (A̲ T̲oroidal L̲HC A̲pparatu̲S). This is a photo of ATLAS before its construction was completed. We'll discuss the particular parts of a particle physics experiment in Chapter 5, but this figure should illustrate the sizes involved. ATLAS, and its sister experiment **CMS** (C̲ompact M̲uon S̲olenoid) at the LHC, each are about the size and weight of a five-story building!

For an idea of what happens in the proton collisions, Fig. 1.4 shows an **event display**. The proton beams come in from either side of the figure, and collide in the center. All of the lines emanating from the center correspond to individual particles produced in the collision. Different parts of the detector are sensitive to different physics. For example, see the lines at the center of the figure, called **tracks**? There's a magnetic field in that region of the detector, and so charged particles bend when passing through. The charge of the particles can be identified by the direction of the bending by using the right-hand rule. The shaded background of the figure represents the different detector components. Outside of the region with the tracks is the **calorimetry**, which measures the energy of particles.

Fig. 1.3 Photo of the ATLAS detector before construction was completed. The large metal tubes contain the electromagnet that sources its namesake toroidal magnetic field. The interior region is now filled with detector electronics. Note the person for scale. Credit: CERN © CERN.

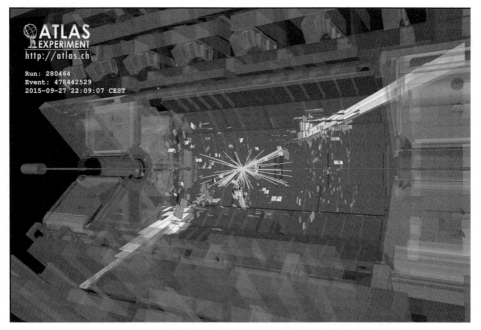

Fig. 1.4 Image of a proton collision event at the ATLAS experiment. The proton beams collide at the center of the figure, and tracks and calorimeter deposits represent particles detected by the experiment. Credit: CERN © CERN.

Deposits of particles in the calorimeters are denoted by the rectangular bars in the figure. At the top and bottom of the background image is another component called the **muon system** which is responsible for detecting muons.

1.4 Units of Particle Physics and Dimensional Analysis

The first thing we need to do in order to begin studying particle physics is to establish the appropriate system of units in which to express the outcome of experimental measurements. Good units should represent the realm in which they are being used. Because particle physics is the realm of short distances and high energies, we need to use units that naturally and usefully express quantities in this domain.

Both relativity and quantum mechanics are necessary to describe particle physics phenomena. The particles we will consider will be traveling at or near the speed of light c, and so c will appear in equations everywhere. For example, the particles we consider satisfy the relativistic energy–momentum relation,

$$E^2 = m^2 c^4 + |\vec{p}|^2 c^2 \,, \tag{1.5}$$

where E is the energy, m is the mass, and $\vec{p}$ is the momentum of the particle of interest. In SI units, c is

$$c = 3 \times 10^8 \text{ m} \cdot \text{s}^{-1} \,, \tag{1.6}$$

so every time we have to use the energy–momentum relationship, we have to lug around this huge number.

Particle physics is also the realm of quantum mechanics, the description of Nature at the shortest distances. The fundamental unit in quantum mechanics is $\hbar$, Planck's reduced constant, which quantifies units of angular momentum. It also appears in the Schrödinger equation,

$$i\hbar \frac{\partial \psi}{\partial t} = -\frac{\hbar^2}{2m} \nabla^2 \psi + V\psi \,. \tag{1.7}$$

We'll discuss how to generalize the Schrödinger equation to account for relativity in Chapter 2, but any time we want to describe quantum phenomena, we need an $\hbar$. In SI units, $\hbar$ is

$$\hbar = 1.05 \times 10^{-34} \text{ J} \cdot \text{s} \,. \tag{1.8}$$

This is a teensy-tiny number in SI units.

Additionally, the masses or other properties of individual particles are exceptionally small. In SI units, the mass of the electron is

$$m_e = 9.11 \times 10^{-31} \text{ kg} \,. \tag{1.9}$$

The mass of the proton, while much larger than the electron, is still minuscule in SI units:

$$m_p = 1.67 \times 10^{-27} \text{ kg} \,. \tag{1.10}$$

Even the most massive elementary particle, the **top quark**, has a mass in SI units of

$$m_t = 3.1 \times 10^{-25} \text{ kg} . \tag{1.11}$$

Whenever we talk about the electron traveling at relativistic speeds, we need to keep track of numbers that are spread over about 40 orders of magnitude! This is inconvenient.

There's another, philosophical, reason to abandon SI units in particle physics. We believe, perhaps with a bit of hubris, that particle physics is truly fundamental. The speed of light as measured by a distant civilization would be the same as what we have measured on Earth. However, why would they use SI units? The second is defined as a part of the day, a very Earth-centric notion, and the meter was originally one ten-millionth the distance from the North Pole to the Equator. Later, the meter was defined from a platinum-iridium alloy bar in France, which then depends on the precision to which such a bar can be machined.

For these reasons and to express the fundamental-ness of particle physics, we introduce **natural units**, or "God's units," in which we set

$$\hbar = c = 1 \quad \text{(unitless)} . \tag{1.12}$$

Correspondingly, the permittivity and permeability of free space, ϵ_0 and μ_0, are also set to 1. Note that the units of $\hbar$ and c are

$$[\hbar] = [\text{mass}][\text{length}]^2[\text{time}]^{-1} , \qquad [c] = [\text{length}][\text{time}]^{-1} . \tag{1.13}$$

Because we set $\hbar = c = 1$, this defines two relationships between the three fundamental measurement units of mass, length, and time. Therefore, natural units can be completely expressed in terms of one unique combination of mass, length, and time. We take the measurement unit of natural units to be energy, and everything in natural units can be expressed solely in terms of energy. The reason to use energy is that it is a conserved quantity, so once the energy of a system is defined, that fixes intrinsic mass, length, and time scales for that system.

In particle physics, we typically use the **electron volt** (eV) as the energy unit of choice as this is naturally (closer to) the scale at which we work. In SI units, one electron volt is

$$1 \text{ eV} = 1.6 \times 10^{-19} \text{ J} . \tag{1.14}$$

One electron volt is the energy that a particle with the fundamental unit of electric charge $e = 1.6 \times 10^{-19}$ coulombs acquires in an electric potential of 1 volt. For example, let's see how this works for the electron mass. To express the electron mass as an energy, we multiply by c^2:

$$m_e c^2 = 8.19 \times 10^{-14} \text{ J} = 5.11 \times 10^5 \text{ eV} = 511 \text{ keV (kilo-eV)} . \tag{1.15}$$

The mass of the proton is

$$m_p c^2 = 1.5 \times 10^{-10} \text{ J} = 9.38 \times 10^8 \text{ eV} = 938 \text{ MeV (mega-eV)} . \tag{1.16}$$

For comparison, the LHC collides protons that have kinetic energies of $6.5 \times 10^{12} \text{ eV} = 6.5$ TeV (tera-eV), almost 7000 times the mass of the proton.

Using natural units, we can turn distances into energies, as well. Recall the Heisenberg uncertainty principle for momentum Δp and position Δx standard deviations:

$$\Delta p \cdot \Delta x \geq \frac{\hbar}{2}. \tag{1.17}$$

This tells us how to convert to natural units for distances. A distance x has the same units as $\hbar/p$, which you might also recall as the de Broglie wavelength divided by 2π. Momentum p can be related to energy via the relativistic energy–momentum relation

$$E = pc, \tag{1.18}$$

which holds for massless particles. Then, the quantity

$$\frac{x}{\hbar c} = \frac{1}{E} \tag{1.19}$$

is a distance x expressed in natural units. Let's see how this works in an example.

Example 1.1 The Bohr radius is the average distance between an electron and the proton nucleus in the hydrogen atom. What is the Bohr radius expressed in natural units?

Solution

The Bohr radius a_0 in SI units is

$$a_0 = 5.3 \times 10^{-11} \text{ m}. \tag{1.20}$$

Converting to natural units, this is

$$\frac{a_0}{\hbar c} = 1.7 \times 10^{15} \text{ J}^{-1} = \frac{1}{3.7 \text{ keV}}. \tag{1.21}$$

Note that this corresponding energy is much larger than the magnitude of the ground state energy of hydrogen, which is 13.6 eV.

Throughout this book, we will employ natural units as they will make expressions and algebra much easier. From natural units, one can always uniquely go to any other unit system by restoring the factors of c and $\hbar$. You just have to remember what the quantity is (a length, time, or mass, for example).

Exercises

1.1   *Energy of a Mosquito.* The mass of a mosquito is approximately 2.5×10^{-6} kg. Estimate the kinetic energy of a flying mosquito and express it in eV. What is the approximate kinetic energy per **nucleon** (proton or neutron) for a mosquito? How does this compare to the energy of protons at the LHC?

1.2   *Yukawa's Theory.* In the 1930s, Hideki Yukawa predicted the existence of a new particle, now called the **pion**. It was theorized to be responsible for binding protons and neutrons together in atomic nuclei. Based on the size of an atomic nucleus,

Yukawa was able to estimate the mass of the pion.[1] Estimate the mass of the pion from the assumption that the relevant distance scale is the radius of an atomic nucleus of about 1 femtometer (10^{-15} meters). Express the mass in eV.

1.3   *Mass of the Photon.* The photon, the force carrier of electromagnetism, is predicted to be massless. We can test this by observations of electromagnetic phenomena in the universe over galactic (or extragalactic) distances. One such bound on the mass of the photon comes from measurements of the Milky Way's magnetic field.[2] Assuming that the properties of the magnetic field are exactly as predicted by Maxwell's equations, estimate an upper bound on the mass of the photon in eV and kg. How much smaller is this than the mass of the electron? The diameter of the Milky Way galaxy is approximately 100,000 light-years.

1.4   *Planck Units.* In this chapter, we discussed natural units in which we set $\hbar = c = 1$. Then, we expressed everything in terms of energies. In 1899, after the definition of h, Max Planck defined a set of units now called **Planck units**, in which $\hbar = c = G_N = 1$, where G_N is Newton's constant.[3] In these units, everything is dimensionless.

    (a) How long is 1 unit of Planck time, in seconds?

        Hint: Express the Planck time t_P as a product of powers of G_N, $\hbar$, and c as

$$t_P = G_N^{\alpha} \hbar^{\beta} c^{\gamma} , \tag{1.22}$$

    and determine the powers α, β, γ by matching the dimensions on both sides.

    (b) What is 1 unit of Planck mass, in kg and eV? How does this compare to the mass of the proton? The enormous difference between the mass of the proton (or any "normal" mass scale of the Standard Model) and the Planck mass is called the **hierarchy problem**.

    (c) How much larger is the electric force between a proton and an electron than the gravitational force?

1.5   *Expansion of the Universe.* The **cosmic microwave background**, or CMB, is remnant electromagnetic radiation from the early universe. It was discovered by Arno Penzias and Robert Wilson in the 1960s as background static that they could not explain in their radio telescope data.[4] Before an explanation of it, they thought it might have been caused by pigeon droppings in the telescope!

    (a) The energy of the photons in the CMB is observed to be almost a perfect blackbody spectrum. As such, we typically reference the characteristic energy as a temperature, which for the CMB is 2.7 K. What does this correspond to in eV?

        Hint: Recall that Boltzmann's constant is $k_B = 1.38 \times 10^{-23}$ J $\cdot$ K.

[1] H. Yukawa, "On the interaction of elementary particles I," Proc. Phys. Math. Soc. Jap. **17**, 48 (1935) [Prog. Theor. Phys. Suppl. **1**, 1].

[2] G. V. Chibisov, "Astrophysical upper limits on the photon rest mass," Sov. Phys. Usp. **19**, 624 (1976) [Usp. Fiz. Nauk **119**, 551 (1976)].

[3] M. Planck, "Über irreversible Strahlungsrorgänge," *Sitzungsberichte der Königlich Preußischen Akademie der Wissenschaften zu Berlin* **5**, 440 (1899), and Ann. Phys. **306**, no. 1, 69 (1900).

[4] A. A. Penzias and R. W. Wilson, "A measurement of excess antenna temperature at 4080 Mc/s," Astrophys. J. **142**, 419 (1965).

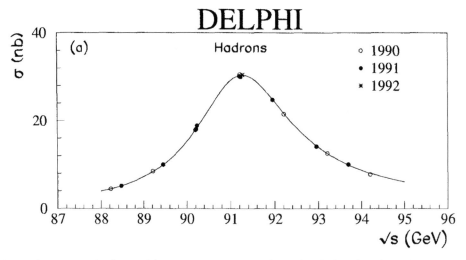

Fig. 1.5 Center-of-mass energy distribution of electron–positron scattering that produces hadrons from the DELPHI experiment. Reprinted from Nucl. Phys. B **418**, P. Abreu *et al.* [DELPHI Collaboration], "Improved measurements of cross-sections and asymmetries at the Z0 resonance," 403 (1994), with permission from Elsevier.

(b) In the early universe, electrons and protons had kinetic energies that were too large to electrically bind and so formed an opaque plasma. When the temperature of the universe was cool enough, they then formed an electrically neutral bound state, called hydrogen. This period in the history of the universe is called **recombination**. From the ground state energy of the bound state of hydrogen, estimate the temperature at which recombination occurred.

(c) At the time of recombination, the energy of photons was set by the recombination temperature. It is these photons that we now observe as the CMB. Using the results of parts (a) and (b) of this exercise, estimate the ratio of the wavelength of CMB photons that we observe now to the wavelength of photons at the time of recombination. This is called the **redshift factor**, and is corroborating evidence that the universe is expanding. The expansion of the universe stretches distances, including wavelengths of photons!

Note that this is only a very rough estimate of the redshift factor; to determine it precisely requires a much more detailed analysis including the thermodynamics of the recombination of electrons and protons into hydrogen.[5]

1.6 *Decay Width of the Z boson.* Figure 1.5 is the distribution of the probability that electrons and their anti-particle positrons interacted and produced hadrons when collided head-on with the corresponding center-of-mass energy on the horizontal axis. The dots represent the data points from the DELPHI experiment at the **Large Electron–Positron Collider** (LEP). The peak of this distribution at about 91 GeV corresponds to the Z boson. The location of the peak corresponds to the mass of the Z

---

[5] P. J. E. Peebles, "Recombination of the primeval plasma," Astrophys. J. **153**, 1 (1968); Y. B. Zeldovich, V. G. Kurt and R. A. Sunyaev, "Recombination of hydrogen in the hot model of the universe," Sov. Phys. JETP **28**, 146 (1969) [Zh. Eksp. Teor. Fiz. **55**, 278 (1968)].

boson (in natural units). The characteristic size of this peak is called the **decay width** of the Z boson, or just the width. More precisely, the decay width can be determined by the full width of the distribution at half of the maximum, measured in GeV.

(a) Estimate the decay width of the Z boson from this plot. Express your answer in GeV. Give an estimate to the tenth of a GeV.

(b) The decay width represents a fundamental quantum mechanical uncertainty on the mass energy of the Z boson. By the energy–time uncertainty principle, the decay width corresponds to a finite lifetime of the Z boson. What does this decay width correspond to, in seconds? This is called the **lifetime**, which is approximately the half-life, of the Z boson.

(c) If the decay width of a particle is very small (approaching 0), what lifetime does that correspond to? What if the decay width gets very large (approaching ∞)? A particle with a very small decay width is called **stable**.

1.7 *Decay of Strange Hadrons.* Look at Fig. 1.6 of the decay chain of particles. In this decay chain, an Ω^- particle is produced and decays to π^- and Ξ^0 particles (which subsequently decay). This is the trace of data from a **bubble chamber**, which is a supersaturated chamber of alcohol vapor. Charged particles that pass through the bubble chamber seed condensation of the vapor, and can be observed by trails of condensed alcohol. The flight path of the Ξ^0 particle is about 3 cm. Using this, estimate the lifetime of the Ω^- particle in seconds and express this time in eV to determine the decay width.

1.8 *PDG Review.* For the following questions, you'll need to use the PDG's *Review of Particle Physics*, located at `http://pdg.lbl.gov`.

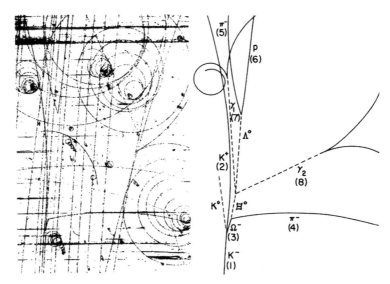

Fig. 1.6   Trace of the decay chain of the negatively charged kaon K^- hadron observed in a bubble chamber experiment at Brookhaven National Laboratory. The Ω^- hadron is observed as a decay product of the K^-. Reprinted figure with permission from V. E. Barnes *et al.*, Phys. Rev. Lett. **12**, 204 (1964). Copyright 1964 by the American Physical Society.

(a) From the "Particle Listings" section of the PDG, look up the properties of the proton. What is the lower bound of the mean lifetime of the proton? This is determined by watching a huge amount of water for a long time and not observing any protons decay. Now, assume that the proton's lifetime is just above this bound. About how much water would you need in order to observe one proton decay in one year? Express your answer in cubic meters.

   Hint: On average, the lifetime corresponds to how long an individual particle exists. How many particles would you need to ensure that one of them decayed in a year?

(b) The heaviest particles of the Standard Model are the W, Z, and Higgs bosons, and the top quark. Find the masses of these particles and determine the atomic elements that are approximately the same mass as each particle.

(c) From the PDG, what are the masses of various particles that appear in the decay represented in Fig. 1.6? What is the width of the Ω^- according to the PDG, and how does that compare to your estimate in Exercise 1.7?

1.9 *InSpire and arXiv.* Two invaluable tools for searching the particle physics literature are **InSpire** (`http://inspirehep.net`) and the preprint **arXiv** (pronounced "archive," `https://arxiv.org`). InSpire is a searchable database of essentially every paper ever written that is relevant for particle physics. The arXiv updates with new research papers posted each weekday. It's a daily task to "check the arXiv" to see what other people in the field are thinking about.

(a) Go to InSpire and search for Emmy Noether's papers. The search field in InSpire is different than a typical search engine. To search for an author, like Noether, you would enter: "find a Noether". The specification "a" denotes author. What is her most highly cited paper?

(b) You can also search by date on InSpire. Find all papers from 1967. What are the two most highly cited papers from that year? You can order by citation count using one of the drop-down menus below the search bar.

(c) Now, go to the arXiv. The arXiv is arranged by subject; originally when it started in 1991, only theoretical physics papers were hosted there. Now there is a wide range of physics and non-physics subjects represented. The most relevant subjects for this book are High Energy Physics - Phenomenology (hep-ph) and High Energy Physics - Experiment (hep-ex).

   For a few days, look through the new papers in these subjects. Are there common themes in the papers that are posted? Which experiments are posting results on arXiv? Are there any papers with titles or abstracts that seem interesting?

1.10 *Research Problem.* From Exercise 1.4, we introduced the hierarchy problem as the enormous difference between the Planck mass and, say, the mass of the proton. Why is there such a difference? Why is gravity so weak?

Special Relativity

The review of special relativity for this book will be very practical: we won't think about shining a light on a train or falling in an elevator. We will, on the other hand, emphasize the importance of symmetries in particle physics as a guiding principle. In any field of physics, **how** something works is straightforward and requires just measuring appropriate quantities. For example, how two blocks collided just requires measuring their initial and final velocities. **Why** the blocks collided in the manner that they did is much more interesting and explains the mechanism of their collision; i.e., what laws of Nature govern their collision. In this case, we know what laws govern their collision: conservation of energy and conservation of momentum. With these two simple principles, we can uniquely predict what will happen when any two blocks collide.

The connection between conservation laws and symmetries is provided by Noether's theorem. You are likely familiar with consequences of Noether's theorem. For example, to ensure conservation of angular momentum of a system, we can express every relevant quantity as vector dot products, which are invariant to rotations. We use this analogy to form relativistic vectors and dot products that enable us to construct Lorentz-invariant quantities. When analyzing a system, it is best to work with Lorentz-invariant quantities as much as possible, as they can be evaluated in any reference frame of convenience.

Additionally, we want to describe dynamical relativistic systems, i.e., systems that change in time. We construct equations of motion analogous to the Schrödinger equation of non-relativistic quantum mechanics. While there are many similarities with the Schrödinger equation, we find multiple relativistic equations that are distinguished by the spin of particle that they describe. As discussed in the Introduction, the Standard Model consists of bosons with spin 0 and 1, and fermions with spin 1/2. The spin-1 equation of motion is very familiar as it is just Maxwell's equations of electromagnetism written in an explicitly Lorentz-invariant form. The spin-0 and spin-1/2 equations of motion are new and will introduce a number of features for which we will develop a physical intuition.

2.1 Symmetries and Their Consequences

The goal of particle physics is to explain what happens at the shortest distance scales. Explanation is provided by conservation laws that highly restrict the possible outcomes of an experiment. To proceed, we then need two things: we need to identify the conservation laws of particle physics and we need to determine how those conservation laws restrict

Box 2.1

Historical Profile: Emmy Noether

Emmy Noether, in addition to her theorem, made fundamental contributions to abstract algebra and the calculus of variations. Noether faced significant obstacles as a woman mathematician, including not being paid for teaching or research for many years while at the university in Göttingen, Germany. When she was formally forbidden from teaching, she found an ally in David Hilbert, who would sign up to teach a course in which Noether would actually lecture. She was aware of how she was able to succeed nevertheless, as she wrote in a letter to fellow mathematician Helmut Hasse:[1]

My methods are really methods of working and thinking; this is why they have crept in everywhere anonymously.

It was in Göttingen that she proved her famous theorem. However, her time there was short-lived because of the rise of the National Socialist party in the 1930s. Noether, who was Jewish, was expelled from Göttingen and relocated to Bryn Mawr College in Pennsylvania. She died in 1935.

the possible physics. Determining the conservation laws can be done experimentally. We can measure quantities before and after particles interact to identify what remains the same. Determining how those conservation laws restrict physics is much more subtle and exciting. So, how do we do it?

How conservation laws restrict physics is so important it is elevated to a theorem in physics, and is possibly the most important and general result in all of theoretical physics. It is called **Noether's theorem**,[2] after Emmy Noether, a German mathematician. Using the Lagrangian formulation of classical mechanics, Noether proved that a symmetry of the Lagrangian has a corresponding conservation law. A **symmetry** is a transformation that can be performed on a system that leaves its physical description unchanged. The mathematical language of symmetries is group theory, and Chapter 3 is devoted to exploiting group theory in particle physics.

Examples of symmetries and their corresponding conservation laws from Noether's theorem include:

- energy conservation from time translation invariance
- momentum conservation from spatial translation invariance
- angular momentum from rotational invariance
- charge conservation from phase invariance of charge density.

The power of Noether's theorem is in the way it restricts the possible description of a system. Given a set of conservation laws determined empirically, we can write down the Lagrangian of the system simply by demanding that it satisfies the appropriate symmetries. We'll review Lagrangians for relativistic systems later in this chapter. Now, let's develop an understanding for how symmetries are constraining and review relativistic kinematics.

---

[1] Reprinted by permission from Springer Nature: Springer Nature *Emmy Noether 1882–1935* by A. Dick (1981).
[2] E. Noether, "Invariant variation problems," Gott. Nachr. **1918**, 235 (1918) [Transp. Theory Statist. Phys. **1**, 186 (1971)] [arXiv:physics/0503066]. This result is properly known as Noether's first theorem as, in that same paper, Noether also proves what is now known as Noether's second theorem. Both theorems are profound statements about Lagrangian mechanics.

Let's see how we can use the consequences of Noether's theorem in our formulation of special relativity and its utility for particle physics. Before we work with special relativity, let's work with the more familiar example of vectors in three dimensions. Let's say we want to describe the physics of a system in which angular momentum is conserved. As mentioned above, angular momentum conservation corresponds to invariance of the system under rotations. For example, we might be considering a particle in a central potential, like an electron orbiting a proton because of electromagnetism. Positions and velocities of the particles are represented by vectors $\vec{x}$ and $\vec{v}$, respectively. If angular momentum is conserved, then how are these vectors allowed to appear in the calculation of properties of the system? That is, what function of the vectors is consistent with rotational invariance?

2.1.1 Rotational Invariance

To understand this, let's be concrete and consider two two-dimensional vectors $\vec{a}$ and $\vec{b}$. To understand rotational invariance we need to figure out how they transform under a rotation. A **rotation** is a linear operation; that is, it can be represented as a matrix $\mathbb{M}$ acting on the vector:

$$\vec{a} \to \mathbb{M}\vec{a}, \tag{2.1}$$

where

$$\mathbb{M} = \begin{pmatrix} \cos\theta & -\sin\theta \\ \sin\theta & \cos\theta \end{pmatrix}, \tag{2.2}$$

with θ as the angle of rotation. The rotation changes the vector $\vec{a}$, and so the vector is not invariant to rotations. However, note that the rotation of the transpose of the vector is

$$\vec{a}^{\mathsf{T}} \to (\mathbb{M}\vec{a})^{\mathsf{T}} = \vec{a}^{\mathsf{T}}\mathbb{M}^{\mathsf{T}}, \tag{2.3}$$

where

$$\mathbb{M}^{\mathsf{T}} = \begin{pmatrix} \cos\theta & \sin\theta \\ -\sin\theta & \cos\theta \end{pmatrix}. \tag{2.4}$$

Then, the rotation matrix times its transpose is the identity matrix $\mathbb{I}$:

$$\mathbb{M}^{\mathsf{T}}\mathbb{M} = \begin{pmatrix} \cos\theta & \sin\theta \\ -\sin\theta & \cos\theta \end{pmatrix} \begin{pmatrix} \cos\theta & -\sin\theta \\ \sin\theta & \cos\theta \end{pmatrix} = \begin{pmatrix} 1 & 0 \\ 0 & 1 \end{pmatrix} = \mathbb{I}. \tag{2.5}$$

This then tells us that the quantity $\vec{b}^{\mathsf{T}}\vec{a}$ is invariant or unchanged under a rotation. That is, under a rotation implemented by the matrix $\mathbb{M}$, $\vec{b}^{\mathsf{T}}\vec{a}$ transforms to

$$\vec{b}^{\mathsf{T}}\vec{a} \to (\mathbb{M}b)^{\mathsf{T}}\,\mathbb{M}\vec{a} = \vec{b}^{\mathsf{T}}\mathbb{M}^{\mathsf{T}}\mathbb{M}\vec{a} = \vec{b}^{\mathsf{T}}\mathbb{I}\vec{a} = \vec{b}^{\mathsf{T}}\vec{a}. \tag{2.6}$$

Therefore, for angular momentum to be conserved, the system must be described by quantities like $\vec{b}^{\mathsf{T}}\vec{a}$. Of course, this is nothing more than the dot product:

$$\vec{b}^{\mathsf{T}}\vec{a} = \vec{b} \cdot \vec{a} = \sum_{i=1}^{2} b_i a_i, \tag{2.7}$$

where we use the summation notation over the elements of the vectors. This can be equivalently written as

$$\vec{b} \cdot \vec{a} = \sum_{i=1}^{2} b_i a_i = \sum_{i,j=1}^{2} b_i \delta_{ij} a_j . \tag{2.8}$$

Here, δ_{ij} is the Kronecker-δ, where

$$\delta_{ij} = \begin{cases} 1, & \text{if } i = j, \\ 0, & \text{otherwise.} \end{cases} \tag{2.9}$$

This may seem a bit tautological, but now consider the rotation of this expression:

$$\sum_{i,j=1}^{2} b_i \delta_{ij} a_j \rightarrow \sum_{i,j=1}^{2} b_i \mathbb{M}_{ik} \delta_{kl} \mathbb{M}_{lj} a_j . \tag{2.10}$$

There are a couple of things we did to write this. $\mathbb{M}_{ij}$ is the component of the matrix $\mathbb{M}$ in the ith row and jth column. On the right, we use the **Einstein summation notation** in which repeated indices are summed over. That is,

$$\mathbb{M}_{lj} a_j = \mathbb{M}_{l1} a_1 + \mathbb{M}_{l2} a_2 , \tag{2.11}$$

which implements the rotation on the lth component of the vector $\vec{a}$. Note that the dot product is invariant to rotations for general vectors $\vec{a}$ and $\vec{b}$ if and only if

$$\mathbb{M}_{ik} \delta_{kl} \mathbb{M}_{lj} = \delta_{ij} . \tag{2.12}$$

However, this is nothing more than the (i, j)th entry of the matrix

$$\left(\mathbb{M}^{\mathsf{T}} \mathbb{M}\right)_{ij} = \delta_{ij} , \tag{2.13}$$

by the form of the rotation matrix. Therefore, an equivalent way to define the rotation matrix is as the set of all matrices that leave the identity matrix invariant:

$$\mathbb{M} \text{ is a rotation matrix if and only if } \mathbb{M}^{\mathsf{T}} \mathbb{I} \mathbb{M} = \mathbb{I} . \tag{2.14}$$

This is what we mean by invariance under rotations: vectors are combined with the Kronecker-δ/dot product, and rotation matrices do not change this.

We'll dive into much more detail about rotation matrices and their mathematical structure as a group in Chapter 3, but note a few things at this stage. Because we haven't restricted the size of the matrices in Eq. 2.14, this actually defines a rotation matrix in any dimension D. We aren't restricted to $D = 2$, where our analysis started. The set of all matrices of dimension D that satisfy Eq. 2.14 is called the **orthogonal group in dimension D**, denoted as O(D). The result of Eq. 2.14 and generalization of it will be helpful in working in special relativity. If we want to describe a system in which relativistic energy, momentum, and angular momentum are conserved, then, just as for rotations, we want to identify vectors and take dot products appropriately.

2.1.2 Relativistic Invariance

Starting from the relativistic conservation laws, we expect that we are able to construct vectors and the notion of a dot product. The relativistic energy–momentum relationship is

$$E^2 = m^2 + |\vec{p}|^2 \qquad \text{or} \qquad m^2 = E^2 - |\vec{p}|^2 \,. \tag{2.15}$$

On the right, we have rearranged the expression so that both sides of the equation are Lorentz invariant. $E^2 - |\vec{p}|^2$ looks like a kind of dot product with a kind of vector. If we define a **four-vector**

$$p = (E, \vec{p}) = (E, p_x, p_y, p_z)\,, \tag{2.16}$$

then $p \cdot p \equiv p^2 = E^2 - |\vec{p}|^2$ is Lorentz invariant. This defines the four-vector dot product.

As with a familiar space-vector or "three-vector," we can represent the four-vector dot product as matrix multiplication. We will denote individual elements of a four-vector with Greek indices, as p_μ or p^μ, which is the μth element of p. Then, the four-vector dot product can be expressed as

$$p \cdot p \equiv p_\mu \eta^{\mu\nu} p_\nu \equiv p_\mu p^\mu = \sum_{\substack{\mu=0 \\ \nu=0}}^{3} p_\mu \eta^{\mu\nu} p_\nu = p^\mu \eta_{\mu\nu} p^\nu \,. \tag{2.17}$$

Here, we use the standard notation that $p_0 = E$ and $p_1 = p_x$, $p_2 = p_y$, $p_3 = p_z$. We also employ the Einstein summation notation, where repeated indices are summed over. The matrix $\eta^{\mu\nu}$ implements the dot product for four-vectors:

$$\eta^{\mu\nu} = \begin{pmatrix} 1 & 0 & 0 & 0 \\ 0 & -1 & 0 & 0 \\ 0 & 0 & -1 & 0 \\ 0 & 0 & 0 & -1 \end{pmatrix}^{\mu\nu} \,. \tag{2.18}$$

On this matrix expression, we have included the indices $\mu\nu$ superscript. $\eta^{\mu\nu}$ is a single entry of the matrix η, and indices on the matrix object means that we are considering just the entry $\mu\nu$, though we leave which particular entry undetermined. We can use η to change upper to lower indices as

$$p_\mu = \eta_{\mu\nu} p^\nu \,, \tag{2.19}$$

via matrix multiplication.

So what did we gain from this? If we want to preserve relativistic energy, momentum, and angular momentum, then we should express the physical system in terms of four-vector dot products, as they are Lorentz invariant. Just like rotations, **Lorentz transformations** are linear transformations and so are implemented by a matrix Λ on a four-vector as

$$p_\mu \to \Lambda_\mu{}^\nu p_\nu \,, \tag{2.20}$$

using Einstein summation notation. If we demand that the four-vector dot product of two four-vectors p and q is Lorentz invariant, we have

$$p \cdot q = p_\mu \eta^{\mu\nu} q_\nu \to p_\rho \Lambda^\rho{}_\mu \eta^{\mu\nu} \Lambda_\nu{}^\sigma q_\sigma = p_\mu \eta^{\mu\nu} q_\nu \,. \tag{2.21}$$

That is, just by relabeling indices

$$p_\mu \Lambda^\mu{}_\rho \eta^{\rho\sigma} \Lambda_\sigma{}^\nu q_\nu = p_\mu \eta^{\mu\nu} q_\nu \,, \tag{2.22}$$

for any four-vectors p and q. Therefore, a Lorentz transformation implemented by a matrix Λ leaves η invariant:

$$\Lambda^\mu{}_\rho \eta^{\rho\sigma} \Lambda_\sigma{}^\nu = \eta^{\mu\nu} \,. \tag{2.23}$$

The set of matrices Λ that satisfy this constraint are called O(3,1). Also, we will often call η the "flat-space" or Minkowski metric, or typically just the **metric**. Let's see how these Lorentz transformations work in a couple of examples.

Example 2.1 One type of Lorentz transformation is just a rotation about a fixed axis. Let's consider a rotation about the $\hat{x}$-axis by an angle θ. What is the matrix Λ that implements this rotation on a four-vector?

Solution

Let's first identify the rotation acting on the (three-)momentum vector $\vec{p}$. This rotation is implemented by a matrix $\mathbb{M}$, where

$$\vec{p} \to \mathbb{M}\vec{p} = \begin{pmatrix} 1 & 0 & 0 \\ 0 & \cos\theta & -\sin\theta \\ 0 & \sin\theta & \cos\theta \end{pmatrix} \begin{pmatrix} p_x \\ p_y \\ p_z \end{pmatrix} = \begin{pmatrix} p_x \\ p_y \cos\theta - p_z \sin\theta \\ p_y \sin\theta + p_z \cos\theta \end{pmatrix} \,. \tag{2.24}$$

The energy depends only on the magnitude of the momentum, and so remains unchanged under a rotation. Therefore, the matrix Λ that implements a rotation on the four-vector p is

$$p_\mu \to \Lambda^\nu_\mu p_\nu = \begin{pmatrix} 1 & 0 & 0 & 0 \\ 0 & 1 & 0 & 0 \\ 0 & 0 & \cos\theta & -\sin\theta \\ 0 & 0 & \sin\theta & \cos\theta \end{pmatrix}^\nu_\mu \begin{pmatrix} E \\ p_x \\ p_y \\ p_z \end{pmatrix}_\nu = \begin{pmatrix} E \\ p_x \\ p_y \cos\theta - p_z \sin\theta \\ p_y \sin\theta + p_z \cos\theta \end{pmatrix}_\mu \,. \tag{2.25}$$

The rotation can also be implemented by acting from the right as

$$p_\mu \to p_\nu \Lambda^\nu_\mu = (E \ p_x \ p_y \ p_z)_\nu \begin{pmatrix} 1 & 0 & 0 & 0 \\ 0 & 1 & 0 & 0 \\ 0 & 0 & \cos\theta & \sin\theta \\ 0 & 0 & -\sin\theta & \cos\theta \end{pmatrix}^\nu_\mu \,. \tag{2.26}$$

Note that the matrix Λ does indeed satisfy

$$\Lambda^\mu_\rho \eta^{\rho\sigma} \Lambda^\nu_\sigma \tag{2.27}$$

$$= \begin{pmatrix} 1 & 0 & 0 & 0 \\ 0 & 1 & 0 & 0 \\ 0 & 0 & \cos\theta & \sin\theta \\ 0 & 0 & -\sin\theta & \cos\theta \end{pmatrix}^\mu_\rho \begin{pmatrix} 1 & 0 & 0 & 0 \\ 0 & -1 & 0 & 0 \\ 0 & 0 & -1 & 0 \\ 0 & 0 & 0 & -1 \end{pmatrix}^{\rho\sigma} \begin{pmatrix} 1 & 0 & 0 & 0 \\ 0 & 1 & 0 & 0 \\ 0 & 0 & \cos\theta & -\sin\theta \\ 0 & 0 & \sin\theta & \cos\theta \end{pmatrix}^\nu_\sigma$$

$$= \eta^{\mu\nu} \,,$$

and so is a Lorentz transformation.

Example 2.2 Let's now consider a different type of Lorentz transformation: a boost. A **Lorentz boost** is a change of inertial reference frame implemented by moving with a relative velocity. Let's consider boosting along the $\hat{z}$-axis by changing the relative velocity of the inertial frame by a velocity $\vec{\beta} = \beta\hat{z}$. In natural units, the speed β is a fraction of the speed of light c, and so $-1 < \beta < 1$ (negative because we could boost in the opposite direction). $|\beta| = 1$ means that the boost is by the speed of light c, which of course isn't possible. What is the matrix Λ that implements this boost?

Solution

Boosting along the $\hat{z}$-axis means that p_x and p_y are unchanged by the boost:

$$p_x \to p_x \qquad\qquad\qquad p_y \to p_y, \qquad\qquad (2.28)$$

while the energy E and p_z change. Under this boost, the energy becomes

$$E \to \gamma(E + \beta p_z), \qquad\qquad (2.29)$$

where γ is called the **boost factor** and is

$$\gamma = \frac{1}{\sqrt{1 - \beta^2}}. \qquad\qquad (2.30)$$

(In SI units, this would be $\gamma^{-1} = \sqrt{1 - v^2/c^2}$, where v is the velocity of the boost.)
    The z-component of the momentum transforms as

$$p_z \to \gamma(p_z + \beta E). \qquad\qquad (2.31)$$

Note that both of these transformations are linear: they are represented by a linear combination of energy E and momentum p_z. This can therefore be implemented by a matrix Λ acting on the four-vector p. Acting from the left, we have

$$p_\mu \to \Lambda^\nu_\mu p_\nu = \begin{pmatrix} \gamma & 0 & 0 & \gamma\beta \\ 0 & 1 & 0 & 0 \\ 0 & 0 & 1 & 0 \\ \gamma\beta & 0 & 0 & \gamma \end{pmatrix}^\nu_\mu \begin{pmatrix} E \\ p_x \\ p_y \\ p_z \end{pmatrix}_\nu = \begin{pmatrix} \gamma E + \gamma\beta p_z \\ p_x \\ p_y \\ \gamma\beta E + \gamma p_z \end{pmatrix}_\mu. \qquad (2.32)$$

Acting instead from the right, we have

$$p_\mu \to p_\nu \Lambda^\nu_\mu = (E\ p_x\ p_y\ p_z)_\nu \begin{pmatrix} \gamma & 0 & 0 & \gamma\beta \\ 0 & 1 & 0 & 0 \\ 0 & 0 & 1 & 0 \\ \gamma\beta & 0 & 0 & \gamma \end{pmatrix}^\nu_\mu. \qquad (2.33)$$

In Exercise 2.1, you will show that the matrix Λ does satisfy the constraint that all Lorentz transformations must satisfy:

$$\Lambda^\mu_\rho \eta^{\rho\sigma} \Lambda^\nu_\sigma = \eta^{\mu\nu}. \qquad\qquad (2.34)$$

2.1.3 Applying Relativity

A very common problem that we want to analyze in particle physics is the decay of an unstable particle. The vast majority of particles decay to two or more particles. The only particles for which there is no evidence that they decay are electrons and protons. The pion, for example, a particle composed of up and down quarks, decays to two photons. The pion has a mass of about 135 MeV and in its decay to photons energy and momentum are conserved. For now, we won't worry about the consequences of angular momentum conservation. How can we understand this decay using four-vectors?

Of course, four-vectors depend on the frame in which they are evaluated, so we need to pick a frame to analyze this decay. The pion is massive so we can always boost to the frame where the pion is at rest. In this frame, its momentum four-vector, or **four-momentum**, is

$$p_\pi = (m_\pi, 0, 0, 0).$$ (2.35)

That is, when the pion is at rest, its energy is just set by the pion mass, m_π. Now, we want to determine the four-momentum of the two photons in the pion decay. We can again perform a rotation (= Lorentz transformation) to have the photons travel along the $\hat{z}$-axis. By momentum conservation, the photons must be traveling back-to-back from the pion decay:

$$\gamma \longleftarrow \boxed{\pi} \longrightarrow \gamma$$

In this frame, we can express the four-vectors of the two photons as

$$p_{\gamma_1} = (E_{\gamma_1}, 0, 0, p_{z,\gamma_1}), \qquad\qquad p_{\gamma_2} = (E_{\gamma_2}, 0, 0, p_{z,\gamma_2}).$$ (2.36)

By energy and momentum conservation the sum of the four-vectors of the photons must add up to the pion's four-momentum:

$$p_\pi = p_{\gamma_1} + p_{\gamma_2}.$$ (2.37)

This is actually a system of four equations, two of which are just $0 = 0$. The non-trivial equations are conservation of energy and conservation of the z-component of momentum:

$$m_\pi = E_{\gamma_1} + E_{\gamma_2}, \qquad\qquad 0 = p_{z,\gamma_1} + p_{z,\gamma_2}.$$ (2.38)

Denoting $p_{z,\gamma_1} \equiv p_z$ and $E_{\gamma_1} \equiv E$, the four-vectors of the photons are then

$$p_{\gamma_1} = (E, 0, 0, p_z), \qquad\qquad p_{\gamma_2} = (m_\pi - E, 0, 0, -p_z).$$ (2.39)

Photons are massless particles and so the dot product of a photon's four-momentum with itself must be 0. Enforcing this for both of the decay product photons, we have

$$p_{\gamma_1}^2 = 0 = E^2 - p_z^2, \qquad\qquad p_{\gamma_2}^2 = 0 = (m_\pi - E)^2 - p_z^2.$$ (2.40)

It then follows that $E = p_z$ and $E = m_\pi/2$. Then, the four-vectors of the photons are

$$p_{\gamma_1} = \left(\frac{m_\pi}{2}, 0, 0, \frac{m_\pi}{2}\right), \qquad\qquad p_{\gamma_2} = \left(\frac{m_\pi}{2}, 0, 0, -\frac{m_\pi}{2}\right).$$ (2.41)

Note that the four-vectors we have found satisfy

$$p_\pi^2 = (p_{\gamma_1} + p_{\gamma_2})^2 = p_{\gamma_1}^2 + p_{\gamma_2}^2 + 2p_{\gamma_1} \cdot p_{\gamma_2} = 2p_{\gamma_1} \cdot p_{\gamma_2}. \qquad (2.42)$$

We will refer to a four-vector for which $p^2 = m^2$, as for the pion four-momentum p_π, as "on the mass shell" or just **on-shell** for short. This terminology derives from the fact that the energy–momentum relationship in relativity defines a hyperbolic curve:

$$m^2 = E^2 - |\vec{p}|^2. \qquad (2.43)$$

The closest approach of this curve to the origin is when $|\vec{p}| = 0$, where $E = m$. The asymptotes of this hyperbolic curve correspond to the lines $E = |\vec{p}|$, which are only reached in the limit of infinite momentum.

To find the four-vector in any other frame, we just Lorentz transform appropriately. While we just considered the case when the decay products are massless, one can also consider the case when the decay products are massive, which adds some complication. The following example will demonstrate the power of working with Lorentz-invariant quantities.

Example 2.3 In the 1960s, Kenneth Greisen, Vadim Kuzmin, and Georgiy Zatsepin predicted that through interactions with the cosmic microwave background (CMB), extragalactic **cosmic rays** would have an upper bound to their energy.[3] They proposed that cosmic rays (high-energy protons) would lose energy in interacting with the CMB photons by producing a neutral pion:

$$p + \gamma_{\text{CMB}} \to p + \pi^0. \qquad (2.44)$$

The proton energy at which this process can occur is called the **GZK cutoff**. In this example, we will analyze this reaction and estimate the GZK cutoff energy.

The simplest way to analyze this reaction is to express the reaction exclusively in Lorentz-invariant four-vector dot products. The sum of the initial momentum four-vectors must be equal to the sum of the final four-vectors, which will greatly simplify the analysis. Let p_p and p_γ be the initial proton and CMB photon four-vectors, and $p_{p'}$ and p_π be the final proton and pion four-vectors.

Solution

By conservation of energy and momentum, we must have

$$(p_p + p_\gamma)^2 = (p_{p'} + p_\pi)^2. \qquad (2.45)$$

Because both sides of this equation are Lorentz invariant, we can evaluate them in any frame. Let's then compute $(p_{p'} + p_\pi)^2$ in the frame in which the proton and pion are both at

---

[3] K. Greisen, "End to the cosmic ray spectrum?," Phys. Rev. Lett. **16**, 748 (1966); G. T. Zatsepin and V. A. Kuzmin, "Upper limit of the spectrum of cosmic rays," JETP Lett. **4**, 78 (1966) [Pisma Zh. Eksp. Teor. Fiz. **4**, 114 (1966)].

rest, which corresponds to the minimum energy at which this process could occur. In that case, the four-vectors are just

$$p_{p'} = (m_p, 0, 0, 0), \qquad\qquad p_\pi = (m_\pi, 0, 0, 0), \qquad\qquad (2.46)$$

where m_p and m_π are the masses of the proton and pion, respectively. The square of their sum is then

$$(p_{p'} + p_\pi)^2 = (m_p + m_\pi)^2. \qquad\qquad (2.47)$$

Masses are Lorentz invariant, so indeed this expression is independent of frame.

Now, let's evaluate the other side of the momentum-conservation equation, Eq. 2.45. We can expand

$$(p_p + p_\gamma)^2 = m_p^2 + 2p_p \cdot p_\gamma, \qquad\qquad (2.48)$$

where we note that the photon is massless. Then, we have to evaluate the dot product $p_p \cdot p_\gamma$. To do this, we can work in the frame in which the proton and CMB photon collide head-on. The energy of the proton will be enormously larger than the mass of the proton, so we can safely approximate the proton as massless in evaluating $p_p \cdot p_\gamma$. With these assumptions, the four-vectors are

$$p_p = (E_p, 0, 0, E_p), \qquad\qquad p_\gamma = (E_{\text{CMB}}, 0, 0, -E_{\text{CMB}}). \qquad\qquad (2.49)$$

Here, E_p is the energy of the proton and E_{CMB} is the energy of the CMB photon. Colliding head-on means that their three-momenta are in opposite directions, and we choose to align them along the $\hat{z}$-axis. Their four-vector dot product is thus

$$p_p \cdot p_\gamma = 2E_p E_{\text{CMB}}. \qquad\qquad (2.50)$$

Now, we just need to assemble the pieces and solve for E_p in Eq. 2.45. We then find that

$$E_p = \frac{2m_p m_\pi + m_\pi^2}{4E_{\text{CMB}}}. \qquad\qquad (2.51)$$

The masses of the proton and pion are $m_p = 938$ MeV and $m_\pi = 135$ MeV, respectively, while the energy of CMB photons is approximately $E_{\text{CMB}} \simeq 3 \times 10^{-4}$ eV (which follows from the CMB temperature of about 2.7 K). The proton energy in this process is

$$E_p = \frac{2m_p m_\pi + m_\pi^2}{4E_{\text{CMB}}} \simeq 2 \times 10^{20} \text{ eV} \simeq 30 \text{ J}. \qquad\qquad (2.52)$$

30 J is a huge amount of energy! That's approximately the same amount of energy as a baseball traveling at 50 miles per hour (23 meters per second). Extragalactic protons above this energy can lose energy by interaction with the CMB. Because the CMB exists throughout the visible universe, this suggests that there cannot be protons with energies larger than this value, hence the name "GZK cutoff."

Figure 2.1 shows the distribution of cosmic ray energies from numerous experiments, tabulated in the PDG. The horizontal axis is the energy of cosmic rays in eV, and the vertical axis is a measure of the number of cosmic rays with that energy. The labeled "Knee," "2nd Knee," and "Ankle" are features in the distribution that correspond to different astrophysical sources of cosmic rays.

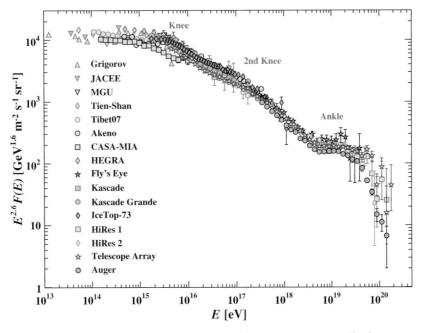

Observed energy distribution of cosmic rays (extragalactic protons) from various experiments. The data are determined by measuring the energy of particles from air showers due cosmic rays hitting the upper atmosphere of Earth. Credit: M. Tanabashi *et al.* [Particle Data Group], "Review of particle physics," Phys. Rev. D **98**, 030001 (2018).

Now that we have computed the GZK cutoff energy, let's compare this to cosmic ray data. The maximum energy that cosmic rays are recorded on this plot is at about (or just below) our calculated value of the GZK cutoff. So, it doesn't appear that there are any cosmic rays observed with energies above the GZK cutoff. A more careful analysis (including the thermal nature of the CMB) actually shows that the GZK cutoff is about 5×10^{19} eV, and so there have been a few ultra-high-energy cosmic rays observed that violate the GZK limit! It remains an open problem to explain the source of cosmic rays that violate the GZK cutoff.

2.2 Relativistic Wave Equations

As particle physics is the realm of both quantum mechanics and special relativity, we should have (or would expect to have) a wavefunction equation like the Schrödinger equation that describes the time evolution of a system. For a brief refresher, a review of quantum mechanics is provided in Appendix B. The Schrödinger equation for a wavefunction ψ of a particle of mass m in a potential V is

$$i\hbar \frac{\partial}{\partial t}\psi = -\frac{\hbar^2}{2m}\nabla^2 \psi + V\psi. \tag{2.53}$$

There are a couple of problems in attempting to use the Schrödinger equation to describe particle physics. The first and biggest problem is that the Schrödinger equation represents conservation of non-relativistic energy. To write the Schrödinger equation, we have identified energy E and momentum $\vec{p}$ with derivatives:

$$E \leftrightarrow i\hbar \frac{\partial}{\partial t}, \qquad\qquad \vec{p} \leftrightarrow -i\hbar \vec{\nabla}, \qquad\qquad (2.54)$$

and so the Schrödinger equation represents

$$E = \frac{|\vec{p}|^2}{2m} + V, \qquad\qquad (2.55)$$

which is not invariant to Lorentz transformations. Apparently, the Schrödinger equation treats space and time differently; there are two spatial derivatives but only one time derivative in the Schrödinger equation. As a final point, what is a potential in relativity?

2.2.1 The Klein–Gordon Equation

It appears that, in order to develop quantum mechanics relativistically, we need to abandon the Schrödinger equation. Just as the Schrödinger equation encoded non-relativistic energy conservation, we should find a wave equation that encodes relativistic energy conservation. So, starting from the relativistic energy–momentum relation (putting back the cs for now)

$$E^2 - |\vec{p}|^2 c^2 - m^2 c^4 = 0, \qquad\qquad (2.56)$$

we replace E and $\vec{p}$ with the canonical quantum mechanical operators:

$$E \leftrightarrow i\hbar \frac{\partial}{\partial t}, \qquad\qquad \vec{p} \leftrightarrow -i\hbar \vec{\nabla}, \qquad\qquad (2.57)$$

and let these act on a wavefunction $\phi(\vec{x}, t)$. We then find the **Klein–Gordon equation**,[4]

$$\left(-\hbar^2 \frac{\partial^2}{\partial t^2} + \hbar^2 c^2 \nabla^2 - m^2 c^4 \right) \phi(\vec{x}, t) = 0. \qquad\qquad (2.58)$$

Because this equation has both c and $\hbar$, it is indeed a relativistic quantum mechanical wave equation. Going back to natural units, this is

$$\left(\frac{\partial^2}{\partial t^2} - \nabla^2 + m^2 \right) \phi(\vec{x}, t) = 0. \qquad\qquad (2.59)$$

The wavefunction $\phi(\vec{x}, t)$ is also an object called a **field** because its arguments extend over all space and time. Another way to say this is that $\phi(\vec{x}, t)$ exists throughout spacetime and its fluctuations are described by the Klein–Gordon equation. This is analogous to a field of wheat that fluctuates according to the movement of the wind.

---

[4] W. Gordon, "Der Comptoneffekt nach der Schrödingerschen theorie," Z. Phys. **40**, 117 (1926); O. Klein, "Elektrodynamik und wellenmechanik vom standpunkt des korrespondenzprinzips," Z. Phys. **41**, 407 (1927); V. Fock, "On the invariant form of the wave equation and the equations of motion for a charged point mass," Z. Phys. **39**, 226 (1926) [Surveys High Energ. Phys. **5**, 245 (1986)].

The Klein–Gordon equation has the solution

$$\phi(\vec{x}, t) = e^{-ip \cdot x}, \tag{2.60}$$

where $p \cdot x = Et - \vec{p} \cdot \vec{x}$ is the four-vector dot product. To see that this is indeed a solution, note that

$$\frac{\partial^2}{\partial^2 t} \phi(\vec{x}, t) = -E^2 \phi(\vec{x}, t), \qquad\qquad -\nabla^2 \phi(\vec{x}, t) = |\vec{p}|^2 \phi(\vec{x}, t). \tag{2.61}$$

The Klein–Gordon equation implies that

$$(E^2 - |\vec{p}|^2 - m^2)\phi(\vec{x}, t) = 0. \tag{2.62}$$

This is indeed true if the four-vector p is on-shell: $p^2 = m^2$. Thus we often say that solutions to the Klein–Gordon equation are on-shell solutions. Note also that the Klein–Gordon equation is Lorentz invariant. The solution $\phi(\vec{x}, t) \equiv \phi(x) = \exp(-ip \cdot x)$ only depends on a four-vector dot product.

There are other interesting features of the Klein–Gordon equation. First, $p \cdot x$ has the same dimensions as angular momentum, $\hbar$, and a classical action. The energy conservation encoded in the Klein–Gordon equation is the quadratic relationship

$$E^2 = m^2 + |\vec{p}|^2. \tag{2.63}$$

There are two solutions to this equation for the energy, differing in their sign:

$$E = \pm\sqrt{m^2 + |\vec{p}|^2}. \tag{2.64}$$

Another way to say this is that there are two solutions to the Klein–Gordon equation because it is a second-order differential equation. The existence of negative-energy solutions is weird, and unfamiliar for free particles from the Schrödinger equation. This is a recurring feature of relativistic wave equations, and we will address it carefully in Section 2.2.2.

The Klein–Gordon Lagrangian

The Klein–Gordon equation isn't the most general way to express the dynamics of a relativistic wavefunction ϕ. We can express all of the information encoded in the Klein–Gordon equation and even more as a Lagrangian. Let's look back at the Klein–Gordon equation, rewritten in the suggestive form

$$\frac{\partial^2}{\partial t^2}\phi = \nabla^2\phi - m^2\phi. \tag{2.65}$$

This is a second-order (in time) differential equation for a field ϕ. As a field, $\phi(x)$ permeates space and time and its configuration must satisfy Eq. 2.65. Contrast this with a particle: a particle is defined by a position $\vec{x}(t)$; that is, a particle is at a unique location $\vec{x}$ at time t. In classical mechanics, the equation that governs $\vec{x}(t)$ would be Newton's second law,

$$\frac{d^2}{dt^2}\vec{x}(t) = -\nabla U, \tag{2.66}$$

where $U(\vec{x})$ is the potential energy. Equation 2.65 looks just like this, though for a field. We have the second time derivative piece

$$\frac{\partial^2}{\partial t^2}\phi\,, \tag{2.67}$$

which is like acceleration for a particle. Now, look at the right side of Eq. 2.65: $\nabla^2\phi - m^2\phi$. The effective "force" from the mass term,

$$-m^2\phi, \tag{2.68}$$

is just like the restoring force for a harmonic oscillator with spring constant $k = m^2$. The Laplacian term

$$\nabla^2\phi \tag{2.69}$$

is a shear force: if the difference between values of ϕ at nearby spatial points is large, then effectively there is a large force. This makes some sense: if you shear a fabric, then there is a force; that is, the fabric becomes warped. This is illustrated in Fig. 2.2.

Okay, we now have an intuition for what the Klein–Gordon wave equation is telling us: it is just Newton's law for a field that experiences shear forces in a harmonic oscillator potential. Further exploiting the analogy with Newton's law, we can determine the corresponding potential energy density, $u(\phi)$. Newton's second law implies that

$$-\frac{\partial u}{\partial\phi} = \nabla^2\phi - m^2\phi\,. \tag{2.70}$$

The solution u to this differential equation is a potential energy density (potential energy per unit volume) because the field $\phi(x)$ itself depends on location in space. This is a bit different than the case with Newton's second law for a particle, Eq. 2.66, where U there is just the potential energy.

Integrating the corresponding force to determine the potential energy density, we find:

$$-\frac{\partial u}{\partial\phi} = \nabla^2\phi - m^2\phi \quad \text{with solution} \quad u = \frac{1}{2}\left(\vec{\nabla}\phi\right)\cdot\left(\vec{\nabla}\phi\right) + \frac{m^2}{2}\phi^2\,. \tag{2.71}$$

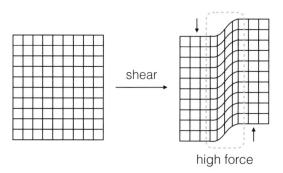

shear

high force

Fig. 2.2 Illustration of shear force on a warped fabric. The shear force is largest in the region where the grid has the highest curvature.

We can safely ignore any constant energy contribution as it can be eliminated by appropriate relabeling.[5] The kinetic energy density k is found from the generalization of $\frac{1}{2}\dot{x}^2$ to fields. This is just

$$k = \frac{1}{2}\left(\frac{\partial}{\partial t}\phi\right)^2 . \tag{2.72}$$

Then, the total energy density of the field at the point $(\vec{x}, t)$ is

$$k + u = \frac{1}{2}\left(\frac{\partial}{\partial t}\phi\right)^2 + \frac{1}{2}\vec{\nabla}\phi \cdot \vec{\nabla}\phi + \frac{m^2}{2}\phi^2 . \tag{2.73}$$

The total energy or **Hamiltonian** H of the field ϕ requires integrating over all space, i.e., summing over all energy contributions at time t:

$$H = \int d\vec{x}\left[\frac{1}{2}\left(\frac{\partial}{\partial t}\phi\right)^2 + \frac{1}{2}\vec{\nabla}\phi \cdot \vec{\nabla}\phi + \frac{m^2}{2}\phi^2\right] . \tag{2.74}$$

The integral extends over all space in the x, y, and z directions. The integration measure is

$$d\vec{x} = dx\,dy\,dz . \tag{2.75}$$

We will also often denote this as $d^3x = d\vec{x}$. Demanding that the Hamiltonian is independent of time (that is, energy is conserved) requires that the field ϕ satisfies the Klein–Gordon equation, Eq. 2.65. With this Hamiltonian, we can formulate the relativistic field equations in a totally different, and ultimately more useful, way.

The classical mechanics of a point particle can be reformulated with a **Lagrangian** and the **principle of least action**. The Lagrangian L is the difference between the kinetic and potential energies of a particle with trajectory $\vec{x}(t)$ (with unit mass):

$$L = \frac{1}{2}\dot{\vec{x}}^2 - U(\vec{x}) . \tag{2.76}$$

The **action** $S[\vec{x}(t)]$ is defined as a function of the function of time $\vec{x}(t)$, or a functional of the Lagrangian integrated over time:

$$S[\vec{x}(t)] = \int dt\left[\frac{1}{2}\dot{\vec{x}}^2 - U(\vec{x})\right] . \tag{2.77}$$

Newton's second law is a consequence of the principle of least action: classical trajectories $\vec{x}(t)$ are those that minimize the action. Minimizing the action (taking the

---

[5] By the way, to see that the potential energy density u contains the term $\frac{1}{2}\left(\vec{\nabla}\phi\right) \cdot \left(\vec{\nabla}\phi\right)$, we can take a derivative by the finite difference method and Taylor expand:

$$\frac{\partial u}{\partial \phi} = \lim_{\epsilon \to 0}\frac{u(\phi + \epsilon) - u(\phi)}{\epsilon} = \lim_{\epsilon \to 0}\frac{1}{2\epsilon}\left(\vec{\nabla}(\phi + \epsilon) \cdot \vec{\nabla}(\phi + \epsilon) - \vec{\nabla}\phi \cdot \vec{\nabla}\phi\right) \tag{2.78}$$

$$= \lim_{\epsilon \to 0}\frac{1}{\epsilon}\left(\vec{\nabla}\epsilon \cdot \vec{\nabla}\phi + \frac{1}{2}\vec{\nabla}\epsilon \cdot \vec{\nabla}\epsilon\right) = \frac{1}{\epsilon}\vec{\nabla} \cdot (\epsilon\vec{\nabla}\phi) - \nabla^2\phi .$$

We can safely ignore the $\vec{\nabla} \cdot (\epsilon\vec{\nabla}\phi)$ term, called a **total derivative**, because we always integrate over all space and this just contributes a constant energy offset.

derivative of $S[\vec{x}(t)]$ with respect to $\vec{x}(t)$ and setting it to 0), one finds that $\vec{x}(t)$ must satisfy

$$\frac{d^2}{dt^2}\vec{x}(t) = -\nabla U, \qquad (2.79)$$

exactly Newton's second law.

We can formulate the Lagrangian and action for the relativistic field ϕ in an analogous way. The Lagrangian L is the difference of the total kinetic and potential energies:

$$L = K - U = \int d^3x \left[\frac{1}{2}\left(\frac{\partial}{\partial t}\phi\right)^2 - \frac{1}{2}\vec{\nabla}\phi \cdot \vec{\nabla}\phi - \frac{m^2}{2}\phi^2 \right]. \qquad (2.80)$$

The action is then the time integral of the Lagrangian:

$$S[\phi(\vec{x},t)] = \int dt\, d^3x \left[\frac{1}{2}\left(\frac{\partial}{\partial t}\phi\right)^2 - \frac{1}{2}\vec{\nabla}\phi \cdot \vec{\nabla}\phi - \frac{m^2}{2}\phi^2 \right]. \qquad (2.81)$$

The wave equation that we started with, the Klein–Gordon equation, can be found by minimizing this action with respect to $\phi(x)$.

We can massage this into a nice form. We will denote

$$dt\, d^3x = d^4x, \qquad (2.82)$$

the Lorentz-invariant integration measure. Further, we can form the four-vector

$$\partial_\mu = \left(\frac{\partial}{\partial t}, -\vec{\nabla}\right)_\mu \qquad (2.83)$$

from the time and space derivatives. Then, we can nicely express the action as

$$S[\phi(\vec{x},t)] = \int d^4x \left[\frac{1}{2}(\partial_\mu\phi)(\partial^\mu\phi) - \frac{m^2}{2}\phi^2 \right]. \qquad (2.84)$$

This is **manifestly** Lorentz invariant: all four-vectors are only present in dot products and so the Lorentz invariance is easy to see by eye. The object

$$\mathcal{L} \equiv \frac{1}{2}(\partial_\mu\phi)(\partial^\mu\phi) - \frac{m^2}{2}\phi^2 \qquad (2.85)$$

is called the Lagrangian density, or typically (though technically incorrectly) just the Lagrangian. Given this Lagrangian, we have all of the information of the wave equation and actually much more!

While this was a long, roundabout way to get there, this now tells us how to construct, in general, descriptions of relativistic, quantum mechanical systems. We just need to write down the corresponding Lorentz-invariant Lagrangian and we are "done." "Done" is in quotes because much of the rest of this book will be devoted to attempting to unpackage the information in the Lagrangian.

Physical Interpretation of the Klein–Gordon Lagrangian

With this Lagrangian in hand, it is useful to pause for a bit and discuss the physical configuration that it describes. We have discussed the analogy of this Lagrangian to a harmonic oscillator, with a spring constant that is set by the mass of the Klein–Gordon field. The way in which the Lagrangian of Eq. 2.85 is Lorentz invariant, especially the term with derivatives, has a nice physical interpretation. In quantum mechanics, we associate the derivative to the momentum operator. That is, a derivative acting on a wavefunction imparts momentum on that wavefunction. In special relativity, we can't distinguish between spatial and temporal dimensions, and so the derivative four-vector ∂_μ imparts momentum and energy when acting on a wavefunction.

What does imparting momentum on a wavefunction mean? If a particle has momentum and energy, that means that it is moving through space; that is, its spacetime position has changed. This can be made more precise through the Taylor expansion. A field ϕ evaluated at the spacetime position $x + \Delta x$ can be expanded about its value at $x = (t, \vec{x})$ as

$$\phi(x + \Delta x) = \phi(x) + \Delta x^\mu \partial_\mu \phi(x) + \cdots . \tag{2.86}$$

That is, the derivative or momentum operator is responsible for displacing the field from its initial position x. Let's continue understanding the consequences of this displacement. If a particle has momentum and is displaced from some position x, then that particle will have a non-zero angular momentum about that position x. The exact same thing happens with our field. Because the field $\phi(x)$ has been displaced from position x by the derivative ∂_μ, it has a non-zero angular momentum about x.

The next natural question to ask is how much angular momentum the object $\partial_\mu \phi(x)$ carries about position x. Restoring the factors of $\hbar$, the momentum operator $\hat{P}_\mu$ is related to the derivative as

$$\hat{P}_\mu = i\hbar \partial_\mu = \left(i\hbar \frac{\partial}{\partial t}, -i\hbar \vec{\nabla} \right)_\mu . \tag{2.87}$$

Note that there is one factor of $\hbar$ and one free Lorentz index μ. A Lorentz transformation, like a rotation, of this operator is implemented by a single matrix Λ as discussed earlier:

$$\hat{P}_\mu \to \Lambda_\mu^\nu \hat{P}_\nu . \tag{2.88}$$

Because there is one free Lorentz index and a rotation is implemented by a single matrix, the derivative imparts one $\hbar$ unit of orbital angular momentum on the field ϕ. A system with non-zero total angular momentum is not rotationally invariant, and $\partial_\mu \phi(x)$ is not Lorentz invariant, so we need to do a little more digging to completely understand the Lagrangian.

The term with derivatives in the Klein–Gordon Lagrangian is

$$\mathcal{L} \supset \frac{1}{2} \left(\partial_\mu \phi \right) \left(\partial^\mu \phi \right) , \tag{2.89}$$

and we now have an interpretation for $\partial_\mu \phi$ and $\partial^\mu \phi$. Each of these corresponds to imparting one unit of angular momentum on the field ϕ. Two four-vectors with upper and lower indices, but otherwise the same, effectively differ in the sign of the spatial components, as indices are raised and lowered with the metric η. That is, if we choose the convention that

$$\partial_\mu = \left(\frac{\partial}{\partial t}, -\vec{\nabla} \right)_\mu , \qquad (2.90)$$

then

$$\partial^\mu = \left(\frac{\partial}{\partial t}, \vec{\nabla} \right)^\mu . \qquad (2.91)$$

Which derivative has the minus sign is irrelevant; all that matters is that there is a relative minus sign. Because the spatial components differ by a minus sign, ∂_μ and ∂^μ displace the field ϕ by the same amount, but impart opposite momenta. Because angular momentum $\vec{L} = \Delta \vec{x} \times \vec{p}$, for some displacement $\Delta \vec{x}$, opposite momentum $\vec{p}$ means opposite angular momentum. To find the total angular momentum of the Lagrangian term with derivatives we just sum together the angular momenta of parts $\partial_\mu \phi$ and $\partial^\mu \phi$. These are equal in magnitude and opposite in sign, and so the angular momentum of $(\partial_\mu \phi)(\partial^\mu \phi)$ is 0, consistent with the requirement of Lorentz invariance.

The physical picture of the configuration of fields described by the term of the Lagrangian in Eq. 2.89 is:

$$\partial^\mu \phi(x) \longleftarrow \underset{\underset{x}{\Delta x \vdots}}{\hspace{3cm}} \longrightarrow \partial_\mu \phi(x)$$

The fields are both displaced by an amount Δx from the spacetime position x, but have equal and opposite momenta. There is indeed 0 net angular momentum of the $(\partial_\mu \phi)(\partial^\mu \phi)$ system because their individual momenta lie along the same line in space. Also, the total momentum of the two fields is zero, another requirement for Lorentz invariance.

2.2.2 The Dirac Equation

The Klein–Gordon equation just enforces conservation of relativistic energy and momentum for the field $\phi(x)$. As a second-order differential equation, it contains no information about intrinsic angular momentum; that is, there are no free Lorentz indices in the Klein–Gordon equation. Therefore, the field $\phi(x)$ has no intrinsic angular momentum or it is a spin-0 field. Under a Lorentz transformation, the field $\phi(x)$ does not transform. If we are to describe the spin-1/2 and spin-1 particles of the Standard Model, the Klein–Gordon equation isn't sufficient. We need to develop relativistic equations of motion that have free Lorentz indices and so transform **non-trivially**, enabling a non-zero spin. In this section, we will develop the Dirac equation which describes spin-1/2 particles like the electron. In Section 2.2.3, we will develop the spin-1 equation of motion that describes the photon from the familiar Maxwell's equations.

Our starting point for the Dirac equation will be innocent enough. The Klein–Gordon equation is a second-order differential equation; can we find a first-order relativistic equation of motion? Let's assume that there exists a wave equation linear in time and spatial derivatives, for some field ψ. Let's write this in a suggestive form as

$$\left(\alpha\frac{\partial}{\partial t} + \vec{\beta}\cdot\vec{\nabla}\right)\psi = m\psi\,, \tag{2.92}$$

for some constant α, a constant vector $\vec{\beta}$, and mass m. If this is relativistically invariant, it must imply the Klein–Gordon equation. In particular, we will assume that this equation is the square-root of the Klein–Gordon equation. So, squaring the operators acting on ψ on both sides of this equation, we have

$$\left(\alpha\frac{\partial}{\partial t} + \vec{\beta}\cdot\vec{\nabla}\right)\left(\alpha\frac{\partial}{\partial t} + \vec{\beta}\cdot\vec{\nabla}\right)\psi = m^2\psi\,. \tag{2.93}$$

Expanding this equation out, we have

$$\left[\alpha^2\frac{\partial^2}{\partial t^2} + \alpha\vec{\beta}\cdot\vec{\nabla}\frac{\partial}{\partial t} + \vec{\beta}\cdot\vec{\nabla}\alpha\frac{\partial}{\partial t} + (\vec{\beta}\cdot\vec{\nabla})^2\right]\psi = m^2\psi\,. \tag{2.94}$$

For this to produce the Klein–Gordon equation,

$$\left(-\frac{\partial^2}{\partial t^2} + \nabla^2\right)\psi = m^2\psi\,, \tag{2.95}$$

we must have

$$\alpha^2 = -1\,, \qquad \beta_i\beta_j = \delta_{ij}\,, \qquad \alpha\beta_i + \beta_i\alpha = 0\,. \tag{2.96}$$

Here, we have denoted the components of the vector $\vec{\beta}$ by β_i. The requirement $\alpha^2 = -1$ is simple and can be satisfied by α being just a number. By contrast, the requirement $\beta_i\beta_j = \delta_{ij}$ is weird, and cannot be satisfied by the β_i being just numbers! However, let's just keep going and see where it takes us.

Let's denote $\alpha = i\gamma_0$, $\beta_i = i\gamma_i$, for $i = 1, 2, 3$ and some things $\gamma_0, \gamma_1, \gamma_2, \gamma_3$. Then, these things satisfy

$$\gamma_0\gamma_0 = 1\,, \qquad \gamma_i\gamma_j = -\delta_{ij}\,, \qquad \gamma_0\gamma_i + \gamma_i\gamma_0 = 0\,. \tag{2.97}$$

That is, for $\mu, \nu = 0, 1, 2, 3$, we have

$$\gamma_\mu\gamma_\nu + \gamma_\nu\gamma_\mu = \begin{pmatrix} 2 & 0 & 0 & 0 \\ 0 & -2 & 0 & 0 \\ 0 & 0 & -2 & 0 \\ 0 & 0 & 0 & -2 \end{pmatrix}_{\mu\nu} = 2\eta_{\mu\nu}\,. \tag{2.98}$$

We can denote this with curly braces $\{\gamma_\mu, \gamma_\nu\} = 2\eta_{\mu\nu}$, where the operation $\{,\}$ denotes the **anti-commutator**. With this notation, our original linear differential equation is

$$\left(i\gamma_0\frac{\partial}{\partial t} + i\vec{\gamma}\cdot\vec{\nabla} - m\right)\psi = 0\,, \tag{2.99}$$

Paul Dirac was perhaps the oddest of theoretical physicists, a group that is synonymous with aloof, absent-minded, and out-of-touch. Dirac was famously awkward in conversation. He married Margit Wigner, the sister of Eugene Wigner, another famous theoretical physicist. He would sometimes introduce his wife to guests at his house by saying "This is Wigner's sister, who is also my wife." When he met Richard Feynman at a conference, his first words were, "I have an equation. Do you have one too?" During the relatively rare occasions when he spoke at length, Dirac was also critical of religion. At the 1927 Solvay conference, Dirac went on a particularly long tirade against religion, which was surprising to some of his fellow conference attendees. When asked for his opinion of Dirac's view, the Catholic-raised Wolfgang Pauli summarized Dirac's personality well, saying "There is no God and Paul Dirac is his prophet." Everyone, including Dirac, laughed.

or, with the derivative four-vector ∂_μ,

$$(i\gamma_\mu \partial^\mu - m)\, \psi = 0 \,. \tag{2.100}$$

This is known as the **Dirac equation**, discovered by Paul Dirac, a British physicist, in 1928.[6]

So what are these γ_μ things? While ordinary numbers cannot satisfy the anti-commutation relations of Eq. 2.98, matrices can. These are referred to as the γ **matrices**, and the smallest matrices that satisfy $\{\gamma_\mu, \gamma_\nu\} = 2\eta_{\mu\nu}\mathbb{I}$ are 4×4. Therefore, the identity matrix $\mathbb{I}$ here is 4×4. One set of matrices that satisfies this requirement is

$$\gamma_\mu = \begin{pmatrix} 0 & \sigma_\mu \\ \bar{\sigma}_\mu & 0 \end{pmatrix} \,. \tag{2.101}$$

In Exercise 2.6, you will relate this choice of γ matrices, called the **Weyl basis**, to other choices of basis. Here, σ_μ and $\bar{\sigma}_\mu$ are vectors of 2×2 matrices:

$$\sigma_\mu = (\mathbb{I}, \sigma_1, \sigma_2, \sigma_3) \,, \tag{2.102}$$

$$\bar{\sigma}_\mu = (\mathbb{I}, -\sigma_1, -\sigma_2, -\sigma_3) \,. \tag{2.103}$$

$\mathbb{I}$ in these expressions is the 2×2 identity matrix and the σ_i are the **Pauli spin matrices**:

$$\sigma_1 = \begin{pmatrix} 0 & 1 \\ 1 & 0 \end{pmatrix}, \qquad \sigma_2 = \begin{pmatrix} 0 & -i \\ i & 0 \end{pmatrix}, \qquad \sigma_3 = \begin{pmatrix} 1 & 0 \\ 0 & -1 \end{pmatrix}. \tag{2.104}$$

Because the Dirac equation has Pauli spin matrices in it, it describes spin-1/2 particles, like the electron. Because the γ_μ are matrices, the solution ψ will be a vector; properly, it is called a four-component **spinor**. In the following example, we will make this concrete by explicitly solving the Dirac equation for the spinor ψ.

Example 2.4 Using the Weyl basis of the γ matrices, let's solve the Dirac equation for a spinor ψ.

---

[6] P. A. M. Dirac, "The quantum theory of the electron," Proc. Roy. Soc. Lond. A **117**, 610 (1928).

Solution

We will make the ansatz that a solution to the Dirac equation can be written in the form

$$\psi(x) = u(p)e^{-ip\cdot x}, \tag{2.105}$$

where the four-momentum of the particle is p and $u(p)$ is a four-component spinor that only depends on p. Plugging this into the Dirac equation, we have

$$(i\gamma_\mu\partial^\mu - m)\,\psi = (i\gamma_\mu(-ip^\mu) - m)\,ue^{-ip\cdot x} = 0. \tag{2.106}$$

The matrix equation for u is therefore

$$(\gamma\cdot p - m)u = 0. \tag{2.107}$$

For a particle with mass m, we can work in the frame in which $\vec{p} = 0$ so that $p = (E, 0, 0, 0)$. Then, the Dirac equation in the Weyl basis becomes

$$\begin{pmatrix} -m\mathbb{I} & E\mathbb{I} \\ E\mathbb{I} & -m\mathbb{I} \end{pmatrix} u = 0. \tag{2.108}$$

Because the matrix is formed from 2×2 sub-blocks, we can write the spinor u as a concatenation of two two-component spinors ξ and ζ:

$$u = \begin{pmatrix} \xi \\ \zeta \end{pmatrix}. \tag{2.109}$$

Plugging this in for u, we find two linear equations for ξ and ζ:

$$-m\xi + E\zeta = 0, \qquad\qquad E\xi - m\zeta = 0. \tag{2.110}$$

Solving for ζ from the first equation and plugging it into the second, we find that the energy E satisfies

$$E^2 = m^2. \tag{2.111}$$

Therefore, the energy $E = \pm m$. The existence of both positive- and negative-energy solutions is something we also observed with the Klein–Gordon equation.

First, setting $E = m$, we find that $\zeta = \xi$. Correspondingly, setting $E = -m$, this fixes $\zeta = -\xi$. Therefore, the two solutions we find to the Dirac equation are

$$\psi_+ = \begin{pmatrix} \xi \\ \xi \end{pmatrix} e^{-imt}, \tag{2.112}$$

$$\psi_- = \begin{pmatrix} \xi \\ -\xi \end{pmatrix} e^{imt}.$$

The $+$ and $-$ subscripts correspond to the positive- and negative-energy solutions, respectively. Note that the exponential time-evolution factors are opposite in sign. That is, if the positive-energy solution corresponds to forward time evolution, the negative-energy solution corresponds to backward time evolution! While this "traveling backward in time" may seem exotic, we identify this as an **anti-particle** of the original particle: it has all the same properties, but the opposite electric charge.

The two-component spinor ξ represents two possible spin states: spin-up and spin-down. The relative sizes of the components of ξ represent the relative probability amplitudes for the particle to have a particular spin. Therefore, as advertised, the Dirac equation describes a spin-1/2 particle with mass m.

When Paul Dirac discovered his equation, he found these two solutions and therefore predicted the existence of anti-particles. J. J. Thomson had discovered the electron in 1897,[7] but no corresponding anti-particle of the electron had yet been observed. Only four years after the Dirac equation was introduced, the electron's anti-particle, called the **positron**, a spin-1/2 particle with positive electric charge and mass of 511 keV, was discovered in 1932 by Carl Anderson.[8]

Physical Interpretation of the Dirac Lagrangian

As we did with the Klein–Gordon equation, it is useful to formulate a Lagrangian from which the Dirac equation follows by minimization of the action. Note that the solution to the Dirac equation ψ is a complex spinor and as such has a conjugate spinor that we will denote as $\overline{\psi}$. As complex spinors, we can consistently take ψ and $\overline{\psi}$ to be independent of one another. Therefore, the action from which the Dirac equation can be derived is

$$S_{\text{Dirac}}[\overline{\psi}, \psi] = \int d^4x\, \overline{\psi}(i\gamma_\mu \partial^\mu - m)\psi\,. \tag{2.113}$$

The derivative in the action acts to the right. Because ψ and $\overline{\psi}$ are independent, we can vary the action with respect to either separately. Varying the action with respect to $\overline{\psi}$ and setting it to zero indeed produces the Dirac equation:

$$(i\gamma_\mu \partial^\mu - m)\psi = 0\,. \tag{2.114}$$

The Lorentz-invariant Dirac Lagrangian is therefore

$$\mathcal{L}_{\text{Dirac}} = \overline{\psi}(i\gamma_\mu \partial^\mu - m)\psi\,. \tag{2.115}$$

The Lorentz invariance of this Lagrangian is fascinating. First, let's focus on the mass term in the Lagrangian:

$$\mathcal{L}_{\text{Dirac}} \supset -m\overline{\psi}\psi\,. \tag{2.116}$$

---

[7] J. J. Thomson, "Cathode rays," Phil. Mag. Ser. 5 **44**, 293 (1897).

[8] C. D. Anderson, "The positive electron," Phys. Rev. **43**, 491 (1933). See also C-Y Chao, "The absorption coefficient of hard γ-rays," Proc. Natl. Acad. Sci. U. S. A. **16**, no. 6, 431 (1930). Chung-Yao Chao was a fellow student at California Institute of Technology with Anderson and the results of Chao's experiments were explained by the existence of the positron, though not known to him at the time.

As it describes a spin-1/2 field (like the electron), ψ carries spin-1/2. Non-zero spin is not Lorentz invariant because it will transform under a rotation. For this term to be Lorentz invariant, its total spin must be zero. The total angular momentum of $\overline{\psi}\psi$ is found from summing the spins of $\overline{\psi}$ and ψ, and therefore the spins of $\overline{\psi}$ and ψ must be anti-aligned. Note also that because this term has no derivatives, the spinors are not displaced from the spacetime point x. The physical configuration of this term is therefore:

This term indeed has total angular momentum 0 and so is Lorentz invariant.

With this understanding, we can then interpret the term in the Lagrangian with the derivative

$$\mathcal{L}_{\text{Dirac}} \supset \overline{\psi}i\gamma_\mu\partial^\mu\psi. \qquad (2.117)$$

In the discussion of the Lagrangian of the Klein–Gordon equation, we argued that the derivative displaces the field by an amount that provides it with one $\hbar$ unit of orbital angular momentum. In the Klein–Gordon case, there were two derivatives whose contributions to the orbital angular momentum canceled, producing a Lagrangian which had total spin 0 and was Lorentz invariant. In the Dirac Lagrangian case, we have only a single derivative, and so to make the total spin 0, it must be canceled in some other way. Indeed, because the spinor ψ carries spin-1/2, we can see how this happens. Because the derivative acts only on ψ, it is displaced in space from $\overline{\psi}$. If the spins of $\overline{\psi}$ and ψ were anti-aligned, as they were for the mass term, then this configuration would have total spin 1, and not be Lorentz invariant.

Thus, the γ matrices play a crucial role. With the derivative imparting one unit of angular momentum, the total angular momentum can be zero if the spins of $\overline{\psi}$ and ψ are aligned and oriented opposite that to the orbital angular momentum. Therefore, the effect of the γ matrix in Eq. 2.117 is to align the spins of $\overline{\psi}$ and ψ. Then, the angular momentum in the spin will exactly cancel the orbital angular momentum, and render this term Lorentz invariant. The physical configuration of this term is therefore:

For illustration, we have drawn the spins to be parallel to the momentum of spinor ψ, but they are of course in the direction opposite to the total orbital angular momentum.

2.2.3 Electromagnetism

From the spin-1/2 Dirac equation, let's now move to spin-1. Interestingly, the spin-1 equations of motion are those with which you are likely the most familiar. Electromagnetism was the first physical theory to be constructed that was Lorentz invariant. Maxwell's equations for electric field $\vec{E}$ and magnetic field $\vec{B}$, in natural units where $\mu_0 = \epsilon_0 = c = 1$, are

$$\vec{\nabla} \cdot \vec{E} = \rho \,, \qquad\qquad \vec{\nabla} \times \vec{B} - \frac{\partial \vec{E}}{\partial t} = \vec{J} \,, \qquad (2.118)$$

$$\vec{\nabla} \cdot \vec{B} = 0 \,, \qquad\qquad \frac{\partial \vec{B}}{\partial t} + \vec{\nabla} \times \vec{E} = 0 \,. \qquad (2.119)$$

As written, these are not manifestly Lorentz invariant; nothing is expressed in terms of four-vectors. The charge density ρ and the current density $\vec{J}$ can be combined into a current four-vector:

$$J^\mu = (\rho, \vec{J})^\mu \,. \qquad (2.120)$$

This transforms under Lorentz transformations as a four-vector.

With a total of six components, the electric and magnetic fields can't combine into a four-vector. However, they can combine into an anti-symmetric tensor $F_{\mu\nu}$ which transforms linearly under Lorentz transformations. The anti-symmetric tensor $F_{\mu\nu}$ satisfies

$$F_{\mu\nu} = -F_{\nu\mu} \,, \qquad (2.121)$$

and can be written as a matrix with electric and magnetic fields as entries of the matrix:

$$F_{\mu\nu} = \begin{pmatrix} 0 & E_x & E_y & E_z \\ -E_x & 0 & -B_z & B_y \\ -E_y & B_z & 0 & -B_x \\ -E_z & -B_y & B_x & 0 \end{pmatrix}_{\mu\nu} . \qquad (2.122)$$

In this form, it is an anti-symmetric matrix. As it has two Lorentz indices μ and ν, it transforms with two Lorentz matrices Λ:

$$F_{\mu\nu} \xrightarrow{\text{Lorentz transformation}} \Lambda^\rho_\mu \Lambda^\sigma_\nu F_{\rho\sigma} \,. \qquad (2.123)$$

$F_{\mu\nu}$ is called the **field strength tensor** of electromagnetism.

The two Maxwell's equations that involve charge (Gauss's law and Ampère's law) can be expressed very compactly:

$$\partial_\mu F^{\mu\nu} = J^\nu \,. \qquad (2.124)$$

Gauss's law, for example, corresponds to taking the index $\nu = 0$. Then, $J^0 = \rho$ and

$$\partial_\mu F^{\mu 0} = \partial_0 F_{00} - \partial_1 F_{10} - \partial_2 F_{20} - \partial_3 F_{30} = \vec{\nabla} \cdot \vec{E} = \rho \,. \qquad (2.125)$$

In general, Eq. 2.124 is nicely Lorentz invariant. We can express Eq. 2.124 as

$$\partial_\mu F^{\mu\nu} - J^\nu = 0 \,, \qquad (2.126)$$

and this still equals 0 after Lorentz transforming:

$$\partial_\mu F^{\mu\nu} - J^\nu = 0 \rightarrow \Lambda^\sigma_\nu (\partial_\mu F^{\mu\nu} - J^\nu) = 0 \,. \qquad (2.127)$$

The other two of Maxwell's equations, those with no charges, follow from the **Bianchi identity**

$$\partial_\mu F_{\nu\rho} + \partial_\rho F_{\mu\nu} + \partial_\nu F_{\rho\mu} = 0 \,. \qquad (2.128)$$

Note that the Bianchi identity is also Lorentz invariant.

There's an aspect of Maxwell's equations that we haven't accounted for yet. Maxwell's equations can be equivalently expressed in terms of a scalar potential V and a vector potential $\vec{A}$, with the definitions

$$\vec{E} = -\vec{\nabla}V - \frac{\partial \vec{A}}{\partial t}, \qquad \vec{B} = \vec{\nabla} \times \vec{A}. \tag{2.129}$$

With these potentials, Maxwell's equations exhibit a **gauge symmetry**. The potentials can be changed in a way that does not modify the physical electric and magnetic fields. For an arbitrary function λ of space and time, the gauge transformations

$$V \to V + \frac{\partial}{\partial t}\lambda, \qquad \vec{A} \to \vec{A} - \vec{\nabla}\lambda \tag{2.130}$$

produce the identical electric and magnetic fields. For example, for the gauge transformation of $\vec{A}$, note that the curl of a gradient is always 0:

$$\vec{\nabla} \times \vec{\nabla}\lambda = 0. \tag{2.131}$$

Therefore, the magnetic field is unchanged under this gauge transformation. The introduction of these scalar and vector potentials motivates the introduction of the **four-vector potential** A_μ:

$$A_\mu = (V, \vec{A})_\mu. \tag{2.132}$$

The gauge transformation of this object can be compactly represented as

$$A_\mu \to A_\mu + \partial_\mu \lambda. \tag{2.133}$$

The field strength tensor also has a nice representation in terms of A_μ. Appropriately identifying the electric and magnetic field components, we can express $F_{\mu\nu}$ as

$$F_{\mu\nu} = \partial_\mu A_\nu - \partial_\nu A_\mu. \tag{2.134}$$

Note that this is invariant to gauge transformations:

$$F_{\mu\nu} \to \partial_\mu(A_\nu + \partial_\nu \lambda) - \partial_\nu(A_\mu + \partial_\mu \lambda) = F_{\mu\nu}. \tag{2.135}$$

Also, in this form, the Bianchi identity is a trivial consequence of the fact that partial derivatives commute with one another: $\partial_\mu \partial_\nu = \partial_\nu \partial_\mu$.

This will be discussed at various points in this book, but these gauge transformations, despite their name, aren't transformations, per se. In particular, under a gauge transformation, the physical electric and magnetic fields are unchanged. Therefore, there is no observable consequence of a gauge transformation of the vector potential. More precisely, the gauge transformation should be thought of as an identification of a class of vector potentials. Any two vector potentials A_μ and A'_μ that are related via a gauge transformation

$$A'_\mu = A_\mu + \partial_\mu \lambda \tag{2.136}$$

correspond to the exact same physical system. In expressing electromagnetism in terms of the vector potential, we have traded the six components of the physical electric and magnetic fields for four components of A_μ, but at the cost of introducing a redundancy of the description of physics through the gauge transformation. While this may seem like a bad trade-off, it will actually turn out to be extremely powerful. Requiring the physical description to be gauge invariant will enable us to essentially uniquely write down the Lagrangians for the particle physics systems we consider in this book.

Lagrangian of Electromagnetism

Okay, so using this new four-vector potential and field strength tensor formalism, how do we write a Lagrangian that is Lorentz invariant and expresses Maxwell's equations as a consequence the principle of least action? We will provide some motivation here, and a more detailed intuitive construction will be the topic of Chapter 8. The Lagrangian needs to have kinetic energy, which involves two derivatives. The field strength has one derivative and we also need to include the current J^μ that couples to the potential. This motivates the Lagrangian

$$\mathcal{L}_{\text{E\&M}} = -\frac{1}{4}F_{\mu\nu}F^{\mu\nu} - J^\mu A_\mu. \tag{2.137}$$

The $-1/4$ is a canonical normalization factor that is necessary to match with the normalization of electric and magnetic fields that we have used in Maxwell's equations. The corresponding action of electromagnetism is

$$S[A_\mu] = \int d^4x \left[-\frac{1}{4}F_{\mu\nu}F^{\mu\nu} - J^\mu A_\mu \right]. \tag{2.138}$$

Varying the action with respect to A_μ and setting it to 0 produces two of Maxwell's equations, Eq. 2.124. The remaining two Maxwell's equations follow from the Bianchi identity.

This action is both Lorentz invariant and gauge invariant. Lorentz invariance is manifest, as all Lorentz indices are contracted. Gauge invariance is a bit more subtle. Under a gauge transformation, $F_{\mu\nu}$ is invariant, and so the only thing that might change is the last term of the action, with the current four-vector coupled to the vector potential. Under a gauge transformation, this turns into

$$S[A_\mu] \supset -\int d^4x\, J^\mu A_\mu \rightarrow -\int d^4x\, [J^\mu A_\mu + J^\mu \partial_\mu \lambda] \tag{2.139}$$

$$= -\int d^4x\, J^\mu A_\mu - \int d^4x\, \partial_\mu(J^\mu \lambda) + \int d^4x\, \lambda \partial_\mu J^\mu.$$

On the second line, we have used the derivative product rule to move the derivative off of the function λ. Assuming that the current J^μ vanishes at the boundary of spacetime, the total derivative term vanishes:

$$\int d^4x\, \partial_\mu(J^\mu \lambda) = 0. \tag{2.140}$$

Then, gauge invariance follows from demanding that the divergence of the current is 0, $\partial_\mu J^\mu = 0$, which is just conservation of charge:

$$\partial_\mu J^\mu = \frac{\partial \rho}{\partial t} - \vec{\nabla} \cdot \vec{J} = 0 \,. \tag{2.141}$$

Therefore, if electric charge is conserved the electromagnetic action is gauge invariant.

Photon Equations of Motion

For studying electromagnetism and its consequences in particle physics, it is most useful to work with the vector potential A_μ directly. We will often refer to A_μ as the **photon field**, as it is the object that directly describes the photon. In the case where there are no sources, $J_\mu = 0$, the equation of motion of A_μ that follows from Eq. 2.124 is

$$\partial_\mu F^{\mu\nu} = \partial^2 A_\mu - \partial_\mu \partial \cdot A = (\eta_{\mu\nu} \partial^2 - \partial_\mu \partial_\nu) A^\nu = 0. \tag{2.142}$$

To solve this equation, we do the now-familiar thing of including the spatial dependence as an exponential phase, and introducing a function that depends on the momentum of the photon. For the photon, we have the ansatz

$$A_\mu = \epsilon_\mu(p) e^{-ip \cdot x} \,, \tag{2.143}$$

where $\epsilon_\mu(p)$ is called the **polarization vector** of the photon. Plugging this expression into the equation of motion, we find

$$(\eta_{\mu\nu} \partial^2 - \partial_\mu \partial_\nu) A^\nu = 0 \qquad \Longrightarrow \qquad (\eta_{\mu\nu} p^2 - p_\mu p_\nu) \epsilon^\nu = 0 \,. \tag{2.144}$$

The photon is massless so, when it is on-shell, $p^2 = 0$, in which case the equation of motion for ϵ_μ reduces to $p \cdot \epsilon = 0$. We can choose the frame in which the photon's momentum is aligned along the $\hat{z}$-axis and so

$$p = E(1, 0, 0, 1) \,, \tag{2.145}$$

where E is the energy of the photon. The most general expression for a polarization vector that satisfies the equation of motion is

$$\epsilon(p) = (a, b, c, a) \,, \tag{2.146}$$

for some values a, b, c. Explicitly, note that

$$p \cdot \epsilon = E(1, 0, 0, 1) \cdot (a, b, c, a) = Ea - Ea = 0 \,. \tag{2.147}$$

The three constants a, b, c correspond to three different photon configurations. Focusing first on the components b and c that are perpendicular to the photon's three-momentum $\vec{p}$, it is convenient to decompose them into two components corresponding to different directions of the photon's spin. A **right-handed circularly polarized** photon has a polarization vector

$$\epsilon_R(p) = \frac{1}{\sqrt{2}} (0, 1, i, 0) \,. \tag{2.148}$$

This describes a photon whose spin is aligned parallel to its momentum, by the right-hand rule. (Recall that the cross-product of the electric and magnetic fields is the Poynting vector, which describes the momentum carried by the electromagnetic field.) A **left-handed circularly polarized** photon is correspondingly represented by

$$\epsilon_L(p) = \frac{1}{\sqrt{2}}(0, 1, -i, 0). \tag{2.149}$$

This describes a photon whose spin is anti-parallel to its momentum. Note that $\epsilon_R = \epsilon_L^*$ and $\epsilon_R \cdot \epsilon_R = \epsilon_L \cdot \epsilon_L = 0$, and $\epsilon_R \cdot \epsilon_L = -1$. The polarization vector for a photon traveling in any other direction can be found by rotating these vectors.

We have constructed the photon to have only two degrees of freedom: right- and left-handed circular polarization. However, the vector potential A_μ has four components corresponding to $\mu = 0, 1, 2, 3$. Additionally, we still haven't addressed the components of the polarization vector denoted by a in Eq. 2.146. What's going on? It turns out that two of the components of the vector potential do not correspond to degrees of freedom; that is, to physical states of the photon. To understand this, first take a look at the equation of motion for the 0th component of A_μ with no sources:

$$\partial^2 A_0 - \partial_0 \partial \cdot A = -\nabla^2 A_0 + \partial_0 \vec{\nabla} \cdot \vec{A} = 0. \tag{2.150}$$

There is no time derivative ∂_0 that acts on A_0 in its equation of motion. As such, A_0 is not allowed to change with time. Another way to say this is the following. Because the equation of motion of A_0 corresponds to Gauss's law, we say that Gauss's law is actually not an equation of motion (allowing for time dependence), but rather an equation that imposes a constraint on the component A_0, or the electric field $\vec{E}$. So, if A_0 is not allowed to have time dependence or **propagate** in time, then the vector potential A_μ at most has three degrees of freedom.

We still need to eliminate another degree of freedom; this is accomplished by the gauge invariance of the electromagnetic action. As mentioned earlier, the gauge transformation really should be thought of as an identification of different vector potentials. The 0 vector potential, $A_\mu = 0$, is identified with (i.e., has the equivalent physical consequences of) the vector potential A'_μ, where

$$A'_\mu = \partial_\mu \lambda. \tag{2.151}$$

Such a vector potential is called **pure gauge** as it is exclusively a function of the gauge parameter λ. The zero vector potential produces zero electric and magnetic fields and as such does not correspond to any propagating degrees of freedom. Through the expression for the solution to the equations of motion in Eq. 2.143, the derivative ∂_μ corresponds to the momentum of the photon p_μ. That is, the pure gauge potential of Eq. 2.151 corresponds to the component of the vector potential that lies along the direction of photon four-momentum. This exactly corresponds to the components of the polarization vector defined by a from Eq. 2.146, with our choice of direction of the photon three-momentum. By the argument above, this component cannot have any physical consequences, and therefore does not correspond to a degree of freedom of the photon. Out of the four components of the vector potential A_μ, neither A_0 nor the component along the direction of momentum p_μ

corresponds to a physical degree of freedom. The photon only has two degrees of freedom, or only two physical polarizations. Another argument for why the photon has only two polarizations will be discussed later in Section 3.4.

In our discussion of the Klein–Gordon equation (spin-0) or the Dirac equation (spin-1/2), this is where we would have ended the discussion, until we saw the application of the solutions later. However, the photon is a spin-1 particle, and as such its equation of motion is quite weird. Let's consider the equation of motion with $p^2 \neq 0$, also called **off-shell**. That equation of motion was

$$(\eta_{\mu\nu}p^2 - p_\mu p_\nu)\epsilon^\nu = 0. \qquad (2.152)$$

The matrix that corresponds to the object in the parentheses, $D_{\mu\nu} \equiv \eta_{\mu\nu}p^2 - p_\mu p_\nu$, is degenerate and has determinant 0. To see the degeneracy, note that anything contributing to the polarization vector that is proportional to the photon momentum is annihilated:

$$(\eta_{\mu\nu}p^2 - p_\mu p_\nu)p^\nu = p_\mu p^2 - p_\mu p^2 = 0. \qquad (2.153)$$

That is, the matrix $D_{\mu\nu}$ has an eigenvector (p_μ) with eigenvalue 0. As such, $D_{\mu\nu}$ has no inverse.

This feature is a manifestation of the gauge invariance of electromagnetism. We are free to add with impunity the derivative of any scalar function to the vector potential, and we get back the same Maxwell's equations. When we encode the position dependence in the complex phase, the derivative turns into the photon momentum. That is, the gauge invariance of electromagnetism ensures that there are no electromagnetic fields in the direction of photon propagation. Electric and magnetic fields are purely transverse; this is encoded in the on-shell equation of motion,

$$p \cdot \epsilon = 0. \qquad (2.154)$$

This gauge invariance will actually cause problems if we want to define the Green's function of the photon by inverting the matrix $D_{\mu\nu}$. As it is singular, we cannot invert this matrix. However, once we fix a gauge by choosing a particular λ in Eq. 2.133, then the Green's function for the photon (in a particular gauge) is well defined. We won't discuss issues with gauge choice much in this book. Universally, we will use the **Feynman–'t Hooft gauge** as it renders the calculations that we do simplest. Feynman–'t Hooft gauge is a generalization of **Lorenz gauge**, which is defined by the requirement

$$\partial \cdot A = 0. \text{ (Lorenz gauge)} \qquad (2.155)$$

This corresponds to enforcing that the gauge function λ is harmonic:

$$\partial^2 \lambda = 0. \qquad (2.156)$$

In Feynman–'t Hooft gauge, the source-free equation of motion of the photon reduces to

$$\partial^2 A_\mu = 0, \qquad (2.157)$$

which is just the (massless) Klein–Gordon equation.

Exercises

2.1 *Properties of Lorentz Transformations.* By explicit matrix multiplication, show that the matrix Λ that implements a Lorentz boost along the z-axis with velocity β

$$\Lambda^{\nu}_{\mu} = \begin{pmatrix} \gamma & 0 & 0 & \gamma\beta \\ 0 & 1 & 0 & 0 \\ 0 & 0 & 1 & 0 \\ \gamma\beta & 0 & 0 & \gamma \end{pmatrix}^{\nu}_{\mu}, \tag{2.158}$$

leaves the metric invariant:

$$\Lambda^{\mu}_{\ \rho}\eta^{\rho\sigma}\Lambda_{\sigma}^{\ \nu} = \eta^{\mu\nu}. \tag{2.159}$$

Recall that the boost factor γ is

$$\gamma = \frac{1}{\sqrt{1-\beta^2}}. \tag{2.160}$$

2.2 *Rapidity.* In experimental particle physics, it is often very useful to express the direction of motion of a particle in terms of its **rapidity** y. The rapidity is defined as

$$y = \frac{1}{2}\log\frac{E+p_z}{E-p_z}, \tag{2.161}$$

for a particle with energy E and z-component of momentum p_z. What makes rapidity so nice is its simple properties under Lorentz transformation. Perform a Lorentz boost of the energy and momentum along the $\hat{z}$-axis with velocity β. How does the rapidity transform under this boost? You should be able to write the Lorentz-boosted rapidity as a simple function of the original rapidity.

2.3 *Lorentz-Invariant Measure.* Why is d^4x Lorentz invariant? Make the change of variables to a new frame and determine the Jacobian of the change of variables. You'll need to use properties of Lorentz transformations as defined by the relation Eq. 2.23.

   Hint: What is the Jacobian if d^4x is Lorentz invariant?

2.4 *Properties of Klein–Gordon Equation.* In the solution of the Klein–Gordon equation, we introduced the exponential phase factor $\exp[-ip \cdot x]$, where p is the four-momentum and x is the spacetime position. What is the frequency of oscillation of this solution to the Klein–Gordon equation? What is the wavelength? What is the phase velocity?

2.5 *Maxwell's Equations.* In Eq. 2.125, we showed how Gauss's law follows from taking the 0 component of the equations of motion of the electromagnetic Lagrangian. Show explicitly that the three other of Maxwell's equations follow from Eqs. 2.124 and 2.128. To do this, take individual components and identify the corresponding Maxwell's equation.

2.6 *Properties of the Clifford Algebra.* In writing the Dirac equation, we chose a particular representation of the γ matrices that satisfied $\{\gamma_\mu, \gamma_\nu\} = 2\eta_{\mu\nu}$, which is called the **Clifford algebra**. The choice we used is called the Weyl basis, in Eq. 2.101. In this exercise, we will study the Clifford algebra and the Weyl basis.

(a) By explicit multiplication, show that the γ matrices in the Weyl basis satisfy the Clifford algebra.

(b) The Weyl basis isn't the unique choice of matrices that satisfy the Clifford algebra. Another set of matrices that does so is

$$\gamma_0 = \begin{pmatrix} \mathbb{I} & 0 \\ 0 & -\mathbb{I} \end{pmatrix}, \qquad \gamma_i = \begin{pmatrix} 0 & \sigma_i \\ -\sigma_i & 0 \end{pmatrix}. \qquad (2.162)$$

This choice of γ matrices is related to the Weyl basis by a **similarity transformation**:

$$\gamma_\mu = S\gamma_\mu^{\text{Weyl}} S^\dagger, \qquad (2.163)$$

where S is a unitary matrix such that $SS^\dagger = \mathbb{I}$. A similarity transformation of the γ matrices respects the Clifford algebra. Determine the matrix S that relates the Weyl basis to this new basis of γ matrices.

(c) Applying the similarity transformation to the Dirac equation, we find

$$(iS\gamma_\mu S^\dagger \partial^\mu - m)\psi = S(i\gamma \cdot \partial - m)S^\dagger \psi = 0. \qquad (2.164)$$

Given the solution to the Dirac equation that we found in Example 2.4, use the similarity transformation S you found in the previous part to find the solutions to the Dirac equation using the γ matrix basis of Eq. 2.162.

2.7 *Relativity of Spin-1/2.* The Dirac equation involves the γ matrices, which themselves are constructed out of the Pauli spin matrices, σ_i. The Pauli matrices are

$$\sigma_1 = \begin{pmatrix} 0 & 1 \\ 1 & 0 \end{pmatrix}, \qquad \sigma_2 = \begin{pmatrix} 0 & -i \\ i & 0 \end{pmatrix}, \qquad \sigma_3 = \begin{pmatrix} 1 & 0 \\ 0 & -1 \end{pmatrix}. \qquad (2.165)$$

Construct a set of matrices σ_μ where

$$\sigma_\mu = (\mathbb{I}, \sigma_1, \sigma_2, \sigma_3), \qquad (2.166)$$

where $\mathbb{I}$ is the 2×2 identity matrix.

(a) For a momentum four-vector $p_\mu = (p_0, p_1, p_2, p_3)_\mu$, compute the 2×2 matrix $p \cdot \sigma$. What is $\det(p \cdot \sigma)$?

(b) What is the trace of this matrix, $\text{tr}(p \cdot \sigma)$? Recall that the trace is the sum of diagonal entries. What are the two eigenvalues of the matrix $p \cdot \sigma$? Use the fact that the determinant is the product of eigenvalues and the trace is the sum.

(c) A general Lorentz transformation of the matrix $p \cdot \sigma$ can be implemented by multiplying on the left and right by appropriate 2×2 matrices $\mathbb{A}$ and $\mathbb{B}$:

$$p \cdot \sigma \to \mathbb{A}p \cdot \sigma\mathbb{B}. \qquad (2.167)$$

Note that the matrix $p \cdot \sigma$ is Hermitian: $(p \cdot \sigma)^\dagger = p \cdot \sigma$. Use this to prove that $\mathbb{B} = \mathbb{A}^\dagger$.

(d) Using the result in part (a), what is $\det(\mathbb{A}\mathbb{A}^\dagger)$?

(e) Construct the matrices $\mathbb{A}$ that implement the following Lorentz transformations on the four-vector p:

(i) A rotation by an angle ϕ in the $\hat{x}\hat{y}$ plane.

(ii) A boost along the $\hat{z}$-axis with velocity β.

To do this, it is useful to first construct the corresponding matrix $p \cdot \sigma$ after Lorentz transformation, and then work out the components of the matrix $\mathbb{A}$ from that.

2.8 *Dark Matter Searches.* Numerous experiments are searching for **dark matter**, a component of energy in the universe necessary for the observed features of early universe cosmology, large-scale structure, and galactic dynamics. There has been no direct evidence for a particle nature of dark matter as of yet, but there are very strong constraints on its properties. Some of the strongest constraints were imposed by null observations of the **LUX** (**L**arge **U**nderground **X**enon) **dark matter experiment**. LUX consists of a vat of 250 kg of ultra-pure liquid xenon situated about 1 mile (1.6 kilometers) underground in the Sanford Underground Research Facility in Lead, South Dakota, USA. In this exercise, we will assume that the xenon is pure ^{131}Xe.

A major result from the LUX experiment is shown in Fig. 2.3. This is a plot of the constraints that it has placed on possible dark matter particles called **WIMPs** (**W**eakly **I**nteracting **M**assive **P**articles). LUX searches for WIMPs by their scattering off of the ^{131}Xe nuclei. If a significant amount of kinetic energy is transferred to the ^{131}Xe, then it excites a scintillator and is observed as a "hit." The amount of kinetic energy transfer is a function of the WIMP mass (the abscissa on the plot) and the rate of hits is a function of the strength of interaction of the WIMPs with xenon

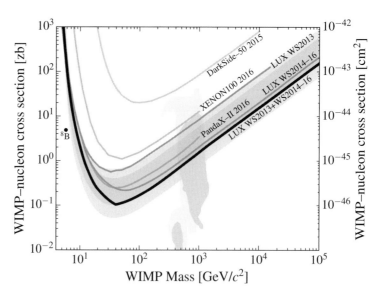

Fig. 2.3   A plot of the WIMP mass and interaction strength limits from the LUX experiment. Reprinted figure with permission from D. S. Akerib *et al.* [LUX Collaboration], Phys. Rev. Lett. **118**, no. 2, 021303 (2017). Copyright 2017 by the American Physical Society.

(the ordinate on the plot). The interaction strength is measured in a unit of cross section called a zeptobarn. We'll introduce the "barn" unit in Chapter 4. By the way, such a plot of dark matter mass versus interaction rate is called a **Goodman–Witten plot**.[9]

WIMP masses and interaction strengths have been **ruled out** (i.e., not observed to exist) by the LUX experiment above the thick black curve. In this exercise, we will study where the lower bound on the WIMP mass comes from (the nearly vertical bound on the left of the plot). You will study the upper bound on the mass in Exercise 4.8.

(a) A model for WIMP dark matter in the Milky Way galaxy is as a spherically distributed halo of particles that is at rest. What is the WIMP's apparent speed through the LUX detector on Earth? The radius of the solar system's orbit around the Milky Way is about 25,000 light-years, with an orbital period of about 230,000,000 years. (The fact that the Earth orbits the Sun affects this on a yearly timescale, but we will ignore this small effect.)

(b) The scattering process that LUX is sensitive to is

$$\chi + {}^{131}\text{Xe} \to \chi + {}^{131}\text{Xe}, \tag{2.168}$$

where χ denotes a WIMP particle. Write down the four-vectors for all particles involved in this collision. Assume that the initial xenon nucleus is at rest, and the initial dark matter particle χ's momentum is aligned along the $\hat{z}$-axis. Call these four-vectors p_{Xe} and p_χ, respectively. After collision, assume that both χ and xenon only have a non-zero $\hat{z}$-component of momentum. Call the four-vectors after collision p'_χ and p'_{Xe}. Make sure all four-vectors are on-shell and enforce three-momentum conservation (don't worry about energy conservation yet). Express the four-vectors in terms of the mass of the WIMP m_χ, the mass of the xenon nucleus m_{Xe}, the initial $\hat{z}$ momentum of the WIMP p_z, and the final $\hat{z}$-component of momentum of the xenon, p_z^{Xe}. (Assuming that the scattering occurs in a line means that the energy transfer to the xenon is maximized.)

(c) Now, using the conservation of the momentum four-vectors,

$$p_\chi + p_{\text{Xe}} = p'_\chi + p'_{\text{Xe}}, \tag{2.169}$$

solve for the recoiling xenon's momentum p_z^{Xe}. A nice way to do this is to write $p_z^{\text{Xe}} = \alpha p_z$, and solve for α. You will find two solutions; only report the solution corresponding to the larger value of momentum. You should find

$$p_z^{\text{Xe}} = 2p_z \frac{m_{\text{Xe}}^2 + m_{\text{Xe}}\sqrt{m_\chi^2 + p_z^2}}{m_{\text{Xe}}^2 + m_\chi^2 + 2m_{\text{Xe}}\sqrt{m_\chi^2 + p_z^2}}. \tag{2.170}$$

(d) LUX measures the recoiling xenon nucleus's kinetic energy K. Relativistically, this is defined as $K = E - m_{\text{Xe}}$. What is the kinetic energy of the recoiling xenon nucleus as a function of WIMP mass m_χ and momentum p_z? Because the

---

[9] M. W. Goodman and E. Witten, "Detectability of certain dark matter candidates," Phys. Rev. D **31**, 3059 (1985).

velocity of the dark matter halo is small, Taylor expand your result to lowest non-zero order in p_z and set $p_z = m_\chi v$, where v is the WIMP velocity with respect to Earth. You should find

$$K = \frac{2m_\chi^2 m_{Xe}}{(m_\chi + m_{Xe})^2} v^2. \tag{2.171}$$

(e) Below a kinetic energy of about 1 keV, LUX cannot detect the recoiling xenon nucleus. Using the result of part (d), what is the minimum dark matter mass that can provide this kinetic energy kick? You'll also need the WIMP velocity v calculated in part (a) of this exercise. Compare this mass to the lower mass bound in Fig. 2.3. (We've made many simplifications, so you'll see that LUX is sensitive to masses a bit below what you will find in this exercise.)

2.9 *Top Quark Decay.* This exercise studies the decay of the top quark, the heaviest particle of the Standard Model. Because the top quark decays almost instantly, its properties have to be inferred from its decay products. At the LHC, a plot from a study by CMS to determine the top quark mass is presented in Fig. 2.4. This plot shows the invariant mass of the four-momenta of two of the decay products of the top quark (a bottom quark and a lepton) versus the number of observed events at the respective mass. The features of this distribution enable a determination of the top quark mass. Here, we will identify the endpoint of this distribution and its connection to the top quark. Later, in Chapter 11, we will compute this distribution in the

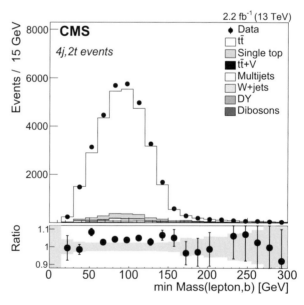

Fig. 2.4 Distribution of the invariant mass of the bottom quark and the charged lepton from the decay of a top quark produced in proton collisions at the CMS experiment. From A. M. Sirunyan *et al.* [CMS Collaboration], "Measurement of the $t\bar{t}$ production cross section using events with one lepton and at least one jet in pp collisions at $\sqrt{s} = 13$ TeV," J. High Energy Phys. **1709**, 051 (2017), doi:10.1007/JHEP09(2017)051 [arXiv:1701.06228 [hep-ex]].

context of the electroweak theory in Example 11.2 and you will test that prediction in Exercise 11.8.

(a) The top quark decays almost 100% of the time to a W boson and a bottom quark: $t \to W^+ + b$. It is easiest to analyze this decay using Lorentz-invariant four-vector dot products. Let p_t, p_W, and p_b be the top quark, W boson, and bottom quark momentum four-vectors, respectively. By conservation of momentum we have

$$p_t = p_W + p_b \,. \tag{2.172}$$

The bottom quark has a mass that is much smaller than the W boson, and so, to a good approximation, we can assume that it is massless. With this assumption, evaluate the dot product $2p_W \cdot p_b$. Your answer should be in terms of the mass of the top quark m_t and the mass of the W boson m_W.

(b) The W boson is also an unstable particle that decays. One of the ways that the W boson can decay is via a positron and a neutrino: $W^+ \to e^+ + \nu_e$. Assuming that both the positron and the neutrino are massless, evaluate the four-vector dot product $2p_l \cdot p_\nu$, where p_l is the four-vector of the positron and p_ν is the four-vector of the neutrino.

(c) Now, we want to combine these decays. The top quark decays sequentially as

$$t \to W^+ + b \to e^+ + \nu_e + b \,. \tag{2.173}$$

Using the conservation of four-momentum,

$$p_t = p_l + p_\nu + p_b \,, \tag{2.174}$$

express the four-vector dot product $2p_l \cdot p_b$ in terms of m_t, m_W, and the dot product $p_\nu \cdot p_b$.

(d) As we will discuss later in Chapter 5, neutrinos interact very weakly with other matter, and so are not observed in particle collision experiments. In the top quark decay studied in this exercise, we only observe the momentum of the bottom quark and the electron. This means that we cannot determine the top quark mass by just summing the four-momenta of its decay products. However, we can exploit appropriate limits. If the neutrino and bottom quark are both massless, what is the minimum value of the dot product $p_\nu \cdot p_b$? Therefore, what is the maximum value of $2p_l \cdot p_b$? The four-momentum of both the positron and the bottom quark are measurable, and so this endpoint can be found and used to determine the mass of the top quark. In the following, call the dot product $2p_l \cdot p_b \equiv m_{lb}^2$.

(e) Figure 2.4 is the distribution of m_{lb} measured by CMS. Using this plot and the result of the previous part, estimate the top quark mass, given the W mass $m_W = 80$ GeV. Note that because of the imperfections in identifying the bottom quark, the endpoint of this distribution is smeared out over a range of tens of GeV. To estimate the endpoint of the distribution, just extrapolate the steepest part of the distribution to the abscissa. The method used by CMS to determine the top quark mass is more sophisticated: information about the entire shape of the distribution is used, and not just the endpoint.

2.10 *Research Problem.* In particle physics, we assume that the universe is Lorentz invariant at short distances. Every observation ever made is consistent with this assumption. However, is this actually true? Is it possible for Lorentz invariance to be violated at shorter distances than we have probed?

3   A Little Group Theory

Beauty is typically associated with symmetry. We like things that have balance, a repeating structure, or look the same when viewed from various directions. We notice almost immediately if the subject of a photograph is uncentered, if a shelf isn't level, or if the pattern of a mosaic is abruptly broken. Beyond these desirable aesthetic properties, symmetries are also endowed with a rich mathematical description in the theory of groups. The connection of symmetries to physics is provided by Noether's theorem through their manifestation of conservation laws. Groups and their relevance in quantum mechanics were firmly established by Eugene Wigner in 1931 in a result now known as Wigner's theorem.[1] If someone says that a theory of Nature is "beautiful," that means that it is experimentally verified and exhibits an extensive symmetry.

In this chapter, we review groups and symmetries and their application to particle physics. We've already hinted at symmetries from Lorentz transformations in the previous chapter, but this likely raised more questions. We just postulated particles of different spin, but where does that come from? Why only those spins? Additionally, in quantum mechanics, commutation relations of Hermitian operators are central to identification of commensurate observables. Why Hermitian and why commutators? It all follows from applying the group theory to quantum mechanics, and we will develop this from the ground up in this chapter.

3.1 Groups as Symmetries

Here is a familiar object:

You might recognize it; it's a triangle. There's nothing identifying about the triangle, but it is special: it is an equilateral triangle. You might be wondering "What does a triangle have to do with particle physics?" and rightly so. Call the vertex at the top of the triangle

---

[1] E. P. Wigner, *Gruppentheorie und ihre Anwendung auf die Quantenmechanik der Atomspektren*, Vieweg (1931).

vertex 1 and going clockwise call the next vertex 2, and the final vertex 3. You'll have to remember this because we won't make any identifying marks on the triangle yet.

Now, imagine closing your eyes and while your eyes are closed, the vertices of the triangle are moved around. We'll keep the orientation of the triangle the same so that a vertex always points up. After the vertices are moved around and you open your eyes, can you identify which one is vertex 1? The triangle is oriented in the same way as before you closed your eyes, and it is equilateral, so there is no way you could know (unless the triangle had identifying marks otherwise). Any action while your eyes were closed just permuted the vertices. Operations that leave an object or system unchanged, like permutation of vertices of an equilateral triangle, are called **symmetries**. Symmetries are elevated to a central guiding principle through their consequences for conservation laws, by Noether's theorem.

Let's work systematically to determine what we could do to the triangle that makes it look the same before and after action of a symmetry transformation. To make the action of these symmetries more clear, we now write numbers on the vertices. One thing we can do is nothing at all. Another is rotation by 120° clockwise:

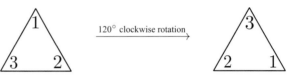

Another is rotation by 240°:

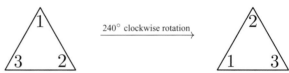

We could also rotate by 360°, but this is the same as not doing anything. Additionally, we could rotate by 120° or 240° counterclockwise as:

but this is identical to the 240° clockwise rotation. Therefore, there are only three possible rotations. Is that all we can do? As a rule of thumb, the answer to every rhetorical yes/no question is "No"; see Hinchliffe's rule.[2]

There's nothing that distinguishes the back from the front of the triangle, so we can also flip the triangle about a vertex. Let's flip about vertex 1:

---

[2] B. Peon, "Is Hinchliffe's rule true?," submitted to Annals Gnosis.

What if we flip about 1 again? This goes back to the original orientation; "nothing" was
done. We can flip about any vertex:

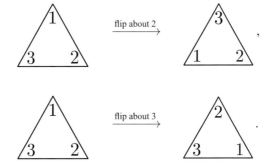

What if we do two different flips consecutively? Let's flip about 1 and then 2:

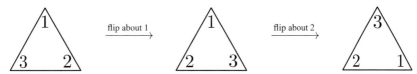

This is just identical to a 120° clockwise rotation. One can try this with any two operations
we have listed. The composition (= subsequent application) of two symmetries of a triangle
is still a symmetry. These six symmetries that we identified are everything.

These examples should be sufficient to illustrate that the action of these symmetries has
a rich mathematical structure. Indeed, they form a mathematical object called a **group**. A
group is a set of actions (like 120° rotation) that multiply (or compose) with an operation
denoted by ·. The set of these actions satisfies the four properties:

1 In the group, there is an identity element, 1. For any element a in the group,

$$a \cdot 1 = 1 \cdot a = a. \tag{3.1}$$

2 Every element of the group has an inverse. For an element a, we denote its inverse as
a^{-1} and it satisfies

$$a \cdot a^{-1} = a^{-1} \cdot a = 1. \tag{3.2}$$

3 The group is closed. For any two elements a and b in the group, their product c is in the
group:

$$a \cdot b = c \text{ is in the group}. \tag{3.3}$$

4 The multiplication/composition operation is associative. For three elements in the group
a, b, and c, the association of terms in a product is irrelevant:

$$(a \cdot b) \cdot c = a \cdot (b \cdot c). \tag{3.4}$$

These four properties define a group. The symmetry group of the equilateral triangle is
called the symmetric group of degree 3 and is denoted by S_3. It's a very useful exercise to
verify that the elements of S_3 we have identified here form a group, though we won't do
that here.

One thing to note here is that while these properties seem familiar from typical multiplication of numbers (which forms a group when 0 is not included), there is a very important property that is not implied by the group requirements. It is not required that a group be commutative: for two elements of the group a and b, we do not require

$$a \cdot b = b \cdot a. \tag{3.5}$$

For groups in which all elements do commute (like multiplication with numbers) we say that the group is **Abelian**. For groups where this is not true, the group is called **non-Abelian**, named after the Norwegian mathematician Niels Abel. The group of symmetries of the triangle are non-Abelian. Two different flips result in a different orientation if they are applied in opposite order:

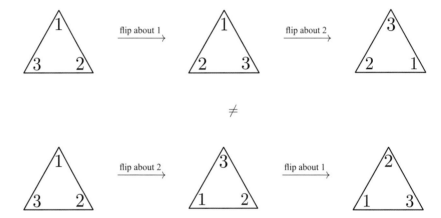

While we will consider and study Abelian groups in this book, most symmetries in particle physics are implemented by non-Abelian groups. To get a better intuition for groups, in the next sections we study the group of rotations in two and three dimensions. We've already seen this group in our discussion of the construction of the dot product in Section 2.1.1.

3.2 The Rotation Group

Recall that a matrix $\mathbb{M}$ that rotates a vector $\vec{v}$ leaves the identity matrix invariant:

$$\mathbb{M}^{\mathsf{T}} \mathbb{I} \mathbb{M} = \mathbb{I}. \tag{3.6}$$

The vector $\vec{v}$ is rotated as $\vec{v} \rightarrow \mathbb{M}\vec{v}$. The set of matrices that leave the identity invariant form a group. To see this, we just check all four properties. First, note that the identity matrix $\mathbb{I}$ is an element of the group:

$$\mathbb{I}^{\mathsf{T}} \mathbb{I} \mathbb{I} = \mathbb{I}. \tag{3.7}$$

The matrix $\mathbb{I}$ multiplied by any rotation matrix $\mathbb{M}$ is just $\mathbb{M}$.

The rotation matrix has an inverse: $\mathbb{M}^\mathsf{T}\mathbb{M} = \mathbb{I}$, and so $\mathbb{M}^{-1} = \mathbb{M}^\mathsf{T}$. If $\mathbb{M}$ is in the group, then so is its transpose $\mathbb{M}^\mathsf{T}$, as that implements a rotation in the opposite direction as $\mathbb{M}$. Matrix multiplication is associative: for three matrices $\mathbb{A}$, $\mathbb{B}$, and $\mathbb{C}$, we have

$$\mathbb{A}(\mathbb{B}\mathbb{C}) = (\mathbb{A}\mathbb{B})\mathbb{C}. \tag{3.8}$$

Finally, the set of matrices is closed. Let $\mathbb{M}$ and $\mathbb{N}$ be two rotation matrices:

$$\mathbb{M}^\mathsf{T}\mathbb{I}\mathbb{M} = \mathbb{I}, \qquad \mathbb{N}^\mathsf{T}\mathbb{I}\mathbb{N} = \mathbb{I}. \tag{3.9}$$

Then, consider the matrix $\mathbb{M}\mathbb{N}$. We have

$$(\mathbb{M}\mathbb{N})^\mathsf{T} = \mathbb{N}^\mathsf{T}\mathbb{M}^\mathsf{T}, \tag{3.10}$$

and so

$$(\mathbb{M}\mathbb{N})^\mathsf{T}\mathbb{I}(\mathbb{M}\mathbb{N}) = \mathbb{N}^\mathsf{T}\mathbb{M}^\mathsf{T}\mathbb{I}\mathbb{M}\mathbb{N} = \mathbb{N}^\mathsf{T}\mathbb{I}\mathbb{N} = \mathbb{I}. \tag{3.11}$$

This proves that the set of rotation matrices forms a group. The set of N-dimensional rotation matrices is called the orthogonal group, denoted as O(N).

3.2.1 Two-Dimensional Rotations: SO(2)

With the group nature of rotation matrices established, let's see how this manifests for rotations in two and three dimensions. Starting with rotations in two dimensions we consider those 2×2 matrices $\mathbb{M}$ that satisfy

$$\mathbb{M}^\mathsf{T} \begin{pmatrix} 1 & 0 \\ 0 & 1 \end{pmatrix} \mathbb{M} = \begin{pmatrix} 1 & 0 \\ 0 & 1 \end{pmatrix}. \tag{3.12}$$

Let's figure out some properties of this group. First, note that

$$\det(\mathbb{M}^\mathsf{T}\mathbb{M}) = \det \mathbb{M} \cdot \det \mathbb{M} = 1, \tag{3.13}$$

and so $\det \mathbb{M} = \pm 1$. What we typically think of as rotations preserve relative orientation of vectors. This corresponds to matrices with determinant 1 and are referred to as proper rotations. A matrix with a determinant of -1 flips entries of a vector, changing orientation. For example, the matrix

$$\begin{pmatrix} 0 & 1 \\ 1 & 0 \end{pmatrix} \tag{3.14}$$

flips $v_1 \leftrightarrow v_2$ when acting on $\vec{v} = (v_1, v_2)^\mathsf{T}$, while the matrix

$$\begin{pmatrix} 1 & 0 \\ 0 & -1 \end{pmatrix} \tag{3.15}$$

turns $v_1 \to v_1$, and $v_2 \to -v_2$. These negative determinant matrices have moved the $+\hat{x}$-axis from right of the $\hat{y}$-axis to the left. We will discuss this more in Chapter 10, but these matrices with determinant of -1 are examples of **parity transformations**: they flip entire axes, without rotation.

Proper rotations are implemented by matrices with $\det \mathbb{M} = 1$. Note that this restriction is a group: two matrices $\mathbb{M}$ and $\mathbb{N}$ each with determinant 1 have a product that also has determinant 1:

$$\det(\mathbb{M}\mathbb{N}) = \det \mathbb{M} \cdot \det \mathbb{N} = 1. \tag{3.16}$$

For the rest of this section, we will restrict to studying matrices with unit determinant.

The set of all N-dimensional matrices $\mathbb{M}$ with determinant 1 that satisfy

$$\mathbb{M}^{\mathsf{T}}\mathbb{M} = \mathbb{I} \tag{3.17}$$

is called the **special orthogonal group of dimension** N, denoted as SO(N). We're studying SO(2) now. All 2×2 matrices that are elements of SO(2) can be written as a function of one rotation angle θ:

$$\mathbb{M}(\theta) = \begin{pmatrix} \cos\theta & -\sin\theta \\ \sin\theta & \cos\theta \end{pmatrix}. \tag{3.18}$$

Note the product of two matrices with rotation angles θ and ϕ is simple:

$$\begin{aligned} \mathbb{M}(\theta)\mathbb{M}(\phi) &= \begin{pmatrix} \cos\theta & -\sin\theta \\ \sin\theta & \cos\theta \end{pmatrix} \begin{pmatrix} \cos\phi & -\sin\phi \\ \sin\phi & \cos\phi \end{pmatrix} \\ &= \begin{pmatrix} \cos\theta\cos\phi - \sin\theta\sin\phi & -\cos\theta\sin\phi - \sin\theta\cos\phi \\ \sin\theta\cos\phi + \cos\theta\sin\phi & -\sin\theta\sin\phi + \cos\theta\cos\phi \end{pmatrix} \\ &= \begin{pmatrix} \cos(\theta+\phi) & -\sin(\theta+\phi) \\ \sin(\theta+\phi) & \cos(\theta+\phi) \end{pmatrix}. \end{aligned} \tag{3.19}$$

This result implies that the SO(2) group is Abelian:

$$\mathbb{M}(\theta)\mathbb{M}(\phi) = \mathbb{M}(\phi)\mathbb{M}(\theta) = \mathbb{M}(\theta+\phi). \tag{3.20}$$

This makes sense: SO(2) is the group of symmetries of a circle. We can rotate the circle by any angle θ and the circle looks unchanged. An additional rotation by an angle ϕ is just another rotation, with a total rotation angle of $\theta + \phi$.

A circle can be represented in many ways, and this enables us to express its symmetries in many ways. One way to represent a circle is on the complex plane. The circle with radius r consists of those points (x, y) such that

$$|z|^2 = x^2 + y^2 = r^2. \tag{3.21}$$

A complex number z can be expressed in terms of a magnitude r and phase ϕ:

$$z = re^{i\phi}. \tag{3.22}$$

The absolute value squared is then $|z|^2 = re^{i\phi} \cdot re^{-i\phi} = r^2$. This is unchanged if the phase ϕ is rotated by any angle θ. That is, if $\phi \to \phi + \theta$, then

$$z \to re^{i(\phi+\theta)} \text{ but } |z|^2 \to |z|^2 = r^2. \tag{3.23}$$

A rotation of the circle in the complex plane by an angle θ is accomplished by multiplying complex numbers by $e^{i\theta}$. This angle $\theta \in [0, 2\pi)$, because angles larger than 2π

or smaller than 0 can be mapped onto $[0, 2\pi)$. Note also that the product of these factors, with angles of rotation of θ and ϕ, is

$$e^{i\theta} e^{i\phi} = e^{i(\theta + \phi)}. \tag{3.24}$$

This is the exact same multiplication law as we found for SO(2)! These rotations of a circle in the complex plane are called the **unitary group of one complex dimension**, or U(1). What we have shown is

$$\text{SO}(2) \simeq \text{U}(1), \tag{3.25}$$

where $\simeq$ means that these two groups are identical (isomorphic) as groups. U(1) is called **unitary** because an element times its complex conjugate is unity:

$$\left(e^{i\phi} \right) \left(e^{i\phi} \right)^* = e^{i\phi} e^{-i\phi} = 1. \tag{3.26}$$

We say that the set of rotations implemented by $e^{i\theta}$ are a **unitary representation** of SO(2). That is, they are unitary and satisfy or "represent" the multiplication law of SO(2).

3.2.2 Three-Dimensional Rotations: SO(3)

Our universe is not two-dimensional, it is three-dimensional, so SO(2) will be of limited use for understanding the properties of rotations. So, let's move on to discussing SO(3), the symmetries of a two-dimensional sphere. The matrices $\mathbb{M}$ that are elements of SO(3) satisfy

$$\mathbb{M}^\mathsf{T} \begin{pmatrix} 1 & 0 & 0 \\ 0 & 1 & 0 \\ 0 & 0 & 1 \end{pmatrix} \mathbb{M} = \begin{pmatrix} 1 & 0 & 0 \\ 0 & 1 & 0 \\ 0 & 0 & 1 \end{pmatrix}, \tag{3.27}$$

and $\det \mathbb{M} = 1$. Unlike SO(2), SO(3) is non-Abelian. We can visualize this through rotations of a three-dimensional object, like a coffee cup. Let's consider rotation of the handle, and then about the cup; and vice-versa. We find something different depending on the order of rotations, as shown in Fig. 3.1. These rotations can be expressed by the three **Euler angles**, which might be familiar from classical mechanics. There are three Euler angles because there are three orthogonal planes to rotate in three dimensions.

For applications to quantum mechanics and particle physics, we want to identify the unitary representations of SO(3). This is not just a novelty; unitary representations preserve probabilities in quantum mechanics. For concreteness, let's consider a spinor ψ that describes the electron, a spin-1/2 particle. A rotation of the spinor is implemented by a matrix $\mathbb{M}$ as

$$\psi \to \mathbb{M}\psi, \tag{3.28}$$

where the matrix $\mathbb{M}$ rotates the different components of the spinor into one another. ψ is a quantum mechanical object: it represents a complex probability amplitude for observing the electron with a given spin. The probability density ρ is formed from multiplying ψ by its conjugate $\bar{\psi}$:

$$\rho = \bar{\psi}\psi. \tag{3.29}$$

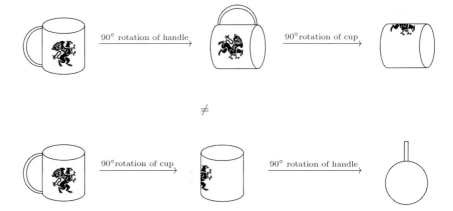

Fig. 3.1 An illustration of the non-Abelian nature of rotations in three dimensions, SO(3). Composing rotations of a coffee cup in different orders results in a different final orientation of the cup.

As a probability density, ρ is an observable and so cannot change under a rotation implemented by $\mathbb{M}$. Explicitly, ρ transforms under a rotation as

$$\rho = \bar{\psi}\psi \rightarrow (\bar{\psi}\mathbb{M}^\dagger)(\mathbb{M}\psi) = \bar{\psi}\psi = \rho. \tag{3.30}$$

$\bar{\psi}$ transforms with the Hermitian conjugate $\mathbb{M}^\dagger$ from the right because it is itself the transpose conjugate of ψ. For ρ to be invariant and $\mathbb{M}$ to be a symmetry we must enforce

$$\mathbb{M}^\dagger\mathbb{M} = \mathbb{I}. \tag{3.31}$$

This defines $\mathbb{M}$ to be a unitary matrix that implements a rotation. More generally, a symmetry in quantum mechanics is implemented by a unitary matrix or a unitary representation of the symmetry group, as required by probability conservation.[3]

Enforcing unitarity of $\mathbb{M}$ is actually quite easy. The procedure was essentially introduced in our discussion of SO(2) and U(1). While SO(2) elements are 2×2 matrices, U(1) elements are just exponential phases, with absolute value equal to unity. We can generalize this procedure to SO(3) or any group of which we want a unitary representation. Let's write the matrix $\mathbb{M}$ in the group SO(3) as an exponential:

$$\mathbb{M} = e^{iX}, \tag{3.32}$$

where X is itself a matrix. Exponentiating a matrix may look weird, but it is just defined by the Taylor expansion

$$e^{iX} = \mathbb{I} + iX - \frac{X^2}{2} - i\frac{X^3}{6} + \cdots. \tag{3.33}$$

Multiplying $\mathbb{M}$ by its Hermitian conjugate then establishes what we mean by unitary representation:

$$\mathbb{M}^\dagger\mathbb{M} = e^{-iX^\dagger}e^{iX} = \mathbb{I} = e^{i(X-X^\dagger)}. \tag{3.34}$$

---

[3] This is almost true. A symmetry in quantum mechanics could also be implemented by an anti-unitary operator. Such an operator still preserves probability, but also complex conjugates the object that it acts on. We'll study this in Chapter 10.

For the final equality to hold, we must enforce $X = X^\dagger$, or that the exponentiated matrix is **Hermitian**.

A particularly nice way to represent the exponentiated matrix X is by a linear combination of a basis of matrices T^a, where a ranges over the basis matrices. That is, we express

$$X = \alpha^a T^a , \tag{3.35}$$

where the α^a are real constants and Einstein summation is employed. We call the basis matrices T^a the **Lie algebra** of the group, named after another Norwegian mathematician, Sophus Lie. A general unitary matrix $\mathbb{M}$ is then expressed in terms of its Lie algebra as

$$\mathbb{M} = e^{i\alpha^a T^a} . \tag{3.36}$$

In the case of SO(3), we denote its Lie algebra with lowercase Gothic script $\mathfrak{so}(3)$. The Lie algebra $\mathfrak{so}(3)$ consists of three matrices T_x, T_y, T_z which implement rotations about the x, y, and z axes, respectively. The real coefficients that appear in the exponentiated group element correspond to the rotation angle about each axis. Quantum mechanically, because the Lie algebra matrices are Hermitian, we interpret them as measurement operators whose real eigenvalues correspond to possible outcomes of experiments. For $\mathfrak{so}(3)$, for example, T_x measures the x-component of angular momentum of a wavefunction.

The Lie algebra isn't just a vector space. It has a very rich structure which is enforced by closure of the group. Let's consider two group elements $\mathbb{M}$ and $\mathbb{N}$ expressed as

$$\mathbb{M} = e^{i\alpha^a T^a} , \qquad\qquad \mathbb{N} = e^{i\beta^a T^a} , \tag{3.37}$$

for some real constants α^a and β^a. If the group is closed, then the product of $\mathbb{M}$ and $\mathbb{N}$ must be able to be written as an exponentiated linear combination of the Lie algebra:

$$\mathbb{M}\mathbb{N} = e^{i\alpha^a T^a} e^{i\beta^a T^a} = e^{i\gamma^a T^a} , \tag{3.38}$$

for some real constants γ^a. If the T^a were just numbers, as they are for U(1), γ^a would just be the sum of α^a and β^a. However, matrices in general do not commute, and so one must be very careful with the order of multiplication. Using the Taylor expansion definition of the exponentiated matrices, carefully multiplying $\mathbb{M}$ by $\mathbb{N}$, one finds

$$\mathbb{M}\mathbb{N} = e^{i\alpha^a T^a} e^{i\beta^a T^a} = e^{i(\alpha^a + \beta^a)T^a + \frac{\alpha^a \beta^b}{2}[T^a, T^b] + \cdots} = e^{i\gamma^a T^a} . \tag{3.39}$$

$[T^a, T^b]$ represents the **commutator** of the Lie algebra matrices:

$$[T^a, T^b] \equiv T^a T^b - T^b T^a . \tag{3.40}$$

The dots in the exponent correspond to nested commutators involving three, four, and more matrices. This product of exponentiated matrices is called the **Baker–Campbell–Hausdorff (BCH) formula**, and you will show this structure in Exercise 3.4.

For the result of the BCH formula to be compatible with the unitary representation of group elements, we must enforce the very non-trivial quadratic relationship of the Lie algebra:

$$[T^a, T^b] = if^{abc} T^c . \tag{3.41}$$

The f^{abc} objects are called the **structure constants** and are just a collection of real numbers. This commutation relation *defines* the Lie algebra. Given a basis T^a and the structure constants f^{abc}, the group elements and their multiplication rules are uniquely defined by exponentiation. While this seems quite abstract, you have likely seen the $\mathfrak{so}(3)$ commutation relation before, when studying angular momentum in quantum mechanics. For an angular momentum operator $\hat{L}_i$ that implements a rotation about axis i, where $i = x, y, z$, the commutation relation is

$$[\hat{L}_i, \hat{L}_j] = i\epsilon_{ijk}\hat{L}_k \,. \tag{3.42}$$

ϵ_{ijk} is called the Levi–Civita tensor or **totally anti-symmetric symbol** which is defined to be 1 if ijk are in order (xyz, zxy, or yzx), -1 if they are in the opposite order (zyx, xzy, or yxz), and 0 if any two of i, j, and k are identical. For example, we have

$$[\hat{L}_x, \hat{L}_y] = i\hat{L}_z \,. \tag{3.43}$$

Note that this structure of the Lie algebra implies that SO(3) is a non-Abelian group. A group is Abelian if and only if the structure constants of its Lie algebra are all 0.

3.2.3 SO(3), SU(2), and Spin

As SO(3) is the rotation group of symmetries of a sphere, we naturally expect its Lie algebra to consist of 3×3 matrices. Just as we saw with SO(2) and U(1), however, we can represent a group in many different ways and all that is required is that the Lie algebra $\mathfrak{so}(3)$ is satisfied. Any set of matrices that satisfies Eq. 3.42 is $\mathfrak{so}(3)$ and defines quantum mechanical observables. One set of matrices that satisfies the $\mathfrak{so}(3)$ Lie algebra are the Pauli spin matrices, divided by 2. By explicit multiplication, you can show that

$$\left[\frac{\sigma_i}{2}, \frac{\sigma_j}{2}\right] = i\epsilon_{ijk}\frac{\sigma_k}{2} \,, \tag{3.44}$$

where the σ_i are defined in Eq. 2.104. Oddly, the Pauli spin matrices are 2×2 matrices, and define the Lie algebra $\mathfrak{su}(2)$ for the group SU(2). SU(2) is the **special unitary group of 2×2 unitary matrices with determinant 1**. Apparently, the $\mathfrak{su}(2)$ and $\mathfrak{so}(3)$ Lie algebras are identical:

$$\mathfrak{su}(2) \simeq \mathfrak{so}(3) \,. \tag{3.45}$$

The consequences of this identification for particle properties are profound. SU(2) matrices enact rotations on two-component spinors, as we saw in the solutions of the Dirac equation. The properties of these rotations, however, are very strange. Let's just consider a rotation by an angle ϕ about the $\hat{z}$-axis implemented by the Lie algebra $\mathfrak{su}(2)$. The matrix that implements this rotation is found by exponentiating σ_3 appropriately:

$$\mathbb{M}(\phi) = e^{i\phi\frac{\sigma_3}{2}} = \begin{pmatrix} e^{i\frac{\phi}{2}} & 0 \\ 0 & e^{-i\frac{\phi}{2}} \end{pmatrix} \,. \tag{3.46}$$

This compact form of the matrix can be found by explicitly summing the Taylor expansion definition of the exponential. Now, we see the importance of the factors of $1/2$ in the $\mathfrak{su}(2)$ Lie algebra with the Pauli matrices. When we rotate about the $\hat{z}$-axis by 2π, we might expect to get back to where we started. However, this matrix does not become the identity matrix if $\phi = 2\pi$:

$$\mathbb{M}(2\pi) = \begin{pmatrix} e^{i\pi} & 0 \\ 0 & e^{-i\pi} \end{pmatrix} = \begin{pmatrix} -1 & 0 \\ 0 & -1 \end{pmatrix}. \tag{3.47}$$

A rotation by 2π is equivalent to multiplying by -1! You only get back to the identity after rotating by 4π:

$$\mathbb{M}(4\pi) = \begin{pmatrix} 1 & 0 \\ 0 & 1 \end{pmatrix}. \tag{3.48}$$

This feature of rotations of spinors is why we call them "spin-1/2." A rotation by 2π only rotates the spinor by $1/2$ of that, or by an angle π.

Symmetries of the sphere don't describe all possible rotations in three dimensions. From our analogy of rotations as symmetries of a sphere, the fact that only rotating by 4π gets you back to the identity may seem extremely counterintuitive. To make sense of this feature of rotations, there is a simple demonstration that you can do. The procedure is illustrated in Fig. 3.2. Put your arm out straight with your palm up and imagine that you are carrying a plate on your hand. Now, rotate your hand by 2π, passing your hand above your arm, keeping your palm up so the plate doesn't fall off. You'll find that your hand is in the original orientation, but your arm is twisted. The remarkable thing about rotations in three dimensions is that you can rotate your hand in the same direction and completely untwist your arm. Now, rotate your hand by 2π keeping the plate on your palm, but this time pass your hand *under* your arm. If you did it right, your arm should be untwisted! Only after a total rotation of 4π can you untwist your arm. Note the importance of working in three dimensions: you have to rotate *above* and *below* your arm to untwist it. It's therefore impossible to untwist in two dimensions by continuing to rotate. This demonstration is referred to as the **plate trick**, but it is also called Dirac's belt trick, the Balinese cup trick, and a number of other names.

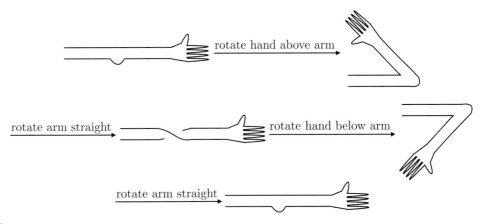

Fig. 3.2 Illustration how to rotate your hand to demonstrate that rotations by 4π in three dimensions are identical to no rotation.

Even though we started with what we thought of as the rotation group in three dimensions, SO(3), by following our noses guided by probability conservation in quantum mechanics, we are led to the group SU(2). Elements of SU(2) implement rotations in three dimensions. The particular type of matrix that implements the rotation of a given object defines its spin. Therefore, we also call the group SU(2) the **spin group** as its representations define different spins. For the first few representations:

1 The one-dimensional representation consists of one-dimensional matrices with determinant 1 or just the number 1. That is, the one-dimensional representation consists of a single spin state that is unchanged by a rotation; that is, a spin-0 object.

2 We've discussed the two-dimensional representation a bit already. This consists of 2×2 matrices that rotate two-component spinors. That is, they rotate an object that has two spin states: spin-up and spin-down. This corresponds to a spin-1/2 object.

3 The three-dimensional representation consists of 3×3 matrices that implement rotations on three-component vectors. This describes a spin-1 object, which has three possible spin states: spin 1, spin 0, and spin -1.

Representations of SU(2) exist with dimension equal to any natural number. A spin ℓ object rotates with SU(2) matrices in the $2\ell + 1$ dimensional representation, where ℓ is a half- or whole-integer value.

To close this section, let's apply this insight to calculating the rotation of a spinor about an axis perpendicular to the direction of spin.

Example 3.1 Consider a spinor in the spin-up state along the $\hat{z}$-axis. What is the spinor after a rotation by an angle ϕ about the $\hat{x}$-axis?

Solution

The first thing we need to do is to construct the spinor that represents spin-up. For spin-1/2, the $\hat{z}$-component of angular momentum corresponds to the matrix

$$\hat{L}_z = \frac{\sigma_3}{2} = \frac{1}{2} \begin{pmatrix} 1 & 0 \\ 0 & -1 \end{pmatrix}. \tag{3.49}$$

Spin-up about the $\hat{z}$-axis is represented by a two-component spinor that has eigenvalue $+1/2$ when acted on by $\hat{L}_z$. By appropriate normalization, this spinor just has 1 in the top component and 0 in the lower component:

$$\hat{L}_z \begin{pmatrix} 1 \\ 0 \end{pmatrix} = \frac{1}{2} \begin{pmatrix} 1 \\ 0 \end{pmatrix} = \frac{1}{2} \begin{pmatrix} 1 & 0 \\ 0 & -1 \end{pmatrix} \begin{pmatrix} 1 \\ 0 \end{pmatrix}. \tag{3.50}$$

A rotation about the $\hat{x}$-axis is implemented by exponentiation of σ_1, where

$$\sigma_1 = \begin{pmatrix} 0 & 1 \\ 1 & 0 \end{pmatrix}. \tag{3.51}$$

Note that the square of this matrix is just the identity:

$$\sigma_1^2 = \begin{pmatrix} 0 & 1 \\ 1 & 0 \end{pmatrix} \begin{pmatrix} 0 & 1 \\ 1 & 0 \end{pmatrix} = \begin{pmatrix} 1 & 0 \\ 0 & 1 \end{pmatrix}. \tag{3.52}$$

Therefore, the matrix that implements the rotation is

$$\mathbb{M}(\phi) = e^{i\phi\frac{\sigma_1}{2}} = \mathbb{I} + i\frac{\phi}{2}\sigma_1 - \frac{1}{2}\left(\frac{\phi}{2}\right)^2\sigma_1^2 - i\frac{1}{6}\left(\frac{\phi}{2}\right)^3\sigma_1^3 + \cdots \tag{3.53}$$

$$= \mathbb{I} + i\frac{\phi}{2}\sigma_1 - \frac{1}{2}\left(\frac{\phi}{2}\right)^2\mathbb{I} - i\frac{1}{6}\left(\frac{\phi}{2}\right)^3\sigma_1 + \cdots$$

$$= \mathbb{I}\cos\frac{\phi}{2} + i\sigma_2\sin\frac{\phi}{2}.$$

That is,

$$\mathbb{M}(\phi) = \begin{pmatrix} \cos\frac{\phi}{2} & i\sin\frac{\phi}{2} \\ i\sin\frac{\phi}{2} & \cos\frac{\phi}{2} \end{pmatrix}. \tag{3.54}$$

Acting this matrix on the spinor, we find

$$\mathbb{M}(\phi)\begin{pmatrix} 1 \\ 0 \end{pmatrix} = \begin{pmatrix} \cos\frac{\phi}{2} & i\sin\frac{\phi}{2} \\ i\sin\frac{\phi}{2} & \cos\frac{\phi}{2} \end{pmatrix}\begin{pmatrix} 1 \\ 0 \end{pmatrix} = \begin{pmatrix} \cos\frac{\phi}{2} \\ i\sin\frac{\phi}{2} \end{pmatrix}. \tag{3.55}$$

For a rotation angle of $\phi = 0$, this of course does nothing to the spinor. However, a rotation by $\phi = \pi$ flips the spin from spin-up to spin-down. As observed earlier, the spinor only returns to itself after rotation by $\phi = 4\pi$.

3.3 Isospin and the Quark Model

Particle physics, such as it was, was becoming a more mature field in the 1930s. The neutron was discovered by James Chadwick in 1932,[4] about 20 years after the discovery of the proton by Ernest Rutherford.[5] The discovery of the neutron enabled the first quantitative understanding of atomic nuclei and consequences of extracting energy from them. Scientists at the time noted that there were some very interesting similarities between the proton and the neutron. From the PDG, some properties of protons and neutrons are summarized in Table 3.1.

Some of these things are eerily similar: both protons and neutrons are spin-1/2 particles and their masses differ by about 2 MeV (about 1 part in 500). Some of these things are not: to the best of our knowledge, the proton is stable; its lifetime is at least 10^{29} years, while the neutron decays in about 900 seconds, or 15 minutes. (When bound in a nucleus, the neutron is stable; when it is isolated, it can decay.) These times differ by a factor of

[4] J. Chadwick, "Possible existence of a neutron," Nature **129**, 312 (1932).
[5] E. Rutherford, "The scattering of alpha and beta particles by matter and the structure of the atom," Phil. Mag. Ser. 6 **21**, 669 (1911).

Table 3.1 Proton and Neutron Properties		
	proton	neutron
mass	938 MeV	940 MeV
lifetime	$> 2.1 \times 10^{29}$ yrs	~ 900 s
charge	$+1\,e$	$0\,e$
spin	1/2	1/2

about 10^{34}. However, for times relevant for particle physics, 900 seconds is a really long time. For a neutron traveling near the speed of light, this corresponds to a distance of about 3×10^{11} meters, or about the distance between the Earth and the Sun. So, for particle physics, 900 seconds is essentially an infinite amount of time.

The other thing that differs between protons and neutrons is their electric charge. This is of course very important for chemistry, but depending on the questions we ask in particle physics, this may be irrelevant. For now, let's ignore the difference of charges of protons and neutrons. Another way to say this is to imagine a universe where electromagnetism doesn't exist. In this universe, protons and neutrons would be identical. That is, we could change all protons to neutrons and vice-versa and everything would be the same. While we don't live in this universe, we approximately live in this universe. It will be useful to study this approximate symmetry between protons and neutrons.

3.3.1 Isospin

With these observations after the discovery of the neutron, Werner Heisenberg proposed an approximate symmetry between protons and neutrons, later called isotopic spin, or **isospin**.[6] (Isospin, despite the name, has nothing to do with the spins of protons and neutrons.) Let's figure out what this isospin is. Protons and neutrons are described quantum mechanically as wavefunctions, which we will denote with the bra-ket notation as $|p\rangle$ and $|n\rangle$, respectively. These ket wavefunctions are normalized with the corresponding Hermitian conjugate bra:

$$\langle p|p\rangle = \langle n|n\rangle = 1\,, \tag{3.56}$$
$$\langle p|n\rangle = \langle n|p\rangle = 0\,.$$

The symmetry between protons and neutrons means that any linear combination of their wavefunctions describes the same physics.

We will call a generic linear combination a **nucleon** state, denoted by $|N\rangle$, where

$$|N\rangle = a|p\rangle + b|n\rangle\,. \tag{3.57}$$

a and b are complex numbers which are constrained by demanding that the nucleon wavefunction is also normalized:

$$\langle N|N\rangle = ((\langle p|a^* + \langle n|b^*)(a|p\rangle + b|n\rangle)) = |a|^2 + |b|^2 = 1\,. \tag{3.58}$$

---

[6] W. Heisenberg, "On the structure of atomic nuclei," Z. Phys. **77**, 1 (1932).

That is, a and b represent the probability amplitudes for the nucleon to be in the proton and neutron states, respectively. Any other linear combination of the proton and neutron can be implemented by a linear operator; let's call it $\hat{U}$. As it is a linear operator it can be represented by an outer product of proton and neutrons kets with bras:

$$\hat{U} = \alpha|p\rangle\langle p| + \beta|n\rangle\langle n| + \gamma|p\rangle\langle n| + \delta|n\rangle\langle p| , \qquad (3.59)$$

where $\alpha, \beta, \gamma, \delta$ are some complex numbers. To see the power of this notation, let's act $\hat{U}$ on the proton:

$$\hat{U}|p\rangle = (\alpha|p\rangle\langle p| + \beta|n\rangle\langle n| + \gamma|p\rangle\langle n| + \delta|n\rangle\langle p|)|p\rangle = \alpha|p\rangle + \delta|n\rangle . \qquad (3.60)$$

Identifying α with a and δ with b, we see that the operator $\hat{U}$ can produce an arbitrary nucleon state by acting on the proton. A similar result is found by acting on the neutron.

So far, the $\alpha, \beta, \gamma, \delta$ coefficients in $\hat{U}$ are unconstrained; let's see what constrains them. $\hat{U}$ must preserve the normalization of the nucleon state $|N\rangle$ to ensure that the total probability is unity:

$$\langle N|\hat{U}^\dagger\hat{U}|N\rangle = \langle N|N\rangle = 1 , \qquad (3.61)$$

or that $\hat{U}^\dagger\hat{U} = \hat{I}$ is just the identity operator. Acting on any state, the identity operator $\hat{I}$ just returns that state, and so is

$$\hat{I} = |p\rangle\langle p| + |n\rangle\langle n| . \qquad (3.62)$$

This requirement forces $\hat{U}$ to be unitary and, because it acts on a linear combination of two states, the set of all such $\hat{U}$ forms the group U(2),[7] the set of all 2×2 unitary matrices. As should be usual by now, we will further restrict the $\hat{U}$ operator to have determinant 1; in terms of the coefficients $\alpha, \beta, \gamma, \delta$, this is the restriction that $\alpha\beta - \gamma\delta = 1$. The set of all such $\hat{U}$ then form the group SU(2) (which is the reason for the name isospin).

In this universe where there is an exact SU(2) isospin symmetry between protons and neutrons, Noether's theorem tells us that in reactions involving protons and neutrons, isospin is conserved. Of course, this isn't our universe, but in processes where the differences between protons and neutrons are irrelevant (or not dominant), we do expect isospin conservation. In particular, as long as we study protons and neutrons in a way that doesn't involve electromagnetism or don't care about the neutron decay, then isospin should be conserved. In addition to other conservation laws we have identified, isospin will help constrain particle interactions. You'll see how this works in Exercise 3.8.

Now, let's go in a different direction from the groups we've studied so far. Most of our attention has been focused around representations of symmetry groups of the smallest dimension. For example, we analyzed rotations of spin-1/2 spinors, which transform under the two-dimensional representation of the rotation group. Correspondingly, the proton and neutron transform as a two-dimensional representation of isospin, appropriately referred to as an isospin **doublet**. Higher-dimensional representations of these groups exist, but they have the property that they can be formed from the smallest-dimensional

---

[7] We still haven't found what we're looking for, however.

representation. As such, the smallest-dimensional representation of a group is called the **fundamental representation**. To see how this works, let's study how a state with two nucleons transforms under isospin.

Example 3.2 Let's consider the system that consists of two nucleons (protons or neutrons), which we denote generically as N. We will denote the wavefunction of two nucleons as $|NN\rangle$. How does this di-nucleon state transform under SU(2) isospin transformations?

Solution

So, what are we working with? The possible nucleon combinations are

$$|pp\rangle, \quad |pn\rangle, \quad |np\rangle, \quad |nn\rangle, \tag{3.63}$$

where the first entry is nucleon 1 and the second is nucleon 2. A generic di-nucleon state will be some linear combination of these basis states. These states are orthonormal and the notation employed here is somewhat of a shorthand. The di-proton state $|pp\rangle$, for example, is more fully represented as

$$|pp\rangle = |p\rangle_1 |p\rangle_2, \tag{3.64}$$

where the subscripts denote the appropriate nucleon. When taking an inner product with this notation, we only contract nucleon 1 with nucleon 1, and nucleon 2 with nucleon 2. Explicitly, we have, for example,

$$\langle pp|pn\rangle = {}_1\langle p|\,{}_2\langle p||p\rangle_1|n\rangle_2 = {}_1\langle p|p\rangle_1\,{}_2\langle p|n\rangle_2 = 0. \tag{3.65}$$

The isospin operator $\hat{U}$ accordingly acts on each nucleon individually.

Let's see what happens if we isospin transform the linear combination $a|pn\rangle + b|np\rangle$. This transforms according to $\hat{U}$ defined in Eq. 3.59 as

$$\hat{U}(a|pn\rangle + b|np\rangle) = a|(\alpha p + \delta n)(\gamma p + \beta n)\rangle + b|(\gamma p + \beta n)(\alpha p + \delta n)\rangle \tag{3.66}$$

$$= \alpha\gamma(a+b)|pp\rangle + \beta\delta(a+b)|nn\rangle + (a\alpha\beta + b\gamma\delta)|pn\rangle + (a\gamma\delta + b\alpha\beta)|np\rangle.$$

By appropriate choice of a and b, this transformation is extremely simple. Let's choose $b = -a$, which then eliminates the $|pp\rangle$ and $|nn\rangle$ states:

$$\hat{U}(a|pn\rangle - a|np\rangle) = a(\alpha\beta - \gamma\delta)|pn\rangle - a(\alpha\beta - \gamma\delta)|np\rangle.$$

The quantity $\alpha\beta - \gamma\delta$ is just the determinant of $\hat{U}$, which is 1 in the group SU(2). Further, demanding that the state $a|pn\rangle - a|np\rangle$ be normalized fixes $a = 1/\sqrt{2}$. Therefore, this transformation law is very simple:

$$\hat{U}\frac{1}{\sqrt{2}}(|pn\rangle - |np\rangle) = -\frac{1}{\sqrt{2}}(|pn\rangle - |np\rangle). \tag{3.67}$$

The fact that there is a "$-$" sign after the transformation means that we refer to this antisymmetric combination as **odd**. Because it transforms to itself, it is a one-dimensional representation, also called a **singlet**.

Table 3.2 Isospin Decompositions	
singlet	triplet
	$\lvert pp\rangle$
$\frac{1}{\sqrt{2}}(\lvert pn\rangle - \lvert np\rangle)$	$\frac{1}{\sqrt{2}}(\lvert pn\rangle + \lvert np\rangle)$
	$\lvert nn\rangle$

There is another linear combination of $\lvert pn\rangle$ and $\lvert np\rangle$ that can be formed that is orthogonal to the singlet. It is

$$\frac{1}{\sqrt{2}}(\lvert pn\rangle + \lvert np\rangle)\,. \tag{3.68}$$

Let's see how this transforms under an isospin transformation:

$$\hat{U}\frac{1}{\sqrt{2}}(\lvert pn\rangle + \lvert np\rangle) = \sqrt{2}\alpha\gamma\lvert pp\rangle + \sqrt{2}\beta\delta\lvert nn\rangle + \frac{\alpha\beta + \gamma\delta}{\sqrt{2}}(\lvert pn\rangle + \lvert np\rangle)\,. \tag{3.69}$$

While this doesn't transform into itself, it does transform to the combinations

$$\lvert pp\rangle\,, \quad \lvert nn\rangle\,, \quad \frac{1}{\sqrt{2}}(\lvert pn\rangle + \lvert np\rangle)\,. \tag{3.70}$$

Under any isospin transformation these three combinations transform into linear combinations of one another. That is, they are closed under isospin and form another representation. Because there are three combinations of nucleons, this is a three-dimensional representation, or a **triplet**. Note that this is symmetric under exchange of nucleons 1 and 2.

Therefore, we have shown that the state that consists of two nucleons $\lvert NN\rangle$ decomposes into the singlet and triplet representations as shown in Table 3.2.

By the way, deuterium (hydrogen with a proton and a neutron in the nucleus) corresponds to either the singlet or the symmetric combination of $\lvert pn\rangle$ and $\lvert np\rangle$, depending on the spin of the deuterium nucleus.

This example shows how we can construct higher-dimensional representations from products of the fundamental representation. This is often expressed in direct product and direct sum notation. Because the fundamental representation of SU(2) isospin is two-dimensional it is often denoted as **2**. Correspondingly, the one- and three-dimensional representations are **1** and **3**, respectively. We have shown that

$$\mathbf{2} \otimes \mathbf{2} = \mathbf{3} \oplus \mathbf{1}\,. \tag{3.71}$$

In this expression $\otimes$ means direct product (as in multiplying two nucleon wavefunctions together) and $\oplus$ is the direct sum (as in a general linear combination of the singlet and the triplet). The factors of $\pm 1/\sqrt{2}$ that appear in the decomposition of Table 3.2 are called **Clebsch–Gordan coefficients**. They tell you how to take linear combinations of the original states to construct other representations.

Further, this example illustrates **irreducible representations**, or irreps for short. In acting with an isospin transformation, we separated out the singlet and triplet contributions.

This was well defined because each transformed exclusively into itself, or was closed under isospin transformations. Irreps are the smallest set of states of a representation of a given dimension that is closed under the action of the symmetry group. Other examples of irreps that we've seen already are representations of the spin group, SU(2), which correspond to all those states with a total given spin.

3.3.2 What is a "Particle"?

With the title of this book containing the word "particle," you might think that the meaning of this term is well understood. Indeed, this word has already been used extensively in this text, but only now are we ready to define it properly. A "particle" in particle physics has a technical, quantum mechanical definition, and requires knowledge of the representations of symmetry groups under which it transforms. To see what this means, let's list a few desirable properties of the definition of a particle:

- The definition of a particle should not depend on its velocity. That is, a particle at rest and one that is moving (but otherwise identical) are the same particle.
- The definition of a particle should not depend on the choice of coordinate axes. That is, a particle with a given direction of its spin and one whose spin is at a relative angle (but otherwise identical) are the same particle.

While the considerations listed here are just related to the Lorentz transformation properties of a particle, this suggests a more general definition. We'll provide the definition first, and then work to unpack and understand it in the rest of this section.

Box 3.1	Definition of a particle

A **particle** (in particle physics) is an object that is localized in space and whose intrinsic properties are unchanged under the action of any symmetry group. That is, a particle is defined by the irreps under which it transforms.

Going back to the very beginning of this chapter, where we discussed the symmetries of an equilateral triangle, this definition of a particle is consistent with the colloquial definition of "equilateral triangle." An equilateral triangle doesn't depend on how you label the vertices, and anyway the action of S_3 mixes the vertices. Because the equilateral triangle has three vertices that are mixed with S_3, it transforms under the three-dimensional representation of S_3. Therefore, one unique way to define an equilateral triangle is as the object that transforms under the three-dimensional irrep of S_3. You'll explicitly construct this irrep in Exercise 3.1, as well as studying another irrep of S_3.

To see how this definition works for honest particles, let's consider its consequences for the electron, e^-. We'll denote the wavefunction of the electron as $|e^-\rangle$. We can measure the energy and momentum of the electron by acting with the momentum operator, $\hat{P}_\mu = i\partial_\mu$. The electron is an eigenstate of the momentum operator, with eigenvalue equal to its four-momentum p_μ:

$$\hat{P}_\mu |e^-\rangle = p_\mu |e^-\rangle. \tag{3.72}$$

The four-momentum p_μ is of course not Lorentz invariant, so we can't define an electron by its particular four-momentum. However, we can construct a Lorentz-invariant operator by acting again with $\hat{P}_\mu$:

$$\hat{P}_\mu \hat{P}^\mu |e^-\rangle = p^2 |e^-\rangle = m_e^2 |e^-\rangle. \tag{3.73}$$

The mass of the electron m_e tells you how the electron's energy and momentum are affected by a Lorentz boost with velocity β. So, the eigenvalue under the action of $\hat{P}_\mu \hat{P}^\mu$ defines a property of the electron particle. Equation 3.73 is of course nothing more than the Klein–Gordon equation.

The electron also has spin, and the measured value of spin depends on the choice of coordinate axes. Regardless of the choice of coordinates, the electron, as a spin-1/2 particle, always has two possible spin states. This number of possible spin states of the electron is unchanged by rotations. What operator can we construct that measures the number of spin states? This situation isn't so different from a typical vector like position, for example. A position vector $\vec{x} = (x, y, z)$ of course isn't rotation invariant; however, its magnitude is. The magnitude is defined by the Pythagorean theorem:

$$|\vec{x}|^2 = x^2 + y^2 + z^2. \tag{3.74}$$

For spin operators $\hat{S}_x$, $\hat{S}_y$, $\hat{S}_z$, this suggests that the squared-spin operator

$$\hat{S}^2 \equiv \hat{S}_x^2 + \hat{S}_y^2 + \hat{S}_z^2 \tag{3.75}$$

is rotation invariant and measures the "magnitude" of the electron spin.

If $\hat{S}^2$ is invariant to rotations, it must commute with all spin operators. For the electron, the spin operators are just the Pauli spin matrices, and it is easy to explicitly construct $\hat{S}^2$. As mentioned earlier, the square of any of the Pauli spin matrices is just the 2×2 identity matrix $\mathbb{I}$, so

$$\hat{S}^2 = \frac{\sigma_1^2}{4} + \frac{\sigma_2^2}{4} + \frac{\sigma_3^2}{4} = \frac{3}{4}\mathbb{I}. \tag{3.76}$$

The identity matrix of course commutes with any matrix. This sum-of-squares operator is called the **Casimir** of the representation and is unique to the particular representation of the symmetry group. Acting the spin Casimir on the electron wavefunction, this produces

$$\hat{S}^2 |e^-\rangle = s(s+1)|e^-\rangle = \frac{3}{4}|e^-\rangle. \tag{3.77}$$

Here, s is the value of the electron spin, $s = 1/2$. That the value of the Casimir is $s(s+1)$ we won't prove here, but this will be studied in more detail in Chapter 8. Another interpretation of the fact that the Casimir is proportional to the identity is that, regardless of whether the electron has spin-up or spin-down, the value of the Casimir is always the same.

Continuing, one can construct a charge operator $\hat{C}$ which measures the charge of the electron. We won't do that here, but we note that its eigenvalue when acting on the electron wavefunction is

$$\hat{C}|e^-\rangle = -e|e^-\rangle, \tag{3.78}$$

where e is the fundamental unit of charge. This is invariant to the action of any symmetry.

Table 3.3 Electron Quantum Numbers

property	Casimir	value
mass	$\hat{P}_\mu \hat{P}^\mu$	511 keV
spin	$\hat{S}^2$	1/2
charge	$\hat{C}$	$-e$

To summarize, we have identified operators whose eigenvalues identify the irreps of symmetry groups under which the electron transforms. The operators are called the Casimirs, and commute with every element of the Lie algebra of the symmetry group. The eigenvalues of the Casimirs are referred to as a particle's **quantum numbers** and the collection of them for all symmetry groups uniquely identifies the particle. For the electron, we present the quantum numbers corresponding to the mass, spin, and charge assignments in Table 3.3. Eugene Wigner initiated this definition of a particle, by identifying the irreps and quantum numbers of the momentum operator and Lorentz transformations. This is now called **Wigner's classification**.[8]

3.3.3 The Quark Model

The power of this group theory construction of particle physics is that it enables concrete predictions for the fundamental constituents of matter. Perhaps the most profound prediction from this group theory approach was with the construction of the **quark model**. In Exercise 1.2 in Chapter 1, we estimated the mass of the pion. There are actually three pions, denoted as π^+, π^-, and π^0, with $+e$, $-e$, and 0 electric charge, respectively. Like the proton and neutron, these pions have eerily close masses:

$$m_{\pi^+} = 139\,\text{MeV}\,, \qquad m_{\pi^-} = 139\,\text{MeV}\,, \qquad m_{\pi^0} = 135\,\text{MeV}\,. \qquad (3.79)$$

Their charge does distinguish them, but we can imagine a universe where electricity and magnetism are turned off. This is suggestive of the pions forming a triplet of isospin, as we saw with pairs of protons and neutrons. However, because the masses of protons and neutrons are much larger than those of the pions, the pions cannot consist of pairs of nucleons. Therefore, if they do form a triplet of isospin, they must consist of things more fundamental than protons and neutrons. This expectation is consistent with the triplet arising from a product of doublet representations of isospin. So, pions could consist of pairs of particles that are doublets of isospin.

In the 1950s and 1960s more and more particles like the pions, collectively called hadrons, were discovered and interesting relationships between them were identified. Beyond SU(2) isospin, it was realized that there was a larger SU(3) group structure to the pattern of hadrons. This symmetry is called SU(3) flavor. Murray Gell-Mann (though he was not aware of it at the time) had organized the measured hadrons in what he called

---

[8] E. P. Wigner, "On unitary representations of the inhomogeneous Lorentz group," Annals Math. **40**, 149 (1939) [Nucl. Phys. Proc. Suppl. **6**, 9 (1989)].

the *Eightfold Way*, which corresponded to the irreps of SU(3) flavor.[9] For example, one set of hadrons is called the **baryon octet** and consists of:

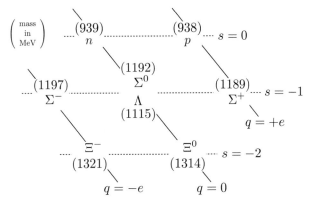

The masses of particles in MeV are in parentheses, with increasing masses going down. All particles along negatively sloped diagonals have the same electric charge q. A quantity called **strangeness** s also changes going down, with the top row having strangeness $s = 0$, the middle row strangeness $s = -1$, and the bottom row strangeness $s = -2$.

This octet is an eight-dimensional representation of flavor SU(3). Like the pion triplet of SU(2) isospin, it is not the fundamental representation of SU(3), as that would be three-dimensional. In fact, representations of flavor SU(3) corresponding to sets of hadrons of all stripes were observed, such as the octet and the ten-dimensional representation called the decuplet, for example. However, the fundamental representation of this symmetry was never observed. With this and other evidence, Gell-Mann and George Zweig theorized that there were fundamental particles whose different combinations produced the zoo of hadrons.[10] Because the observed flavor group was SU(3), they predicted that there were just three fundamental particles that were responsible for everything. Gell-Mann called them quarks,[11] after the poem in James Joyce's novel *Finnegans Wake*:[12]

> –Three quarks for Muster Mark!
> Sure he hasn't got much of a bark
> And sure any he has it's all beside the mark.

---

[9] M. Gell-Mann, "The Eightfold Way: A theory of strong interaction symmetry," CTSL-20, TID-12608. See also Y. Ne'eman, "Derivation of strong interactions from a gauge invariance," Nucl. Phys. **26**, 222 (1961).

[10] M. Gell-Mann, "A schematic model of baryons and mesons," Phys. Lett. **8**, 214 (1964); G. Zweig, *An SU(3) Model for Strong Interaction Symmetry and Its Breaking. Version 1*, CERN-TH-401, CERN (1964); G. Zweig, "An SU(3) model for strong interaction symmetry and its breaking: Version 2," in Lichtenberg, D. and Rosen, S. (eds), *Developments in the Quark Theory of Hadrons*, Vol. 1, pp. 22–101.

[11] Quark is also a fresh acid-set cheese, but not the eponym of the fundamental particle.

[12] Excerpt from FINNEGANS WAKE: CENTENNIAL EDITION by James Joyce, copyright © 1939 by James Joyce, copyright renewed © 1967 by Giorgio Joyce and Lucia Joyce. Used by permission of Viking Books, an imprint of Penguin Publishing Group, a division of Penguin Random House LLC. All rights reserved. Any third party use of this material, outside of this publication, is prohibited. Interested parties must apply directly to Penguin Random House LLC for permission.

These three quarks were named up (u), down (d), and strange (s). The pions, for example, consist of up and down quarks in the following combinations:

$$|\pi^+\rangle = |u\bar{d}\rangle\,, \quad |\pi^-\rangle = |\bar{u}d\rangle\,, \quad |\pi^0\rangle = \frac{1}{\sqrt{2}}(|u\bar{u}\rangle - |d\bar{d}\rangle)\,, \tag{3.80}$$

where $\bar{u}$ is the anti-up quark. A beautiful application of fundamental mathematics leading to discovery of new physics!

We now know of six quarks, the evidence for which we'll discuss throughout the rest of this book.

3.4 Why the Photon Has Two Polarizations

We'll end this chapter on group theory in particle physics by addressing a question that was raised in the previous chapter, but from a new perspective. In Section 2.2.3, we discussed electromagnetism, but couched in a language centered around the photon and its properties. There, we mentioned that the photon has two polarizations, or, equivalently, two spin states. This is the same situation as for the electron: the electron has two spin states (spin-up and spin-down), and yet the electron is spin-1/2 while the photon is spin-1. However, as the photon is spin-1, it naïvely transforms according to the three-dimensional representation of the spin group. Thus, for a spin 1 particle like the photon, we should expect three spin states, not two. So, what gives?

The reasoning behind this requires an understanding of the consequences of special relativity for the properties of the photon. The photon is a massless particle, unlike the electron, and it is this masslessness that is responsible for the effective elimination of one spin state. Let's first consider the case of a massive particle, like the electron. We will also need to assume that the particle is point-like, and has no spatial extent. This is true for the electron, and in fact true for all particles of the Standard Model; no intrinsic length scale has been measured for any of the particles of the Standard Model. For composite particles, like the proton, we work in the regime in which the proton appears point-like. That is, we imagine only probing the proton with wavelengths that are long as compared to its Compton wavelength.

With these assumptions, the only thing around to describe the dynamics of a massive particle is its momentum four-vector. As we are considering massive particles, we can boost to a frame in which the particle is at rest:

$$p = (m, 0, 0, 0)\,. \tag{3.81}$$

Here, m is the mass of the particle. This frame makes the analysis that follows clear, but Lorentz invariance implies that it must hold in any other frame. In this representation, this four-vector has a large amount of symmetry. Because the three-momentum of this particle at rest is just the $\vec{0}$ vector, we can, with impunity, rotate the spatial coordinates into one another with a 3×3 matrix $\mathbb{M}$. Because $\mathbb{M}$ implements a symmetry transformation on the particle's four-momentum, it must be an appropriate Lorentz transformation. The matrix

$$p = (m, 0, 0, 0)$$

Fig. 3.3 The rotational symmetry of a massive particle. Because we can Lorentz boost to its rest frame, there are three independent axes about which we can rotate.

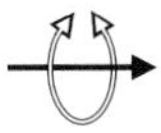

$$p = (E, 0, 0, E)$$

Fig. 3.4 The rotational symmetry of a massless particle. Because we cannot Lorentz boost to its rest frame, there is only one independent axis about which we can rotate.

$\mathbb{M}$ can therefore be expressed in terms of the Lie algebra of the rotation group, $\mathfrak{so}(3)$. The three elements of the Lie algebra corresponding to the three independent rotational directions of a point-like massive particle are illustrated in Fig. 3.3.

As discussed earlier, if this is a particle, it transforms as an irrep of the spin group, SU(2), the members of which are classified by an integer or half-integer value of spin: 0, 1/2, 1, 3/2, 2, etc. All massive particles that we have measured indeed have a definite spin which defines its properties under rotation. For example, the Higgs boson is a spin-0 or **scalar** particle, the electron is a spin-1/2 particle with two spin states, and the rho meson has spin-1 with three spin states. For massive particles, our understanding of spin states indeed aligns with our more familiar non-relativistic quantum mechanics intuition.

Now for a massless particle like the photon. Unlike with massive particles, there is no Lorentz transformation that can be performed to boost a massless particle into a frame where its four-momentum is of the form

$$p = (E, 0, 0, 0). \tag{3.82}$$

This would imply that the particle has mass E, which must be non-zero if the particle has non-zero energy. For a massless particle, the frame in which the four-momentum is simplest is when the three-momentum is aligned along the $\hat{z}$-direction:

$$p = (E, 0, 0, E). \tag{3.83}$$

This four-vector is indeed massless with energy E. Going through a similar analysis as with the massive particle, we find that the symmetries of this four-vector correspond to the group SO(2), as we can only rotate two axes, $\hat{x}$ and $\hat{y}$, into one another and leave this four-vector unchanged. SO(2) is identical to U(1) as a group and, unlike SU(2), it is also an Abelian group. Additionally, an element of SO(2) is specified by a single number which represents the corresponding rotation angle. For these reasons, the irreps of SO(2), which would correspond to different particles, are all one-dimensional. By the way, this SO(2) symmetry that leaves massless four-vectors unchanged is called the **little group**. The single rotational direction of a massless particle that leaves its four-vector invariant is illustrated in Fig. 3.4.

This has profound consequences for the physical polarization states of the photon. Because all irreducible representations of the SO(2) little group are one-dimensional, this means that massless particles have only one accessible spin state, regardless of the intrinsic spin of the particle. A massless particle always travels at the speed of light, and so its three-momentum is non-zero for any non-zero energy. Therefore, a massless particle has an intrinsic axis about which to quantize spin. That is, this one spin state can be specified by the projection of spin onto the axis defined by the three-momentum of the massless particle. This spin projection is called the particle's **helicity**. The magnitude of this projection of the spin onto the three-momentum axis is not so important; it is just determined by the intrinsic spin of the particle. However, the sign of the projection can be positive or negative; that is, the spin can be aligned with the three-momentum or it can be anti-aligned with the three-momentum. The sign of helicity of a massless particle is Lorentz invariant: there is no Lorentz transformation that you can perform to change the sign of helicity. This is unlike a massive particle, for which a rotation will change the projection of spin on any axis.

We say that the particle has positive helicity if the spin and three-momentum are aligned, and negative helicity if they are anti-aligned. These two spin or polarization states are all that are possible for a massless particle. When we discussed the polarization states for the photon we referred to them as right- and left-handed. Right-handed helicity just means that the spin of the photon is aligned with the three-momentum according to the right-hand rule, and analogously for left-handed.

It may seem that we could in principle just consider massless particles with only one helicity. In fact, this turns out to be inconsistent with quantum mechanics. First, according to the rules of quantum mechanics, everything that is not expressly forbidden is mandatory, and unless there is a mechanism for forbidding one helicity, then both necessarily are present. There is a deep and fundamental result in quantum field theory called the **CPT theorem**, which states that only if the Lorentz-invariant Hamiltonian of a system is invariant under the combined action of charge conjugation (C), parity transformation (P), and time reversal (T) is that Hamiltonian actually Hermitian, and so has real eigenvalues. The combined action of CPT on a right-handed helicity photon transforms it into a left-handed helicity photon, and vice-versa. Therefore, for the Hamiltonian of a quantum system to be Hermitian, we must require that both left- and right-handed helicities of massless particles be present. We will discuss charge conjugation, parity transformation, and time reversal in more detail in Section 10.2.

Effectively, the action of CPT is complex conjugation. Demanding that the combined action of CPT on the Hamiltonian $\hat{H}$ is a symmetry then imposes that $\hat{H} = \hat{H}^{\dagger}$, which is indeed the statement of Hermiticity. If $\hat{H}$ is Hermitian, then every state and its complex conjugate must be present in the Hilbert space of the theory. Recall that when we introduced photon polarization states in Section 2.2.3, we showed that right-handed polarization is the complex conjugate of left-handed polarization (and vice-versa):

$$\epsilon_R = \epsilon_L^* . \tag{3.84}$$

Therefore, if a right-handed photon state exists in the Hilbert space, then by the CPT theorem or equivalently the Hermitivity of the Hamiltonian, the left-handed photon state must exist as well.

Exercises

3.1 *Representations of the Symmetric Group.* The group of symmetries of the equilateral triangle discussed at the beginning of this chapter can be represented with matrices. Because the transformations of the triangle were just manipulations of a two-dimensional plane, this symmetry group is a subgroup of O(2). In this exercise, we will study the representations of this group, S_3.

(a) Consider a three-dimensional vector with entries a, b, and c:

$$\begin{pmatrix} a \\ b \\ c \end{pmatrix}. \tag{3.85}$$

Let a correspond to vertex 1 of the triangle, b correspond to vertex 2, and c correspond to vertex 3. Construct the six matrices that implement the six unique symmetries of the triangle. For example, the matrix

$$\begin{pmatrix} 0 & 0 & 1 \\ 1 & 0 & 0 \\ 0 & 1 & 0 \end{pmatrix} \tag{3.86}$$

implements a 120° clockwise rotation:

$$\begin{pmatrix} 0 & 0 & 1 \\ 1 & 0 & 0 \\ 0 & 1 & 0 \end{pmatrix} \begin{pmatrix} a \\ b \\ c \end{pmatrix} = \begin{pmatrix} c \\ a \\ b \end{pmatrix}. \tag{3.87}$$

(b) In part (a), you constructed the three-dimensional representation of the symmetric group, S_3. However, this is not the fundamental representation. Construct the six 2×2 matrices that correspond to the exact same group operations, S_3, and therefore to the two-dimensional representation. These matrices act on a two-dimensional vector,

$$\begin{pmatrix} x \\ y \end{pmatrix}, \tag{3.88}$$

where x and y represent the horizontal and vertical components, respectively. To construct the matrices, note the following properties. Three of the matrices correspond to rotations in the plane by 0°, 120°, and 240°. Another matrix flips about the vertical axis and sends x to $-x$, or, in terms of the triangle, flips about vertex 1. The other two flip matrices can be found by composing an appropriate rotation with the flip about vertex 1.

3.2 *Lorentz Group.* From the defining requirement that a Lorentz transformation implemented by a matrix Λ leave the metric invariant,

$$\Lambda^\mu_\rho \eta^{\rho\sigma} \Lambda^\nu_\sigma = \eta^{\mu\nu}, \tag{3.89}$$

prove that the set of such matrices $\{\Lambda\}$ form a group.

3.3 *Hermitian Matrices.* A Hermitian matrix $\mathbb{M}$ is one for which it is equal to its Hermitian conjugate:

$$\mathbb{M} = \mathbb{M}^\dagger = (\mathbb{M}^\mathsf{T})^* \,. \tag{3.90}$$

(a) Prove that Hermitian matrices have real eigenvalues. That is, for an eigenvector $\vec{v}$ of a Hermitian matrix $\mathbb{M}$ prove that eigenvalue λ is always real:

$$\mathbb{M}\vec{v} = \lambda\vec{v} \,. \tag{3.91}$$

(b) Assume that a Hermitian matrix $\mathbb{M}$ has two eigenvectors $\vec{v}_1$ and $\vec{v}_2$ with eigenvalues λ_1 and λ_2, with $\lambda_1 \neq \lambda_2$. Prove that $\vec{v}_1$ and $\vec{v}_2$ are orthogonal; that is, $\vec{v}_1^\dagger \vec{v}_2 = \vec{v}_2^\dagger \vec{v}_1 = 0$.

3.4 *Baker–Campbell–Hausdorff Formula.* In this chapter we introduced the Lie algebra as the basis of unitary representations of a group. An element of a group is constructed by exponentiating a matrix T in the Lie algebra:

$$\mathbb{M} = e^{iT} = \mathbb{I} + iT - \frac{T^2}{2} - i\frac{T^3}{6} + \cdots \,. \tag{3.92}$$

What is the multiplication law of exponentiated matrices? That is, for two matrices T_1 and T_2, determine the matrix T_3 such that

$$e^{iT_1} e^{iT_2} = e^{iT_3} \,. \tag{3.93}$$

Find T_3 up through quadratic order in matrices T_1 and T_2.

Hint: Express T_3 as

$$iT_3 = \log\left(e^{iT_1} e^{iT_2}\right) \,, \tag{3.94}$$

and Taylor expand to quadratic order in the matrices T_1 and T_2.

3.5 *Casimir Operator.* In this chapter, we introduced the Casimir operator as the object that measures the irrep under which a particle transforms. For the spin group, the Casimir $\hat{S}^2$ is defined as

$$\hat{S}^2 \equiv \hat{S}_x^2 + \hat{S}_y^2 + \hat{S}_z^2 \,. \tag{3.95}$$

Using the Lie algebra of the spin group

$$[\hat{S}_i, \hat{S}_j] = i\epsilon_{ijk}\hat{S}_k \,, \tag{3.96}$$

show that the Casimir commutes with all elements of the Lie algebra:

$$[\hat{S}^2, \hat{S}_i] = 0 \,. \tag{3.97}$$

3.6 *Helicity.* In Section 2.2.3, we constructed the right- and left-handed polarization vectors for the photon. For momentum $p = (E, 0, 0, E)$, they were

$$\epsilon_R(p) = \frac{1}{\sqrt{2}}(0, 1, i, 0) \,, \qquad \epsilon_L(p) = \frac{1}{\sqrt{2}}(0, 1, -i, 0) \,. \tag{3.98}$$

These polarization vectors are eigenstates of the little group, SO(2). Perform an SO(2) rotation in the $\hat{x}\hat{y}$ plane by an angle ϕ and determine the eigenvalues of these polarization vectors. What are the eigenvalues? Which polarization vector rotates with ϕ? Which one rotates oppositely? What happens to the polarization vectors after rotation by $\phi = 2\pi$?

3.7 *Symplectic Group.* The symplectic group Sp(N) consist of those $N \times N$ matrices $\mathbb{M}$ with determinant 1 that leave an off-diagonal matrix invariant:

$$\mathbb{M}^\mathsf{T} \begin{pmatrix} 0 & \mathbb{I} \\ -\mathbb{I} & 0 \end{pmatrix} \mathbb{M} = \begin{pmatrix} 0 & \mathbb{I} \\ -\mathbb{I} & 0 \end{pmatrix}. \tag{3.99}$$

Here, $\mathbb{I}$ is the $\frac{N}{2} \times \frac{N}{2}$ identity matrix; therefore, N must be even.

Sp(2) are just those 2×2 matrices $\mathbb{M}$ that satisfy

$$\mathbb{M}^\mathsf{T} \begin{pmatrix} 0 & 1 \\ -1 & 0 \end{pmatrix} \mathbb{M} = \begin{pmatrix} 0 & 1 \\ -1 & 0 \end{pmatrix}. \tag{3.100}$$

(a) Prove that Sp(N) is a group.

(b) Now, study the group Sp(2). What are the properties of the matrices $\mathbb{M}$ in Sp(2)? Have we seen this group before?

3.8 *π-p Scattering.* We are able to estimate the relative rates of interactions between different isospin multiplets with the corresponding Clebsch–Gordan coefficients. In this chapter, we introduced Clebsch–Gordan coefficients for determining the decomposition of product isospin states into irreps of isospin. In practice, the procedure for determining the Clebsch–Gordan coefficients is to simply look them up in a table. A nice table of Clebsch–Gordan coefficients is located in the PDG. Go to the PDG website, click on "Reviews, Tables, Plots" and then scroll down to "Mathematical Tools." From there, you can click on the link to the Clebsch–Gordan coefficient tables. In the following questions, we will use these tables to estimate relative pion–proton scattering rates.

To see how these tables work, look at the $1/2 \times 1/2$ table. Horizontally, the rows correspond to the coefficients of combining an isospin-1/2 state with an isospin-1/2 state with the corresponding eigenvalue of the third component of isospin at the left. An isospin doublet, like the nucleons, is an isospin-1/2 state, while the isospin triplet is an isospin-1 state. For example, consider looking for the coefficient of the isospin-1 and $I_3 = 0$ combination of a proton ($I_3 = 1/2$) and a neutron ($I_3 = -1/2$). This is found by looking at the following row and columns:

$$
\begin{array}{cc|c}
 & & 1 \\
 & & 0 \\
\hline
1/2 & -1/2 & 1/2
\end{array}
\tag{3.101}
$$

The $I_3 = 0$ state is the symmetric combination of the proton and neutron in the isospin triplet (the middle entry in the isospin triplet from Table 3.2). There is an implicit square-root for every entry of the table, so the coefficient of this term is $1/\sqrt{2}$, which is exactly what we found in Example 3.2.

(a) The proton and neutron are in the isospin $I = 1/2$ doublet with $I_3 = 1/2$ and $I_3 = -1/2$, respectively, while the pions are isospin $I = 1$ with $I_3 = 1$ for π^+,

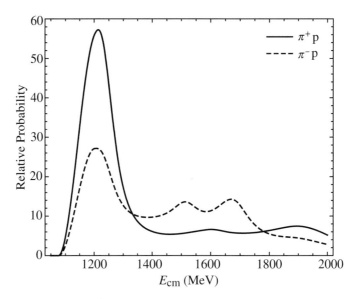

Fig. 3.5   Relative probability for scattering of $\pi^+ p$ and $\pi^- p$ as a function of the pion–proton center-of-mass collision energy, E_{cm}. This plot was constructed from hadron resonance data from the PDG.

$I_3 = 0$ for π^0, and $I_3 = -1$ for π^-. Determine the Clebsch–Gordan coefficients for the combinations:

(i) $\pi^0 n$ in the $I = 3/2$, $I_3 = -1/2$ state
(ii) $\pi^- p$ in the $I = 1/2$, $I_3 = -1/2$ state
(iii) $\pi^+ p$ in the $I = 3/2$, $I_3 = 1/2$ state.

(b) The Δ baryon is a particle that is similar to a proton or a neutron, but in an excited state configuration of up and down quarks. It has isospin $I = 3/2$. From the Clebsch–Gordan table, estimate the ratio of the probability that the scattering of a proton and a π^+ produces a Δ baryon versus a proton and a π^-.

(c) Figure 3.5 is a plot of the relative scattering probability for pion–proton collisions for a range of pion–proton center-of-mass collision energies, measured in MeV. $\pi^+ p$ results are solid and $\pi^- p$ results are dashed.

What is the minimum value that the mass of the pion–proton system can have? Is this consistent with the plot?

(d) On this plot, the Δ baryon corresponds to the peak (resonance) at 1232 MeV. From this plot, estimate the ratio of the probability for $\pi^+ p$ scattering to produce a Δ versus $\pi^- p$ scattering. How does this compare to your result in part (b)?

3.9   *Research Problem.* Why do groups and their structure describe physical phenomena so well? You might want to start with reading E. P. Wigner, "The unreasonable effectiveness of mathematics in the natural sciences," Comm. Pure Appl. Math. **13**, no. 1, 1 (1960).

4 Fermi's Golden Rule and Feynman Diagrams

Golden Rule #1: *Do unto others as you would have them do unto you.*

The foundation of civilization.

Golden Rule #2: *The transition rate from an initial to a final state in quantum mechanics is the matrix element squared of the Hamiltonian times the density of states.*

The foundation of predictions in particle physics.[1]

Special relativity and applications of group theory aren't sufficient alone to study particle physics. We need to couch them within the framework of quantum mechanics, especially as applied to experiments like the LHC. Our primary experimental tool in particle physics is colliding particles and observing the detritus that comes from it. Quantum mechanically, we cannot say for certain what the outcome of any given particle collision will be, but we can determine the probability of a particular collision. Just as in non-relativistic quantum mechanics, the probability that a given pair of particles interacts and produces a collection of particles from the collision is controlled by the overlap of the initial state with the final-state wavefunctions. So, we need to develop a method for calculating these wavefunction overlaps and extracting corresponding probabilities.

Probability is a dimensionless quantity; it is just a number between 0 and 1. However, this makes it extremely subtle because it is absolute. To calculate the probability of the outcome of any given particle collision means that we need to know all possible outcomes a priori. In quantum mechanics, this is essentially the statement that the eigenstate basis for a given potential is complete: it can describe *any* potential outcome of an experiment. Unfortunately, this is effectively impossible in particle physics because the number of final states is uncountably infinite and (in most cases) the set of final states is not even fully known. With this motivation, we want to find a quantity whose value represents relative probability, but is not strictly connected to an absolute scale. We'll find the notion of cross section and the corresponding barn unit to be a particularly physically appealing solution to this problem.

Starting from the introduction of the barn in this chapter, we work backward from the observation of outcomes of collision events, to counting the possible final states consistent with our particular measurement, to calculating wavefunction overlaps in particle physics.

---

[1] The term "Fermi's Golden Rule" comes from colorful names that Fermi used in a nuclear physics class he taught at University of Chicago; see E. Fermi, *A Course Given by Enrico Fermi at the University of Chicago, 1949*, University of Chicago Press (1950). The Golden Rule #1 that he cites in his lectures is another formula for transition rates in quantum mechanics, which, oddly, is less general than what we now know as Fermi's Golden Rule.

Along the way, we'll develop Fermi's Golden Rule and Feynman diagrams, two cornerstone techniques for making predictions in particle physics.

4.1 Invitation: The Barn

Let's consider what happens in collisions at the Large Hadron Collider. Protons are collided at high energies and we observe what comes out. This is a quantum mechanical process, so all we can predict is the probability that protons will interact and produce a particular final state, and not exactly what will result from each proton collision. So, we want to define a unit that is a measure of this scattering probability. Let's imagine watching the protons collide. Immediately before they collide, it would look something like this:

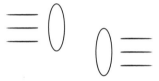

Protons are Lorentz contracted into pancake shapes from their spherical shape at rest. The lines trailing each proton are an admittedly poor representation of the blur as the protons zoom past near the speed of light.

What property of the proton controls the probability of interaction? This drawing makes it clear that the size of the proton is clearly important. However, the radius of the proton is hard to define; we can't just measure it with a ruler. One very useful way to define the radius of the proton is as (effectively) the radius at which the electric field of the proton is largest. This definition is called the **charge radius** and has a nice analogy in classical electromagnetism. The electric field of a spherical shell of charge is maximized when you are on the surface of the shell, by Gauss's law. The charge radius of a spherical shell is therefore just its radius.

Another way to define the size/radius of the proton is through its rate of interacting with itself or other particles. For example, if you strapped yourself on a proton that was traveling toward another proton for collision, you would see something like this:

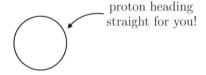
proton heading straight for you!

That is, you would see a cross-sectional area of the proton. The larger this area is, the more likely it is that you will interact with it. The smaller the area, the less likely to interact.

This motivates a connection between proton size and scattering probability. In particle physics, a collision or interaction rate is expressed in effective cross-sectional area, typically just called **cross section**. As an "area," we can measure scattering cross sections as the square of some relevant length scale. Interestingly, this is not what is typically done in particle physics.

Box 4.1
Historical Profile: The Unit of the Barn

The origin of the name **barn** is quite amusing and its history is recounted in a letter to the editor in *Physics Today*:[2]

> The tradition of naming a unit after some great man closely associated with the field ran into difficulties since no such person could be brought to mind. Failing this, the names Oppenheimer and Bethe were tried, since these men had suggested and made possible the work on the problem with which the Purdue project was concerned. The "Oppenheimer" was discarded because of its length, though in retrospect "Oppy" or "Oppie" would seem to be short enough. The "Bethe" was thought to lend itself to confusion because of the widespread use of the Greek letter. Since John Manley was directing the work at Purdue, his name was tried, but the "Manley" was thought to be too long. The "John" was considered, but was discarded because of the use of the term for purposes other than as the name of a person. The rural background of one of the authors then led to the bridging of the gap between the "John" and the "barn." This immediately seemed good and further it was pointed out that a cross section of 10^{-24} cm^2 for nuclear processes was really as big as a barn. Such was the birth of the barn.

If we want to define the cross section of something at human scales, we would express it in square meters or perhaps square centimeters. For example, the cross section of a baseball is about 45 cm^2, while the "meat" of a baseball bat is about 150 cm^2, which are both nice numbers. Square centimeters are less useful in particle physics. For the proton, with a radius of about 10^{-15} m (or 10^{-13} cm), it has a cross section of about 10^{-26} cm^2. This is a tiny exponent that can be annoying to lug around. As with the development of natural units from Chapter 1, we want a measure of cross section that respects the scales of subatomic physics.

The standard unit of cross section in particle physics is called the **barn**. During World War II, when scientists were working on the atomic bomb, it was very important to understand the cross section of uranium (i.e., the probability that uranium interacted with itself), and they wanted to name an appropriate unit for this cross-sectional area. Correspondingly, the barn unit is approximately the cross-sectional area of a uranium nucleus, 10^{-24} cm^2. In these units, the cross-sectional area of a proton is about a hundredth of a barn or so. In elementary particle physics, the cross sections that we consider are typically much smaller than a barn, so we often use nanobarns (nb, 10^{-9}), picobarns (pb, 10^{-12}), femtobarns (fb, 10^{-15}), or even attobarns (ab, 10^{-18}) to express them. The barn is one of the few quantities in particle physics that is not within the natural unit system. The history of settling on the word "barn" for cross section is described in the Historical Profile of Box 4.1.

4.2 Scattering Systematics

Let's see how we can take this idea of cross section and turn it into a concrete prediction for the rate of production of a particular final state from particle collisions. For concreteness, let's consider colliding protons at the LHC. Accelerating and colliding individual protons

---

[2] Reproduced from M. G. Holloway and C. P. Baker, "How the barn was born," Phys. Today **25**, no. 7, 9 (1972), with the permission of the American Institute of Physics.

is extremely inefficient; for two protons to interact you would have to get them to within about a femtometer of one another. At the LHC, **bunches** that each contain about 10^{11} protons are collided and out of these bunch crossings maybe 20 pairs of protons actually collide. Let's analyze the collision of two of these bunches; call one bunch A and the other bunch B. We can visualize this as:

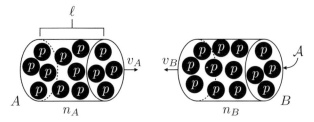

The velocity of bunch A is v_A, and correspondingly v_B for bunch B. The length of each bunch is ℓ and the cross-sectional area of each bunch is $\mathcal{A}$. The number of protons per unit volume in the bunches (the number density) is n_A and n_B for bunches A and B, respectively. Note that the cross-sectional area $\mathcal{A}$ is the size of the bunch, and not the size of an individual proton. For colliding protons at the LHC, the relevant quantity is the number of collision events per second, as this will tell us how often protons will collide over the time that we run the machine. Let's see how to define this.

As the bunches pass through one another, every proton of bunch B travels through a region of length ℓ where there are protons (and similarly for bunch A). The total number of protons N_A in bunch A that a slice of the protons in bunch B sees in one unit of time is

$$N_A/t = n_A \cdot \mathcal{A} \cdot |v_A - v_B|, \tag{4.1}$$

where $|v_A - v_B|$ is the relative speed of the bunches. Note that $|v_A - v_B|$ is the length per unit time of bunch A that passes through bunch B. Multiplying by $\mathcal{A}$ gives the volume per unit time of A that passes through B, and then multiplying by the number density n_A counts the total number of protons that pass through in a unit time.

The total number of protons in bunch B that could possibly interact with the protons of bunch A is

$$N_B^{\text{eff}} = n_B \cdot \ell \cdot \sigma. \tag{4.2}$$

That is, the number of B protons that can possibly interact with the protons of bunch A is given by the number density n_B times the total volume of pure protons in bunch B. This volume is just the length of the bunch multiplied by the cross-sectional area of each proton, which we denote as σ. The total number of scattering events per unit time is then the product of these two factors:

$$\text{events/time} = \frac{N_A N_B^{\text{eff}}}{t} = n_A n_B \mathcal{A} \ell |v_A - v_B| \sigma. \tag{4.3}$$

The prefactor $n_A n_B \mathcal{A} \ell |v_A - v_B|$ is called the **flux factor**, and depends on the precise parameters of the LHC accelerator. The proton scattering cross section σ is intrinsic to individual proton–proton interactions, and is the same for protons scattering in the LHC as in the center of the galaxy. Note indeed that it has units of area. Let's see what the consequences of the flux factor are at the LHC.

Example 4.1 What is the approximate value of the flux factor in collisions at the LHC?

Solution

The flux factor is also called the instantaneous luminosity or just the **luminosity**, denoted by $\mathcal{L}$. (Yes, this is the same symbol as for Lagrangian density but one never uses them in close proximity so there's no chance for confusion.) The value of $\mathcal{L}$ is

$$\mathcal{L} = n_A n_B \mathcal{A} \ell |v_A - v_B| = \frac{N_A N_B |v_A - v_B|}{\text{Vol}}, \tag{4.4}$$

where N_A and N_B are the numbers of protons in a bunch and Vol is the volume of a bunch, $\text{Vol} = \mathcal{A}\ell$. The proton–proton scattering cross section σ is fixed, so to produce as many collisions per second as possible, we need to make the luminosity as large as possible. This is done in just one way at the LHC. The number of protons per bunch N is a fixed number:

$$N = N_A = N_B = 10^{11}. \tag{4.5}$$

Further, the velocity of the two bunches of protons is very close to the speed of light because the energy of protons at the LHC is much, much larger than the proton mass. As the bunches travel in opposite directions, their relative speed is essentially twice the speed of light:

$$|v_A - v_B| = 2c. \tag{4.6}$$

Finally, there is the volume factor in the luminosity. Protons at the LHC are accelerated in bunches in a radio-frequency (RF) electromagnetic field that has a frequency of 400 MHz. Protons effectively ride along in the "troughs" of this field and its wavelength therefore sets the length of a bunch. We'll analyze how this is done more carefully in Chapter 5, but for now, we will just set the bunch length ℓ to be half of the wavelength of the RF field:

$$\ell = \frac{c}{2 \times 400 \text{ MHz}}. \tag{4.7}$$

So far, all of this is fixed. The one thing that is controlled at the LHC to increase the luminosity is the cross-sectional area of the bunch, $\mathcal{A}$. At collision points like those in the center of the ATLAS and CMS experiments, quadrupole focusing magnets are used to squeeze the bunches into an extremely small region. These focusing magnets squeeze the bunches down to a radius of 10 micrometers! This corresponds to a cross-sectional area of

$$\mathcal{A} = \pi \times (10 \text{ microns})^2 \simeq 3 \times 10^{-10} \text{ m}^2. \tag{4.8}$$

Putting this all together, the luminosity of the LHC collisions is

$$\mathcal{L} = 2c\frac{N^2}{\ell \mathcal{A}} \simeq 2 \times 10^{22} \times 2 \times 400 \text{ MHz} \frac{1}{3 \times 10^{-10} \text{ m}^2} \simeq 10^{36} \text{ cm}^{-2}\text{s}^{-1}. \tag{4.9}$$

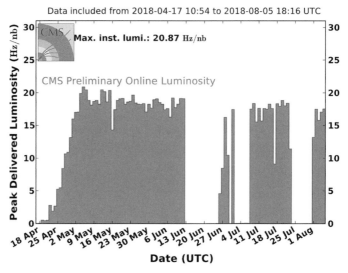

CMS Peak Luminosity Per Day, pp, 2018, $\sqrt{s} = 13$ TeV

Plot of the proton luminosity as measured at the CMS experiment during 2018. Note that the luminosity is about $20 \, \text{nb}^{-1}\text{s}^{-1}$ throughout the time when the machine was running. Credit: CMS Experiment, © CERN. Figure reused with permission from `https://twiki.cern.ch/twiki/bin/view/CMSPublic/LumiPublicResults`.

Luminosity is often quoted in units of $\text{cm}^{-2}\text{s}^{-1}$, as done above. It is also expressed in inverse barns per second. A barn is $10^{-24} \, \text{cm}^2$, and so the calculated luminosity is also

$$\mathcal{L} \simeq 10^{12} \, \text{b}^{-1}\text{s}^{-1} = 1 \, \text{pb}^{-1}\text{s}^{-1}. \tag{4.10}$$

This is a measure of the inverse cross section per second produced in the experiments at the LHC, and this approximation is a bit of an overestimate of the luminosity at the LHC. At its peak, the maximum luminosity is only about $20 \, \text{nb}^{-1}\text{s}^{-1} = 2 \times 10^{34} \, \text{cm}^{-2}\text{s}^{-1}$ at the LHC, as shown in the luminosity measurements at CMS in Fig. 4.1. The reason for this overestimate is that proton bunches are only collided every 25 nanoseconds, or 40 million times a second. That is, only every tenth trough of the 400 MHz RF field is populated by a bunch of protons, which decreases our estimate of the luminosity by a factor of ten.

In astrophysics or other fields of physics, the term "luminosity" references the rate of emission or intrinsic brightness of a stellar object. In particle physics, this connotation still holds, as the luminosity is the intrinsic "brightness" of the proton collisions. How often they actually do something, however, is controlled by their scattering cross section.

For the rest of this chapter we will work to define and learn how to calculate the proton scattering cross section, σ. A careful and detailed derivation of the form of the cross section σ requires quantum field theory and its axioms; here, we will justify the form of the cross section by its expected properties.

4.2.1 The Scattering Cross Section

When colliding protons, there are two things that we need to define to determine the probability or cross section for the collision to take place. First, we need to define what the initial state is; that is, what momenta the colliding protons have. Second, we need to define what final state we are considering; that is, what particles are produced in the collision and what their momenta are. These states are described by wavefunctions with the initial state of

$$|p_A p_B\rangle_{\text{in}} \tag{4.11}$$

and the final state of

$$_{\text{out}}\langle p_1 p_2 \ldots p_n| . \tag{4.12}$$

Here, p_A and p_B are the momenta of the protons in the A and B bunches that collide, and $p_1, p_2, \ldots, p_n$ are the momenta of the n particles produced from the collision. Initial states, or "in" states, exist from time $t = -\infty$ and propagate forward in time. Final states, or "out" states, are produced from the collision and propagate out to time $t = +\infty$. Of course, no initial state is prepared infinitely far in the past, and no final state travels infinitely far into the future, but these time scales are exponentially longer than the time of collision. The time over which the collision takes place is approximately the time that it takes a proton at the LHC to travel the width of a proton, or about 10^{-23} s. This is trillions of times smaller than the time it takes for final-state particles to travel from the collision point to the detectors of the LHC, and so the infinite past and future limits are not bad approximations.

Examples of scattering processes that we might be interested in include the collision of protons that produces two positrons,

$$pp \to e^+ e^+ , \tag{4.13}$$

or two positrons and three photons,

$$pp \to e^+ e^+ \gamma\gamma\gamma , \tag{4.14}$$

or any collection of particles. In these example reactions, note that electric charge is explicitly conserved, and energy and momentum are implicitly conserved. The probability that two protons collided to produce n final-state particles with momenta $p_1, \ldots, p_n$ is the absolute squared inner product of the wavefunctions:

$$\text{Prob. of scattering of protons to } p_1, \ldots, p_n = |_{\text{out}}\langle p_1 p_2 \ldots p_n | p_A p_B\rangle_{\text{in}}|^2 . \tag{4.15}$$

This inner product has a lot of moving parts that we will work to identify. For compactness in what follows, we will denote the final state as f, but we'll consider explicit final states in a bit.

As discussed earlier, the proton scattering cross section is proportional to this probability:

$$\sigma \propto |_{\text{out}}\langle f | p_A p_B\rangle_{\text{in}}|^2 . \tag{4.16}$$

Probabilities are dimensionless and cross sections have units of area, so we need to include the appropriate scales of the problem to get the dimensions right. This is determined by thinking about how the protons can scatter and their relevant size, which is sensitive to their **de Broglie wavelength**.

The de Broglie wavelength of a particle is the distance over which the particle is coherent; that is, the distance within which the particle's position is most likely to be. The de Broglie wavelength λ_{dB} is inversely proportional to the magnitude of three-momentum of a particle:

$$\lambda_{\mathrm{dB}} = \frac{\hbar}{|\vec{p}|} . \tag{4.17}$$

That is, higher-momentum particles are localized in a smaller region of space, while low-momentum particles are delocalized (can be "anywhere"). It is harder for protons with short de Broglie wavelengths to collide because their wavefunctions only have support in a small region of space, while it is much easier for protons with long de Broglie wavelengths to interact. Therefore, we expect the cross section to be proportional to the de Broglie wavelengths of the two protons:

$$\sigma \propto \lambda_A \lambda_B \left|_{\mathrm{out}}\langle f | p_A p_B \rangle_{\mathrm{in}}\right|^2 = \frac{1}{|\vec{p}_A||\vec{p}_B|} \left|_{\mathrm{out}}\langle f | p_A p_B \rangle_{\mathrm{in}}\right|^2 . \tag{4.18}$$

Now, the right side of Eq. 4.18 has units of area (in natural units), like the cross section, though we aren't done yet.

One issue to clear up is what happens as momentum goes to 0: $|\vec{p}| \to 0$. This would suggest that the cross section diverges, which is unphysical behavior. In particular, a proton at rest has an intrinsic size determined by the **Compton wavelength** λ_{C}:

$$\lambda_{\mathrm{C}} = \frac{\hbar}{mc} , \tag{4.19}$$

where m is the mass of the particle. At high momentum, the relevant size of a particle is its de Broglie wavelength, while at low momentum, its Compton wavelength is its size. Neither the de Broglie nor the Compton wavelength interpolates between these regimes, but a distance scale inversely proportional to the energy of the particle does. This then suggests that the cross section can be expressed as

$$\sigma \propto \frac{1}{2E_A} \frac{1}{2E_B} \left|_{\mathrm{out}}\langle f | p_A p_B \rangle_{\mathrm{in}}\right|^2 . \tag{4.20}$$

The factors of $1/2$ come from careful normalization of the initial-state wavefunctions in quantum field theory.

As a cross-sectional area, σ should also have particular properties under Lorentz transformations. Let's remind ourselves about the picture of proton collision:

Let's call the axis along which the scattering occurs $\hat{z}$. The cross-sectional areas of the protons then lie in the plane of the $\hat{x}$ and $\hat{y}$ axes. Looking at a proton in a head-on collision we see:

This is clearly rotationally invariant about the $\hat{z}$-axis, and it is Lorentz-boost invariant along the $\hat{z}$-axis. That is, the area that you see does not change with these transformations. We need to make sure that this is true for the expression for the cross section. The probability

$$\left| {}_{\text{out}}\langle f | p_A p_B \rangle_{\text{in}} \right|^2 \tag{4.21}$$

is fully Lorentz invariant (essentially by definition), so it remains unchanged under rotations and boosts. The overall factors of energy are rotationally invariant, but change if the system is boosted along the $\hat{z}$-axis. So, we need to fix this.

Under a Lorentz boost by velocity β along the $\hat{z}$-direction, momentum and energy transform as

$$p^z \to \gamma(p^z + \beta E), \qquad\qquad E \to \gamma(E + \beta p^z), \tag{4.22}$$

and so the factors of energy in the cross section from Eq. 4.20 transform as

$$\frac{1}{E_A E_B} \to \frac{1}{\gamma(E_A + \beta p_A^z)\gamma(E_B + \beta p_B^z)} = \frac{1}{E_A E_B} \frac{1}{\gamma^2(1 + \beta v_A)(1 + \beta v_B)}. \tag{4.23}$$

In writing this, we have used the relationship that the ratio of momentum to energy is the velocity of the particle:

$$\frac{p^z}{E} = v. \tag{4.24}$$

In general, this Lorentz transformation factor is not 1, so, as written, the cross section isn't invariant to boosts along the $\hat{z}$-axis. So, we'll need to multiply by another factor to cancel this transformation.

This factor must be dimensionless in natural units, because the cross section of Eq. 4.20 already has dimensions of area. Also, it must actually transform under a Lorentz boost. The only objects that satisfy these criteria are the protons' velocities, as defined in Eq. 4.24. Under a Lorentz boost along the $\hat{z}$-axis, the velocity transforms as

$$v = \frac{p^z}{E} \to \frac{\gamma(p^z + \beta E)}{\gamma(E + \beta p^z)} = v \frac{1 + \frac{\beta}{v}}{1 + \beta v}. \tag{4.25}$$

The relative speed $|v_A - v_B|$ of the colliding protons therefore transforms as

$$|v_A - v_B| \to \left| v_A \frac{1 + \frac{\beta}{v_A}}{1 + \beta v_A} - v_B \frac{1 + \frac{\beta}{v_B}}{1 + \beta v_B} \right| \tag{4.26}$$

$$= \frac{|v_A - v_B|}{\gamma^2(1 + \beta v_A)(1 + \beta v_B)}. \tag{4.27}$$

This is exactly the inverse transformation of the product of energies, Eq. 4.23! Their product is invariant to boosts along the $\hat{z}$-axis, and therefore the expression for the proton–proton scattering cross section σ is

$$\sigma = \frac{1}{2E_A} \frac{1}{2E_B} \frac{1}{|v_A - v_B|} \left| _{\text{out}}\langle f | p_A p_B \rangle_{\text{in}} \right|^2 . \tag{4.28}$$

This has the correct dimensions of a cross-sectional area (energy^{-2} in natural units) and the correct Lorentz transformation properties under rotations and boosts about/along the collision axis.

4.2.2 Fermi's Golden Rule

We've so far been a bit cavalier about what the object

$$\left| _{\text{out}}\langle f | p_A p_B \rangle_{\text{in}} \right|^2 \tag{4.29}$$

actually is. It's the final piece standing between us and predicting how often protons will collide and produce the final state f in our experiment. To define this, we need to think about what is measured in a particle physics experiment. Experiments like ATLAS and CMS at the LHC are exceptionally good at measuring energy and momentum of particles, but are not designed to measure the positions of particles at collision. In the calculation of this probability, we should therefore work with momenta and be completely ignorant about positions of particles. From an expression with particular momenta of the final-state particles, we must sum over all sets of momenta that are consistent with the measurements that we make. For example, let's say we observe two positrons produced from proton collisions. For a consistent calculation of the cross section, we must sum over all possible momenta of the positrons which have positive energy, conserve four-momentum, and correspond to real, on-shell positrons.

For concreteness, let's consider a final state with n particles:

$$\underbrace{A + B}_{\text{initial protons}} \rightarrow \underbrace{1 + 2 + \cdots + n}_{\text{final particles}} . \tag{4.30}$$

We will just demand that we measure the existence of the n final-state particles, with any physical momenta. We denote the probability amplitude for the initial-state protons A and B with given momenta to collide and produce the final-state particles $1, 2, \ldots, n$ with particular momenta as

$$\mathcal{M}(A + B \rightarrow 1 + 2 + \cdots + n), \tag{4.31}$$

which is called the Lorentz-invariant matrix element or just the **matrix element**. This will just be a placeholder until Section 4.3 when we will define this object precisely. Because we only care about the existence of the final-state particles and not their momenta, we need to sum over all possible momenta, consistent with conservation laws.

Let's work through how to do this. We start schematically, with

$$\left| _{\text{out}}\langle p_1 p_2 \cdots p_n | p_A p_B \rangle_{\text{in}} \right|^2 = \sum_{\substack{\text{final particle} \\ \text{momenta}}} \left| \mathcal{M}(A + B \rightarrow 1 + 2 + \cdots + n) \right|^2 . \tag{4.32}$$

For a final-state particle i, its four-momentum is p_i and is four continuous components. Therefore, this sum should consist of n four-dimensional integrals over each component of each final-state particle's momentum. We will express this as

$$\sum_{\substack{\text{final particle} \\ \text{momenta}}} \rightarrow \int \prod_{i=1}^{n} \frac{d^4 p_i}{(2\pi)^4} \,. \tag{4.33}$$

For now, the integrals extend over all possible values of each momentum component: non-negative energies and three-momentum components that are any real number. Here, the product symbol $\prod_{i=1}^{n}$ means we multiply the momentum integration measures for each particle together:

$$\prod_{i=1}^{n} \frac{d^4 p_i}{(2\pi)^4} = \frac{d^4 p_1}{(2\pi)^4} \frac{d^4 p_2}{(2\pi)^4} \cdots \frac{d^4 p_n}{(2\pi)^4} \,. \tag{4.34}$$

The factors of 2π come from Fourier transforming from position to momentum space representations of the wavefunctions for the initial and final state.

As it stands, this representation is much too general and includes configurations of final-state particles that violate conservation laws. First, the relative domain of the energy and three-momentum integrals is unrelated. However, the final-state particles must each be on-shell. We must enforce that the integrals are 0 if $p_i^2 \neq m_i^2$, where m_i is the mass of particle i, for all $i = 1, 2, \ldots, n$. This can be accomplished by including a Dirac δ-function in the integrand:

$$\sum_{\substack{\text{final particle} \\ \text{momenta}}} \rightarrow \int \prod_{i=1}^{n} \left[\frac{d^4 p_i}{(2\pi)^4} 2\pi \delta(p_i^2 - m_i^2) \right] \,. \tag{4.35}$$

Again, the factors of 2π originate from a Fourier transform. The Dirac δ-function $\delta(x)$ is 0 if $x \neq 0$ and infinite if $x = 0$. However, the infinity is constrained to produce a finite integral:

$$\int_{-\epsilon}^{\epsilon} \delta(x) \, dx = 1 \,, \tag{4.36}$$

for any $\epsilon > 0$. So, this indeed restricts the particles to be on the mass shell.

There are still unphysical configurations of the particles included in the momentum integrals. So far, there's no connection between the initial-state and final-state momenta; however, these are of course equal by momentum conservation. The integrals must only be non-zero if the total four-momentum of the initial protons is equal to the total four-momentum of all n final-state particles. This can also be enforced by a Dirac δ-function; we need four of them to account for conservation of energy and three components of momentum. Thus, including this factor, we have

$$\sum_{\substack{\text{final particle} \\ \text{momenta}}} = \int \prod_{i=1}^{n} \left[\frac{d^4 p_i}{(2\pi)^4} 2\pi \delta(p_i^2 - m_i^2) \right] (2\pi)^4 \delta^{(4)} \left(p_A + p_B - \sum_{i=1}^{n} p_i \right) \,. \tag{4.37}$$

The δ-function that imposes momentum conservation is shorthand for conservation of each component:

$$\delta^{(4)}\left(p_A + p_B - \sum_{i=1}^{n} p_i\right) = \delta\left(E_A + E_B - \sum_{i=1}^{n} E_i\right) \tag{4.38}$$

$$\times \delta\left(p_A^x + p_B^x - \sum_{i=1}^{n} p_i^x\right) \delta\left(p_A^y + p_B^y - \sum_{i=1}^{n} p_i^y\right) \delta\left(p_A^z + p_B^z - \sum_{i=1}^{n} p_i^z\right).$$

This integral over the n final particle momenta is Lorentz invariant, and is called **n-body Lorentz-invariant phase space**, and is denoted by $d\Pi_n$ or $d\text{LIPS}_n$:

$$\int d\Pi_n = \int \prod_{i=1}^{n} \left[\frac{d^4 p_i}{(2\pi)^4} 2\pi\delta(p_i^2 - m_i^2)\right] (2\pi)^4\delta^{(4)}\left(p_A + p_B - \sum_{i=1}^{n} p_i\right). \tag{4.39}$$

Lorentz-invariant phase space is the relativistic density of states of n on-shell particles in momentum space with a fixed total four-momentum.

Using the expression for Lorentz-invariant phase space and plugging it into Eq. 4.28, our final expression for the proton scattering cross section is

$$\sigma = \frac{1}{2E_A} \frac{1}{2E_B} \frac{1}{|v_A - v_B|} \int \prod_{i=1}^{n} \left[\frac{d^4 p_i}{(2\pi)^4} 2\pi\delta(p_i^2 - m_i^2)\right] \tag{4.40}$$

$$\times |\mathcal{M}(A + B \to 1 + 2 + \cdots + n)|^2 (2\pi)^4 \delta^{(4)}\left(p_A + p_B - \sum_{i=1}^{n} p_i\right).$$

This result is called **Fermi's Golden Rule**.[3] In the next section, we will discuss how to calculate the Lorentz-invariant matrix element, $\mathcal{M}(A + B \to 1 + 2 + \cdots + n)$. We'll end this section with an evaluation of two-body phase space, which we will use numerous times throughout the rest of this book.

Example 4.2 Performing as many integrals in Eq. 4.39 as possible, what is the Lorentz-invariant, two-body phase space?

Solution

The two-body phase space integral that we want to evaluate is

$$\int d\Pi_2 = \int \frac{d^4 p_1}{(2\pi)^4} 2\pi\delta(p_1^2 - m_1^2) \frac{d^4 p_2}{(2\pi)^4} 2\pi\delta(p_2^2 - m_2^2) (2\pi)^4\delta^{(4)}(Q - p_1 - p_2). \tag{4.41}$$

Here, Q is the total energy–momentum four-vector, and we choose to work in the center-of-mass frame where the total three-momentum is 0:

$$Q = (E_{\text{cm}}, 0, 0, 0), \tag{4.42}$$

---

[3] Though called "Fermi's Golden Rule," it was derived by Dirac in 1927; see P. A. M. Dirac, "Quantum theory of emission and absorption of radiation," Proc. Roy. Soc. Lond. A **114**, 243 (1927). This is an example of **Stigler's law of eponymy**; see S. M. Stigler, "Stigler's law of eponymy," Trans. N. Y. Acad. Sci. **39**, 147 (1980).

where E_{cm} is the energy in the center-of-mass frame. The phase space integral can then be written as

$$\int d\Pi_2 = \frac{1}{4\pi^2} \int d^4 p_1 \, d^4 p_2 \, \delta(p_1^2 - m_1^2) \, \delta(p_2^2 - m_2^2) \, \delta(E_{cm} - E_1 - E_2) \, \delta^{(3)} (\vec{p}_1 + \vec{p}_2) \,.$$

(4.43)

E_1 (E_2), $\vec{p}_1$ ($\vec{p}_2$), and m_1 (m_2) are the energy, three-momentum, and mass of particle 1 (2).

Now, we use the on-shell δ-functions to do the integral over the energies of the two particles. First, the integration measure is

$$d^4 p = dE \, d^3 \vec{p},$$

(4.44)

for a four-vector p. The integral we want to do is

$$\int dE \, \delta(p^2 - m^2) = \int dE \, \delta(E^2 - |\vec{p}|^2 - m^2) \,.$$

(4.45)

To do this integral, it's almost in the form of Eq. 4.36, but in the δ-function, the energy is squared. So, we'll make the change of variables to

$$x = E^2 \,,$$

(4.46)

so that

$$dE = \frac{dx}{2\sqrt{x}} \,.$$

(4.47)

Then, the integral over the on-shell δ-function is

$$\int dE \, \delta(E^2 - |\vec{p}|^2 - m^2) = \int dx \frac{1}{2\sqrt{x}} \delta(x - |\vec{p}|^2 - m^2) = \frac{1}{2\sqrt{|\vec{p}|^2 + m^2}} \,.$$

(4.48)

Using this result, we can then write the phase space integral as

$$\int d\Pi_2 = \frac{1}{16\pi^2} \int \frac{d^3 \vec{p}_1}{\sqrt{|\vec{p}_1|^2 + m_1^2}} \frac{d^3 \vec{p}_2}{\sqrt{|\vec{p}_2|^2 + m_2^2}}$$

(4.49)

$$\times \delta \left(E_{cm} - \sqrt{|\vec{p}_1|^2 + m_1^2} - \sqrt{|\vec{p}_2|^2 + m_2^2} \right) \delta^{(3)} (\vec{p}_1 + \vec{p}_2) \,.$$

Next, the integrals over all three components of $\vec{p}_2$ can be done with the last δ-function. This δ-function fixes $\vec{p}_2 = -\vec{p}_1$:

$$\int d\Pi_2$$

(4.50)

$$= \frac{1}{16\pi^2} \int \frac{d^3 \vec{p}_1}{\sqrt{|\vec{p}_1|^2 + m_1^2}} \frac{1}{\sqrt{|\vec{p}_2|^2 + m_2^2}} \delta \left(E_{cm} - \sqrt{|\vec{p}_1|^2 + m_1^2} - \sqrt{|\vec{p}_1|^2 + m_2^2} \right) \,.$$

For the remaining integrals over $\vec{p}_1$, we express the integral in spherical coordinates. The integration measure is

$$d^3 \vec{p}_1 = |\vec{p}_1|^2 \, d|\vec{p}_1| \, d\cos\theta \, d\phi \,.$$

(4.51)

Here, θ is the polar angle with $\cos\theta \in [-1, 1]$ and ϕ is the azimuthal angle where $\phi \in [0, 2\pi)$. Typically, and in all situations considered in this book, we will not study matrix elements with azimuthal angle dependence. So, we can just integrate over ϕ:

$$\int d\Pi_2 \tag{4.52}$$

$$= \frac{1}{8\pi} \int \frac{|\vec{p}_1|^2 \, d|\vec{p}_1| \, d\cos\theta}{\sqrt{|\vec{p}_1|^2 + m_1^2}} \frac{1}{\sqrt{|\vec{p}_2|^2 + m_2^2}} \delta\left(E_{\text{cm}} - \sqrt{|\vec{p}_1|^2 + m_1^2} - \sqrt{|\vec{p}_1|^2 + m_2^2}\right).$$

To integrate over the final δ-function, we change variables from the magnitude of momentum $|\vec{p}_1|$ to E, where

$$E = \sqrt{|\vec{p}_1|^2 + m_1^2} + \sqrt{|\vec{p}_1|^2 + m_2^2}. \tag{4.53}$$

Then, the differential element is

$$d|\vec{p}_1| = \frac{\sqrt{|\vec{p}_1|^2 + m_1^2}\sqrt{|\vec{p}_1|^2 + m_2^2}}{|\vec{p}_1|(\sqrt{|\vec{p}_1|^2 + m_1^2} + \sqrt{|\vec{p}_1|^2 + m_2^2})} dE = \frac{\sqrt{|\vec{p}_1|^2 + m_1^2}\sqrt{|\vec{p}_1|^2 + m_2^2}}{|\vec{p}_1|E_{\text{cm}}} dE, \tag{4.54}$$

where, on the right, we have used the δ-function which sets $E = E_{\text{cm}}$. Plugging this into the integral, we then find

$$\int d\Pi_2 = \frac{1}{8\pi} \frac{|\vec{p}|}{E_{\text{cm}}} \int_{-1}^{1} d\cos\theta. \tag{4.55}$$

The magnitude of one of the final-state particles' momenta $|\vec{p}|$ is the solution of the equation

$$E_{\text{cm}} = \sqrt{|\vec{p}|^2 + m_1^2} + \sqrt{|\vec{p}|^2 + m_2^2}, \tag{4.56}$$

which, for compactness, we leave implicit. Two-body phase space is apparently dimensionless, and consists of a single integral over the angle θ, called the **scattering angle**.

4.3 Feynman Diagrams

The last piece in Fermi's Golden Rule of Eq. 4.40 that we need in order to calculate the cross section is the Lorentz-invariant matrix element

$$\mathcal{M}(A + B \to 1 + 2 + \cdots + n), \tag{4.57}$$

which is the probability amplitude for protons A and B to collide and produce particles $1, 2, 3, \ldots, n$ after collision. It represents the overlap of the wavefunction of the initial and final states:

$$\mathcal{M}(A + B \to 1 + 2 + \cdots + n) = {}_{\text{out}}\langle p_1 p_2 \ldots p_n | p_A p_B \rangle_{\text{in}}, \tag{4.58}$$

where the momenta of the final-state particles $p_1, p_2, \ldots, p_n$ and the momenta of the initial-state particles p_A, p_B are specified. The object $\mathcal{M}$ must be Lorentz invariant for the cross section to have the correct transformation properties. It is called a "matrix element" because it represents an entry of the scattering matrix, or **S-matrix**. The S-matrix of a quantum system encodes all possible transitions from an initial to a final state.

4.3.1 Diagrams in Physics: Circuits

This is all still incredibly abstract. What we want is a procedure or algorithm for calculating the Lorentz-invariant matrix element. It will be exceptionally convenient to introduce graphs or diagrams to represent the transition from an initial to a final state. For the $A + B \to 1 + 2 + \cdots + n$ process, or compactly, "2-to-n" scattering, we can draw the diagram:

$$\mathcal{M}(A + B \to 1 + 2 + \cdots + n) = \quad\quad\quad\quad\quad \text{(4.59)}$$

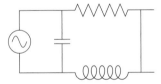

$$\text{past} \xrightarrow[\text{time}]{} \text{future}$$

Time in this diagram runs left to right: in the far past, protons A and B were accelerated and then collided. After collision, n final-state particles were then produced, and subsequently measured in the far future. The blob in the middle represents the magic that happened during collision. We would like to unpack that blob and have a way of calculating it explicitly. Somehow that blob should be represented by lines and vertices that connect the initial to the final state and correspond to a mathematical expression.

As an example of another instance in physics where we use diagrams, and to motivate the diagrams in particle physics, consider circuit diagrams. For a circuit diagram, we define different lines to represent different circuit elements, and are often asked to determine a voltage across some external wires. For example, consider the following circuit diagram:

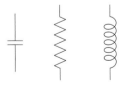

(This is meaningless; it's just an illustration.) The circuit is sourced by an AC voltage denoted by ⊛. The different symbols

denote the circuit elements: capacitors, resistors, and inductors, respectively. Really, they represent different functions for the voltage across that element in terms of the charge flowing through the element. For example, the voltage across a capacitor is $V = Q/C$, where C is the capacitance and Q is the total charge on the capacitor. The other voltages are

$$\Rightarrow V_C = \frac{Q}{C}, \qquad \Rightarrow V_R = R\frac{dQ}{dt}, \qquad \Rightarrow V_I = L\frac{d^2Q}{dt^2}. \qquad (4.60)$$

Here, R is the resistance of the resistor and L is the inductance of the inductor. So, we can compactly denote voltage/charge relationships with symbols in circuit diagrams. Additionally, we need information about what happens at nodes in the circuit:

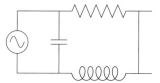

Of course, we have Kirchoff's rules, which are just the statements of energy and charge conservation applied to circuits. Kirchoff's first rule is that the charge flowing into a node is equal to the charge flowing out of a node. Kirchoff's second rule is the identical statement, but applied to energy. A consequence of this is that the net voltage around any closed loop in a circuit is zero.

Given mathematical definitions of the symbols of lines and Kirchoff's rules, we can uniquely determine the voltage across the two open wires on the right in the diagram:

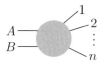

We want to develop similar diagrammatics for particle physics to calculate the probability that an initial collection of particles turns into a final collection of particles:

Just like circuit diagrams, these particle physics diagrams should have the following properties:

(a) Different particles are represented by different shapes of lines.

(b) At nodes (or vertices) in the diagram, relativistic energy and momentum are conserved.

(c) The charge flowing into a vertex is equal to the charge flowing out of a vertex.

Unlike circuit diagrams, in particle physics we also need to specify what happens at vertices to conserve angular momentum. It is best to see how this works in an example.

4.3.2 Diagrams in Physics: Electron–Muon Scattering

For concreteness, let's consider one of the simplest processes in particle physics: the collision of an electron and a positron that annihilate and produce a muon and an anti-muon, $e^+e^- \to \mu^+\mu^-$. We can visualize this as

$$\mathcal{M}(e^+e^- \to \mu^+\mu^-) = \qquad\qquad\qquad . \qquad\qquad (4.61)$$

The electron and muon are both electrically charged and so interact via electromagnetism. The simplest way they can interact is by exchange of a photon, the force carrier of electromagnetism. We can denote a photon by a wavy line $\wedge\wedge\wedge\wedge$ called a **propagator**. Then, we can draw this scattering process as

$$\mathcal{M}(e^+e^- \to \mu^+\mu^-) = \qquad\qquad\qquad . \qquad\qquad (4.62)$$

We've denoted particles with arrows that point with time (e^-, μ^-) and anti-particles with arrows that point against the flow of time (e^+, μ^+). Such a diagram is called a "spacetime diagram" or **Feynman diagram** after Richard Feynman, who introduced them in the 1940s.[4]

For this diagram to be useful, it should (a) tell us a physical picture of what is happening and (b) assist in calculation. The beauty (some may say curse[5]) of Feynman diagrams is their immediate physical interpretation. From this diagram, an electron and positron collide and annihilate into a photon. The photon travels some distance and then transmogrifies into a $\mu^+\mu^-$ pair that sails off into the sunset. This is a nice picture, but is not what is actually happening, for reasons we will discuss in Section 4.3.4. Feynman diagrams are our mathematical representation of an approximation to a particle physics process.

Momentum Conservation

To make Feynman diagrams useful, we need to define what all of its parts are. These are called the **Feynman rules**. We will step through the description of the rules needed

---

[4] R. P. Feynman, "Space-time approach to quantum electrodynamics," Phys. Rev. **76**, 769 (1949). While Feynman introduced his diagrams, Freeman Dyson was the one who shortly after provided the solid mathematical basis for Feynman diagrams. See F. J. Dyson, "The radiation theories of Tomonaga, Schwinger, and Feynman," Phys. Rev. **75**, 486 (1949); F. J. Dyson, "The S matrix in quantum electrodynamics," Phys. Rev. **75**, 1736 (1949). Freeman Dyson is the eponym of the character in the "HλLF-LIFE" video game series. Dyson is also perhaps the most well-known physicist who does not hold a Ph.D.

[5] Julian Schwinger derisively stated that with his diagrams, "Feynman brought field theory to the masses." Schwinger's sentiment from a modern viewpoint is harsh, but nevertheless one must be careful with the physical interpretation of Feynman diagrams.

to evaluate a Feynman diagram and then summarize them in Section 4.3.3. First, as we mentioned earlier, relativistic energy and momentum is conserved at each vertex. To impose this, let's go back to our diagram and write the momentum four-vectors for each particle:

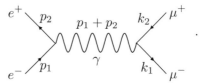

Momentum conservation requires that $p_1 + p_2 = k_1 + k_2$. Note also that the electrons and muons are **external particles** and as such are on-shell; for example, $p_1^2 = m_e^2$, the mass of the electron. The photon is an internal particle: its ends are stuck on other particles. The momentum of the photon is also fixed to be the sum of the electron and positron momenta. Because of this, it is impossible for the photon to be on-shell!

To see this, let's fix ourselves to scattering $e^+ e^-$ along the $\hat{z}$-axis, with equal energy. Then, the four-vectors of the electron and positron can be written as

$$p_1 = \left(\sqrt{p_z^2 + m_e^2}, 0, 0, p_z \right), \qquad p_2 = \left(\sqrt{p_z^2 + m_e^2}, 0, 0, -p_z \right), \qquad (4.63)$$

for some $\hat{z}$-component of momentum p_z. The four-vector of the photon is then

$$p_\gamma = p_1 + p_2 = \left(2\sqrt{p_z^2 + m_e^2}, 0, 0, 0 \right). \qquad (4.64)$$

Note that $p_\gamma^2 = 4(p_z^2 + m_e^2) \neq m_\gamma^2 = 0$. This can never equal 0, for any momentum p_z, and therefore the internal photon is not on-shell. Particles that are not on-shell are said to be **off-shell** or virtual.

External Particle Wavefunctions

Okay, what else is going on here? The external particles e^+, e^-, μ^+, μ^- are on-shell and all are spin-1/2 fermions. Therefore, they are described by solutions of the Dirac equation. As spin-1/2 particles, their spin can be up or down, or some linear combination of up and down. Recall that we could express the solution ψ to the Dirac equation as

$$(i\gamma^\mu \partial_\mu - m)\psi = 0 \qquad \Longrightarrow \qquad \psi = u_s(p)e^{-ip\cdot x} \text{ and } v_s(p)e^{-ip\cdot x}, \qquad (4.65)$$

where $p = (E_p, \vec{p})$, with $E_p = \pm\sqrt{|\vec{p}|^2 + m^2}$. $u_s(p)$ and $v_s(p)$ are the four-component spinors that describe the spin of external particles (u) and anti-particles (v). They are complex (as they are wavefunctions) and so have conjugates $u^\dagger$ and $v^\dagger$, respectively. The spinors $u(p)$ and $v^\dagger(p)$ describe the spin of an initial spin-1/2 particle and anti-particle, respectively, while $u^\dagger(p)$ and $v(p)$ describe the spin of a final spin-1/2 particle and anti-particle. The subscript s in u_s denotes the spin state (up or down), which is defined relative to an appropriate axis.

In the case when the mass of the spin-1/2 particle is zero (or in the extreme relativistic limit $|\vec{p}| \gg m$), the spin is described by the helicity of the particle. The helicity is the

Table 4.1 External Spinor Assignments		
	particle	anti-particle
$\overrightarrow{\text{time}}$	▶	◀
initial state	$u(p)$	$v^{\dagger}(p)$
final state	$u^{\dagger}(p)$	$v(p)$

projection of spin along the direction of motion. Because massless particles must travel at the speed of light, helicity is Lorentz invariant: there is no Lorentz transformation one can perform to change the helicity. We call helicity right- or left-handed based on the direction of spin. For example, a (massless) electron with right-handed helicity would be

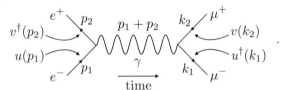

while left-handed would be

To understand these names, curl your right or left hand about the axis of momentum and look where your thumb points. The spin state s that defines the external wavefunction spinor for massless fermions is either R or L for right- or left-handed helicity, respectively.

Because they describe on-shell fermions, we identify the external lines of the Feynman diagram with the spinors we defined:

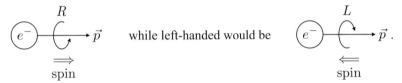

The relative direction of the fermion arrow compared to the arrow of time tells us if it is a particle or an anti-particle, and whether it is initial or final state tells us how we should treat it. The rules are summarized in Table 4.1.

Electron–Photon Vertex

Getting closer! We just need to figure out what the squiggle $\vee\!\!\vee\!\!\vee\!\!\vee$ means and what a vertex $\!\!\prec$ is. First, let's focus on the vertex:

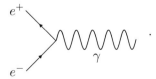

Look at what this is symbolizing: initial electrically charged particles emit or radiate a photon. The likelihood of emission of a photon is controlled by the electric charge e of the electron and positron, since, if they were neutral, they could never emit a photon. The vertex also must conserve angular momentum. The electron and positron are spin-1/2 particles and the photon is spin-1. So, this vertex must align the spins of e^+ and e^- into total spin 1.

Let's again align the e^+e^- pair along the $\hat{z}$-axis and assume that the electron and positron are massless. This is a good approximation if their energies are much larger than their masses. For the electron and positron to annihilate into a photon, their helicities must sum appropriately to spin-1. Consider right-handed e^+ and e^- colliding:

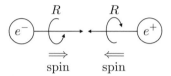

In this configuration, the spins of the e^+ and e^- are in opposite directions, which sum to 0 total spin. This spin configuration is therefore not allowed in this process. For production of a spin-1 photon, we must align the spins of the e^+ and e^-:

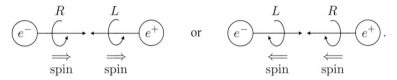

Therefore, the vertex aligns the spins of the electron and positron. This is accomplished by multiplying the spinors together with a γ matrix, as discussed in Section 2.2.2. For the initial-state electron and positron spinors, this would correspond to

$$v^\dagger(p_2)\gamma_\mu u(p_1)\,. \tag{4.66}$$

We then assign the vertex a value of the product of the electric charge e and a γ matrix:

$$\text{\raisebox{-1ex}{\includegraphics}} = e\gamma_\mu\,. \tag{4.67}$$

In the Weyl basis, the γ matrix can be written in terms of the Pauli spin matrices, and the four-component spinors u and v decompose into two two-component spinors of definite helicity. Calculating this spinor product in two-component notation will be done in Chapter 6.

Photon Propagator

The last part of the Feynman diagram is the photon propagator $\sim\!\!\sim\!\!\sim$. The photon propagator expresses the momentum dependence of the strength of the electromagnetic field. As such, it will depend on the four-momentum that flows through it, which we will denote as q. The strength of the electromagnetic force on charged particles doesn't depend on the spin of those particles, so we should expect that the propagator is exclusively a function of the Lorentz-invariant q^2.[6] That is, with momentum q flowing through the photon, we assume that the squiggle can only be some function f of q^2:

$$\text{\reflectbox{}} = f(q^2). \tag{4.68}$$

What function?

Because the internal photon is off-shell, $q^2 \neq 0$, we can think of q^2 as its effective mass. For e^+e^- collision in the center-of-mass frame, the four-vector q is

$$q = \left(2\sqrt{|\vec{p}|^2 + m_e^2}, 0, 0, 0\right), \tag{4.69}$$

where $\vec{p}$ is the initial electron momentum. This photon has 0 momentum, so it is at rest. Then, the "size" of this photon is determined by the Compton wavelength:

$$\lambda_C = \frac{\hbar}{mc} = \frac{\hbar}{\sqrt{q^2}c}, \tag{4.70}$$

where factors of $\hbar$ and c have been included for illustration. If this wavelength is very small, then it is unlikely that the electron and muon will interact; they have to be very close to know about one another. On the other hand, if the Compton wavelength is large, then there is greater likelihood that the electron and muon interact; the photon can "see" both of them at the same time, even if they are widely separated. This motivates the propagator to be inversely proportional to q^2:

$$\text{\reflectbox{}} \propto \frac{1}{q^2}. \tag{4.71}$$

Result

Putting all these pieces together, we can express the mathematical content of the Feynman diagram. Explicitly, the Feynman diagram for $e^+e^- \to \mu^+\mu^-$ with a particular choice of helicities for the external particles is

$$\mathcal{M}(e_R^- e_L^+ \to \mu_R^- \mu_L^+) =$$

$$= v_L^\dagger(p_2)\gamma^\mu u_R(p_1)\frac{e^2}{(p_1+p_2)^2}u_R^\dagger(k_1)\gamma_\mu v_L(k_2). \tag{4.72}$$

---

[6] This isn't quite true, but the details depend on the choice of gauge that one makes to write the propagator. In this book, we will exclusively use the Feynman–'t Hooft gauge, for which the propagator only depends on q^2.

In analogy with Kirchoff's rules, the Feynman rules we have discussed here (what ⋀⋀⋀ means, what happens at a vertex, etc.) represent a unique map from the Feynman diagram to a concrete mathematical expression that represents the complex probability amplitude for the interaction. Feynman diagrams are a powerful tool for predicting and interpreting experimental results in particle physics.

The results presented in this section can be directly derived from the Lagrangian of the theory called **quantum electrodynamics (QED)**:

$$\mathcal{L} = -\frac{1}{4}F_{\mu\nu}F^{\mu\nu} + \bar{\psi}(i\gamma \cdot \partial - m - e\gamma \cdot A)\psi, \tag{4.73}$$

where A_μ is the vector potential of the photon and $F_{\mu\nu}$ is the field strength tensor of electromagnetism,

$$F_{\mu\nu} = \partial_\mu A_\nu - \partial_\nu A_\mu. \tag{4.74}$$

While we won't discuss QED in this book much, it is the most precise physical theory humanity has constructed. It makes predictions that agree to better than 1 part per billion with experiment.

4.3.3 Feynman Diagrams: Summary

Feynman diagrams and their construction from the Feynman rules are a central tool for analyzing particle physics processes. While we have developed them somewhat piecemeal in this chapter, here we summarize their content applied to electron–muon scattering. We will see numerous applications throughout the rest of the book, so it will be important to have a central repository to reference later.

- Identify all initial- and final-state particles; collectively, they are called external particles. External particles are described by solutions to the on-shell equations of motion (the Klein–Gordon, Dirac, or electromagnetic equations of motion). Draw an arrow of time that identifies the transition from initial to final state. Arrows on external fermions identify them as particles or anti-particles. The spinor assignments of external particles and anti-particles is summarized in Table 4.1.

- Connect external fermion lines together with photon propagators denoted by a wavy line and assign its value as

$$\text{⋀⋀⋀}_{\overrightarrow{q}} = \frac{1}{q^2}, \tag{4.75}$$

where q is the four-momentum that flows through the propagator. The momentum q is determined by momentum conservation at each vertex in the diagram. Make every possible connection of fermion lines with photon propagators. Diagrams that are topologically distinct should be summed together.

- At every vertex in the diagram, multiply spinors together with the electric charge e and a γ matrix. That is, a vertex with electrically charged fermions and a photon is assigned the value

$$\text{(diagram)} = e\gamma_\mu . \tag{4.76}$$

- Multiply spinors along the same fermion line together with vertices. Multiply different fermion lines together and contract the Lorentz indices of vertices that are connected by a photon propagator. With explicit momenta and spin assignments, the diagram can then be evaluated as a complex number.
- Repeat with all distinct Feynman diagrams for the process of interest and sum all diagrams together.

In this chapter, we just introduce the procedure of constructing Feynman diagrams. We leave the Feynman diagram of Eq. 4.72 unevaluated, with no explicit spinors. In Chapter 6, we will evaluate this Feynman diagram and construct the cross section for the process $e^+e^- \to \mu^+\mu^-$.

4.3.4 Feynman Diagrams: Caveat Emptor

While Feynman diagrams have a satisfying physical interpretation, as mentioned earlier one must be careful to not take them too seriously. Additionally, Feynman diagrams are useful to answer some questions of interest in particle physics, but are not useful for other questions. Feynman diagrams are a tool of particle physics, but particle physics is much more than the content of Feynman diagrams.

For instance, in the process $e^+e^- \to \mu^+\mu^-$ the interaction proceeds through electromagnetism. In the Feynman diagram of Eq. 4.72, the electromagnetic field is illustrated as a wavy line representing a photon. So, we often say that the interaction proceeds via a photon. This is a bit sloppy, however. The photon that mediates the interaction is off-shell and virtual. Virtual particles are not real: they can never be measured directly. The reason why they cannot be measured directly is that the p_γ^2 of the intermediate photon does not correspond to the eigenvalue of any Hermitian operator. For a real photon described by a wavefunction $|\gamma\rangle$, the Hermitian operator $\hat{P}_\mu\hat{P}^\mu$ has eigenvalue

$$\hat{P}_\mu\hat{P}^\mu|\gamma\rangle = 0|\gamma\rangle . \tag{4.77}$$

This is why Feynman diagrams are useful mathematical tools, but as physical descriptions of the scattering process, should be interpreted with care. Precisely, the wavy line in the Feynman diagram of Eq. 4.72 is the graphical representation of an approximation to the quantum electromagnetic field.

The questions that Feynman diagrams can address in particle physics are relatively limited, as well. In Chapter 6, we will introduce the notion of an inclusive cross section, for which few or no restrictions are made on the final state. Inclusive cross sections are well described by Feynman diagrams and their input into Fermi's Golden Rule. However, when there are constraints on the final state that restrict particle energies or angles, Feynman diagrams are no longer an accurate representation of the process. An example of this is studied in Chapter 9 in which an infinite number of final-state particles are necessary for an accurate description. As the representation of the wavefunction overlap of an initial with a final state, Feynman diagrams do not exhibit any time evolution. An initial state is one that (formally) exists infinitely far into the past, while a final state is infinitely far into the future. Time evolution will be important in Chapter 12 when we discuss neutrino oscillation, but our analysis will effectively just use familiar methods from quantum mechanics. Finally, as an approximation technique, Feynman diagrams actually do not converge. We present a simple argument for why this is true in Chapter 14. This is never a problem in practice as calculations are only ever done to low orders, like that presented in Eq. 4.72. However, a consequence of this is that Feynman diagrams cannot be used to predict properties of bound states like the proton, for example. Using Feynman diagrams to make deep statements about particle physics or quantum field theory requires care, and one should only do it at one's own risk.

Exercises

4.1 *Galactic Collisions.* The Milky Way and Andromeda (M31) galaxies will collide in about 4 billion years. Both galaxies are spiral-type and contain about 100 billion stars each, which is comparable to the number of protons in a bunch at the LHC. In this exercise, we will study this collision and estimate the number of individual stars that interact.

(a) The two galaxies are similar in size and disk-shaped, with a diameter of about 10^5 light-years and a thickness of about 10^3 light-years. If the relative velocity of the Milky Way and Andromeda galaxies is about 10^5 m $\cdot$ s^{-1}, estimate the flux factor (luminosity) at collision. How does this compare to the luminosity of proton collisions at the LHC?

(b) What is the average distance between two stars in these galaxies? Assuming that an average star is about the size of our Sun (radius 7×10^8 m), what is the ratio of the average distance to star radius? Correspondingly, what is the ratio of the average distance of protons in a bunch at the LHC to the proton radius? Use 10^{-15} m for the proton radius.

(c) If the process of collision of the Milky Way and Andromeda galaxies takes about a billion years, approximately how many individual stars will collide? Is this similar to the number of proton collisions per bunch crossing at the LHC?

4.2 *Integrating δ-functions.* In the definition of n-body phase space, there were a lot of δ-functions that imposed on-shell conditions for particles' four-momenta or energy–

momentum conservation. So, when evaluating Fermi's Golden Rule, we need to know how to evaluate integrals involving δ-functions. In this exercise, we will see how to do this.

(a) Suppose you have an integral to evaluate of the form

$$\int dx\, \delta\left(y - f(x)\right) ,\qquad (4.78)$$

where $f(x)$ is some function of x. Show that this is equal to

$$\int dx\, \delta\left(y - f(x)\right) = \int dx\, \frac{1}{\left|\frac{df}{dx}\right|} \delta\left(x - f^{-1}(y)\right) .\qquad (4.79)$$

Hint: You'll need to make a change of variables in the integral.

(b) Using part (a), evaluate the integral

$$\int_{0}^{2} dx\, \delta(x^2 - 1) .\qquad (4.80)$$

4.3   *Three-Body Phase Space.* In this exercise, we will simplify three-body phase space ($n = 3$), which will be useful when we discuss processes in QCD and decays mediated by the weak force. We will denote the four-vectors of the three final-state particles as p_i, for $i = 1, 2, 3$, and they have masses m_i, for $i = 1, 2, 3$. We will work in the center-of-mass frame where the total momentum is 0 and the total energy is E_{cm}. The total four-vector of the process will be denoted by $Q = (E_{\text{cm}}, 0, 0, 0)$.

(a) Very useful quantities for expressing three-body phase space are the x_i variables, for $x = 1, 2, 3$. Define

$$x_i = \frac{2Q \cdot p_i}{Q^2} .\qquad (4.81)$$

Show that

$$x_1 + x_2 + x_3 = 2 .\qquad (4.82)$$

(b) In the center-of-mass frame, determine expressions for the energy of particle i, E_i, and the magnitude of the three-momentum $|\vec{p}_i|$ in terms of the x_i and the masses of the particles.

(c) As in Example 4.2, we can integrate over the on-shell δ-functions, producing the integral for three-body phase space:

$$\int d\Pi_3 = \int \frac{d^3 p_1}{(2\pi)^3} \frac{d^3 p_2}{(2\pi)^3} \frac{d^3 p_3}{(2\pi)^3} \frac{1}{2E_1} \frac{1}{2E_2} \frac{1}{2E_3} (2\pi)^4 \delta^{(4)}\left(Q - p_1 - p_2 - p_3\right) .$$

$$(4.83)$$

Here, $E_i = \sqrt{|\vec{p}_i|^2 + m_i^2}$. We can eliminate the integral over $\vec{p}_3$ by enforcing the three-momentum conserving δ-functions. Do this; you should find

$$\int d\Pi_3 = \int \frac{d^3 p_1}{(2\pi)^3} \frac{d^3 p_2}{(2\pi)^3} \frac{1}{2E_1} \frac{1}{2E_2} \frac{1}{2E_3} 2\pi\delta\left(E_{\text{cm}} - E_1 - E_2 - E_3\right). \quad (4.84)$$

(d) Now, express the integrals over the three-momenta $\vec{p}_1$ and $\vec{p}_2$ in spherical coordinates. Note that there are four angular integrals: two azimuthal angles, and two polar angles. By integrating over $\vec{p}_3$, the energy E_3 is a function of $|\vec{p}_1 + \vec{p}_2|$ and therefore a function of the angle between $\vec{p}_1$ and $\vec{p}_2$, $\cos\theta_{12}$. Do the integral over $\cos\theta_{12}$, eliminating the last δ-function. You should find

$$\int d\Pi_3 = \frac{1}{8(2\pi)^5} \int d|\vec{p}_1| \, d\cos\psi \, d\phi_1 d|\vec{p}_2| \, d\phi_2 \frac{|\vec{p}_1|}{E_1} \frac{|\vec{p}_2|}{E_2}. \quad (4.85)$$

ψ is the remaining polar angle, and ϕ_1 and ϕ_2 are the remaining azimuthal angles.

Hint: The angle between the momenta of particles 1 and 2, θ_{12}, can be expressed as one of the polar angles.

(e) The remaining three angles, ψ, ϕ_1, and ϕ_2, simply rotate the three-particle system. Integrate over these angles. You should find

$$\int d\Pi_3 = \frac{1}{32\pi^3} \int d|\vec{p}_1| \, d|\vec{p}_2| \frac{|\vec{p}_1|}{E_1} \frac{|\vec{p}_2|}{E_2}. \quad (4.86)$$

(f) Now, there just remain two integrals over the magnitude of momenta $|\vec{p}_1|$ and $|\vec{p}_2|$. Using part (a), express the remaining integrals in terms of x_1 and x_2. Show that

$$\int d\Pi_3 = \frac{E_{\text{cm}}^2}{128\pi^3} \int dx_1 \, dx_2. \quad (4.87)$$

By momentum conservation the energy fractions x_1 and x_2 are restricted to lie in the domain $[0, 1]$ with the constraint that $x_1 + x_2 > 1$.

4.4 $e^+ e^- \to e^+ e^-$ *Scattering.* In this chapter, we discussed the details of the (lowest-order) Feynman diagram for the process $e^+ e^- \to \mu^+ \mu^-$. In this exercise, we will study the process $e^+ e^- \to e^+ e^-$.

Using the Feynman rules, draw all of the Feynman diagrams for $e^+ e^- \to e^+ e^-$ scattering, called **Bhabha scattering**.[7] Each diagram should contain only one internal photon, and be sure to clearly label the momentum of each external particle.

Hint: Unlike $e^+ e^- \to \mu^+ \mu^-$ scattering, there is more than one Feynman diagram for $e^+ e^- \to e^+ e^-$ scattering.

4.5 *Non-Relativistic Limit of Feynman Diagrams.* Feynman diagrams may seem somewhat magical and disconnected from familiar non-relativistic or classical physics. In this exercise, we will demonstrate how the contents of a Feynman diagram connect to the non-relativistic limit of electromagnetism.

---

[7] H. J. Bhabha, "The scattering of positrons by electrons with exchange on Dirac's theory of the positron," Proc. Roy. Soc. Lond. A **154**, 195 (1936).

Consider the scattering of electrons and muons $e^- \mu^- \rightarrow e^- \mu^-$, which is represented by the following Feynman diagram:

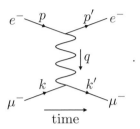

(a) Express the four-momentum q flowing through the photon in terms of the four-momenta of the external particles.

(b) Describe the physical configuration of the electron and muon if the internal photon goes on-shell. That is, as $q^2 \rightarrow 0$, does that correspond to the electron and muon getting closer together or farther apart?

(c) From the photon propagator in this diagram, we can define a corresponding electric potential. Multiplying by the electric charges e of the electron and muon, the momentum electric potential $\tilde{V}(q)$ is

$$\tilde{V}(q) = \frac{-e^2}{q^2} . \tag{4.88}$$

Take the non-relativistic limit of this expression; that is, using the result from part (a), take the limit where the masses of the electron and muon are much larger than their magnitude of (three-)momentum.

(d) Call the non-relativistic limit of the potential from the previous question $\tilde{V}(\vec{q})$; that is, it depends on the three-momentum of the photon. We can write this as

$$\tilde{V}(\vec{q}) = \frac{e^2}{|\vec{q}|^2} . \tag{4.89}$$

We aren't used to dealing with potentials defined by particles' momenta; typically, we deal with potentials that are functions of (relative) position. To determine the electric potential $V(\vec{r})$, we need to Fourier transform the potential $\tilde{V}(\vec{q})$ from a function of momentum to one of position.

The necessary Fourier transform is

$$V(\vec{r}) = \int \frac{d^3q}{(2\pi)^3} \frac{e^2}{|\vec{q}|^2} e^{-i\vec{q}\cdot\vec{r}} . \tag{4.90}$$

Write the integral over the three components of the momentum $\vec{q}$ in spherical coordinates: momentum magnitude $|\vec{q}|$, azimuthal angle ϕ, and polar angle θ. You'll also want to express the dot product $\vec{q} \cdot \vec{r}$ in terms of the magnitude of the $\vec{q}$ and $\vec{r}$ vectors. Integrate over the azimuthal angle ϕ and the polar angle θ. You should find

$$V(\vec{r}) = \frac{ie^2}{4\pi^2} \frac{1}{|\vec{r}|} \int_{-\infty}^{\infty} d|\vec{q}| \, \frac{e^{-i|\vec{q}||\vec{r}|}}{|\vec{q}|} . \tag{4.91}$$

(e) Now, we need to do the integral over $|\vec{q}|$. Note that

$$\int_{-\infty}^{\infty} d|\vec{q}| \, \frac{e^{-i|\vec{q}||\vec{r}|}}{|\vec{q}|} = -i \int_{0}^{|\vec{r}|} dr' \int_{-\infty}^{\infty} d|\vec{q}| \, e^{-i|\vec{q}|r'} . \tag{4.92}$$

Using this and the definition of the δ-function,

$$\int_{-\infty}^{\infty} d|\vec{q}| \, e^{-i|\vec{q}||\vec{r}|} = \pi\delta(|\vec{r}|) , \tag{4.93}$$

show that

$$V(\vec{r}) = \frac{e^2}{4\pi} \frac{1}{|\vec{r}|} . \tag{4.94}$$

That is, the photon propagator that goes like $1/q^2$ in momentum space corresponds to the familiar $1/|\vec{r}|$ potential in position space!

4.6   *Proton–Proton Total Cross Section.* The **TOTEM experiment** (<u>TOT</u>al <u>E</u>lastic and diffractive cross section <u>M</u>easurement) at the Large Hadron Collider measures the total cross section for proton collisions. The TOTEM experiment is located far "downstream" from the collisions at CMS. The basic way that it measures the total cross section of the proton is much the same as how you would measure the cross section of any object.

A very efficient way to measure the cross section of an object is to shine light on it and measure the size of the shadow. The TOTEM experiment sits far from the collisions at CMS so that it can measure the size of the effective "shadow" from the proton collisions. A proton occupies a finite size in space, and so particles produced in the collision must be deflected around the proton, producing a shadow, or lack of particles observed.

Figure 4.2 shows results from TOTEM and various other experiments on the measurements of the total, elastic (kinetic energy preserving), and inelastic cross sections of the proton measured in millibarns, as a function of center-of-mass collision energy $\sqrt{s}$. Estimate the radius of the proton in meters as measured by the TOTEM experiment, at center-of-mass collision energies of 7 and 8 TeV.

4.7   *Proton Collision Beam at ATLAS.* The **integrated luminosity** is the time integral of the luminosity over a time interval $[t_0, t_1]$:

$$\text{Integrated Luminosity} = \int_{t_0}^{t_1} dt \, \mathcal{L} , \tag{4.95}$$

and has units of area^{-1}. The integrated luminosity in collider physics experiments is often measured in units of inverse barns (b^{-1}).

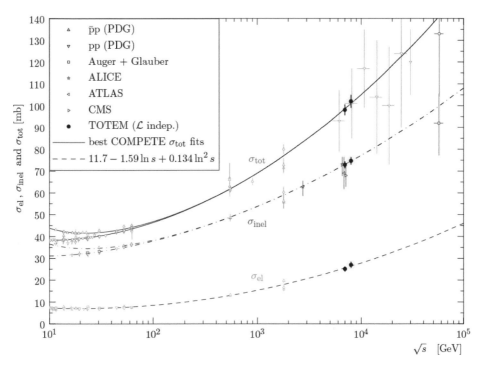

Fig. 4.2 Plots of the elastic σ_{el}, inelastic σ_{inel}, and total σ_{tot} cross sections in millibarns (mb) for proton scattering measured by various experiments, as a function of the center-of-mass collision energy, $\sqrt{s}$. From G. Antchev *et al.* [TOTEM Collaboration], "Luminosity-independent measurement of the proton–proton total cross section at $\sqrt{s} = 8$ TeV," Phys. Rev. Lett. **111**, no. 1, 012001 (2013), doi:10.1103/PhysRevLett.111.012001.

Figure 4.3 is a plot from the ATLAS experiment of the integrated luminosity of proton collisions during data taking in 2016, as a function of date. Note that the date is written in the European standard Day/Month. The integrated luminosity is measured in inverse femtobarns (fb^{-1}).

(a) From this plot, estimate the average luminosity over the data-taking period. Express the result in inverse femtobarns per second and inverse centimeters squared per second.

(b) Using the TOTEM plot Fig. 4.2 from Exercise 4.6, estimate the number of proton collision events that took place in the ATLAS experiment during data taking in 2016. On average, approximately how many proton collision events occurred per second?

(c) Figure 4.4 is nicknamed the **Stairway to Heaven plot** and shows the cross section in picobarns for numerous final states in proton–proton collisions as measured in the ATLAS experiment at the LHC. For example, the cross section for W boson production $pp \to W$ at a center-of-mass collision energy of 13 TeV is about $\sigma_{pp \to W} \simeq 2 \times 10^5$ pb $= 2 \times 10^{-7}$ b.

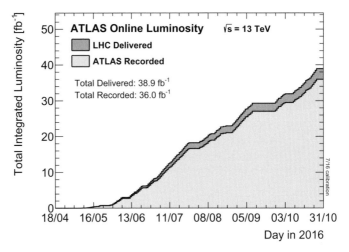

Fig. 4.3   Total integrated luminosity in inverse femtobarns (fb^{-1}) of the ATLAS experiment versus the date in 2016 when data were taken. Credit: ATLAS Experiment © 2018 CERN.

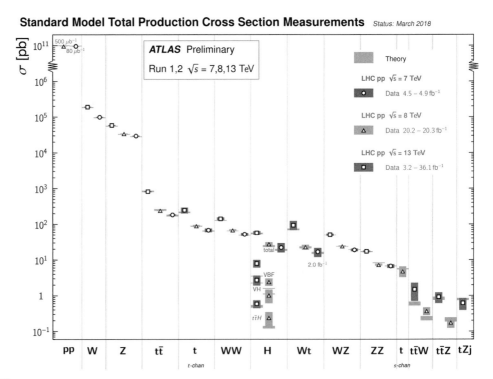

Fig. 4.4   Measured cross sections in picobarns (pb) for the production of numerous Standard Model particles in proton–proton collisions at the LHC. Credit: ATLAS Experiment © 2018 CERN.

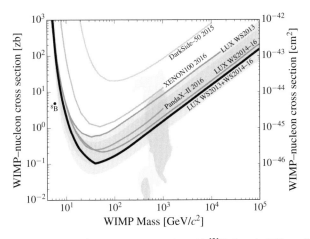

Fig. 4.5 A plot of the limits on the WIMP mass and interaction cross section with ^{131}Xe from the LUX experiment. Reprinted figure with permission from D. S. Akerib *et al.* [LUX Collaboration], Phys. Rev. Lett. **118**, no. 2, 021303 (2017). Copyright 2017 by the American Physical Society.

For the following processes estimate the number of collision events recorded in 2016 (13 TeV collisions):

(i) $pp \to Z$

(ii) $pp \to t\bar{t}$

(iii) $pp \to H$

(iv) $pp \to t\bar{t}Z$.

4.8   *Upper Limits on LUX Bounds.* This exercise is a continuation of Exercise 2.8. In this exercise, we will work to understand the upper limit of the bounds on WIMP dark matter masses established by the LUX experiment. The plot of the WIMP mass versus interaction cross section with xenon nuclei is reprinted in Fig. 4.5. Above a WIMP mass of about 100 GeV, the bound established by LUX is determined by the flux of WIMPs through the LUX detector. We will work to determine how this bound is established.

The LUX detector consists of 250 kg of pure liquid xenon, which you can assume is mostly composed of ^{131}Xe. Liquid xenon has a density of about 3000 kg $\cdot$ m^{-3}. For WIMP dark matter to explain the observed rotational velocity of stars about the galaxy, its halo must have a mass density of about 0.3 GeV $\cdot$ cm^{-3}. Assume that the Earth travels at 240 km $\cdot$ s^{-1} through the halo.

(a) To set the bound on WIMP mass and interaction strength, assume that LUX observed one event in its 332 total days of data collection. With this assumption, determine the cross section σ for WIMPs to interact with ^{131}Xe as a function of the WIMP mass m_χ. Express the cross section in zeptobarns (10^{-21} barns).

(b) Compare this result to the bound plotted in Fig. 4.5, above a WIMP mass of 100 GeV. Note that the cross section is expressed per nucleon (proton or neutron) in the ^{131}Xe nucleus, so make sure to account for this.

(c) What is the maximum mass of a WIMP particle for which one interaction event was possible? That is, what mass corresponds to just one WIMP particle passing through the LUX detector during the entire 332 days?

4.9 *Research Problem.* To study rare (very unlikely) final states at the LHC, we just have to collide a lot of protons together. Is there an upper limit to the proton collision luminosity? If so, is it limited by possible technology or by fundamental physics?

5

Particle Collider Experiment

Physics wouldn't be physics without experiment. We need to concretely test the predictions that we make about how Nature evolves and interacts and search for new phenomena we haven't thought of. Essentially the one and only way in which experiments in particle physics are performed is to collide particles with one another. Expansive high-energy colliders, like the LHC, of course do this, but so do experiments searching for dark matter (looking for collisions of dark matter with atomic nuclei), measuring properties of neutrinos (looking for neutrino collisions with nuclei), probing matter at extreme densities (by colliding nuclei together), or performing detailed studies of the proton (colliding particles with the proton at low energies). Colliders are the bread and butter of particle physics, and we need to understand their properties in detail to know how we can test a hypothesis. We can only test what we can measure, so working within the collider paradigm will highly restrict what is worth calculating at all.

Our discussion of particle collision experiments centers around the LHC and its two all-purpose detectors, ATLAS and CMS. The detectors, while among the most prominent parts of the whole LHC experiment, are just two pieces in a series of machines and equipment that accelerate protons to high energy, collide them, record what comes out, and then ship that data out for analysis. Each item in this chain of operations could occupy whole chapters, but here we just provide a first glimpse at the detail and complexity that go into these experiments. Everything that we introduce here is applicable to all particle physics experiments, but perhaps in different quantities, depending on how the experiment is constructed and to what it is sensitive.

This chapter has three main parts: what happens before protons collide at the LHC, what happens at the point of collision, and what happens after. The ATLAS and CMS experiments occupy the middle part, while in the last part we discuss the data management at the LHC and the standard statistical tools used to claim discovery. We start with what happens before the collision, which requires creating the protons that will be accelerated and collided in the first place.

5.1 Before Collision: Particle Acceleration

After photons and electrons, protons are probably the easiest particles to isolate and create a beam of. Protons are just atomic hydrogen nuclei, so to get a proton you just need to ionize hydrogen. This is pretty easy: take your hydrogen gas and put it in a strong electric field.

In this electric field, the negatively charged electron and the positively charged proton are pulled apart. Once hydrogen is broken up, you can use electric and magnetic fields to do whatever you want to the protons. After isolating protons, the LHC manipulates them with a radio-frequency (RF) electromagnetic field. The protons are organized into collections of about 100 billion protons called **bunches** that sit in the troughs of the field. These bunches have a size fixed by the wavelength of the RF field and are separated in space from one another. The RF field at the LHC has a frequency of 400 MHz, corresponding to a wavelength of about 3/4 of a meter. No proton exists in isolation at the LHC; it is always within its bunch.

Once separated into bunches, the protons are sent through a series of accelerators until they reach the main LHC ring. The first accelerator, Linac 2, is a linear accelerator that consists of a number of dipole electric fields arranged in a line. These electric fields push and pull the proton bunches, accelerating them to an energy of 50 MeV. The bunches are then sent into the Proton Synchrotron Booster, a circular accelerator, to reach an energy of 1.4 GeV. As a circular accelerator, the proton bunches can be sent around the circle many times to reach the desired energy. This is an advantage with respect to a linear collider, which only gets one chance to accelerate the protons. However, this advantage comes at a cost. Even with a constant speed, protons traveling in a circle are accelerating and therefore emit electromagnetic radiation, called **synchrotron radiation**. A circular accelerator must input enough energy to both speed up the protons and maintain their energy, as they lose energy continuously through synchrotron radiation. This is one of the main limitations on the energy to which circular accelerators can reach, which we will discuss shortly.

After the Proton Synchrotron Booster, the proton bunches are accelerated to 25 GeV by the Proton Synchrotron and then to 450 GeV by the Super Proton Synchrotron (SPS). The ring of the SPS is visible in the aerial view of CERN in Fig. 1.2. After the SPS, the protons are injected into the LHC ring, and accelerated to their (current as of 2018) maximum energy of 6.5 TeV. To maintain that energy and to keep the protons in the LHC ring requires continuous energy input through both electric and magnetic fields. For charged particles with mass m and energy E traveling in a circle of radius R, the power P lost through synchrotron radiation is[1]

$$P = \frac{e^2}{6\pi\epsilon_0 m^4 c^{11} R^2} E^4 v^4 .$$

(5.1)

The electric charge of the particle is assumed to be the fundamental unit of charge, e, and the velocity of the particle is v. For protons at the LHC, the velocity is essentially the speed of light, c. We can rewrite this expression in useful units, where the orbital radius R is expressed in km and the particle energy and mass are expressed in the same units (GeV, for example). Then, the power in $eV \cdot s^{-1}$ lost to synchrotron radiation is

$$P \simeq 3 \times 10^{-7} \left(\frac{1 \text{ km}}{R}\right)^2 \left(\frac{E}{m}\right)^4 \text{ eV} \cdot \text{s}^{-1} .$$

(5.2)

---

[1] See, for example, D. J. Griffiths, *Introduction to Electrodynamics, 4th ed.*, Cambridge University Press (2017).

The boost factor E/m of protons at the LHC is about 7000 and the circumference of the LHC is 27 km, and so the power radiated away in synchrotron radiation per proton is

$$P_{\mathrm{LHC}} \simeq 40 \,\mathrm{MeV} \cdot \mathrm{s}^{-1} . \tag{5.3}$$

This is an appreciable rate compared to the energy of protons at the LHC. Accounting for all protons accelerated at the LHC, the power emitted from synchrotron radiation is about a kilowatt. This is a small part of the 200 MW power budget of the entire CERN site. The LHC regularly shuts down for a few months in the winter because electricity usage is largest in winter and costs the most then as well. You'll study the consequences of the synchrotron power losses in Exercise 5.1.

In addition to synchrotron losses from the protons traveling in a circle, we have to keep the protons in the LHC ring. This is accomplished with thousands of superconducting electromagnets that bend the path of the protons, without affecting their kinetic energy. The strength of the magnetic field $|\vec{B}|$ that is required to keep particles in a circle of radius R is a relatively standard calculation in a course on electromagnetism. For a particle with the fundamental electric charge e, this field is

$$|\vec{B}| = \frac{|\vec{p}|}{eR} . \tag{5.4}$$

In this expression, we have assumed that the particle's momentum $\vec{p}$ is perpendicular to the magnetic field $\vec{B}$. For highly relativistic particles, $|\vec{p}| \simeq E/c$ and, expressing the energy in TeV and radius in km, the required magnetic field in tesla is

$$|\vec{B}| \simeq 3 \frac{E\,(\mathrm{TeV})}{R\,(\mathrm{km})} . \tag{5.5}$$

At the LHC with protons with energy of 6.5 TeV, the magnitude of the magnetic field is estimated to be about 5 T. This is quite close to the 8 T magnets that are actually used in the experiment. The strength of the bending magnets is the most restrictive constraint on increasing the collision energy at the LHC. Modern superconducting magnets have a maximum strength of around 20 T or so. Therefore, with current technology, to accelerate protons to significantly higher energies requires a larger ring. Efforts are now underway to plan for a 100 TeV collider located around Geneva. To reach these energies requires an acceleration ring of about 100 km. This and other efforts for future colliders will be presented in Chapter 14.

Now that protons are traveling in the LHC ring at their maximum energy, we need to collide them. The two counterrotating beams consisting of thousands of proton bunches are set to collide at specific locations around the LHC ring. However, even though bunches consist of billions of protons, protons are very small and the vast majority of the volume of a bunch is empty space. The length of a proton bunch is about 40 cm (about half of the RF wavelength that bunches the protons), and the radius of the beam in the LHC ring is about 1 mm. Therefore, the volume of a bunch is

$$\mathrm{Vol}_{\mathrm{bunch}} \simeq 10^{-4} \,\mathrm{m}^3 . \tag{5.6}$$

By contrast, the total volume of the bunch occupied by 100 billion protons is only about

$$\text{Vol}_{\text{protons}} \simeq 10^{-37} \text{ m}^3 \,, \tag{5.7}$$

where we have assumed that the radius of the proton is about 10^{-15} m and one dimension of the protons is Lorentz-contracted by a factor of $\gamma \simeq 7000$. The volume of empty space is more than 30 orders of magnitude larger than the volume of the protons! Such a small fraction of the volume occupied by protons makes it extremely challenging for any of the protons to interact in an interesting way.

To improve the likelihood that protons actually collide, we need to make the beam volume much smaller. This is accomplished at the LHC by quadrupole focusing magnets located around the collision points. Figure 5.1 shows the magnetic field lines of a quadrupole magnet. Consider a beam of protons traveling into the page, into this quadrupole magnet. The beam takes up a finite space, so there will be protons in each of the wedge-shaped regions. The magnetic force $\vec{F}_B$ on the protons is given by the Lorentz force law:

$$\vec{F}_B = e\vec{v} \times \vec{B} \,, \tag{5.8}$$

where e is the fundamental charge, $\vec{v}$ is the velocity of the protons, and $\vec{B}$ is the quadrupole field. By the right-hand rule, protons in the top and bottom wedge regions will be focused; that is, they will move toward the center of the bunch. However, protons in the left and right wedges are defocused: they are pushed away from the center of the bunch. The bunch can be focused in both top–bottom and left–right regions by stacking a series of quadrupole magnets that are rotated $90°$ degrees from one another. At the LHC, these quadrupole

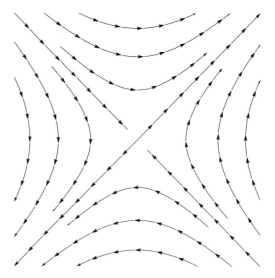

Fig. 5.1   A representation of the magnetic field lines of a quadrupole magnet.

magnets can focus the beam width down to a radius of less than 10 micrometers! Therefore, at the collision point, the volume of the bunch is reduced to

$$\text{Vol}_{\text{bunch,col}} \simeq 10^{-8} \, \text{m}^3 \, . \tag{5.9}$$

Though the bunch volume has only decreased by a factor of about 10,000, this increases the luminosity and therefore the likelihood of proton collision by the same factor, which is vital for interesting physics at the LHC.

5.2 At Collision: Particle Detection

Once the proton beams have been accelerated to their maximum energy and focused at a collision point, they are collided, which just means that two opposite-traveling bunches are passed through one another. At the LHC, bunches are collided every 25 nanoseconds. As we studied in Chapter 4, the bunch parameters and the proton cross section determine how many of the protons in each bunch collision actually interact. When two protons collide and exchange a significantly large amount of momentum, they explode into a shower of particles emanating from the collision point. The dynamics of what happened at the moment of the proton collision is imprinted on these final-state particles, through energy and momentum conservation, through angular momentum conservation, or through charge conservation. To play the detective and be able to reconstruct the physics of the proton collision requires measuring all, or as many as possible, of the particles created and their properties.

At a particle physics detector, we are able to measure many particle properties, but many others we don't even try to measure. For example, measuring energy and momentum are relatively easy, as we will discuss. Also, if you have measured the energy and all three components of three-momentum then you can reconstruct particle masses via

$$m^2 = E^2 - |\vec{p}|^2 \, . \tag{5.10}$$

Electric charge is also relatively easy to measure. While we needed a magnetic field to keep protons in the LHC ring, putting a magnetic field in our detector enables identification of charged particles by their curvature in this field. Angular momentum or spin, on the other hand, is not easy or convenient to measure, and so we don't design our detectors to be sensitive to spin. We will see in Chapter 6 that spin leaves its mark through energy and momentum conservation, so we will be able to infer spin from other measurements.

In this section, we introduce the components of a modern particle physics detector, focusing around the ATLAS and CMS experiments.[2] Schematic illustrations of the ATLAS and CMS experiments are presented in Fig. 5.2. Each illustration has an outline of a person

---

[2] Details about the ATLAS and CMS experiments are available in G. Aad *et al.* [ATLAS Collaboration], "The ATLAS experiment at the CERN Large Hadron Collider," J. Instrum. **3**, S08003 (2008); S. Chatrchyan *et al.* [CMS Collaboration], "The CMS experiment at the CERN LHC," J. Instrum. **3**, S08004 (2008).

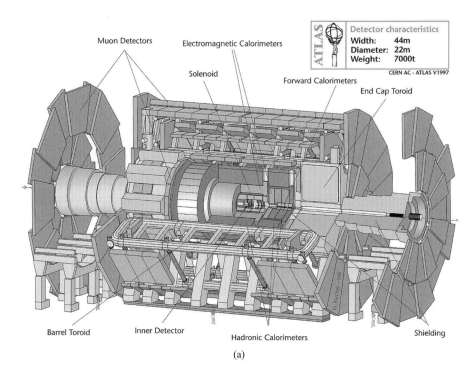

Muon Detectors   Electromagnetic Calorimeters

Solenoid

Forward Calorimeters

End Cap Toroid

Detector characteristics
Width: 44m
Diameter: 22m
Weight: 7000t

ATLAS

CERN AC - ATLAS V1997

Barrel Toroid   Inner Detector   Hadronic Calorimeters   Shielding

(a)

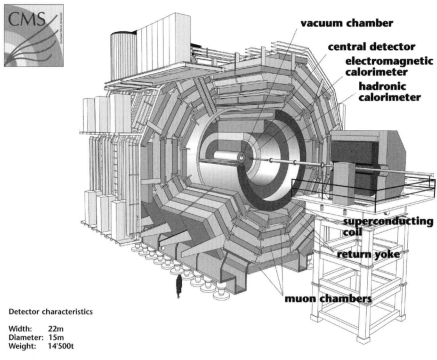

CMS

vacuum chamber

central detector

electromagnetic
calorimeter

hadronic
calorimeter

superconducting
coil

return yoke

muon chambers

Detector characteristics

Width: 22m
Diameter: 15m
Weight: 14'500t

(b)

Fig. 5.2   Illustrations of the (a) ATLAS and (b) CMS detectors. Credit: CERN © CERN.

for scale. While the ATLAS and CMS experiments differ in their precise design, they share many qualities. These are appropriately labeled on the figures as calorimeters or muon detectors, for example; each has a specific purpose and measures specific particles or properties. Before discussing these components, we must introduce a coordinate system by which to describe measurements performed by these experiments.

5.3 Detector Coordinates

To first approximation, both the ATLAS and the CMS detectors are cylinders, with the colliding proton beams along the axis of the cylinder as illustrated in Fig. 5.3. The center of the cylinder is where the proton beams collide, and is called the collision point or interaction point (denoted as "IP"). As these experiments are cylinders, we should use cylindrical coordinates to orient ourselves and express vectors. The cylindrical coordinates we use are azimuthal angle ϕ, pseudorapidity η, and transverse momentum $p_\perp$.

The azimuthal angle ϕ is just the angle about the proton beams, as illustrated in the figure. In cylindrical coordinates, we also need a "z" coordinate that specifies the distance along the proton beams, away from the interaction point. At the LHC, this is measured as the **pseudorapidity** η, which is defined by the polar angle as measured with respect to the proton beam. The pseudorapidity η is

$$\eta = -\ln\tan\frac{\theta}{2}, \tag{5.11}$$

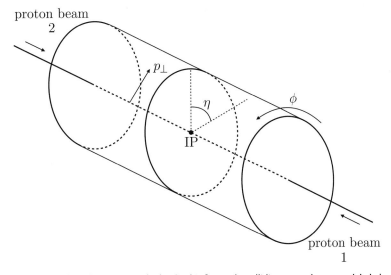

Fig. 5.3 A schematic illustration of the detector as a cylinder. On this figure, the colliding proton beams are labeled, and the point of their collision is denoted as "IP." Coordinates on this cylinder are the azimuthal angle ϕ, the pseudorapidity η, and the transverse momentum $p_\perp$.

where the polar angle is θ. An illustration of this is:

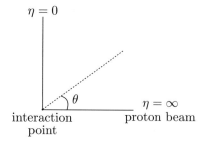

Pseudorapidity η is 0 for central collisions, directly perpendicular to the beam at the interaction point. It is $\pm\infty$ along the proton beams where the polar angle is 0 or π. What makes pseudorapidity useful is that it is true rapidity for massless particles.

In Exercise 2.2 in Chapter 2 we defined the rapidity y as

$$y = \frac{1}{2}\ln\frac{E+p_z}{E-p_z}, \tag{5.12}$$

where the $\hat{z}$-axis coincides with the proton beams. For Lorentz boosts along the proton beam, the rapidity just transforms additively. That is, differences of rapidity are invariant to Lorentz boosts along the $\hat{z}$-direction. This is a very nice and important property that we will exploit when studying the strong force. For a massless particle, energy is equal to the magnitude of momentum: $E = |\vec{p}|$, and so the polar angle is $\cos\theta = p_z/E$. Plugging this into the expression for rapidity, we have

$$y = \frac{1}{2}\ln\frac{E+p_z}{E-p_z} = \frac{1}{2}\ln\frac{1+\cos\theta}{1-\cos\theta} = -\ln\left(\frac{1-\cos\theta}{1+\cos\theta}\right)^{1/2} = -\ln\tan\frac{\theta}{2} = \eta. \tag{5.13}$$

In general, however, rapidity and pseudorapidity do not coincide. In particular, this means that the pseudorapidity of a massive particle does not have a simple transformation property for boosts along the $\hat{z}$-axis.

Transverse momentum, or $p_\perp$, is the effective radial coordinate on the experiment. $p_\perp$ is the magnitude of particle momentum that is transverse to the proton beam, in the $\hat{x}\hat{y}$ plane, as illustrated in Fig. 5.3. Better than rapidity, $p_\perp$ is simply invariant to Lorentz boosts along the proton beam direction. Additionally, the initial momentum of the protons (and their constituent quarks and gluons) is along the $\hat{z}$-axis of the experiment. The initial transverse momentum vector is 0 for the protons as well as for their constituent quarks and gluons, and therefore the final transverse momentum vector must also be 0. That total transverse momentum is 0, is conserved, and is easy to measure at collider experiments are very useful and important properties.

With the $\hat{z}$ coordinate along the proton beam, the $p_\perp$ is

$$p_\perp = \sqrt{p_x^2 + p_y^2}, \tag{5.14}$$

for x- and y-components of momentum. Any massless four-vector p can be expressed in $(\eta, \phi, p_\perp)$ coordinates as

$$p = p_\perp(\cosh\eta, \cos\phi, \sin\phi, \sinh\eta). \tag{5.15}$$

A massive four-vector is similar, we just need to include the mass appropriately. This requires incorporating the mass into the energy of the particle, as well as using true rapidity, rather than pseudorapidity. The four-vector of a massive particle is

$$p = \left(\sqrt{p_\perp^2 + m^2} \cosh y,\ p_\perp \cos\phi,\ p_\perp \sin\phi,\ \sqrt{p_\perp^2 + m^2} \sinh y \right), \qquad (5.16)$$

where the mass is m. The quantity $m_T = \sqrt{p_\perp^2 + m^2}$ is called the transverse mass and is invariant to boosts along the proton beam direction.

5.4 Detector Components

With the detector coordinates established, let's move on to how to measure particle properties from the various components of the detector. While the detailed illustrations of the ATLAS and CMS experiments in Fig. 5.2 are useful for gaining an appreciation for the size and design of the detectors, we will work with a much simplified picture which ATLAS and CMS are to both equivalent. Looking down the cylinder or **barrel** of one of these detectors, they have the schematic form illustrated in Fig. 5.4. At the center is the interaction point, and the proton beams go into and out of the page. Like an onion, these experiments consist of many layers, each of which measures particular properties or is sensitive to particular particles. These different layers are labeled and we will describe each layer in this section, working from the inside out.

Detectors like ATLAS or CMS are often referred to as **4π hermetic** detectors because they capture all particles produced from proton collision throughout 4π steradians of a sphere as best as possible. This task is impossible because, at least, the proton beams must come in somewhere and cables have to go somewhere. However, the coverage is otherwise exceptional. The tracking system extends out to $|\eta| = 2.5$, which is only about an angle of $10°$ above the proton beam. The calorimetry extends further; it goes out to about $|\eta| = 5$, which is less than one degree above the proton beam!

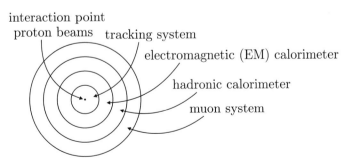

Cross-sectional schematic illustration of the layers of a particle collider detector. The beams of protons go into and out of the page and collide in the center.

5.4.1 Tracking System

Immediately outside the region of the collision point and proton beams is the **tracking system**. The tracking system consists of millions of individual channels that respond when a charged particle hits them. These channels consist of silicon or gas that ionizes when a high-energy particle with charge passes through. This ionization is recorded at numerous points along the trajectory of the charged particle and traces out a **track** of the charged particle. For this reason, charged particles are often just called "tracks" at the LHC experiments.

Observing these tracks provides information about the direction of charged particles, but by itself doesn't provide information about the energy of the particles. What makes the tracker especially useful is that it is embedded in a solenoidal magnetic field. The solenoidal field points along the $\hat{z}$-axis (parallel to the proton beam) and so charged particles' trajectories are affected by the magnitude of their charge and transverse momentum $p_\perp$. Only the component of momentum perpendicular to the magnetic field is affected by the magnetic field. Because of this solenoidal magnetic field, the tracking system is sensitive to both the charge and the momentum of particles. Let's see how this works.

As we discussed in Section 5.1, the radius of curvature R (in meters) of a charged particle is

$$R \simeq 3 \frac{p_\perp \text{ (GeV)}}{Q|\vec{B}| \text{ (T)}}, \tag{5.17}$$

where Q is the electric charge of the particle in units of the fundamental electric charge e, transverse momentum $p_\perp$ is measured in GeV, and the magnitude of magnetic field $|\vec{B}|$ is in T. For example, for a $p_\perp = 1$ GeV electron in the 2 T magnetic field of the ATLAS detector, the radius of curvature is about 1.5 m. Conversely, observing a charged particle with a radius of curvature of 1.5 m means that its $p_\perp$ is 1 GeV. With the radius of curvature, we can measure the magnitude of transverse momentum. With the right-hand rule, the sign of the charge of the particle can be determined as well.

The way in which tracks are constructed by the detector is interesting. The tracking system consists of many layers of material away from the interaction point. As the charged particle passes through, it hits these layers, as illustrated in Fig. 5.5. Curvature can be determined by a minimum of three hits, though the tracking system typically has about 30 layers for redundancy. Importantly, it is the curvature, not the radius of curvature, that is more easily measured. The curvature is just the inverse of the radius of curvature, which is inversely proportional to transverse momentum. Note that the uncertainty on the curvature is

$$\Delta \frac{1}{R} \propto \Delta \frac{1}{p_\perp} = \frac{\Delta p_\perp}{p_\perp^2}. \tag{5.18}$$

That is, the uncertainty on the measurement of momentum in this way increases with increasing $p_\perp$:

$$\Delta p_\perp \simeq p_\perp^2 \left(\Delta \frac{1}{R} \right). \tag{5.19}$$

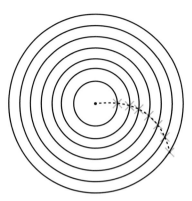

Fig. 5.5 Schematic image of a charged particle passing through the layers of the tracking system. Proton beams go into and out of the page, the interaction point is located at the center, the layers of the tracker are represented by the concentric circles, the trajectory of the particle is illustrated by the dashed curve, and hits in the tracker are the crosses.

Therefore, it is increasingly challenging to measure $p_\perp$ in the tracking system as $p_\perp$ increases.

By hitting elements of the tracking system, charged particles lose some energy due to ionization. When a charged particle, such as a muon, passes close to the electron cloud of an atom in material, like

the charged particle's electric field disrupts the electron cloud and can pull electrons out, ionizing the atom. By ionizing the atom, the muon loses some energy. The mean rate of energy loss per unit length due to ionization was first calculated by Hans Bethe in the 1930s.[3] For a relativistic particle with mass M, energy E, and electric charge Qe, he found the mean energy loss to be

$$\left\langle \frac{dE}{dx} \right\rangle \simeq -16Q^2 \left(\frac{Z}{14} \right) \left(\frac{n}{5 \times 10^{28} \text{ m}^{-3}} \right) \left(\log \frac{4m_e^2 E^2}{E_p^2 (M^2 + 2m_e E)} - 1 \right) \text{ MeV} \cdot \text{m}^{-1}.$$

(5.20)

Z is the atomic number of the ionizing material and n is the number density of atoms per cubic meter. Fiducial values for silicon have been inserted. m_e is the mass of the electron. E_p is the plasma energy of the ionizing material, and $E_p \simeq 30$ eV for most materials.

One potential concern regarding charged particles passing through the tracking system is that they would lose a significant amount of energy to ionization and so their momentum would not be accurately measured. However, with Bethe's formula we can demonstrate

---

[3] H. Bethe, "Theory of the passage of fast corpuscular rays through matter," Annalen Phys. **397**, 325 (1930).

that this isn't much of a concern. For a muon passing through silicon with energy $E = 1$ GeV and mass $M \simeq 106$ MeV, the mean energy loss per meter is

$$\left\langle \frac{dE}{dx} \right\rangle_{E\,=\,1\,\text{GeV}} \simeq -3.9 \text{ MeV} \cdot \text{cm}^{-1} \,. \tag{5.21}$$

In the CMS silicon tracking system, each layer of the tracker is about 500 μm in thickness, and a muon might hit 30 or so such layers. The total thickness of silicon that a such muon would travel through is about 15 cm. The mean energy loss due to ionization in the silicon tracker of a 1 GeV muon would then be approximately

$$\langle E \rangle_{E\,=\,1\,\text{GeV}} \simeq -60 \text{ MeV} \,. \tag{5.22}$$

This is only about 6% of the muon's energy. If the muon had an energy of 100 GeV, the fractional energy loss would be even smaller. A 100 GeV muon would only lose about

$$\langle E \rangle_{E\,=\,100\,\text{GeV}} \simeq -75 \text{ MeV} \tag{5.23}$$

to ionization by passing through the tracking system. This is much less than 1% of the energy of a 100 GeV muon.

5.4.2 Calorimetry

The next layer of the particle detector onion is the **calorimetry**. The electromagnetic and hadronic calorimeters have the same basic function, but are designed to be sensitive to different types of particles. As calorimeters, these parts of the detector measure energy, in much the same way that the calorie content of food is determined. By burning a sandwich and using the expended heat to warm up water, we can determine the calorie, or energy, content of the sandwich by the temperature increase of the water. Similarly, the calorimeters are designed to stop particles and have them explode all of their energy into individual cells of the calorimeters. These cells are referred to as **towers** and are finely segmented in ϕ and η. By the amount that a tower "warms up" or the degree of its response to a particle that hits it, the energy of that particle can be determined. Because of the fine angular segmentation, the direction of the particle can be well measured. Therefore, assuming that the particle that hit the calorimeter tower is massless, its complete four-vector is known.

As their names suggest, the electromagnetic and hadronic calorimeters are sensitive to electromagnetic and hadronic radiation, respectively. In particular, the electromagnetic calorimeter stops electrons and photons, low-mass particles that interact via electromagnetism. The most important way that the electromagnetic calorimeter stops high-energy electrons and photons is through **bremsstrahlung** ("braking radiation" in German).

Bremsstrahlung is the process by which an electron emits a photon which decreases its energy. In a similar way, a photon can split into an electron–positron pair, and each resulting particle has less energy than the initial photon. We can express these processes with the Feynman diagrams

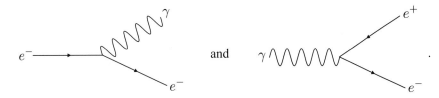

While the rate of bremsstrahlung can be calculated by considering the collinear emission of a photon off of an electron, we will just quote the result here. The rate of energy loss per unit length via bremsstrahlung can be expressed as

$$\frac{dE}{dx} \simeq -\frac{E}{X_0},$$ (5.24)

where X_0 is called the radiation length. The radiation length X_0 for an electron to lose energy is a function of the material through which the electron is moving, just like for ionization radiation. A radiation length X_0 is the distance over which the electron loses a fraction e^{-1} ($e = 2.71828\ldots$) of its energy. Typical radiation lengths are the order of centimeters, and your calorimeter should be many radiation lengths thick to capture all of the energy of the electron.

At ATLAS, for example, the electromagnetic calorimeter consists of lead plates immersed in liquid argon. The liquid argon ionizes, and the energies of low-energy particles can be measured efficiently. The lead plates have strong stopping power for more energetic particles; lead has one of the shortest radiation lengths, of about 5 mm. Together, they form the electromagnetic calorimeter and stop nearly all electrons and photons created in collision.

The hadronic calorimeter acts in much the same way as the electromagnetic calorimeter; however, it must stop particles with a much higher mass than electrons. Hadrons, like pions, interact most strongly with atomic nuclei, and not atomic electrons. As such, their interactions are more complicated to understand, but the same basic principles are at work. Hadrons pass through a material and lose energy by inelastic collisions with atomic nuclei. The rate of collisions, just as with bremsstrahlung, can be characterized by a nuclear interaction length, λ_I. Over one λ_I, a high-energy hadron loses a fraction e^{-1} of its energy to interactions with nuclei. Iron is one of the materials with the smallest λ_I; in ATLAS, for example, the hadronic calorimeter consists mostly of iron. The nuclear interaction length of iron is about 16 cm and the hadronic calorimeter is about $7\lambda_I$ in depth. This captures almost all of the energy of hadrons produced in proton collisions. At CMS, to ensure that the hadronic calorimeter has very low radioactivity, over a million World War II brass shell casings from the Russian army were used to make part of it.

5.4.3 Muon System

Muons are like electrons but 200 times more massive, and so lose much less energy in ionization and bremsstrahlung. The classical intuition for this property is that muons are accelerated less than electrons for the same force. In classical mechanics, the energy loss per unit length is just the force that the material exerts on the particle:

$$|\vec{F}| = -\frac{dE}{dx}.$$

(5.25)

To measure a muon's energy, we first give up hope that we can stop it in a calorimeter.

At both ATLAS and CMS, outside of the hadronic calorimeter there is a **muon detection system**. At ATLAS, it consists of detectors for tracking in a high-tesla, toroidal magnetic field (and is where ATLAS gets its name). CMS, by contrast, uses a high-tesla solenoidal magnet to bend muons (and is where CMS gets its name). The "compact" of "compact muon solenoid" refers to the fact that a CMS detector is about 7 meters smaller in diameter than in ATLAS; see Fig. 5.2. Nevertheless, because of the compactness, CMS is extremely dense, which ensures that the calorimetry can stop particles and the muon system significantly bends the trajectory of muons. Though much smaller than ATLAS, the weight of CMS is more than twice that of ATLAS!

Focusing on the ATLAS muon system, the field lines of the toroidal magnet form concentric circles about the proton beams. The toroidal magnetic field is perpendicular to the solenoidal magnetic field in the tracking system. Therefore, the trajectory of a muon is deflected in two perpendicular planes from the tracking system to the muon system. This enables a direct measurement all three momentum components of the muon. The trajectory of a muon traveling through the detector would look schematically like:

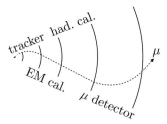

In this figure, the muon travels straight through the calorimeters where there is zero magnetic field, as at the ATLAS experiment, while CMS has a magnetic field throughout the detector. Because the calorimetry at ATLAS and CMS is so good at stopping high-energy hadrons, essentially every charged particle that makes it to the muon system is considered to be a muon.

5.4.4 Unobservable Neutrinos

The tracker, calorimetry, and muon system detects and measures the momentum of almost all detector-stable particles. A **detector-stable** particle is a particle which does not decay before it deposits its energy in the detector. There are very few detector-stable particles: electrons, photons, muons, and a handful of hadrons. These are all well measured by the detector, but there is one more class of particle that is detector-stable but cannot be measured: neutrinos. Like photons, neutrinos are electrically neutral and have very small mass. Unfortunately, neutrinos interact incredibly weakly with matter, and so the vast majority of the time pass right through all detector components without so much as a "hello." Indeed, in collider physics, we assume that neutrinos do not interact with the detector. We will study the interactions of neutrinos and justify their very weak interactions in Chapter 10.

However, this lack of direct measurement does not mean that neutrinos are invisible or have no experimental consequences. We can see the effects and existence of neutrinos indirectly. The total momentum transverse to the proton beam is zero because the initial colliding protons have momentum only along the $\hat{z}$-axis. If you measure a non-zero net $p_\perp$ in the detector from all of the observed particles, then a neutrino (or a few of them) must carry away the unmeasured transverse momentum that is responsible for conserving momentum. This missing momentum is often called **missing transverse energy** and denoted by $\not{E}_T$, or called simply MET (read: "met"). While not necessarily always equivalent, the words "neutrino" and "MET" are often used interchangeably in collider physics.

Now that we've discussed all components of the detectors, let's look at a couple of example event displays and attempt to understand the physics in the proton collision responsible for the collection of particles observed.

Example 5.1 Figure 5.6 shows an event display from the ATLAS experiment recorded in 2015. What happened in this collision of protons?

Solution

The inset panel of Fig. 5.6 in the upper-left corner is a head-on display with the proton beam in the center, and the tracking, calorimetry, and muon system at sequentially larger radii. This figure shows that there were many observed tracks and large energy deposits in both the electromagnetic and hadronic calorimeters. This suggests that this collision event produced numerous hadrons. Note also that the energy deposits in the calorimetry are relatively narrow (small angular spread) and back-to-back in the plane transverse to the beam. This suggests that this event corresponds to the production of two jets, which are collimated, high-energy streams of particles. We'll discuss jets and their physics in Chapter 9.

The other two panels of Fig. 5.6 show different displays of this event. In the upper-right corner, the figure is focused on the tracking system of the detector, and shows the individual **hits** that combine to form tracks. Note that although the tracking system does not extend all the way to the interaction point, the identified tracks can be extended and all overlap, indicating the location of the point where two protons collided. The panel at the bottom of the figure shows the event with the proton beam in the plane of the page. The charged tracks are visible in the center of this panel, and the high-energy deposits in the calorimetry are visible as well.

Example 5.2 Now, let's look at an event display from the CMS experiment, Fig. 5.7, recorded in 2012. What happened in this collision of protons?

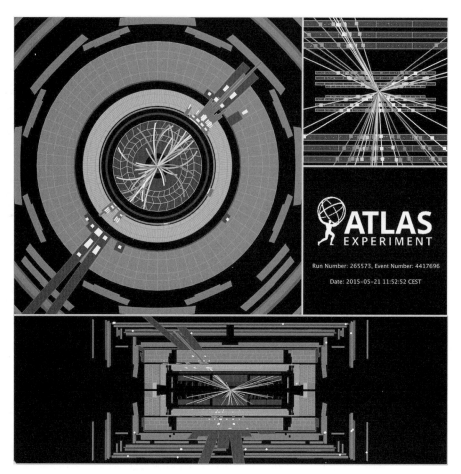

Fig. 5.6 Event display from the ATLAS experiment from May 21, 2015. Credit: ATLAS Experiment © 2015 CERN.

Solution

The CMS detector in this figure is artistically rendered as the light cylinder, just to guide the eye. This event display shows the identified tracks as the curves at the center of the figure, electromagnetic calorimetry deposits as the bars extending to the edge of the figure, and hadronic calorimetry deposits as the thick bars at the end of the cylinder. The most prominent features of this event display are two high-energy deposits in the electromagnetic calorimetry. Note that these high-energy deposits do not have corresponding tracks leading to them; this is distinct from the previous event from ATLAS. Thus, whatever produced these deposits must be neutral, so there are no tracks, but must also interact electromagnetically. The only particle that satisfies these criteria is the photon. Thus, this event corresponds to the production of two high-energy photons from proton collision. In fact, this event is evidence for the existence of the Higgs boson, which was initially produced in the proton collisions and then subsequently decayed to photons. We'll discuss the Higgs boson, its properties, and its discovery in Chapter 13.

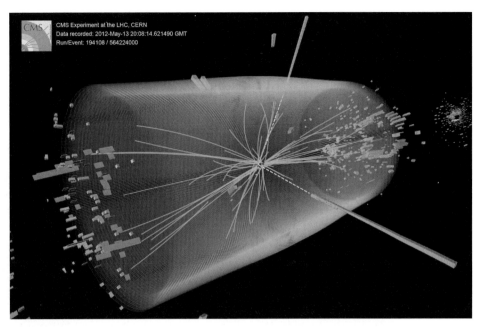

Fig. 5.7 Event display from the CMS experiment from May 13, 2012. Credit: CERN © 2012 CERN, for the benefit of the CMS Collaboration.

Box 5.1   **Historical Profile: Fabiola Gianotti**

Fabiola Gianotti is an experimental particle physicist who earned her Ph.D. from the University of Milan in 1989. Most of her career has been spent on experiments at CERN, including the UA2 experiment, which was on the Super Proton Synchrotron, and the ALEPH experiment, one of the experiments on the Large Electron–Positron Collider. Most recently, Gianotti was a member of the ATLAS experiment, becoming its spokesperson from 2009 to 2013, and presented the ATLAS results of the Higgs boson discovery on July 4, 2012. On January 1, 2016, Gianotti became the first female Director-General of CERN, a position she holds until 2020. She has been recognized with numerous awards, including membership of the Accademia Nazionale dei Lincei ("Academy of the Lynx-Eyed"), a 400-year old Italian science academy of which Galileo Galilei was an early member. Gianotti is also a classically trained pianist.

5.5 After Collision: Triggering and Data Acquisition

After collision, our detector has measured all the particles produced and we then need to store this event for later analysis. We have to do this extremely fast: the LHC collides 40 million proton bunches per second! The amount of information contained in a complete event at the LHC is of the order of a megabyte, so if every collision were recorded, the LHC would produce 40 terabytes of data per second. This is an enormous amount of data, and there's currently no way to record that much data in a time significantly less than a second so that there isn't a backlog of events to record. So, the experiments at the LHC simply cannot store all proton collision events.

Why, then, is there such a high rate for proton collisions at the LHC? As discussed earlier, the vast majority of the volume of a proton bunch is empty space. Even though each bunch contains 100 billion protons, it is extremely unlikely that any protons collide at all. Even if they do collide, most of the time the collision is uninteresting: two protons might just bounce off of one another, without exploding apart into many final-state particles. We therefore need to collide proton bunches at an extremely high rate so that we can hope to observe any interesting proton collisions at all. However, we don't want to record absolutely every proton collision event. ATLAS and CMS employ **triggers**, which are requirements on observed particles that define what it means to be an "interesting event." An example of such a trigger might be the observation of a muon with a transverse momentum $p_\perp > 1$ GeV. If an event contains a muon that passes this requirement, then the event is recorded for later analysis.

The trigger systems at ATLAS and CMS actually involve multiple layers that perform their tests on different time scales, from simple questions about energy deposits in the calorimetry to detailed questions about invariant masses of collections of multiple particles observed in the detector. The total number of events that pass all triggers and are actually recorded and stored for later analysis is only about 100 to 1000 events per second. These hundreds of events per second are sent from the detector to be stored on magnetic tape. For the 39,999,000 events or so that do not pass the triggers each second, it's as if they never happened at all.

Once the events are recorded to tape, they are available to members of the experimental collaborations for analysis. These data are then used to search for new physical phenomena, such as a new particle, that were created in the collision of the protons. Claiming discovery of a new phenomenon has an extremely high bar in particle physics and requires a detailed understanding and interpretation of experimental uncertainties. How is this done?

5.6 Statistical Analyses

The outcome of every physics experiment is a number or collection of numbers that are distributed according to a probability distribution. This is perhaps most apparent in quantum mechanics, where the wavefunction represents the probability amplitude for a particle to be located at a particular position, for example. However, this is true for any experimental result because we can only ever do a finite number of measurements and any measurement apparatus is imperfect. We can never measure energies, masses, or charges with perfect, infinite precision; there is always some uncertainty on the outcome values of an experiment. These can depend on numerous factors and essentially represent our ignorance as to a perfect understanding of the experimental response. A simple example of such a measurement uncertainty is in the measurement of a distance with a tape measure.

On a tape measure, there are only so many ticks, and those ticks have a finite width, so we can only use it to measure to, say, an eighth or sixteenth of an inch (in imperial units). Additionally, it may be that the tape measure is actually slightly off in what it claims. If an inch on the tape is really 1.01 inches, then this could lead to a significant measurement error if you're looking at something that is large enough.

Such uncertainties that correspond to measurement error are called **systematic uncertainties**. In particle physics, determining systematic uncertainties is extremely challenging and requires extensive understanding of the response of the detector to many different particles with many different energies. We've already seen an example of a systematic uncertainty in the discussion of the tracking system. If a charged particle has high enough transverse momentum, then it becomes increasingly challenging to measure its curvature in the tracker, and therefore to determine its transverse momentum. At some high enough transverse momentum, you can no longer rely on the tracking system to accurately measure the curvature. While systematic uncertainties are extremely important to model correctly, we will instead focus on another source of uncertainty which we can understand and model mathematically.

5.6.1 Statistical Uncertainties

This is a figure that represents a fair die; a six-sided cube with each side individually numbered with pips:

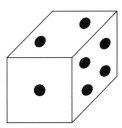

As it is a fair die, we expect that there is probability of 1/6 for it to land with any side up after a roll. That is, there is 1/6 probability that 1 will be up, 1/6 probability that 2 will be up, etc. Consider rolling the die 10 times and tracking the sides that land up. Out of 10 rolls, how many do you expect to be 1? Naïvely, because 1 has 1/6 probability in any roll, we would expect 1 to be up in 10/6 of the rolls. However, this is of course impossible because the number of rolls must be an integer. So our expectation makes only so much sense. In the limit that the number of rolls we make goes to infinity, then we expect that indeed 1/6 of the rolls produce a 1, 1/6 produce a 2, etc., but in any finite number of rolls that isn't necessarily true.

A better question to ask is the following. Let's roll the die 10 times again. Given the outcome of those rolls, how surprised should we be that they were the result of rolling a fair die? That is, with the assumption that the die is fair, are the rolls likely or unlikely? For example, if you rolled 10 4s in a row, you might be very surprised and question whether the die was fair at all. However, if you rolled a healthy mixture of 1, 2, 3, 4, 5, and 6, you wouldn't likely question the fairness of the die. As we can only ever do a finite number of rolls (or a finite number of experimental measurements), it is this question that is fundamental. We need a way to determine how likely an outcome is, given a null hypothesis. In this example, our null hypothesis is that the die is fair, and so we roll it numerous times and calculate the probability that a fair die would produce that outcome. If the probability is relatively large, then we gain confidence that the die is indeed fair. However, if the probability is extremely small, then we gain confidence that the die is not fair; we don't necessarily know how it is weighted, just that it is more consistent with being not fair. Therefore, we need to quantify deviations from expectation in our experiment.

This type of uncertainty, due to the finite number of measurements, is called a **statistical uncertainty**, and represents the fact that any finite number of measurements are expected to deviate from the true result by a well-defined amount. This is true even with perfect experimental resolution, but the effect of statistical uncertainties decreases as the number of events that contribute to the measurement increases. To understand statistical uncertainties, let's assume that we are measuring some continuous quantity that can be represented by x, which is a random variable distributed according to a probability distribution $p(x)$. Probability distributions are normalized:

$$\int_{-\infty}^{\infty} dx \, p(x) = 1 \,, \tag{5.26}$$

where the integral extends over the entire domain of x. Probabilities are also always positive, and so the integral over any subdomain $x \in [a, b]$ is non-negative and at most 1:

$$0 \le \int_{a}^{b} dx \, p(x) \le 1 \,. \tag{5.27}$$

As a concrete example, we might consider measuring the invariant mass of electron–positron pairs m_{ll} in the final state of proton collisions. As a mass, $m_{ll} \ge 0$ and so any value that we measure must respect this. As we collect events, our goal is to determine the probability distribution from which the masses m_{ll} are drawn. We can then compare this to predictions and draw conclusions about the underlying physics. The first step, though, is to approximate the probability distribution, $p(m_{ll})$. As the mass can be any continuous number and we can only perform a finite number of measurements, we need to make a discrete approximation of the continuous probability distribution. That is, we construct a **histogram**, which consists of **bins**, subdomains of the full distribution into which we put the number of events observed that lie in that subdomain. This set-up is illustrated in Fig. 5.8.

As we collect more and more measurements, we fill these bins appropriately. To create a histogram that integrates to 1, just like the fundamental probability distribution, we add an amount

$$\frac{1}{N_{\text{ev}} \Delta x} \tag{5.28}$$

to a bin for each event in it. Here, N_{ev} is the total number of events in our measurement sample and Δx is the width of the bin under question. Note that this does correspond to integrating to 1:

$$1 = \sum_{i \in \text{bins}} \frac{N_i}{N_{\text{ev}} \Delta x} \Delta x \,. \tag{5.29}$$

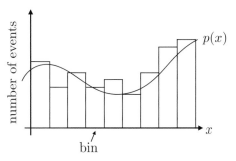

Fig. 5.8 Illustration of a histogram of some measured quantity x. The subdomains of the histogram are bins whose height is proportional to the number of events with a value of x that lies in that particular subdomain. The continuous probability distribution $p(x)$ from which the events are drawn is also illustrated.

N_i is the number of events in bin i, and the sum of all events in all bins is of course N_{ev}. That is, our approximation for the probability distribution that x lies in the region of the bin about the point x_i is

$$p(x_i) \approx \frac{N_i}{N_{ev}\Delta x}.$$ (5.30)

This can be made more precise in the limit that the number of events $N_{ev} \to \infty$. In this limit, and with the bin width $\Delta x \to 0$, the law of large numbers states that

$$\lim_{N_{ev}\to\infty,\Delta x\to 0} \frac{N_i}{N_{ev}\Delta x} = p(x_i).$$ (5.31)

This is a nice result if we can actually perform an infinite number of measurements. What happens in the physically possible case, when N_{ev} is finite? How good an approximation is the histogram to the true probability distribution?

5.6.2 Derivation of Poisson Distribution

It turns out that this question has a well-defined answer and is relatively easy to derive. To do so, we will make a reasonable assumption that is typically true, or at least assumed true, in particle physics analyses. We assume that each event that contributes to a histogram is independent; that is, the measured value from one event has no bearing on the measured value from any other event. The consequence of this is that each event randomly samples the probability distribution $p(x)$ to produce a value x. Let's call P the probability to have a value within dx of x:

$$P = p(x)\,dx.$$ (5.32)

Then, the probability that a measured value is not within dx of x is of course $1 - P$. Out of a total of N_{ev} events, the probability that there are k events within dx of x is therefore

$$p_k = P^k(1 - P)^{N_{ev}-k}\binom{N_{ev}}{k}.$$ (5.33)

The factor on the right,

$$\binom{N_{\text{ev}}}{k} = \frac{N_{\text{ev}}!}{k!\,(N_{\text{ev}} - k)!}\,, \tag{5.34}$$

is called a **binomial number** and is read as "N_{ev} choose k" and is the number of ways to pick k events out of a set of N_{ev} total events. That is, to determine the probability that k events landed near x, we need to choose k of them out of N_{ev}, and multiply by the probabilities that k are near x and $N_{\text{ev}} - k$ are not.

The probability distribution represented by Eq. 5.33 is called the **binomial distribution**, for the following reason. The binomial formula represents the expansion of a binomial expression:

$$(x + y)^N = \sum_{i=0}^{N} x^i y^{N-i} \binom{N}{i}. \tag{5.35}$$

Note that each term of the expansion has the form of Eq. 5.33. Indeed, this proves that the binomial distribution is normalized:

$$\sum_{k=0}^{N_{\text{ev}}} p_k = \sum_{k=0}^{N_{\text{ev}}} P^k (1 - P)^{N_{\text{ev}} - k} \binom{N_{\text{ev}}}{k} = (P + (1 - P))^{N_{\text{ev}}} = 1\,. \tag{5.36}$$

This is therefore what we set out to identify. The binomial distribution tells us the probability distribution of the number of events in a given bin, according to some probability distribution $p(x)$. We could stop here and perform all the analyses that we need, but it turns out that a few more assumptions are useful and make the analysis a bit simpler.

We will assume that $N_{\text{ev}} \to \infty$ with k finite, as well as assuming that $P \ll 1$, which is the limit in which the bins are becoming narrow. In this limit, note that

$$\lim_{N_{\text{ev}} \to \infty} \binom{N_{\text{ev}}}{k} = \lim_{N_{\text{ev}} \to \infty} \frac{N_{\text{ev}}!}{k!\,(N_{\text{ev}} - k)!} = \frac{N_{\text{ev}}^k}{k!}\,, \tag{5.37}$$

and $(1 - P)^k \simeq 1$, for finite k. With these simplifications, the probability for k events in a bin becomes

$$\lim_{N_{\text{ev}} \to \infty} p_k = \lim_{N_{\text{ev}} \to \infty} \frac{(N_{\text{ev}} P)^k}{k!} \left(1 - \frac{N_{\text{ev}} P}{N_{\text{ev}}}\right)^{N_{\text{ev}}} = \frac{(N_{\text{ev}} P)^k}{k!} e^{-N_{\text{ev}} P}\,. \tag{5.38}$$

This probability distribution is called the **Poisson distribution**. Note that it is normalized:

$$1 = \sum_{k=0}^{\infty} \frac{(N_{\text{ev}} P)^k}{k!} e^{-N_{\text{ev}} P}\,. \tag{5.39}$$

The mean of this distribution is

$$\langle k \rangle = \sum_{k=0}^{\infty} k \frac{(N_{\text{ev}} P)^k}{k!} e^{-N_{\text{ev}} P} = N_{\text{ev}} P\,, \tag{5.40}$$

which means that, on average, the number of events k in the bin will be exactly $N_{\text{ev}} P$. The **variance** of this distribution σ^2 is also $N_{\text{ev}} P$:

$$\sigma^2 = \langle k^2 \rangle - \langle k \rangle^2 = N_{\text{ev}} P\,. \tag{5.41}$$

You will prove these statements in Exercise 5.5.

The variance of any probability distribution is a measure of the average squared deviation from the mean. A measure of the average deviation from the mean is called the **standard deviation**, which is the square-root of the variance. In this case of the Poisson distribution, the standard deviation is thus

$$\sigma = \sqrt{N_{\mathrm{ev}}P}\,. \tag{5.42}$$

A colloquial interpretation of the standard deviation is that you "expect" the result of any experiment to deviate by about 1 standard deviation from the true result. More precisely, the standard deviation is a measure of how surprised we should be with the outcome of an experiment. If the outcome is within 1 standard deviation of expectation, then there's no surprise. However, if the outcome is many standard deviations from expectation, then you gain confidence that expectation is not truth. Also note that as more data are collected, the relative size of the standard deviation to the mean decreases:

$$\frac{\sigma}{\langle k \rangle} = \frac{1}{\sqrt{\langle k \rangle}} = \frac{1}{\sqrt{N_{\mathrm{ev}}P}}\,. \tag{5.43}$$

That is, it is more likely that there are large deviations will a small dataset than a large dataset. Deviations from expectation that diminish with more data were likely not interesting.

5.6.3 Significance and Discovery

The Poisson distribution and its standard deviation provide a metric for determining our confidence in the null hypothesis. We typically say that a particular measurement was "within 1 sigma," or a "3-sigma excess," for example, which quantifies the importance of the deviation from expectation. However, in physics in general and particle physics in particular, there are a lot of measurements done, and a potential for significant deviations, even just assuming the null hypothesis. How likely, or with what probability, is there an excess in the data that is at least 3 standard deviations from the mean? One can directly calculate this in principle using the binomial or Poisson distribution, but the standard way to express it is by using a further statistical simplification.

While we won't prove it here, in the limit that the number of events $N_{\mathrm{ev}} \to \infty$ and k is taken to be a continuous random variable, the Poisson distribution simplifies to the **Gaussian distribution**:

$$p_k = \frac{(N_{\mathrm{ev}}P)^k}{k!}e^{-N_{\mathrm{ev}}P} \xrightarrow{N_{\mathrm{ev}} \to \infty} p(x) = \frac{1}{\sqrt{2\pi N_{\mathrm{ev}}P}}\exp\left[-\frac{(x-N_{\mathrm{ev}}P)^2}{2(N_{\mathrm{ev}}P)}\right]. \tag{5.44}$$

This result is known as the **central limit theorem**. This Gaussian distribution is normalized:

$$1 = \int_{-\infty}^{\infty} dx\, \frac{1}{\sqrt{2\pi N_{\mathrm{ev}}P}}\exp\left[-\frac{(x-N_{\mathrm{ev}}P)^2}{2(N_{\mathrm{ev}}P)}\right]. \tag{5.45}$$

For compactness in what follows, we will just denote the expected number of events in a bin as $N \equiv N_{\mathrm{ev}}P$. The Gaussian distribution still has a mean of N events and standard deviation of $\sqrt{N}$, but this form enables us to calculate the probability of deviation from the mean in a universal manner.

Using the Gaussian distribution, let's calculate the probability that there was a deviation that was at least $X\sigma$ above the mean. This is called the **p-value** of the deviation and requires integrating from $N + X\sigma = N + X\sqrt{N}$ to ∞:

$$p_X = \int_{N+X\sqrt{N}}^{\infty} dx \, \frac{1}{\sqrt{2\pi N}} \exp\left[-\frac{(x-N)^2}{2N}\right] \tag{5.46}$$
$$= \int_X^{\infty} dx \, \frac{1}{\sqrt{2\pi}} \exp\left[-\frac{x^2}{2}\right] .$$

In the second line, we have changed variables so that the Gaussian distribution over which we are integrating has a mean of 0 and a standard deviation of 1. Then, this integral can be evaluated in terms of the error function erf:

$$p_X = \frac{1}{2} - \frac{\mathrm{erf}\left(\frac{X}{\sqrt{2}}\right)}{2} . \tag{5.47}$$

As a sense of these values, the probability of a deviation at least at large as 1 sigma is $p_1 \simeq 0.16$, at least 3 sigma is $p_3 \simeq 0.0013$, and at least 5 sigma is $p_5 \simeq 2.9 \times 10^{-7}$. There is an unofficial, yet ubiquitous, standard in particle physics that a claim of discovery can only be made if the deviation from the null hypothesis is at least **5 sigma**. This is the most rigorous such standard in all of science.

Example 5.3 Figure 5.9 is a plot from the CMS experiment of the number of events versus the measured invariant mass of pairs of photons, $m_{\gamma\gamma}$. This plot was used as evidence for the discovery of the Higgs boson, which has a mass of about 125 GeV. What is approximate the statistical significance of this excess?

Solution

Focus on the inset plot in the upper right corner. This plot shows the number of observed events (the dots) and the number of expected events from the null hypothesis of no Higgs boson (the smooth curve). In the three bins near 125 GeV, there are approximately 3400 events expected from the null hypothesis. The standard deviation, assuming Poisson statistics, is therefore $\sqrt{3400} \approx 58$. The number of observed events in those same bins is about 3600, nearly 4 standard deviations away from the null hypothesis! With more detailed analysis and statistical methods, the CMS experiment produced the main plot, which weights the observed events in a particular way to isolate the excess further.

Exercises

5.1   *Synchrotron Losses.* Synchrotron radiation can be a significant source of power loss in a circular collider. The amount of synchrotron radiation depends sensitively

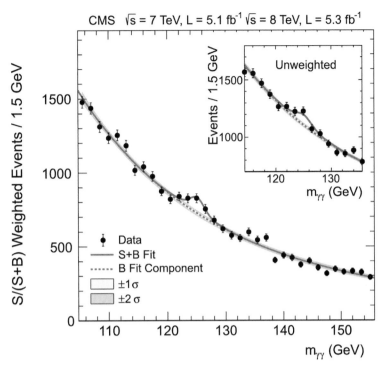

CMS $\sqrt{s}$ = 7 TeV, L = 5.1 fb^{-1} $\sqrt{s}$ = 8 TeV, L = 5.3 fb^{-1}

Fig. 5.9   Invariant mass distribution of photon pairs $m_{\gamma\gamma}$ collected in proton collisions at the CMS experiment up through the end of June, 2012. The bump near a mass of 125 GeV is now known to be the Higgs boson. Reprinted from Phys. Lett. B **716**, S. Chatrchyan *et al.* [CMS Collaboration], "Observation of a new boson at a mass of 125 GeV with the CMS experiment at the LHC," 30 (2012), with permission from Elsevier.

on accelerator parameters, so different experiments have to deal with synchrotron radiation in different ways. In this exercise, we will see how important synchrotron radiation can be.

(a) In the LHC ring at peak performance, there are 2808 bunches of 1.15×10^{11} protons with each proton having an energy of 6.5 TeV. How much total power is emitted in synchrotron radiation when the LHC is running? Express the answer in GeV $\cdot$ s^{-1} and watts (J $\cdot$ s^{-1}). How does this compare to a microwave oven, which typically uses 1000 W?

(b) The Large Electron–Positron Collider (LEP) collided electrons and positrons at a maximum center-of-mass energy of 206 GeV. LEP occupied the same tunnel where the current LHC ring is situated. How much power was emitted in synchrotron radiation from an electron at LEP?

(c) One reason why the LHC can collide protons at much higher energies than LEP collided electrons and positrons is that the synchrotron radiation from protons is much less than for electrons. For the same power in synchrotron radiation, how much larger could the proton energy at the LHC be compared to electrons at LEP?

5.2 *Limits of the Tracking System.* The tracking system at ATLAS is a cylinder with an outer radius of 1.1 meters which contains a 2 T solenoidal magnetic field, i.e., the magnetic field lines are parallel to the axis of the cylinder. Immediately outside of the tracking system is a region of zero magnetic field, where the electromagnetic calorimeter is located.

(a) The energy of charged particles that do not leave the tracking system is poorly measured because those particles do not reach the calorimetry. Estimate the minimum $p_\perp$ in GeV for an electron to reach the calorimetry.

(b) The semiconductor tracker is a subsystem of the tracker that is also a cylinder, but with an outer radius of 0.5 meters. It consists of layers of silicon that can measure particle positions accurate to 17 micrometers. Extremely high-energy charged particles do not bend enough in the magnetic field to have their charge accurately measured. Estimate the maximum $p_\perp$ in GeV of an electron whose path the semiconductor tracker can determine was bent in the magnetic field.

(c) For a high-$p_\perp$ track that just bends in this magnetic field, what is the uncertainty in the measurement of the $p_\perp$? Estimate this using the resolution of the silicon tracking system.

5.3 *Reconstructing Muons.* Because muons are not stopped by the calorimetry, it is harder to determine their momentum four-vector. At ATLAS, for example, the distinct magnetic fields of the tracking system and the muon system come to the rescue. The magnetic field of the tracking system is solenoidal, with a strength of B_{sol} tesla. The magnetic field of the muon system is toroidal, i.e., field lines form concentric circles about the beam, with a strength of B_{tor} tesla. If a muon is observed to have a radius of curvature of R_{sol} in the tracking system and radius of curvature of R_{tor} in the muon system, determine the magnitude of the transverse momentum $p_\perp$ and $\hat{z}$-component p_z of the muon. Express your answer in terms of the elementary charge, the strength of the magnetic fields, and the curvature radii.

Hint: What components of momentum are perpendicular to the magnetic field in the tracking system? What about in the muon system?

5.4 *Data Quantity from the LHC.* Data management at the LHC is a serious issue. Even though only about 100 proton collision events per second are actually recorded to tape, over the time that the experiments run, this adds up to an enormous amount of data. Estimate, in petabytes (10^{15} bytes), the amount of data stored on magnetic tape from the ATLAS and CMS experiments over a year. The experiments are only collecting data for a total of a couple of months per year.

5.5 *Properties of Poisson Statistics.* In this chapter, we derived the Poisson distribution as the distribution of the number of events in a bin about the mean. The probability to observe n events in a Poisson distribution with a mean of λ is

$$p_n = \frac{\lambda^n e^{-\lambda}}{n!}, \tag{5.48}$$

where n is a non-negative integer.

(a) Calculate the mean $\langle n \rangle$ of this distribution and show that it indeed is λ.

(b) Calculate the variance $\sigma^2 = \langle n^2 \rangle - \langle n \rangle^2$ of this distribution and show that it is λ.

Hint: Consider the derivatives with respect to λ of

$$1 = \sum_{n=0}^{\infty} \frac{\lambda^n}{n!} e^{-\lambda}. \tag{5.49}$$

5.6 *Look-Elsewhere Effect.* Often in experimental particle physics it is known where to look for a signal. For example, if you want to study the properties of the Z boson, then you can tune the collision energy of your e^+e^- collider to the mass of the Z boson. However, when searching for possible new physics, where the mass scale is unknown, experimentalists look for excesses over many bins. Purely as a result of finite statistics, one will find excesses when enough bins are considered. Thus, the significance of an excess in any one bin is reduced, simply because that excess could have been anywhere. This is called the **look-elsewhere effect**.

(a) Assume you are looking for excesses in a collection of N_{bins} bins of data. Let's also assume that you only think that excesses are interesting if they deviate by more than $X\sigma$ from the expected number of events in a bin. If the probability for any one bin to have an excess of at least $X\sigma$ is p_X, determine the probability that at least one of the N_{bins} bins has such an excess. In this problem, only assume that the different bins are independent; don't assume anything in particular about the probability p_X.

(b) Now, expand your result from part (a) to lowest order in the limit where $p_X \to 0$. How much larger is the probability for at least one bin to have an excess than p_X? Show that the probability p_X^{global} of an excess at least as large as $X\sigma$ anywhere in your data is

$$p_X^{\text{global}} = N_{\text{bins}} p_X. \tag{5.50}$$

(c) Excesses are considered "interesting" or "evidence" if they are at least a **3σ deviation** from the expected value from the null hypothesis. If the local significance (the significance in one bin) of an excess is 3σ, what is the global significance, which includes the look-elsewhere effect? A typical number of bins in an analysis is about 100, which you can assume for this problem. Including the look-elsewhere effect, do you think such an excess is still interesting?

(d) On December 15, 2015, the ATLAS and CMS experiments held a press conference at which they presented results for the measurement of the invariant mass of pairs of photons, $m_{\gamma\gamma}$.[4] Both experiments observed an excess in their data above the null hypothesis around a mass scale of $m_{\gamma\gamma} = 750$ GeV.

---

[4] M. Aaboud *et al.* [ATLAS Collaboration], "Search for resonances in diphoton events at $\sqrt{s} = 13$ TeV with the ATLAS detector," J. High Energy Phys. **1609**, 001 (2016) [arXiv:1606.03833 [hep-ex]]; V. Khachatryan *et al.* [CMS Collaboration], "Search for resonant production of high-mass photon pairs in proton–proton collisions at $\sqrt{s} = 8$ and 13 TeV," Phys. Rev. Lett. **117**, no. 5, 051802 (2016) [arXiv:1606.04093 [hep-ex]].

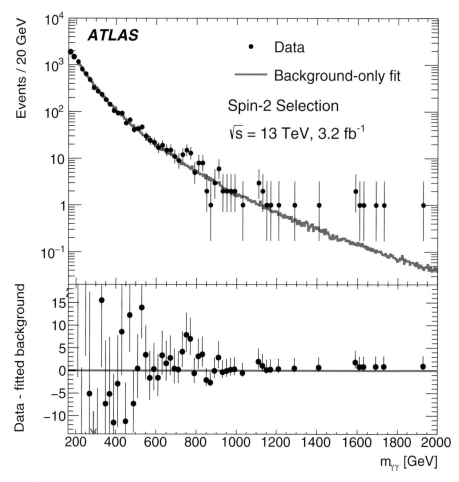

Fig. 5.10 A plot of the invariant mass of pairs of photons $m_{\gamma\gamma}$ in a search for new physics by the ATLAS experiment in 3.2 fb^{-1} of data at 13 TeV proton collision energy. The excess in the three bins around $m_{\gamma\gamma} = 750$ GeV caused a flurry of activity to attempt to describe it. From M. Aaboud *et al.* [ATLAS Collaboration], "Search for resonances in diphoton events at $\sqrt{s}$=13 TeV with the ATLAS detector," J. High Energy Phys. **1609**, 001 (2016), doi:10.1007/JHEP09(2016)001 [arXiv:1606.03833 [hep-ex]].

This excess was later named F (read: "di-gamma"), and inspired a deluge of responses from theorists who wrote more than 500 papers over the next year attempting to explain this excess. A plot of ATLAS's data that demonstrates this excess is presented in Fig. 5.10.

(i) Estimate the local significance σ_{local} of the F excess. To do this, use the three bins that range over 720–780 GeV to calculate the significance and ignore the look-elsewhere effect. Would you be interested in such an excess?

   If you were inspired by the excess in the CMS data, then the excess in ATLAS data would have no look-elsewhere effect because you knew to look around 750 GeV.

(ii) Estimate the global significance σ_{global} of the F excess. Use the same bins as in part (i) and now include the look-elsewhere effect. To include the look-elsewhere effect, you will need to determine how many sets of three neighboring bins there are in these data. You can safely use the approximation you derived in part (b). Would you be interested in such an excess?

   If you were not inspired by the excess in the CMS data, then the excess in ATLAS data would have a look-elsewhere effect because you did not know to look around 750 GeV.

(iii) Assuming that this excess is just a statistical fluctuation of the null hypothesis, how many more events need to be added to these three bins to reduce the local significance of the excess to 1σ?

By summer 2016, the F excess had vanished; apparently it was just a statistical fluctuation.

5.7 *Discovery of the Top Quark.* The top quark was discovered in 1995 at the CDF and DØ experiments on the Tevatron collider located at Fermi National Accelerator Laboratory (Fermilab).[5] Figure 5.11 is a plot presented as evidence for discovery from the CDF (Collider Detector at Fermilab) experiment. Do you believe their claim of discovery?

(a) The null hypothesis predicts a total of $6.9^{+2.5}_{-1.9}$ events to have been measured and to contribute to this plot. How many total events were observed? How significant is this? Assume Poisson statistics and just use 6.9 as the expected number of events.

(b) Assuming the null hypothesis in which there is no top quark, what is the probability for these events to be observed? Is their claim of discovery justified?

5.8 *Missing Energy and Neutrinos.* The most precise method for measuring the mass of the W boson is in its decays to a charged lepton l (an electron or muon) and a neutrino ν:

$$W \to l\nu. \tag{5.51}$$

We'll discuss the properties of this interaction in Chapter 12. Here, we will work to understand this measurement.

As discussed in this chapter, neutrinos are not directly observed in collider physics experiments, so we aren't able to measure the invariant mass of the charged lepton and neutrino. Also, in proton collisions, a W boson can be produced with a range of momenta along the beam axis, called **longitudinal momentum**, in the process

$$pp \to W \to l\nu. \tag{5.52}$$

---

[5] F. Abe *et al.* [CDF Collaboration], "Observation of top quark production in $\bar{p}p$ collisions," Phys. Rev. Lett. **74**, 2626 (1995) [arXiv:hep-ex/9503002]; S. Abachi *et al.* [D0 Collaboration], "Observation of the top quark," Phys. Rev. Lett. **74**, 2632 (1995) [hep-ex/9503003].

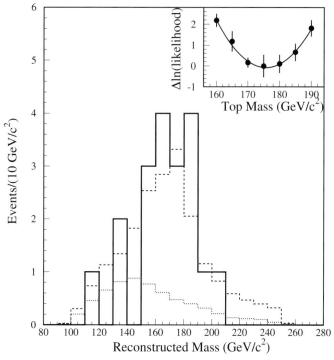

Fig. 5.11   Reconstructed mass distribution from candidate top quark events measured at the CDF experiment. The solid histogram is their recorded data and the dotted histogram is the prediction from the null hypothesis. Reprinted figure with permission from F. Abe *et al.* [CDF Collaboration], Phys. Rev. Lett. **74**, 2626 (1995). Copyright 1995 by the American Physical Society.

Therefore, we are restricted to measuring the transverse momentum of the visible particle, the charged lepton. In the following, assume the W boson is on-shell and the leptons are massless.

(a) In the process $pp \to W \to l\nu$ at the LHC, what is the maximum value of the $p_\perp$ of the charged lepton?

(b) In this process, what is the maximum transverse mass m_T^W of the W boson? This can be defined from the transverse masses of the charged lepton m_T^l and neutrino m_T^ν and their respective transverse momentum vectors $\vec{p}_{\perp l}$ and $\vec{p}_{\perp \nu}$ as

$$m_T^W = \sqrt{2}\sqrt{m_T^l m_T^\nu - \vec{p}_{\perp l} \cdot \vec{p}_{\perp \nu}}. \tag{5.53}$$

(c) Figure 5.12 is a plot from the ATLAS experiment of the transverse momentum distribution of electrons in events with electrons and missing energy. This distribution is used for precision measurement of the W boson mass. From this plot and your result in part (a), can you estimate the W boson mass? Can you identify where on the plot the maximum value of the electron transverse momentum would be? If not, why not?

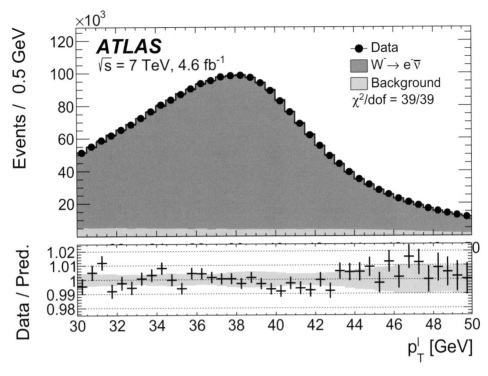

Fig. 5.12 Plot of the transverse momentum of the electron in $pp \rightarrow W^- \rightarrow e^- \bar{\nu}$ events. From M. Aaboud *et al.* [ATLAS Collaboration], "Measurement of the W-boson mass in pp collisions at $\sqrt{s} = 7$ TeV with the ATLAS detector," Eur. Phys. J. C **78**, no. 2, 110 (2018), doi:10.1140/epjc/s10052-017-5475-4 [arXiv:1701.07240 [hep-ex]].

The analysis in which this plot was presented produced one of the most precise measurements of the W boson mass, accurate to a few parts per 10^4.

5.9 *Event Displays.* Figure 5.13 is an event display from the ATLAS experiment. The large image is a head-on view of the ATLAS experiment, with the tracker, calorimeters, and muon system visible. At the center right, underneath the ATLAS logo, is a so-called **Lego plot**. (Can you tell why it is called that?) This "unrolls" the detector, and displays the distribution of $p_\perp$ deposited in the (η, ϕ) plane.

(a) Based on the "hits" in the tracker, EM calorimeter, hadronic calorimeter, and muon system, identify the particle type of the light and dark spikes in the Lego plot. Both light particles are identical as are both dark particles, but light $\neq$ dark.

Hint: The azimuthal angle in the head-on figure is measured with respect to the right, horizontal axis; i.e., $\phi = 0°$ is to the right, $\phi = 90°$ is vertically upward, etc.

(b) From the Lego plot, determine the four-vectors of each of the light and dark particles. A zoomed-in view of the Lego plot is provided in Fig. 5.14. You can safely neglect the particles' masses. With these four vectors, answer the following questions:

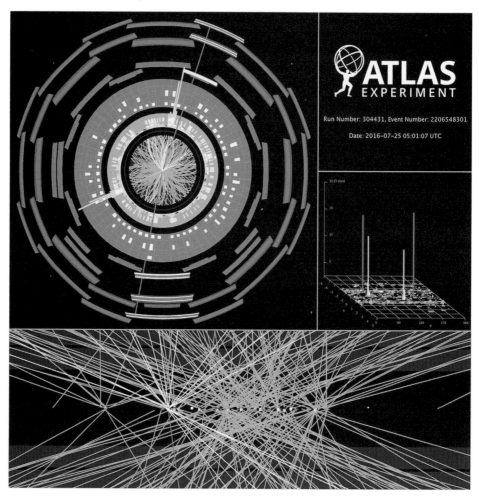

Fig. 5.13 Display of a proton collision event from July 25, 2016 in the ATLAS experiment. Credit: ATLAS Experiment © 2018 CERN.

(i) What is the total transverse momentum vector of the four particles?

(ii) What is the invariant mass of the four light and dark particles together? That is, sum their four-momenta and square it. Which particle of the Standard Model does this mass correspond most closely to?

(iii) Approximately what is the speed of this particle along the proton beam direction (the $\hat{z}$-axis)? To estimate this, approximate the velocity transverse to the beam as 0.

5.10 *Research Problem.* Is there a better way to do particle physics experiments than to collide particles and observe what comes out?

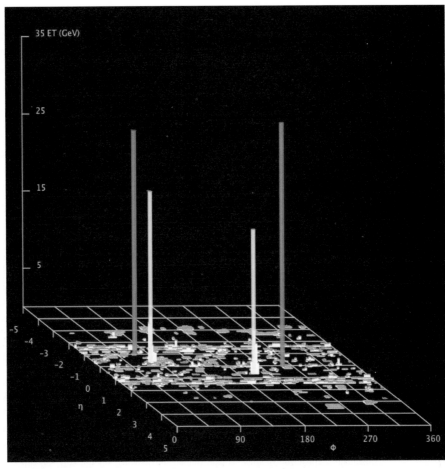

Fig. 5.14 Zoom-in of the Lego plot from the proton collision event from July 25, 2016 in the ATLAS experiment. Credit: ATLAS Experiment © 2016 CERN.

6

Quantum Electrodynamics in e^+e^- Collisions

It's time to put the tools we've acquired over the last several chapters to use. The simplest scattering process that we can consider is the collision of electrons with positrons within the quantum theory of electrodynamics, QED. There are manifold reasons for this simplicity. Electrons and positrons are both fundamental, point particles and so the dynamics of collision is straightforward. Electrons and positrons interact through electromagnetism and so we are able to take a familiar force and leverage it to understand unfamiliar phenomena. Electrons and positrons have relatively small masses compared to the center-of-mass collision energies that we study, so to good approximation we can assume they are massless. This enables a description in terms of helicity states which makes analyses of angular momentum conservation, for example, transparent.

Our first task in this chapter is to finish what we started in Chapter 4 with the calculation of the Feynman diagram for $e^+e^- \to \mu^+\mu^-$ scattering. Through this Feynman diagram and constructing the cross section, we will learn what the cross section tells us about properties, like the intrinsic spin, of the scattering particles. This intuition will be useful when we observe hadron production from electron–positron collisions. To understand, interpret, and predict the process $e^+e^- \to$ hadrons, we will need to develop the notion of an inclusive cross section and construct a mapping to the hadronic final state from the process $e^+e^- \to q\bar{q}$, where q ($\bar{q}$) is a quark (anti-quark). The observation that we can study quarks and their properties in electron–positron collisions will be our entry into studying quantum chromodynamics, or QCD.

6.1 $e^+e^- \to \mu^+\mu^-$

Our first goal is to calculate the matrix element $\mathcal{M}(e^+e^- \to \mu^+\mu^-)$ represented by the Feynman diagram

$$\mathcal{M}(e^+e^- \to \mu^+\mu^-) = \quad \text{[Feynman diagram]} \quad . \tag{6.1}$$

We assume that all external particles are massless, so we are working in the limit in which the center-of-mass collision energy is much larger than the mass of the muon ($m_\mu \simeq 106$ MeV). To complete this calculation, there are several things we need to do:

- determine the massless solutions to the Dirac equation to describe the external electrons and muons
- identify the allowed helicity configurations of the process $e^+e^- \to \mu^+\mu^-$ mediated by electromagnetism
- actually calculate the Feynman diagrams for the helicity configurations that are allowed.

6.1.1 Solutions to the Massless Dirac Equation

Let's begin. The massless Dirac equation is

$$i\gamma \cdot \partial\psi = 0. \tag{6.2}$$

Because the γ matrices are four-dimensional, this one expression represents four first-order differential equations. Therefore, there will be four solutions. To solve it, we will work with γ matrices in the Weyl or chiral representation, where

$$\gamma_\mu = \begin{pmatrix} 0 & \sigma_\mu \\ \bar{\sigma}_\mu & 0 \end{pmatrix}, \tag{6.3}$$

with

$$\sigma_\mu = (\mathbb{I}, \sigma_1, \sigma_2, \sigma_3)_\mu, \qquad\qquad \bar{\sigma}_\mu = (\mathbb{I}, -\sigma_1, -\sigma_2, -\sigma_3)_\mu. \tag{6.4}$$

The σ_i are the Pauli spin matrices,

$$\sigma_1 = \begin{pmatrix} 0 & 1 \\ 1 & 0 \end{pmatrix}, \qquad \sigma_2 = \begin{pmatrix} 0 & -i \\ i & 0 \end{pmatrix}, \qquad \sigma_3 = \begin{pmatrix} 1 & 0 \\ 0 & -1 \end{pmatrix}. \tag{6.5}$$

The power of the Weyl representation is that we can express the four-component spinor ψ in terms of two two-component spinors, ψ_L and ψ_R:

$$\psi = \begin{pmatrix} \psi_L \\ \psi_R \end{pmatrix}. \tag{6.6}$$

Plugging this into the Dirac equation, we find that it separates into two equations:

$$i\sigma \cdot \partial\psi_R = 0, \qquad\qquad i\bar{\sigma} \cdot \partial\psi_L = 0. \tag{6.7}$$

These are called the **Weyl equations**. The solutions to $i\sigma \cdot \partial\psi_R = 0$ are right-handed helicity fermions, while the solutions to $i\bar{\sigma} \cdot \partial\psi_L = 0$ are left-handed helicity fermions. These names will be clear shortly.

Let's solve $i\sigma \cdot \partial\psi_R = 0$ in the standard way. We write

$$\psi_R = u_R e^{-ip \cdot x}, \tag{6.8}$$

for some two-component spinor u_R and four-momentum p. We then have

$$i\sigma \cdot \partial\psi_R = 0 \qquad \to \qquad (\sigma \cdot p)u_R = 0. \tag{6.9}$$

First, for illustration, we consider the particle's three-momentum $\vec{p}$ aligned along the $+\hat{z}$-direction and energy $E > 0$. Then, the dot product of the σ matrix and the momentum is

$$\sigma \cdot p = \mathbb{I}p_0 - \sigma_3 p_z = \begin{pmatrix} E & 0 \\ 0 & E \end{pmatrix} - \begin{pmatrix} p_z & 0 \\ 0 & -p_z \end{pmatrix} = \begin{pmatrix} 0 & 0 \\ 0 & 2E \end{pmatrix}. \tag{6.10}$$

Note that $p_0 = E$ and by the masslessness condition, $E = |\vec{p}| = p_z$. Therefore, the solution to $(\sigma \cdot p)u_R = 0$ is

$$u_R \propto \begin{pmatrix} 1 \\ 0 \end{pmatrix}. \tag{6.11}$$

That is, this fermion is "spin-up," or its spin is aligned with the direction of motion, which is right-handed or $+$-helicity. To ensure that probabilities sum to 1, this spinor is normalized by the energy of the fermion as

$$u_R = \sqrt{2E} \begin{pmatrix} 1 \\ 0 \end{pmatrix}, \qquad \text{or} \qquad \psi_R = \sqrt{2E} \begin{pmatrix} 1 \\ 0 \end{pmatrix} e^{-iEt+iEz}. \tag{6.12}$$

Then, the spinor's inner product with itself is normalized as $u_R^\dagger u_R = 2E$. You will study the origin of this normalization in Exercise 6.2.

In the more general case where the momentum $\vec{p}$ is at an angle θ with respect to the $+\hat{z}$-axis and an angle ϕ about the $+\hat{z}$-axis, we can still determine the solution to the Dirac equation. In this case, the four-momentum is

$$p = E(1, \sin\theta \cos\phi, \sin\theta \sin\phi, \cos\theta), \tag{6.13}$$

and so

$$(\sigma \cdot p)u_R(p) = (\mathbb{I}E - \sigma_1 E \sin\theta \cos\phi - \sigma_2 E \sin\theta \sin\phi - \sigma_3 E \cos\theta)u_R(p) = 0. \tag{6.14}$$

In this expression, we have explicitly included the momentum dependence of the spinor as $u_R(p)$. Written in matrix form, this becomes

$$E \begin{pmatrix} 1 - \cos\theta & -e^{-i\phi}\sin\theta \\ -e^{i\phi}\sin\theta & 1 + \cos\theta \end{pmatrix} u_R(p) = 0. \tag{6.15}$$

The properly normalized solution to this eigenvalue equation is

$$u_R(p) = \sqrt{2E} \begin{pmatrix} e^{-i\phi/2} \cos\frac{\theta}{2} \\ e^{i\phi/2} \sin\frac{\theta}{2} \end{pmatrix}. \tag{6.16}$$

Note that for $\theta \to 0$ and choosing $\phi = 0$,

$$u_R(p) \to \sqrt{2E} \begin{pmatrix} 1 \\ 0 \end{pmatrix}, \tag{6.17}$$

which has the correct limit as the solution for spin-up along the $\hat{z}$-axis. Note also that $u_R(p)^\dagger u_R(p) = 2E$, as required.

The negative-energy solutions (anti-particles) and the left-handed solutions can be found similarly. For momentum p expressed in spherical coordinates, the eigenvalue equation for the negative-energy solution $v_R(p)$ to the right-handed Weyl equation is

$$(\sigma \cdot p)v_R(p) = (-\mathbb{I}E - \sigma_1 E \sin\theta\cos\phi - \sigma_2 E \sin\theta\sin\phi - \sigma_3 E \cos\theta)v_R(p) = 0. \quad (6.18)$$

Now, however, the energy of this particle is $p_0 = -E < 0$. Written in matrix form, we have

$$-E\begin{pmatrix} 1+\cos\theta & e^{-i\phi}\sin\theta \\ e^{i\phi}\sin\theta & 1-\cos\theta \end{pmatrix} v_R(p) = 0. \quad (6.19)$$

This equation is satisfied by the properly normalized solution

$$v_R(p) = \sqrt{2E}\begin{pmatrix} e^{-i\phi/2}\sin\frac{\theta}{2} \\ -e^{i\phi/2}\cos\frac{\theta}{2} \end{pmatrix}. \quad (6.20)$$

Similarly, the left-handed, positive-energy solution $u_L(p)$ satisfies the eigenvalue equation

$$(\bar{\sigma} \cdot p)u_L(p) = (\mathbb{I}E + \sigma_1 E \sin\theta\cos\phi + \sigma_2 E \sin\theta\sin\phi + \sigma_3 E \cos\theta)u_L(p) = 0, \quad (6.21)$$

or that

$$E\begin{pmatrix} 1+\cos\theta & e^{-i\phi}\sin\theta \\ e^{i\phi}\sin\theta & 1-\cos\theta \end{pmatrix} u_L(p) = 0. \quad (6.22)$$

That is, as two-component spinors, v_R and u_L satisfy the same eigenvalue equation! In a similar way, u_R and v_L satisfy the same eigenvalue equation. Therefore, the spinor solutions to the Dirac equation are

$$u_R(p) = v_L(p) = \sqrt{2E}\begin{pmatrix} e^{-i\phi/2}\cos\frac{\theta}{2} \\ e^{i\phi/2}\sin\frac{\theta}{2} \end{pmatrix}, \quad (6.23)$$

$$u_L(p) = v_R(p) = \sqrt{2E}\begin{pmatrix} e^{-i\phi/2}\sin\frac{\theta}{2} \\ -e^{i\phi/2}\cos\frac{\theta}{2} \end{pmatrix}.$$

These spinors are normalized:

$$u_R^\dagger u_R = u_L^\dagger u_L = v_R^\dagger v_R = v_L^\dagger v_L = 2E, \quad (6.24)$$

and different spin solutions are orthogonal:

$$u_R^\dagger u_L = v_L^\dagger v_R = 0. \quad (6.25)$$

These four solutions, u_R, u_L, v_R, and v_L, are the four solutions to the Dirac equation, Eq. 6.2.

6.1.2 Helicity Configurations

With the solutions of the massless Dirac equation in hand, we now turn to determining the helicity configurations of the electrons and muons that are consistent with angular momentum conservation. We introduced how to do this in Chapter 4, but we'll finish that discussion here, especially now that we know the spinors that describe left- and right-handed helicity particles.

From earlier, the Feynman diagram for the process $e^+e^- \to \mu^+\mu^-$ is

The intermediate photon is spin-1, and so the helicities of the e^+e^- pair and the $\mu^+\mu^-$ pair must combine into spin-1. This means that, for electrons colliding in the center-of-mass frame, a configuration like

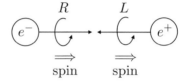

which has a net spin of 1 is allowed. By contrast, a configuration like

is not allowed, as the spins sum to a total spin of 0. This demonstrates that the allowed configurations by angular momentum conservation are all of those for which the e^+ and e^- have opposite helicities and the μ^+ and μ^- have opposite helicities. All non-zero probability helicity configurations of the $e^+e^- \to \mu^+\mu^-$ scattering process are therefore

$$e_L^+ e_R^- \to \mu_L^+ \mu_R^- \,, \qquad\qquad e_R^+ e_L^- \to \mu_L^+ \mu_R^- \,, \qquad (6.26)$$
$$e_L^+ e_R^- \to \mu_R^+ \mu_L^- \,, \qquad\qquad e_R^+ e_L^- \to \mu_R^+ \mu_L^- \,,$$

with the helicities of the particles denoted by the subscripts.

There's a bit more simplification we can do. Let's evaluate the electron–positron–photon vertex with a particular choice of helicities:

$$= v_L^\dagger(p_2)\sigma^\mu u_R(p_1) \,. \qquad (6.27)$$

Because we are working with two-component spinors defined by the Weyl equations of Eq. 6.7 in this chapter, the electron–positron–photon vertex is proportional to a 2×2 Pauli spin matrix, instead of a 4×4 γ matrix. To remember whether the Pauli matrix is σ_μ or $\bar{\sigma}_\mu$ in this vertex, it is the matrix that appears in the appropriate Weyl equation. When this vertex is evaluated for a particular choice of momenta p_1 and p_2, it returns some four-vector with complex number elements. Let's see what happens when we complex conjugate this vertex. We find

$$\left(v_L^\dagger(p_2)\sigma^\mu u_R(p_1)\right)^* = u_R^\dagger(p_1)\sigma^\mu v_L(p_2) = \text{[diagram]} \qquad (6.28)$$

Complex conjugation turns an initial state into a final state! This is a manifestation of the action of CPT that we discussed in Section 3.4, which we'll dive into more in Chapter 10.

For now, this observation relates matrix elements that will simplify our calculation. For example, complex conjugation yields

$$\mathcal{M}(e_L^+ e_R^- \to \mu_L^+ \mu_R^-)^* = \mathcal{M}(\mu_R^+ \mu_L^- \to e_R^+ e_L^-). \qquad (6.29)$$

This may not seem immediately useful because the initial and final states have been flipped. However, we are assuming that all external particles are massless, and so the only difference between electrons and muons is their name, which is just a label. As long as particles and anti-particles of the same label maintain the same label, the matrix element must be unchanged. So, we are free to relabel the muons as electrons and vice-versa in $\mathcal{M}(\mu_R^+ \mu_L^- \to e_R^+ e_L^-)$. Therefore, we have the series of equalities

$$\mathcal{M}(e_L^+ e_R^- \to \mu_L^+ \mu_R^-)^* = \mathcal{M}(\mu_R^+ \mu_L^- \to e_R^+ e_L^-) = \mathcal{M}(e_R^+ e_L^- \to \mu_R^+ \mu_L^-). \qquad (6.30)$$

In the calculation of the cross section from Fermi's Golden Rule, we always take the absolute square of the matrix element. Because of this, we don't care about the fact that complex conjugation relates these processes, as all three matrix elements have the same absolute square. We only have to calculate the absolute squared matrix elements of two non-zero helicity configurations and the other two are identical:

$$|\mathcal{M}(e_L^+ e_R^- \to \mu_L^+ \mu_R^-)|^2 = |\mathcal{M}(e_R^+ e_L^- \to \mu_R^+ \mu_L^-)|^2, \qquad (6.31)$$
$$|\mathcal{M}(e_R^+ e_L^- \to \mu_L^+ \mu_R^-)|^2 = |\mathcal{M}(e_L^+ e_R^- \to \mu_R^+ \mu_L^-)|^2.$$

The only two Feynman diagrams we need to calculate are then

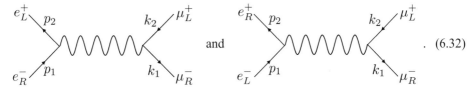

$$\text{and} \qquad (6.32)$$

We will calculate them in our first example of this chapter.

Example 6.1 What are the values of the two Feynman diagrams of Eq. 6.32?

Solution

In Section 4.3.2, we had found that the first diagram evaluated as

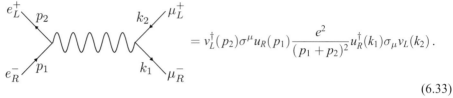

$$= v_L^\dagger(p_2)\sigma^\mu u_R(p_1)\frac{e^2}{(p_1+p_2)^2}u_R^\dagger(k_1)\sigma_\mu v_L(k_2)\,.$$

(6.33)

This is Lorentz invariant, and so can be evaluated in any frame. Let's choose the center-of-mass frame, where the e^+ and e^- have 0 net momentum and collide head-on. Then, we can express their momenta as

$$p_1 = \frac{E_{\rm cm}}{2}(1,0,0,1)\,, \qquad\qquad p_2 = \frac{E_{\rm cm}}{2}(1,0,0,-1)\,,$$

(6.34)

where $E_{\rm cm}$ is the total energy of the e^+e^- system. Their spinors are then

$$v_L^\dagger(p_2) = \sqrt{E_{\rm cm}}(0\;\; -i)\,, \qquad\qquad u_R(p_1) = \sqrt{E_{\rm cm}}\begin{pmatrix}1\\0\end{pmatrix}\,,$$

(6.35)

because $2E = E_{\rm cm}$. To determine the spinor for momenta p_2, note that its momentum is opposite that of p_1. Negating momentum in spherical coordinates requires two steps: transforming the polar angle $\theta \to \pi - \theta$ and transforming the azimuthal angle $\phi \to \phi + \pi$. Using the general expression for momentum in spherical coordinates, Eq. 6.13, one can show that this indeed flips momentum. Such a transformation is called a parity transformation. Then, their spinor product with a σ matrix is

$$v_L^\dagger(p_2)\sigma^\mu u_R(p_1) = E_{\rm cm}(0\;\; -i)(\mathbb{I},\sigma_1,\sigma_2,\sigma_3)^\mu\begin{pmatrix}1\\0\end{pmatrix}$$

(6.36)

$$= E_{\rm cm}(0\;\; -i)\left[\hat{x}\begin{pmatrix}0 & 1\\1 & 0\end{pmatrix} + \hat{y}\begin{pmatrix}0 & -i\\i & 0\end{pmatrix}\right]\begin{pmatrix}1\\0\end{pmatrix}$$

$$= E_{\rm cm}(-i\hat{x}+\hat{y})\,.$$

Only σ_1 and σ_2 have non-zero off-diagonal entries, which is why they remain on the second line of this equation.

To evaluate the Feynman diagram requires also calculating the spinor product $u_R^\dagger(k_1)\sigma_\mu v_L(k_2)$. To do this, we need to know the muon and anti-muon momenta, k_1 and k_2. By energy–momentum conservation, the energies of the muon and anti-muon are both $E_{\rm cm}/2$ and we can choose the frame in which

$$k_1 = \frac{E_{\rm cm}}{2}(1,\sin\theta,0,\cos\theta)\,, \qquad k_2 = \frac{E_{\rm cm}}{2}(1,-\sin\theta,0,-\cos\theta)\,.$$

(6.37)

Then, the spinors are

$$u_R^\dagger(k_1) = \sqrt{E_{\rm cm}}\left(\cos\frac{\theta}{2}\;\; \sin\frac{\theta}{2}\right)\,, \qquad v_L(k_2) = i\sqrt{E_{\rm cm}}\begin{pmatrix}-\sin\frac{\theta}{2}\\\cos\frac{\theta}{2}\end{pmatrix}\,.$$

(6.38)

The spinor product is then

$$u_R^\dagger(k_1)\sigma_\mu v_L(k_2) = iE_{cm}\left(\cos\frac{\theta}{2} \quad \sin\frac{\theta}{2}\right)(\mathbb{I},\sigma_1,\sigma_2,\sigma_3)_\mu\begin{pmatrix} -\sin\frac{\theta}{2} \\ \cos\frac{\theta}{2} \end{pmatrix} \tag{6.39}$$

$$= E_{cm}\left(i\hat{x}\cos\theta + \hat{y} - i\hat{z}\sin\theta\right). \tag{6.40}$$

That is,

$$v_L^\dagger(p_2)\sigma^\mu u_R(p_1) = E_{cm}(0,-i,1,0)^\mu, \tag{6.41}$$
$$u_R^\dagger(k_1)\sigma_\mu v_L(k_2) = E_{cm}(0,i\cos\theta,1,-i\sin\theta)_\mu.$$

The dot product of the spinor products is therefore

$$\left(v_L^\dagger(p_2)\sigma^\mu u_R(p_1)\right)\left(u_R^\dagger(k_1)\sigma_\mu v_L(k_2)\right) = -E_{cm}^2(1+\cos\theta). \tag{6.42}$$

Noting also that $(p_1 + p_2)^2 = E_{cm}^2$ and combining all the factors, we finally find that the Feynman diagram for this helicity configuration is

$$\mathcal{M}(e_R^- e_L^+ \rightarrow \mu_R^- \mu_L^+) = \quad\text{}\quad = -e^2(1+\cos\theta). \tag{6.43}$$

The Feynman diagram for the other spin configuration that we need to calculate is

$$= v_R^\dagger(p_2)\bar{\sigma}^\mu u_L(p_1)\frac{e^2}{(p_1+p_2)^2}u_R^\dagger(k_1)\sigma_\mu v_L(k_2).$$
$$\tag{6.44}$$

The nice thing about this Feynman diagram is that we can reuse the calculation for the muon spinor products from Eq. 6.41. We just need to calculate the spinor product for the new helicity configuration of the electron and positron:

$$v_R^\dagger(p_2)\bar{\sigma}^\mu u_L(p_1) = E_{cm}(i \quad 0)(\mathbb{I},-\sigma_1,-\sigma_2,-\sigma_3)^\mu\begin{pmatrix}0\\1\end{pmatrix} \tag{6.45}$$

$$= E_{cm}(-i\hat{x}-\hat{y}). \tag{6.46}$$

The dot product for this helicity configuration is then

$$\left(v_R^\dagger(p_2)\bar{\sigma}^\mu u_L(p_1)\right)\left(u_R^\dagger(k_1)\sigma_\mu v_L(k_2)\right) = -E_{cm}^2(1-\cos\theta). \tag{6.46}$$

Then, the Feynman diagram is

$$= -e^2(1-\cos\theta) = \mathcal{M}(e_L^- e_R^+ \rightarrow \mu_R^- \mu_L^+). \tag{6.47}$$

We've now calculated our first Feynman diagrams! With these scattering amplitudes in hand, how do we go from these to probabilities? In principle, we can measure whether a particle is an e^+, e^-, μ^+, or μ^- by a charge and mass measurement. As discussed in Chapter 5, by passing a charged particle through a magnetic field, we learn the charge by the direction of trajectory bending and the momentum by the curvature. Additionally, we can, in principle, measure the spin of the electrons and muons via a **Stern–Gerlach experiment**, for example.[1] A Stern–Gerlach experiment is set up with a charged particle with non-zero spin passing through an inhomogeneous magnetic field. Because of its spin and charge, the particle therefore has a magnetic moment, and so will be deflected in the field. Spin-1/2 particles are deflected either up or down, representing the particle in the up- or down-spin state, respectively. Because we can, in principle, therefore conceive of experiments that can distinguish particle type, electric charge, and spin, the processes $e^+_R e^-_L \rightarrow \mu^+_R \mu^-_L$ and $e^+_L e^-_R \rightarrow \mu^+_R \mu^-_L$ are physically distinct. Therefore, they cannot interfere quantum mechanically.

This means that each of these processes distinguished by particle spin or helicity add together at the probability or scattering amplitude squared level. So, all that matters in calculating the cross section for $e^+e^- \rightarrow \mu^+\mu^-$ scattering are the four squared matrix elements:

$$|\mathcal{M}(e^+_R e^-_L \rightarrow \mu^+_R \mu^-_L)|^2 = |\mathcal{M}(e^+_L e^-_R \rightarrow \mu^+_L \mu^-_R)|^2 = e^4(1 + \cos\theta)^2, \qquad (6.48)$$
$$|\mathcal{M}(e^+_R e^-_L \rightarrow \mu^+_L \mu^-_R)|^2 = |\mathcal{M}(e^+_L e^-_R \rightarrow \mu^+_R \mu^-_L)|^2 = e^4(1 - \cos\theta)^2.$$

Let's figure out how to make a cross section from them.

6.1.3 Calculating the Cross Section

To calculate the cross section, we need to think a bit more about our experimental set-up and what we measure. Typically, we accelerate and collide unpolarized electrons and positrons. Unpolarized means that on average the electrons and positrons have probability 1/2 to have right-handed helicity and probability 1/2 to have left-handed helicity. Therefore, we should average over all possible configurations of the initial e^+ and e^- helicities. There are four possible pairs of helicities (*LL*, *LR*, *RL*, and *RR*) and so we should multiply the sum of initial electron–positron helicity configurations by 1/4.

Also, our experiments are not typically sensitive to the final-state muon and anti-muon helicities; they just collect muon four-momenta. Because the experiment does not preferentially detect any particular helicity, we should sum the probabilities of the possible final-state helicities. Just as with the initial state, there are four possible muon–anti-muon helicity configurations, though some of the configurations have zero probability because they do not conserve angular momentum. Accounting for the initial-state averaging and final-state summing of the squared matrix elements, we have

$$\frac{1}{4} \sum_{\mu^+\mu^- \text{ spins}} |\mathcal{M}|^2 = \frac{1}{4}\left(2e^4(1+\cos\theta)^2 + 2e^4(1-\cos\theta)^2\right) = e^4(1+\cos^2\theta). \quad (6.49)$$

---

[1] W. Gerlach and O. Stern, "Experimental test of the applicability of the quantum theory to the magnetic field," Z. Phys. **9**, 349 (1922).

The sum of matrix elements is just the sum of the four non-zero matrix elements given in Eq. 6.48. Now, we can insert this into the expression for the cross section, by Fermi's Golden Rule:

$$\sigma = \frac{1}{2E_{e^+}} \frac{1}{2E_{e^-}} \frac{1}{|v_{e^+} - v_{e^-}|} \tag{6.50}$$

$$\times \int \frac{d^4k_1}{(2\pi)^4} 2\pi\delta(k_1^2) \frac{d^4k_2}{(2\pi)^4} 2\pi\delta(k_2^2) \left(\frac{1}{4}\sum_{\mu\text{ spins}} |\mathcal{M}|^2\right)(2\pi)^4\delta^{(4)}(p_1 + p_2 - k_1 - k_2).$$

Here, E_{e^+} and E_{e^-} are the positron and electron energy, respectively.

The expression for the summed and averaged cross section in Eq. 6.49 is already in terms of the scattering angle θ. We can therefore just take the result from Example 4.2 in Chapter 4 in which we evaluated the phase space integrals, leaving only an integral over the scattering angle. Using the result from that example, the cross section then becomes

$$\sigma = \frac{1}{2E_{e^+}} \frac{1}{2E_{e^-}} \frac{1}{|v_{e^+} - v_{e^-}|} \frac{e^4}{8\pi} \frac{|\vec{p}|}{E_{\text{cm}}} \int_{-1}^{1} d\cos\theta\,(1 + \cos^2\theta). \tag{6.51}$$

In the center-of-mass frame, the positron and electron energies are $2E_{e^+} = 2E_{e^-} = E_{\text{cm}}$, where E_{cm} is the center-of-mass energy. The relative velocity of the electron and positron is just $|v_{e^+} - v_{e^-}| = 2$ because they collide head-on and we approximate both as massless. Finally, the magnitude of final-state momentum $|\vec{p}| = E_{\text{cm}}/2$ because the muon and anti-muon are approximately massless and the total momentum is zero in this frame. Inserting all these pieces, we then find

$$\sigma = \frac{\pi}{2E_{\text{cm}}^2}\left(\frac{e^2}{4\pi}\right)^2 \int d\cos\theta\,(1 + \cos^2\theta). \tag{6.52}$$

The factor that controls the size of this cross section is the **fine structure constant** α:

$$\alpha = \frac{e^2}{4\pi}, \tag{6.53}$$

which is a dimensionless, pure number. In any unit system it has the same value, which is approximately $\alpha \simeq 1/137$. If we remove the integral over the scattering angle in the cross section, we find the cross section differential in the scattering angle θ or just **differential cross section**:

$$\frac{d\sigma(e^+e^- \to \mu^+\mu^-)}{d\cos\theta} = \frac{\pi\alpha^2}{2E_{\text{cm}}^2}(1 + \cos^2\theta). \tag{6.54}$$

This angular distribution can be plotted and compared to data. Figure 6.1 shows data of the scattering angle in $e^+e^- \to \mu^+\mu^-$ collisions measured by the HRS experimental collaboration. Excellent agreement is observed, consistent with our expectation that electrons and muons are spin-1/2 particles. The deviations from the $1 + \cos^2\theta$ prediction can be accounted for by including the effects of the Z boson mediating the scattering.

We can also calculate the **total cross section** by integrating over $\cos\theta$:

$$\sigma(e^+e^- \to \mu^+\mu^-) = \frac{4\pi\alpha^2}{3E_{\text{cm}}^2}. \tag{6.55}$$

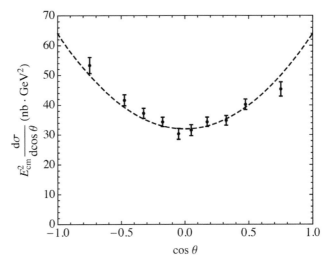

Fig. 6.1   Distribution of the scattering angle $\cos\theta$ in $e^+e^- \to \mu^+\mu^-$ collisions at a center-of-mass energy $E_{cm} = 29$ GeV. The dots represent the data and are compared to the $1 + \cos^2\theta$ shape prediction represented by the dashed curve. The data come from M. Derrick *et al.* [HRS Collaboration], "New results on the reaction $e^+e^- \to \mu^+\mu^-$ at $\sqrt{s} = 29$ GeV," Phys. Rev. D **31**, 2352 (1985).

This cross section can then be used to calculate the total number of $e^+e^- \to \mu^+\mu^-$ events in a collision experiment. Including the value of α in this expression and changing units to barns, the value of the total cross section is approximately

$$\sigma(e^+e^- \to \mu^+\mu^-) \simeq 80 \text{ nb} \cdot \left(\frac{1 \text{ GeV}}{E_{cm}}\right)^2 . \tag{6.56}$$

For a center-of-mass collision energy of order of a GeV, this cross section is about a million times smaller than the total cross section for pp collisions (see Fig. 4.2). So, the process $e^+e^- \to \mu^+\mu^-$ occurs about a millionth as often as a proton–proton collision at high energies.

6.1.4 Inclusive Cross Sections

The cross section we just calculated is called an **inclusive cross section** because it is a good approximation to the cross section for $e^+e^- \to \mu^+\mu^-$ plus anything else in the final state. This inclusive process is denoted by

$$e^+e^- \to \mu^+\mu^- + X, \tag{6.57}$$

where X can be anything. That is, X can be nothing (as we studied here), or it can be a photon, or five photons, or an e^+e^- pair, etc. That is, this inclusive cross section is shorthand for a sum over an infinite number of cross sections:

$$e^+e^- \to \mu^+\mu^- + X = (e^+e^- \to \mu^+\mu^-) + (e^+e^- \to \mu^+\mu^-\gamma) + \cdots . \tag{6.58}$$

The process $e^+e^- \rightarrow \mu^+\mu^-$ is a good approximation to the inclusive cross section because α is small. For example, note that the cross section for $e^+e^- \rightarrow \mu^+\mu^-\gamma$ scales like

$$\sigma(e^+e^- \rightarrow \mu^+\mu^-\gamma) \propto \alpha^3 \,. \tag{6.59}$$

The amplitude for emitting a photon is proportional to the fundamental charge e, which is squared in the cross section. The cross section $\sigma(e^+e^- \rightarrow \mu^+\mu^-\gamma)$ is about 100 times smaller than $\sigma(e^+e^- \rightarrow \mu^+\mu^-)$. In general, the cross section scales like α to the power equal to the number of final-state particles. That is, we can approximate the inclusive cross section as

$$\sigma(e^+e^- \rightarrow \mu^+\mu^- + X) \simeq 80(1 + \mathcal{O}(\alpha)) \text{ nb} \cdot \left(\frac{1 \text{ GeV}}{E_{\text{cm}}}\right)^2 \,, \tag{6.60}$$

where $\mathcal{O}(\alpha)$ means "on the order of α." Then, to an accuracy of about 1% ($\approx \alpha$), $\sigma(e^+e^- \rightarrow \mu^+\mu^- + X) \simeq \sigma(e^+e^- \rightarrow \mu^+\mu^-)$; nevertheless, one can calculate the corrections suppressed by α and find excellent agreement with data.

The utility of inclusive cross sections comes from their generality. It is relatively easy to measure inclusive cross sections in experiment and to collect a large number of events. This is because all we need to identify in the final state are a muon and an anti-muon; we don't care about anything else in the event. Inclusive cross sections are therefore often used for precision extraction of fundamental parameters. For example, in the case of inclusive $e^+e^- \rightarrow \mu^+\mu^-$ collisions, we can make theoretical predictions by calculating Feynman diagrams to whatever order in α that we are strong enough to get to, and then integrate the resulting squared matrix elements over the appropriate phase space to determine the cross section. Correspondingly, we can measure the inclusive cross section for $e^+e^- \rightarrow \mu^+\mu^-$ collisions in experiment and compare the measurement to our prediction. Our theoretical prediction depends on some number of parameters (in the case of the calculation we just completed, it only depends on the value of α), and those parameters can be varied to optimally fit the data. This fit to data then constitutes a "measurement" of the parameters of the cross section. Measurement is in quotes because actually defining the parameters of the cross section precisely takes some care, but can be done consistently. To do so requires renormalization of those parameters; we'll briefly touch on renormalization in Chapters 8 and 9.

6.1.5 Exclusive Cross Sections

While $\sigma(e^+e^- \rightarrow \mu^+\mu^-)$ is a good approximation for the inclusive cross section, it is not a good approximation for the cross section of the exclusive process

$$e^+e^- \rightarrow \mu^+\mu^- + \text{nothing} \,. \tag{6.61}$$

By "nothing," I mean that the final state consists exclusively of a muon and an anti-muon, and nothing else. "Nothing" is an extremely subtle thing in quantum field theory, and so typically we don't require such a strong restriction on the final state. More generally, an **exclusive process** is a process in which there is at least one restriction on the phase space of the final state. For example, we might impose that the energy in final-state particles that are not the muon and anti-muon must be less than 1 GeV. Then, we would want to predict the cross section for the process

$$e^+e^- \to \mu^+\mu^- + (\text{energy} < 1 \text{ GeV}). \tag{6.62}$$

Or, we might require that the muons are relatively isolated. That is, we forbid any photons within an angle of, say, 0.5 radians of the muon or anti-muon. Then, we want to predict the cross section for the process

$$e^+e^- \to \mu^+\mu^- + (\text{angle between } \gamma \text{ and } \mu > 0.5 \text{ rad}). \tag{6.63}$$

The reason why our prediction for $\sigma(e^+e^- \to \mu^+\mu^-)$ is not a good approximation for these or any exclusive processes is that the definition of an exclusive process introduces at least one new energy scale. In the inclusive cross section, we are completely agnostic as to the energy of the muons or the energy of any other particles in the final state. All we care about is the energy scale introduced by the center-of-mass collision energy and the existence of the muon in the final state. Indeed, this is all our prediction specified. By contrast, for the exclusive process of forbidding other particles to have a total energy above 1 GeV, we are now not only sensitive to the center-of-mass energy. We now care about the center-of-mass energy and the energy scale of 1 GeV. Nothing in our prediction for $\sigma(e^+e^- \to \mu^+\mu^-)$ specified this second energy scale, and so we don't expect it to be a good approximation for the exclusive process. To predict the **exclusive cross section** requires consistently accounting for all energy scales defined in the problem.

While most of our focus in this book will be on inclusive cross sections and their importance, we will discuss how to make predictions for a particular exclusive cross section in Chapter 9. Essentially, the existence of multiple energy scales in an exclusive process means that there are dimensionless ratios of energies that can be quite large. These dimensionless ratios will appear in our predictions at every order in the expansion in powers of α, and can potentially spoil the nice convergence properties of the perturbative Feynman diagram expansion.

For example, our calculation of $\sigma(e^+e^- \to \mu^+\mu^-)$ is just one contribution to the prediction of the cross section of the exclusive process for restricting the energy of all particles other than the muons in the final state to be less than 1 GeV. Because there are no other particles in the final state, this somewhat trivially satisfies the requirement. This prediction scales like the square of the fine structure constant, $\sigma^{\text{ex}}(e^+e^- \to \mu^+\mu^-) \propto \alpha^2$, where the "ex" superscript denotes the exclusive cross section. However, it is also possible that there was a photon in the final state whose energy was less than 1 GeV; let's call this prediction $\sigma^{\text{ex}}(e^+e^- \to \mu^+\mu^- + \gamma)$. This prediction indeed scales like a power of α, but also as a function f of the ratio of E_{cm} to 1 GeV:

$$\sigma^{\text{ex}}(e^+e^- \to \mu^+\mu^- + \gamma) \propto \alpha^3 f\left(\frac{E_{\text{cm}}}{1 \text{ GeV}}\right). \tag{6.64}$$

As we will see in Chapter 9, this function f is typically a logarithm, and so if the center-of-mass energy is large enough that it is comparable to the inverse of the fine structure constant,

$$\log \frac{E_{\text{cm}}}{1\,\text{GeV}} \sim \frac{1}{\alpha}, \tag{6.65}$$

then it is not true that the process with a photon in the final state is suppressed by a small number with respect to the process with no photon. To make predictions for exclusive processes, then, we need to account for an arbitrary number of photons in the final state. In many cases, this is easier than it sounds and the sum over all possible numbers of emitted photons can be done explicitly.

6.2 $e^+e^- \to$ Hadrons

6.2.1 Inclusive Hadronic Cross Sections

This discussion of inclusive cross sections leads into the observation and prediction of $e^+e^- \to$ hadrons events. In addition to processes like $e^+e^- \to \mu^+\mu^-$, also observed in the collisions of electrons and positrons is the final-state production of hadrons in the process $e^+e^- \to$ hadrons. Hadrons, such as pions, protons, or neutrons, are complex composite particles. We would like to predict the inclusive cross section for $e^+e^- \to$ hadrons.

How do we do this? One way is to calculate the individual rates for all possible collections of hadrons in the final state. We can calculate the cross sections for the individual processes $e^+e^- \to \pi^0\pi^0$, $e^+e^- \to \pi^+\pi^-$, $e^+e^- \to p\bar{p}$, and every other possible hadronic final state, and then sum them all together. At the very least, this is enormously computationally challenging as we have to consider a huge number of processes. As it turns out, despite significant effort by thousands of theorists, no one has made progress on this "direct" calculation of $e^+e^- \to$ hadrons. So, what do we do? Do we give up?

Of course the answer is no, we just have to think about the problem differently. Let's go back to our understanding of inclusive cross sections and the constituents of hadrons. As we discussed with the quark model in Section 3.3.3, hadrons organize themselves into representations of isospin or SU(3) flavor. All observed representations have a dimension larger than 2 or 3, which would be the dimensionality of the fundamental representation of isospin and SU(3) flavor, respectively. So, this suggests that all representations of hadrons are formed by taking products of smaller representations and the existence of fundamental constituents, called quarks.

We don't observe quarks directly (nor have we ever) for reasons we will discuss in Chapter 8. Nevertheless, with the discovery of quarks as the fundamental constituents of hadrons, we can make progress on understanding and calculating the cross section for $e^+e^- \to$ hadrons. The thing we want to predict is an inclusive cross section; we aren't demanding anything specific about the observed hadrons and there could be anything else produced in the collision. So, with that understanding, considering processes like $e^+e^- \to \pi^+\pi^- + X$ in which specific hadrons are produced is wrong; while this includes some of the possible final states, it misses most of them, like $e^+e^- \to \pi^0\pi^0$, for example.

However, while we may not understand the magic responsible for it, if we consider processes in which quarks are produced, this can include the production of any collection of hadrons in the final state consistent with energy and momentum conservation. For example, the inclusive process $e^+e^- \to u\bar{u} + X$, where u is an up quark, describes both processes where neutral and charged hadrons are produced:

$$e^+e^- \to \pi^0\pi^0 \supset (u\bar{u})(u\bar{u}), \qquad e^+e^- \to \pi^+\pi^- \supset (u\bar{d})(\bar{u}d). \qquad (6.66)$$

Again, we have to be inclusive: $e^+e^- \to u\bar{u} + X$ can produce $\pi^0\pi^0$ and $\pi^+\pi^-$ final states, and many others.

With this insight, we can predict to good approximation the $e^+e^- \to$ hadrons cross section from the inclusive cross section $e^+e^- \to q\bar{q}$, where we sum over all possible final-state quarks q and anti-quarks $\bar{q}$ allowed by energy and momentum conservation. Quarks are spin-1/2 particles, just like muons, so we are able to reuse our results from that study to determine

$$\sigma(e^+e^- \to \text{hadrons}) \simeq \sum_{\text{quarks } q} \sigma(e^+e^- \to q\bar{q} + X). \qquad (6.67)$$

How do we do this?

6.2.2 Properties of the Inclusive Cross Section: Color

By the quark model, quarks must be electrically charged to account for electrical charges of the hadrons. The quark model predicts that the proton consists of two up quarks and a down quark ($p = uud$) and a neutron is two down quarks and an up ($n = ddu$). The electric charge of the proton or the neutron is just the sum of the electric charges of the quarks that compose it. Then,

$$2Q_u + Q_d = 1, \qquad 2Q_d + Q_u = 0, \qquad (6.68)$$

in units of the fundamental charge e. Solving these two equations, we find $Q_u = 2/3$ and $Q_d = -1/3$. So, the quark model predicts that quarks have electric charges that are fractions of the fundamental charge. This is fascinating in its own right, but let's keep going.

Quarks, like electrons and muons, are electrically charged and so they couple to electromagnetism in proportion to their charge. Because of this property, quarks can be produced in the exact same way as muons in the process $e^+e^- \to \mu^+\mu^-$. At this point, the only differences in the calculation of the cross sections $\sigma(e^+e^- \to \mu^+\mu^-)$ and $\sigma(e^+e^- \to \text{hadrons})$ is that we need to (1) include the charge of the quarks in the calculation and (2) sum over all quarks that could be produced. Importantly, this sum is done at the level of the squared matrix element, and not of individual Feynman diagrams. While we have never directly observed individual quarks, in principle we could, and could therefore distinguish different types of quarks by their masses, charges, and other quantum numbers. In the following example, let's include these differences to predict $\sigma(e^+e^- \rightarrow \text{hadrons})$.

Example 6.2 Using the result for the cross section $\sigma(e^+e^- \rightarrow \mu^+\mu^-)$ from Eq. 6.55, what is the inclusive cross section for hadron production, $\sigma(e^+e^- \rightarrow \text{hadrons})$?

Solution

Our insight into inclusive cross sections tells us that the total cross section for $e^+e^- \rightarrow$ hadrons, to good approximation, is just

$$\sigma(e^+e^- \rightarrow \text{hadrons}) \simeq \sum_{\text{quarks } q} \sigma(e^+e^- \rightarrow q\bar{q}) = \sum_{\text{quarks } q} \frac{4\pi\alpha^2}{3E_{\text{cm}}^2} Q_q^2, \tag{6.69}$$

where Q_q is the electric charge of quark q in units of the fundamental charge e. In the last equation, we've used the expression for the $e^+e^- \rightarrow \mu^+\mu^-$ cross section, Eq. 6.55, as the cross section to produce each individual type of quark–anti-quark pair in the final state. This prediction can be compared to experiment. Actually, a better measurement is the ratio of the cross sections, called R,

$$R \equiv \frac{\sigma(e^+e^- \rightarrow \text{hadrons})}{\sigma(e^+e^- \rightarrow \mu^+\mu^-)} \simeq \sum_{\text{quarks } q} Q_q^2, \tag{6.70}$$

which just reduces to the sum over quark charges. To predict the value of R, we just need to know how many quarks there are and what their electric charges are.

By 1974, four quarks were known: up, down, strange, and charm, with charges

$$Q_u = \frac{2}{3}, \qquad Q_d = -\frac{1}{3}, \qquad Q_s = -\frac{1}{3}, \qquad Q_c = \frac{2}{3}. \tag{6.71}$$

Depending on the center-of-mass energies, the types and numbers of quarks that can contribute to R changes. A quark can only be produced in the final state if the center-of-mass energy is greater than twice its mass. At center-of-mass energies below about 4 GeV, only the up, down, and strange quarks are light enough to contribute to R. Thus, we (naïvely) find that R for energies below about 4 GeV is

$$R(E_{\text{cm}} < 4\,\text{GeV}) \simeq \sum_{\text{quarks } u, d, s} Q_q^2 = \frac{4}{9} + \frac{1}{9} + \frac{1}{9} = \frac{2}{3}. \tag{6.72}$$

Above about 4 GeV, the charm quark also contributes and so the value of R will change appropriately:

$$R(E_{\text{cm}} > 4\,\text{GeV}) \simeq \sum_{\text{quarks } u, d, s, c} Q_q^2 = \frac{4}{9} + \frac{1}{9} + \frac{4}{9} + \frac{1}{9} = \frac{10}{9}. \tag{6.73}$$

The observed values in these two regimes are actually close to $R(E_{\text{cm}} < 4\,\text{GeV}) = 2$ and $R(E_{\text{cm}} > 4\,\text{GeV}) = \frac{10}{3}$, respectively, three times larger than our predictions. This is demonstrated in data in the top two panels of Fig. 6.2.

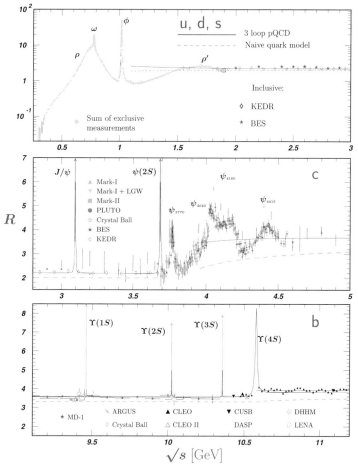

Fig. 6.2 Plot of the ratio R of the cross section for $e^+e^- \rightarrow$ hadrons to the cross section for $e^+e^- \rightarrow \mu^+\mu^-$ in various center-of-mass collision energy windows. The prediction of R in the quark model is shown in the dashed curve, and its value changes depending on the number of quarks with masses less than the collision energy. In the top plot, the up, down, and strange quarks contribute; in the middle plot, the charm quark also contributes about about 4 GeV; and in the bottom plot, the bottom quark also contributes above about 10 GeV. This plot was compiled in M. Tanabashi *et al.* [Particle Data Group], "Review of particle physics," Phys. Rev. D **98**, 030001 (2018).

    Does this mean that our prediction is wrong? Not really, it's just incomplete. Along with other evidence we will discuss in the following chapters, this cross section ratio discrepancy is evidence that there are actually three copies of each type of quark. These three copies are distinguished by a quantity called **color**, typically called red, green, and blue, though has nothing to do with visible light whatsoever. Therefore, there are actually three times as many quark final states available as we naïvely expected. Including this factor of 3, the cross section ratio R becomes

$$R(E_{\text{cm}} < 4\text{ GeV}) = 2\,, \qquad\qquad R(E_{\text{cm}} > 4\text{ GeV}) = \frac{10}{3}\,, \qquad (6.74)$$

for center-of-mass collision energies lower and higher than about twice the charm quark mass ($m_c \simeq 1.3$ GeV). This simple picture and calculation then agrees well with experiment, as shown in the plots of Fig. 6.2. Above a center-of-mass energy of about $E_{cm} \simeq 10.5$ GeV, R is again seen to increase, corresponding to the production of bottom quarks in the final state. Assigning the bottom quark an electric charge of

$$Q_b = \frac{1}{3},\qquad(6.75)$$

the value of R above 10.5 GeV would be

$$R(E_{cm} > 10.5 \text{ GeV}) = \frac{11}{3}.\qquad(6.76)$$

This is in reasonable agreement with data presented in the bottom panel of Fig. 6.2. Improved agreement with data can be accomplished using the fundamental theory of quarks, color, and the strong force, called quantum chromodynamics (the solid curves on the plots of Fig. 6.2).

6.2.3 Properties of the Inclusive Cross Section: Spin

This R measurement is evidence for both the fractional charges in the quark model and for the existence of color. We'll find a lot more evidence for both in later chapters, but for now, let's go back to another property of quarks that we have assumed up to now. To be able to recycle the cross section from $e^+e^- \rightarrow \mu^+\mu^-$ for hadron production, we needed to assume that quarks are spin-1/2 particles, just like muons. However, how do we know? Let's go back to the differential cross section for $e^+e^- \rightarrow \mu^+\mu^-$:

$$\frac{d\sigma}{d\cos\theta} = \frac{\pi\alpha^2}{2E_{cm}^2}(1 + \cos^2\theta).\qquad(6.77)$$

If we translated this to quarks, the overall coefficient would be affected by quark charges and color, but the shape, the $1 + \cos^2\theta$, would remain. This shape, which is peaked for $\theta = 0$ or π, is indicative of the final-state particles being fermions with spin-1/2.

If we can measure the angular distribution of the final-state quarks, then we would have evidence of their spin. For reasons we will study in detail in Chapter 9, hadrons produced in $e^+e^- \rightarrow$ hadrons events are not just uniformly distributed throughout the experiment. At high center-of-mass collision energies, hadrons form collimated streams, called **jets**. These jets are a manifestation of the underlying quarks. That is, in an $e^+e^- \rightarrow$ hadrons event, we will observe something like that illustrated in Fig. 6.3. The two jets are composed of numerous hadrons (pions, protons, etc.) which are schematically denoted by the lines emanating from the collision point. The two jets in this illustration can be thought of as a proxy for their initiating quarks: the momentum of the jets will be very close to the momentum of the initiating quarks. Our modern interpretation of jets initiated by inclusive high-energy quark production was first experimentally verified in 1975.[2] An event display

---

[2] G. Hanson *et al.*, "Evidence for jet structure in hadron production by e^+e^- annihilation," Phys. Rev. Lett. **35**, 1609 (1975).

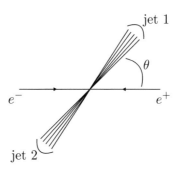

Fig. 6.3   Schematic illustration of two jets of collimated hadrons produced from electron–positron collisions. The scattering angle of the final state is illustrated as the angle of one of the jets from the electron–positron momentum axis.

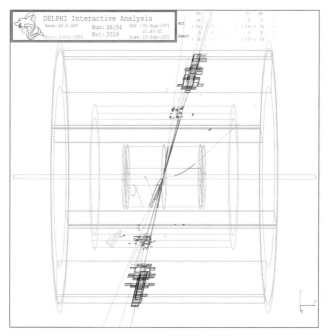

Fig. 6.4   A dijet event display from the DELPHI experiment. Electrons and positrons are collided at $E_{cm} = 91$ GeV and two back-to-back jets are observed in the detector. The electrons and positrons come in from the left and right, respectively, and the jets are the collimated tracks and calorimeter deposits pointing nearly up and down from the collision point in the center. The components of the DELPHI detector are outlined for illustration. Credit: DELPHI experiment © CERN.

of such a **dijet** event from the DELPHI experiment at the **Large Electron–Positron Collider (LEP)** is presented in Fig. 6.4.

The dominant process for hadron production consists of two jets with equal and opposite momentum. This observation suggests the cross section relationships:

$$\sigma(e^+e^- \to \text{hadrons}) \simeq \sigma(e^+e^- \to q\bar{q}) \simeq \sigma(e^+e^- \to \text{two jets}). \qquad (6.78)$$

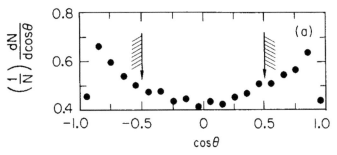

$$\frac{1}{N}\frac{dN}{d\cos\theta}$$

Fig. 6.5 Distribution of the thrust angle $\cos\theta$ in $e^+e^- \to$ hadrons collisions at a center-of-mass energy $E_{cm} = 29$ GeV. The thrust angle is a way to define the scattering angle for the collimated collection of particles in a jet. Reprinted figure with permission from W. T. Ford *et al.*, Phys. Rev. D **40**, 1385 (1989). Copyright 1989 by the American Physical Society.

We can measure the scattering angle θ illustrated in Fig. 6.3 that the jets make with the electron beam axis, and compare to the $1 + \cos^2\theta$ expectation from spin-1/2 quarks. A plot of data in which the scattering angle in $e^+e^- \to$ dijets events was measured is shown in Fig. 6.5. These data exhibit the characteristic $1 + \cos^2\theta$ dependence, providing concrete evidence for the spin-1/2 nature of quarks. Therefore, just from thinking about the consequences of the quark model for $e^+e^- \to$ hadrons, we have evidence for both quark color and spin. In Exercise 6.5, we predict the differential cross section for spin-0 quarks and see whether it can be consistent with data. Spoiler: the spin-0 hypothesis will not be consistent with data!

Exercises

6.1 *Lorentz Transformations of Spinors.* With explicit spinor solutions to the Dirac equation we can then determine how they transform under Lorentz transformations. For a massless momentum four-vector p where

$$p = E(1, \sin\theta\cos\phi, \sin\theta\sin\phi, \cos\theta), \tag{6.79}$$

with energy E, polar angle θ, and azimuthal angle ϕ, recall that the right- and left-handed spinors are

$$u_R(p) = \sqrt{2E}\begin{pmatrix} e^{-i\phi/2}\cos\frac{\theta}{2} \\ e^{i\phi/2}\sin\frac{\theta}{2} \end{pmatrix}, \qquad u_L(p) = \sqrt{2E}\begin{pmatrix} e^{-i\phi/2}\sin\frac{\theta}{2} \\ -e^{i\phi/2}\cos\frac{\theta}{2} \end{pmatrix}. \tag{6.80}$$

Determine the 2×2 matrix $\mathbb{M}$ that implements each of the following Lorentz transformations:

(a) an azimuthal rotation by an angle χ
(b) a polar rotation by an angle ω
(c) a Lorentz boost along the $\hat{z}$-axis by a velocity β.

The matrix $\mathbb{M}$ is defined as

$$\mathbb{M}u = u', \tag{6.81}$$

where u is the original spinor and u' is the Lorentz-transformed spinor.

6.2   *Helicity Spinors.* Spinors of definite helicity have more interesting properties as spin-1/2 representations of the Lorentz group. In this exercise, we will demonstrate some of them and relate them to properties of the Pauli spin matrices. The expressions for the left- and right-handed spinors are rewritten above in Eq. 6.80.

   (a)   The inner product of the spinors is normalized: $u_R^\dagger u_R = 2E$, and orthogonal: $u_R^\dagger u_L = 0$. Evaluate the outer product of spinors $u_R u_R^\dagger$ and $u_L u_L^\dagger$. Your result will be a matrix. How does the matrix $u_R u_R^\dagger$ compare to $p \cdot \bar\sigma$? What matrix is $u_L u_L^\dagger$, in terms of the Pauli σ matrices?

   (b)   Now, take the trace of the outer product matrix. In terms of the original spinors, what did you just compute? Combining this result with part (a) is the justification for the normalization of the spinors.

   (c)   The helicity operator $\hat{h}$ can be expressed as the matrix

$$\hat{h} = \frac{1}{2}\hat{p} \cdot \vec\sigma, \tag{6.82}$$

where $\hat{p}$ is the unit vector in the direction of the three-momentum $\vec{p}$ and $\vec\sigma$ is the vector of the three Pauli spin matrices $\vec\sigma = (\sigma_1, \sigma_2, \sigma_3)$. Evaluate the helicity operator $\hat{h}$ for the momentum represented in spherical coordinates. What are the eigenvalues of $\hat{h}$ acting on the spinors u_R and u_L?

   (d)   In Eq. 6.42, we evaluated the spinor product

$$\left(v_L^\dagger(p_2)\sigma^\mu u_R(p_1)\right)\left(u_R^\dagger(k_1)\bar\sigma_\mu v_L(k_2)\right) = -E_{\mathrm{cm}}^2(1 + \cos\theta) \tag{6.83}$$

in the center-of-mass frame for momenta p_1, p_2, k_1, and k_2, relevant for $e^+e^- \to \mu^+\mu^-$ scattering. In this same frame, evaluate the spinor product

$$\left(v_L^\dagger(p_2)v_L(k_2)\right)\left(u_R^\dagger(k_1)u_R(p_1)\right). \tag{6.84}$$

How does this compare to the other spinor product above? This relationship is an example of a **Fierz identity**.

6.3   *Spin Analysis of $e^+e^- \to \mu^+\mu^-$ Scattering.* In calculating the matrix element for $e^+e^- \to \mu^+\mu^-$, we found zeros at certain regions of phase space. For example, the matrix element

$$\mathcal{M}(e_R^- e_L^+ \to \mu_R^- \mu_L^+) = -e^2(1 + \cos\theta) \tag{6.85}$$

vanishes if $\cos\theta = -1$. Explain this zero probability for scattering by angular momentum conservation.

6.4   *Spin-0 Photon.* We are quite confident that the photon has spin 1, even without testing it in electron–positron collisions. Nevertheless, let's assume the photon were a spin-0 particle. Without doing a calculation, can you determine the scattering angle dependence in the matrix element for the process $e^+e^- \to \mu^+\mu^-$? Is this consistent with data?

Hint: Of course the matrix element is Lorentz invariant. If the photon were spin 0, are there subparts of the matrix element that are additionally Lorentz invariant alone?

6.5 $e^+e^- \to$ *Scalars*. The angular distribution of the jets present in $e^+e^- \to$ hadrons collisions is evidence for the spin-1/2 nature of quarks. But is it possible for quarks to have a different spin? In this exercise, we will calculate the Feynman diagram for the process $e^+e^- \to \phi\phi^*$, where ϕ is an electrically charged massless scalar, a spin-0 particle. As a spin-0 particle, it has no helicity, and its external spin wavefunction is just 1. The Feynman diagram for this process mediated by a photon is

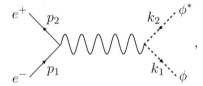

where the external scalars are denoted by the dashed lines.

(a) What helicity configurations of the initial-state e^+e^- pair are allowed by angular momentum conservation?

(b) To evaluate the Feynman diagram, we need to know how the scalar couples to the photon. The unique result consistent with electric charge conservation is

$$= e(k_1 - k_2)^\mu, \tag{6.86}$$

where both momenta k_1 and k_2 leave the vertex. Working in the center-of-mass frame where the initial electron and positron collide head-on with equal energy, compute the Feynman diagrams for $e^+e^- \to \phi\phi^*$ with all possible electron and positron helicities. Assume that the electric charge of the scalar is e. You should find, for example,

$$\mathcal{M}(e_R^+ e_L^- \to \phi\phi^*) = ie^2 \sin\theta. \tag{6.87}$$

(c) Now, square the matrix elements corresponding to different electron and positron helicities, and sum them. What is the cross section differential in the scattering angle θ? Make sure to average over initial spins of the electron and positron. What is the cross section when $\theta \to 0$ or π?

(d) Now, integrate over the scattering angle and determine the total cross section $\sigma(e^+e^- \to \phi\phi^*)$. How does the value of the total cross section compare to that of the process $e^+e^- \to \mu^+\mu^-$?

6.6 *Decays of the Z Boson*. The Z boson is an unstable particle, and as such, decays to less massive particles. The Z boson is electrically neutral and couples to any particle that carries a charge under the weak force. Z boson decays are of the form

$$Z \to f\bar{f}, \tag{6.88}$$

where f is a fermion of the Standard Model, and $\bar{f}$ is its anti-particle. The Z boson couples to the each of the fermions of the Standard Model with (approximately) equal strength. Here, we will understand the relative rates of Z boson decay to different fermions of the Standard Model. We'll discuss many more properties of the Z boson starting in Chapter 11.

(a) The Z boson decays to both electrons and up quarks: $Z \to e^+e^-$ and $Z \to u\bar{u}$. About how much more often does the Z boson decay to up quarks then to electrons?

(b) Using the PDG, what quarks can the Z boson decay to? What charged leptons can the Z decay to?

(c) Combining parts (a) and (b), estimate the ratio of the rate at which Z bosons decay to hadrons to the rate at which they decay to charged leptons.

(d) Using the PDG, determine the measured ratio of the rate of Z decays to hadrons to the rate of Z decays to charged leptons. How does this compare to your estimate in part (c)?

6.7 *Inclusive versus Exclusive Cross Sections.* In this chapter, we discussed the distinction between inclusive and exclusive cross sections, especially in the context of interpreting the process $e^+e^- \to$ hadrons. In this exercise, we will see how different these cross sections are as the restrictions imposed in the exclusive cross section are made extreme.

The ALEPH experiment at the Large Electron–Positron Collider measured the inclusive $e^+e^- \to$ hadrons cross section at a center-of-mass collision energy of $E_{\rm cm} = 206$ GeV. From this inclusive cross section, ALEPH then imposed restrictions on the hadronic final state to extract exclusive cross sections for the processes $e^+e^- \to n$ jets, where $n = 1, 2, 3, 4, 5$, and 6 or more jets. Figure 6.6 plots the results of this study, where each exclusive process is expressed as a fraction of the inclusive process.

To separate the inclusive cross section into individual exclusive cross sections for n-jet production, ALEPH measured the momentum of the jets and imposed restrictions on relations between pairs of jets. ALEPH demanded that a relationship between the energies E_i and E_j of two jets i and j and their relative angle θ_{ij} be greater than a fraction of the total center-of-mass collision energy:

$$2 \min[E_i^2, E_j^2](1 - \cos \theta_{ij}) > y_{\rm cut} E_{\rm cm}^2 . \tag{6.89}$$

Only if this restriction was satisfied did ALEPH count the jets, otherwise the lower-energy jet was ignored. The parameter $y_{\rm cut}$ varies, and ALEPH considered the range $y_{\rm cut} \in [10^{-5}, 1]$. For a fixed number of jets in the final state, smaller values of $y_{\rm cut}$ are stronger restrictions on those jets.

(a) What is the smallest energy scale that is imposed on the hadronic final state when $y_{\rm cut} = 10^{-5}$? Express your answer in GeV. How does this compare to the mass of the proton, for example?

(b) At large values of $y_{\rm cut}$, Fig. 6.6 demonstrates that most of the inclusive cross section for $e^+e^- \to$ hadrons is contained in the exclusive process $e^+e^- \to$ two

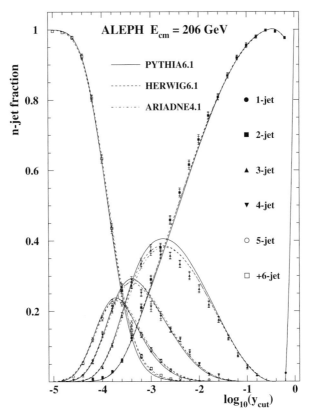

Fig. 6.6   Plot of the *n*-jet fraction in $e^+e^- \rightarrow$ hadrons events from the ALEPH experiment at the Large Electron–Positron Collider as a function of the jet resolution variable y_{cut}. Reprinted by permission from Springer Nature: Springer Nature Eur. Phys. J. C "Studies of QCD at e^+e^- centre-of-mass energies between 91 GeV and 209 GeV," A. Heister *et al.* [ALEPH Collaboration] (2004).

jets. Below approximately what value of y_{cut} is this dijet cross section starting to be a poor approximation to the total cross section of $e^+e^- \rightarrow$ hadrons?

(c) As we will discuss in Chapter 7, the cross section for additional jet production is controlled by the coupling of QCD, α_s, which is called the strong coupling. Because we impose a new energy scale on the final state according to the value of y_{cut}, we expect that the exclusive cross section for three-jet production is related to two-jet production as

$$\sigma(e^+e^- \rightarrow \text{three jets}) \simeq \alpha_s f(y_{cut}) \sigma(e^+e^- \rightarrow \text{two jets}), \qquad (6.90)$$

for some function of y_{cut}, $f(y_{cut})$. This parallels our discussion in Section 6.1.5. In Chapter 9, we will show that this function is

$$f(y_{cut}) = \log^2 y_{cut}. \qquad (6.91)$$

Using Fig. 6.6, what is the approximate value of α_s? Note that the logarithm in Eq. 6.91 is to base e (natural logarithm).

(d) As $y_{\text{cut}} \to 0$, what is the probability that there are only two jets in the final state? In this same limit, how many jets will be observed in the process $e^+e^- \to$ hadrons, and how does this compare to the total number of hadron particles detected by the experiment?

6.8 *Finite Decay Width Effects.* In this chapter, we calculated the cross section for $e^+e^- \to \mu^+\mu^-$ through an intermediate photon. We then modified the predictions of this process to understand the process $e^+e^- \to$ hadrons where individual quarks were still produced through electromagnetism. At very high center-of-mass collision energies, however, the Z boson can also mediate these processes. In this exercise, we will work to incorporate the effects of the Z boson into these predictions as approximately a massive photon. This approximation will be justified in Chapter 11.

(a) The propagator for a massive, unstable particle like the Z boson can be written as

$$\frac{1}{q^2 - m_Z^2 + im_Z\Gamma_Z}. \tag{6.92}$$

q is the momentum flowing through the virtual Z boson, m_Z is the mass of the Z boson, and Γ_Z is the width or decay rate of the Z boson. In the process $e^+e^- \to$ hadrons, $q^2 = E_{\text{cm}}^2$, the squared center-of-mass collision energy. In the evaluation of a cross section, this propagator appears as its absolute value squared. Taking this absolute squared propagator as the relative probability distribution of Z boson production, what is its full width in E_{cm} at half of the maximum value? Does this justify the term "width"?

(b) Including a factor of 3 for quark color, the total cross section for $e^+e^- \to$ hadrons through a photon is

$$\sigma(e^+e^- \to \gamma^* \to \text{hadrons}) = \frac{11}{3}\frac{4\pi\alpha^2}{3E_{\text{cm}}^2}, \tag{6.93}$$

The $*$ superscript denotes that the photon is off-shell and the factor of $11/3$ is the total squared sum of quark charges up through the bottom quark. With the Z boson propagator from part (a), modify this cross section to predict the cross section with an intermediate Z boson, $\sigma(e^+e^- \to Z^* \to \text{hadrons})$. Assume that the Z boson couples to electrons and quarks in the same way as a photon does.

(c) We'll assume that the intermediate photon and Z boson do not interfere quantum mechanically, so we can just sum their cross sections. That is, calculate the total cross section for $e^+e^- \to$ hadrons as

$$\sigma(e^+e^- \to \text{hadrons}) \tag{6.94}$$
$$= \sigma(e^+e^- \to \gamma^* \to \text{hadrons}) + \sigma(e^+e^- \to Z^* \to \text{hadrons}).$$

You should find

$$\sigma(e^+e^- \to \text{hadrons}) = \frac{11}{3}\frac{4\pi\alpha^2}{3E_{\text{cm}}^2}\left(1 + \frac{E_{\text{cm}}^4}{(E_{\text{cm}}^2 - m_Z^2)^2 + m_Z^2\Gamma_Z^2}\right). \tag{6.95}$$

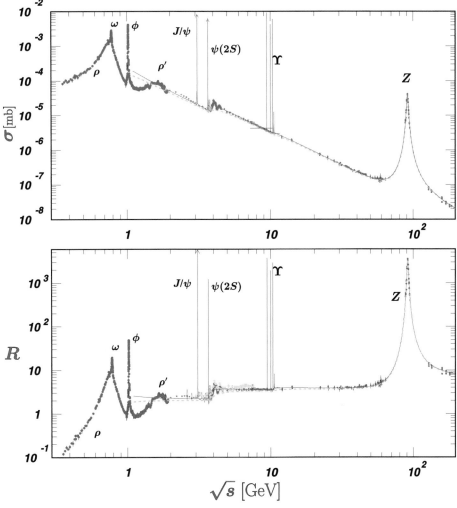

Fig. 6.7   At the top is the cross section for $e^+e^- \rightarrow$ hadrons from various experiments as a function of the center-of-mass energy. The bottom plot is the ratio R of the cross section of $e^+e^- \rightarrow$ hadrons to that of $e^+e^- \rightarrow \mu^+\mu^-$. From M. Tanabashi *et al.* [Particle Data Group], "Review of particle physics," Phys. Rev. D **98**, 030001 (2018).

(d) The top panel of Fig. 6.7 shows the cross section for $e^+e^- \rightarrow$ hadrons in mb as a function of the center-of-mass collision energy $\sqrt{s} = E_{cm}$. From this plot and the functional form of your prediction from part (c), estimate the mass m_Z of the Z boson and its width Γ_Z. You can use $\alpha = 1/137$ and only consider the region where $\sqrt{s} > 10$ GeV. How do your extracted values of m_Z and Γ_Z compare to the properties of the Z boson from the PDG?

6.9 *Research Problem.* In this chapter, we introduced exclusive cross sections and we'll predict one such cross section in Chapter 9. In general, is there a principle for selecting what type of exclusive cross sections can be predicted?

7

Quarks and Gluons

The observation of jet production in e^+e^- collisions and their angular distribution is strongly suggestive of the existence of fundamental quarks. However, it's not quite a smoking gun. If quarks do exist as the fundamental constituents of hadrons then we should be able to observe their point-like structure. Just observing jets isn't really probing the structure of quarks, and there's still quite a logical leap from measuring the angular distribution of jets to concluding the spin of quarks. We need a direct probe of the constituents of hadrons such as the proton, so let's just collide electrons with protons! The process in which electrons and protons are collided at high energies is called deeply inelastic scattering or DIS, and we'll find that DIS is the tool that we need.

Digging a little deeper in the predictions of the quark model, we found that it wasn't sufficient to describe the rate of $e^+e^- \to$ hadrons scattering; we needed to multiply by a factor of 3 to have predictions match data. We called this factor of 3 "color," which so far is totally ad hoc. If, in addition to electric charge, quarks do indeed carry a color charge, then there should be a corresponding force between quarks whose strength is proportional to this color charge. The possibility of such a force is extremely attractive for phenomenology: it could potentially describe why quarks form hadron bound states. As a force, though, it would need a mediating particle just like the photon of electromagnetism. Where is this particle? By revisiting $e^+e^- \to$ hadrons, we'll identify direct evidence for this force carrier that we now call the gluon.

In this chapter, we identify the players of this new force and their properties. We call this force the "strong force," and this will set the stage for construction of a complete theory of the strong force called quantum chromodynamics or QCD in Chapter 8. Along the way, we'll find many surprises and get glimpses of a theory that is very different than familiar electromagnetism.

7.1 Crossing Symmetry

To frame the discussion of this chapter, let's start by discussing the useful kinematic variables to express the experiments we will study. Most of the scattering processes that we consider consist of two initial-state particles that interact and produce two final-state particles, like $e^+e^- \to \mu^+\mu^-$ studied in the previous chapter. Such processes are called **2 → 2** (read: "two-to-two"), and we can completely characterize what happens in such a scattering. The general 2 → 2 scattering process can be visualized as:

time

with initial momenta p_1, p_2 and final momenta p_3, p_4. Momentum conservation imposes

$$p_1 + p_2 = p_3 + p_4 . \tag{7.1}$$

The description of the scattering process must be Lorentz invariant. As such, it can only depend on particle masses (m_1, m_2, m_3, m_4) and Lorentz-invariant four-vector dot products. Naïvely, there are six dot products that can be formed, but momentum conservation relates pairs of them:

$$
\begin{aligned}
s &\equiv (p_1 + p_2)^2 = (p_3 + p_4)^2 , \\
t &\equiv (p_1 - p_4)^2 = (p_2 - p_3)^2 , \\
u &\equiv (p_1 - p_3)^2 = (p_2 - p_4)^2 .
\end{aligned}
\tag{7.2}
$$

The names s, t, and u for these Lorentz-invariant momentum combinations are called **Mandelstam variables**, named after Stanley Mandelstam.[1] Among other work, Mandelstam is known for proving that the exotic theory called $\mathcal{N} = 4$ supersymmetric Yang–Mills exhibits a larger spacetime symmetry than just Lorentz transformations.[2] This theory is actually invariant under conformal transformations: all possible transformations that maintain relative angles. As such, its interactions are extremely highly constrained, more so than the interactions of the Standard Model.

The utility of the Mandelstam variables is that they express different momentum exchanges between the initial and final state. For example, for the process $e^+ e^- \to q\bar{q}$, the Feynman diagram for this process is

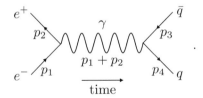

time

Then, the squared momentum of the intermediate photon is $s = (p_1 + p_2)^2$. A scattering process like this is therefore called an **s-channel** process. In the center-of-mass frame, $s = E_{\text{cm}}^2$.

The **t-channel** process is just a different time ordering of the same fundamental scattering process. If time instead flows upward, the Feynman diagram remains the same but corresponds to the process $e^- q \to e^- q$:

[1] S. Mandelstam, "Determination of the pion–nucleon scattering amplitude from dispersion relations and unitarity: General theory," Phys. Rev. **112**, 1344 (1958).

[2] S. Mandelstam, "Light cone superspace and the ultraviolet finiteness of the N=4 model," Nucl. Phys. B **213**, 149 (1983).

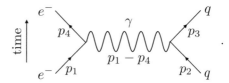

Therefore, the squared momentum of the photon is $t = (p_1 - p_4)^2$. Note that we've relabeled the particle momenta so that p_1 and p_2 are always the initial momenta and p_3 and p_4 are always the final momenta. Specializing to massless particles and working in the center-of-mass frame, we can express the external particle momenta in spherical coordinates as

$$p_1 = \frac{E_{\text{cm}}}{2}(1, 0, 0, 1), \qquad p_3 = \frac{E_{\text{cm}}}{2}(1, -\sin\theta\cos\phi, -\sin\theta\sin\phi, -\cos\theta),$$

$$p_2 = \frac{E_{\text{cm}}}{2}(1, 0, 0, -1), \qquad p_4 = \frac{E_{\text{cm}}}{2}(1, \sin\theta\cos\phi, \sin\theta\sin\phi, \cos\theta). \tag{7.3}$$

Then, the value of t is just

$$t = (p_1 - p_4)^2 = -2p_1 \cdot p_4 = -\frac{E_{\text{cm}}^2}{2}(1 - \cos\theta), \tag{7.4}$$

where θ is the scattering angle.

The **u-channel** would be yet a further different time ordering, but this happens to not exist for electron–quark scattering. The reason for this is that electromagnetism does not allow for electrons to turn into quarks directly. The diagram that would correspond to the u-channel process,

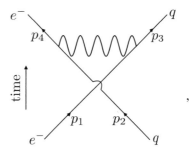

is forbidden by the nature of electromagnetic interactions. The u-channel is allowed when identical particles are scattered, as in the process $e^- e^- \to e^- e^-$, which is called **Møller scattering**.[3] For Møller scattering, the u-channel corresponds to the following diagram:

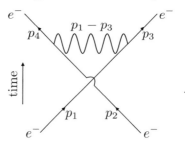

---

[3] C. Møller, "Zur theorie des durchgangs schneller elektronen durch materie," Annalen Phys., **406** no. 5, 531 (1932).

The squared momentum of the photon in this diagram is:

$$u = (p_1 - p_3)^2 = -\frac{E_{cm}^2}{2}(1 + \cos\theta),\tag{7.5}$$

in the center-of-mass frame.

For all massless particle scattering, note that we have the relationship

$$s + t + u = E_{cm}^2 - \frac{E_{cm}^2}{2}(1 - \cos\theta) - \frac{E_{cm}^2}{2}(1 + \cos\theta) = 0,\tag{7.6}$$

which is a Lorentz-invariant constraint. That is, for $2 \to 2$ processes, there are only two quantities that completely characterize the scattering: the center-of-mass collision energy E_{cm} and the scattering angle θ.

7.1.1 Electron–Quark Scattering

Using the Mandelstam variables, we can rewrite the expression for the cross section of the process $e^+e^- \to q\bar{q}$ that we derived in the previous chapter. For one type q of quark with electric charge Q_q, the cross section differential in the scattering angle is

$$\frac{d\sigma(e^+e^- \to q\bar{q})}{d\cos\theta} = \frac{3}{2}\frac{\pi\alpha^2}{E_{cm}^2}Q_q^2(1 + \cos^2\theta).\tag{7.7}$$

This follows from the differential cross section for $e^+e^- \to \mu^+\mu^-$ scattering, Eq. 6.54, where we include the quark electric charge and multiply by a factor of 3 to account for color. Using the relationships $s = E_{cm}^2$ and $t - u = E_{cm}^2 \cos\theta = s\cos\theta$, we can express the scattering angle in terms of the Mandelstam variables. We have

$$\cos\theta = \frac{t - u}{s} = \frac{t - (-s - t)}{s} = 2\frac{t}{s} + 1,\tag{7.8}$$

where we used $s + t + u = 0$ to eliminate u. The angular dependence factor $1 + \cos^2\theta$ is then

$$1 + \cos^2\theta = 1 + \left(\frac{t - u}{s}\right)^2 = \frac{(t + u)^2 + (t - u)^2}{s^2} = 2\frac{t^2 + u^2}{s^2}.\tag{7.9}$$

The last piece that we need to express the differential cross section in terms of the Mandelstam variables is the Jacobian factor. From Eq. 7.8, we have

$$d\cos\theta = \frac{2}{s}dt,\tag{7.10}$$

which enables us to rewrite the cross section differential in $\cos\theta$ to differential in t. Combining these results, we find

$$\frac{d\sigma(e^+e^- \to q\bar{q})}{d\cos\theta} = \frac{d\sigma(e^+e^- \to q\bar{q})}{dt} \cdot \frac{dt}{d\cos\theta},\tag{7.11}$$

or that

$$\frac{d\sigma(e^+e^- \to q\bar{q})}{dt} = \frac{2}{s} \cdot \frac{3}{2} \frac{\pi\alpha^2}{s} Q_q^2 \frac{2}{s^2}(u^2 + t^2) \tag{7.12}$$

$$= \frac{6\pi\alpha^2 Q_q^2}{s^2} \frac{u^2 + t^2}{s^2}.$$

These relationships and this formalism are exceptionally powerful. We can simply write down the cross section for the process $e^-q \to e^-q$ accounting for the change in direction of time. In terms of Feynman diagrams, we start from the process $e^+e^- \to q\bar{q}$ described by

$$\tag{7.13}$$

and transform it to the process $e^-q \to e^-q$:

$$\tag{7.14}$$

Rotating the direction of time by $90°$ in these diagrams is accounted for by exchanging the external momenta as

$$p_1 \to p_1, \qquad p_2 \to -p_4, \qquad p_3 \to p_3, \qquad p_4 \to -p_2. \tag{7.15}$$

In terms of the Mandelstam variables, this change corresponds to

$$s = (p_1 + p_2)^2 \longrightarrow (p_1 - p_4)^2 = t, \tag{7.16}$$
$$t = (p_1 - p_4)^2 \longrightarrow (p_1 + p_2)^2 = s,$$
$$u = (p_1 - p_3)^2 \longrightarrow (p_1 - p_3)^2 = u.$$

By changing the direction of time, we do not change the value of the Feynman diagram in this process. This is not necessarily true for a general $2 \to 2$ process, which we will discuss when studying the weak force in Chapter 10. To determine the cross section for the process $e^-q \to e^-q$, all we need to do is make the appropriate replacements for s, t, and u established above in the process $e^+e^- \to q\bar{q}$.

Making these replacements, we find

$$\frac{d\sigma(e^-q \to e^-q)}{dt} = \frac{2\pi\alpha^2 Q_q^2}{s^2} \frac{s^2 + u^2}{t^2}. \tag{7.17}$$

Note that some things changed while others did not. The factor of 3 accounting for color was removed because there is a quark in the initial state. The Feynman diagram of Eq. 7.14

is only non-zero if the color of the initial-state quark is identical to the color of the final-state quark; the photon cannot change quark colors. The overall factor of $1/s^2$ did not change in transforming from the process $e^+e^- \rightarrow q\bar{q}$ to $e^-q \rightarrow e^-q$. This factor came from the change-of-variables Jacobian from Eq. 7.10 and the factor in Fermi's Golden Rule that accounts for the de Broglie wavelength of the initial colliding particles. In the rest of the expression, we exchanged s, t, and u as specified in Eq. 7.16.

That we can so easily determine the cross section for processes with the same underlying Feynman diagrams by exchanging s, t, and u is called **crossing symmetry**. We have "crossed" the initial-state positron to a final-state electron and a final-state anti-quark to an initial-state quark. We say that the process $e^-q \rightarrow e^-q$ is the crossing of $e^+e^- \rightarrow q\bar{q}$ (and vice-versa). Crossing symmetry was what we identified in Section 6.1.2 as relating matrix elements by complex conjugation. Another way to say this is, for example, that the complex conjugate of an initial-state positron is a final-state electron. Complex conjugation relates particles to their anti-particles.

With a prediction for the cross section of $e^-q \rightarrow e^-q$, we have a new way to test the quark model! If we are able to collide electrons on quarks, then we can test the hypothesis that quarks are point particles; that is, they have no spatial extent. If quarks are point particles, then they should look the same, no matter what energy or wavelength we probe them with. So, is this true and how do we test it?

7.2 Deeply Inelastic Scattering

We need to figure out how to get a sample of quarks with which we can scatter electrons. The quark model itself can help us out here. While we can't produce quarks in isolation with which to collide electrons, we can collide electrons with protons. At sufficiently high energies, the electrons will have a de Broglie wavelength that is shorter than the Compton wavelength of the proton. At these energies, the electron resolves the constituent quarks, according to the quark model. Both electrons and quarks are electrically charged, and so this scattering process is mediated by photons.

In a schematic Feynman-like diagram, the process we are considering is:

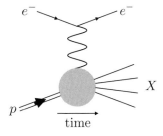

An electron is scattered off of a proton p and the final configuration consists of an electron and X, which is some collection of hadrons. We want to be completely inclusive on the final state (for reasons that will be clear shortly) and so the process we consider is $e^- p \to e^- + X$, where X is anything. If the energy of the electron changes in this interaction, then the collision is inelastic, and the proton explodes into other particles. This inelastic process is called **deeply inelastic scattering** or DIS.[4]

In DIS, the photon is not interacting with the proton, but rather with its constituents, the quarks. A better diagram for DIS might be:

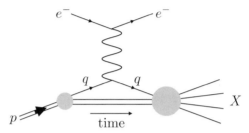

In this diagram, q represents a constituent quark of the proton. Let's break this diagram apart and understand its pieces to make predictions in this model. By the way, the model in which quarks are the "parts" of a proton and therefore govern high-energy interactions is called the **parton model**, introduced by Richard Feynman.[5]

Let's go to the center-of-mass frame for simplicity, though the analysis will be completely Lorentz invariant. Let the momentum of the initial and final electrons be k_e and k'_e, respectively, and the momentum of the initial proton be P. Then, the momentum flowing through the photon is[6]

$$q = k_e - k'_e. \tag{7.18}$$

Note that q is space-like:

$$q^2 = (k_e - k'_e)^2 = -2 k_e \cdot k'_e < 0, \tag{7.19}$$

for $k_e^2 = (k'_e)^2 = 0$. Often, we will denote $-q^2 \equiv Q^2$, which represents the momentum transferred from the electron to the proton. In the parton model picture, the electron actually interacts with an individual quark. What is the momentum of the quark? For a high-energy proton, the quark will just carry a fraction $x \in [0, 1]$ of the total proton momentum. This high-energy limit corresponds to when the proton energy E_p is much larger than the proton mass m_p: $m_p \ll E_p$. In this limit, we can also ignore the mass of the proton. Let's call the momentum of the initial and final quarks k_q and k'_q, respectively. Then,

$$k_q = xP, \qquad \text{and} \qquad k_e + k_q = k'_e + k'_q. \tag{7.20}$$

---

[4] DIS is more commonly referred to as "deep inelastic scattering" but this is a somewhat confusing phrase. The scattering process in DIS is deeply inelastic, i.e., the proton is exploded apart, and not deep (whatever that means) and inelastic.

[5] R. P. Feynman, "The behavior of hadron collisions at extreme energies," in *Proceedings of the 3rd International Conference on High Energy Collisions, Stony Brook, NY*, Gordon and Breach (1969), pp. 237–258; R. P. Feynman, "Very high-energy collisions of hadrons," Phys. Rev. Lett. **23**, 1415 (1969).

[6] Apologies for the repeated use of the letter q to denote both a momentum and a quark. Both notations are standard. From context it should be clear if we mean momentum or quark.

Then, the picture we have of DIS is:

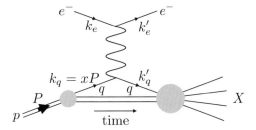

Now, with these kinematics, we need to calculate some things. First, there is the hard electron–quark scattering. We know the differential cross section for this. Exploiting crossing symmetry to relate the process $e^+e^- \to q\bar{q}$ to $e^-q \to e^-q$, the differential cross section is

$$\frac{d\sigma}{d\hat{t}} = \frac{2\pi\alpha^2 Q_q^2}{\hat{s}^2} \frac{\hat{s}^2 + \hat{u}^2}{\hat{t}^2}. \tag{7.21}$$

Here, we introduce $\hat{s}$, $\hat{t}$, and $\hat{u}$, which are the Mandelstam variables for the interacting partons. Note that

$$\hat{s} = (k_e + k_q)^2 = (k_e + xP)^2 = 2xk_e \cdot P = xs, \tag{7.22}$$

where s is the Mandelstam variable for e^-p scattering, approximating the proton as massless. Similarly,

$$\hat{t} = (k_e - k'_e)^2 = q^2 = -Q^2. \tag{7.23}$$

To define $\hat{u}$, we use the linear relationship between $\hat{s}$, $\hat{t}$, and $\hat{u}$:

$$\hat{u} = -\hat{s} - \hat{t} = -xs + Q^2. \tag{7.24}$$

Substituting in these expressions, we can write the differential cross section as

$$\frac{d\sigma(e^-q \to e^-q)}{dQ^2} = 2\pi\alpha^2 Q_q^2 \frac{1 + \left(1 - \frac{Q^2}{xs}\right)^2}{Q^4}. \tag{7.25}$$

Though it seems to depend on the momenta of quarks from the proton, the ratio $\hat{u}/\hat{s}$ is actually observable. To measure this ratio, we only need to know the electron and proton momenta. To see this, note that

$$\frac{\hat{u}}{\hat{s}} = \frac{-2xk'_e \cdot P}{2xk_e \cdot P} = -\frac{k'_e \cdot P}{k_e \cdot P}. \tag{7.26}$$

Then, the combination that appears in the differential cross section,

$$1 - \frac{Q^2}{xs} = 1 + \frac{\hat{t}}{\hat{s}} = -\frac{\hat{u}}{\hat{s}} = \frac{k'_e \cdot P}{k_e \cdot P}, \tag{7.27}$$

is also measurable. This is an extremely important point: though DIS involves interactions between electron and constituent quarks in the proton, all we need to do to completely characterize the scattering is to set the center-of-mass collision energy $E_{cm}^2 = s$ and measure the scattered electron's momentum, k'_e.

Okay, we're getting close. The differential cross section of Eq. 7.25 isn't the whole story; this is just the subprocess $e^-q \to e^-q$ scattering in e^-p collisions. We need to get the quark out of the proton in the first place. In the parton model, the fraction x of the proton's momentum that the quark carries has a probability distribution $f_q(x)$, which is called a **parton distribution function**, or pdf. The probability P_q of extracting a quark parton from a proton with momentum fraction in the range $[x, x + dx]$ is

$$P_q([x, x + dx]) = f_q(x)\, dx\,. \tag{7.28}$$

As written, this is independent of the exchanged momentum Q^2, which is the assumption we will make now. Q^2-independence means that regardless of the energy or wavelength at which they are probed, quarks look the same and have the same pdf. We will study the physical consequences of this assumption and its generalization for the properties of quarks in the proton in the next section.

With these assumptions, the cross section differential in x and Q^2 for a quark q is then

$$\frac{d^2\sigma(e^-q \to e^-q)}{dx\, dQ^2} = 2\pi\alpha^2 Q_q^2 f_q(x) \frac{1 + \left(1 - \frac{Q^2}{xs}\right)^2}{Q^4}\,. \tag{7.29}$$

We can introduce the variable y for which

$$y = 1 + \frac{\hat{u}}{\hat{s}} = 1 + \frac{-\hat{s} - \hat{t}}{\hat{s}} = \frac{Q^2}{xs}\,. \tag{7.30}$$

That is, $Q^2 = xys$. Changing variables from Q^2 to y, the differential cross section becomes

$$\frac{d^2\sigma(e^-q \to e^-q)}{dx\, dy} = 2\pi\alpha^2 s Q_q^2 x f_q(x) \frac{1 + (1 - y)^2}{Q^4}\,. \tag{7.31}$$

The quark or parton model predicts multiple quarks in the proton, so we need to sum over all of them to get the complete cross section:

$$\frac{d^2\sigma}{dx\, dy} = 2\pi\alpha^2 s \left[\sum_{\text{quarks } q} Q_q^2 x f_q(x)\right] \frac{1 + (1 - y)^2}{Q^4} \equiv 2\pi\alpha^2 s F_2(x) \frac{1 + (1 - y)^2}{Q^4}\,. \tag{7.32}$$

Here, $F_2(x)$ is called a **form factor**, which in the parton model is independent of Q^2.

The final thing to note is that the momentum fraction x of the quark is actually observable. While DIS is inelastic scattering because the proton explodes apart, the beautiful thing about the parton model is that the $e^-q \to e^-q$ subprocess is actually elastic scattering. Therefore, the final-state quark is on-shell, which enables us to solve for x:

$$(k_q')^2 = 0 = (q + xP)^2 = q^2 + 2xq \cdot P\,. \tag{7.33}$$

That is, the momentum fraction x is

$$x = \frac{Q^2}{2q \cdot P} = \frac{-k_e \cdot k_e'}{(k_e - k_e') \cdot P}\,. \tag{7.34}$$

So, everything in the differential cross section of Eq. 7.32 is observable!

Again, the parton model assumes that the form factor $F_2(x)$ is independent of Q^2, which can be directly tested. In the late 1960s and early 1970s, experiments at Stanford Linear

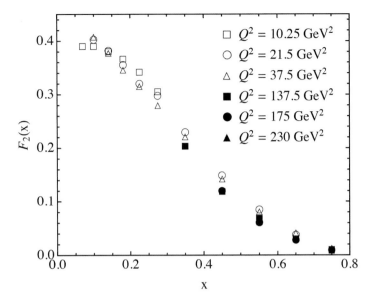

Fig. 7.1 Measured form factor $F_2(x)$ from $\mu^- p$ deeply inelastic scattering data collected by the BCDMS experiment. The data range over more than a decade in Q^2, and yet to good approximation lie on a universal curve. This is a prediction of Bjorken scaling. The data come from A. C. Benvenuti *et al.* [BCDMS Collaboration], "A high statistics measurement of the proton structure functions $F_2(x, Q^2)$ and R from deep inelastic muon scattering at high Q^2," Phys. Lett. B **223**, 485 (1989).

Accelerator Center (SLAC) and elsewhere demonstrated to good approximation that the form factor was independent of Q^2.[7] This feature of the parton model is called **Bjorken scaling**, after James "BJ" Bjorken.[8] Bjorken scaling is evidence for point-like quarks: if quarks had an intrinsic size then there would be an energy scale or wavelength at which this size could be resolved. This would then produce strong dependence of the form factor $F_2(x)$ on Q^2.

Figure 7.1 displays data that demonstrate Bjorken scaling. Here, the form factor $F_2(x)$ is plotted as a function of the momentum fraction x in data from deeply inelastic scattering of muons off of protons in the Bologna–CERN–Dubna–Munich–Saclay (BCDMS) experiment which was located at CERN. In an experiment, the form factor $F_2(x)$ can be determined by measuring the cross section for DIS and then dividing the measurement by the analytic form of the cross section that is independent of x from Eq. 7.32. In this plot, measurements of the form factor over a wide range of Q^2 values agree remarkably well, demonstrating to a large degree that the data lie on a universal curve. However,

[7] The number of experimental results that led to confirmation of the parton model is enormous. Reviews of the experiments and their interpretation for other models of the proton at the time can be found in J. I. Friedman and H. W. Kendall, "Deep inelastic electron scattering," Ann. Rev. Nucl. Part. Sci. **22**, 203 (1972). A nice historical review of the discovery of quarks is E. M. Riordan, "The discovery of quarks," Science **256**, 1287 (1992).

[8] J. D. Bjorken, "Current algebra at small distances," in *Selected Topics in Particle Physics: Proceedings of the International School of Physics "Enrico Fermi,"* Course XLI, Academic Press (1968), pp. 55–81; J. D. Bjorken, "Asymptotic sum rules at infinite momentum," Phys. Rev. **179**, 1547 (1969).

at higher Q^2, it is apparent that the data differ, especially at larger values of x. Bjorken scaling is therefore only an approximate description of the internal dynamics of the proton. As we will discuss in Chapter 9, the deviations from Bjorken scaling are exactly predicted in QCD.

7.2.1 Physical Interpretation of Bjorken Scaling

The prediction and subsequent experimental validation of Bjorken scaling was a major triumph of the parton model. However, the weak observed dependence of the form factor $F_2(x)$ on momentum Q^2 in Fig. 7.1, for example, illustrates that there's something else going on. The observed Q^2 dependence is still suggestive of point-like quarks because we never observe a value of Q^2 beyond which the form factor dramatically changes. Here, we consider the interpretation and consequences of Bjorken scaling and attempt to understand how quarks can still be point particles while at the same time violating Bjorken scaling.

First, let's attempt to understand Bjorken scaling from a different perspective. Bjorken scaling is the statement that the form factor $F_2(x, Q^2)$ is independent of Q^2:

$$F_2(x, Q^2) \underset{\text{Bjorken}}{=} F_2(x). \tag{7.35}$$

Earlier, we interpreted this as a form factor or pdf that is independent of the wavelength that probes the quark in the proton, and therefore the quark is a point particle. How do we see this conclusion more precisely? As of now, we have expressed the form factor in momentum space; that is, as a function of the momentum q of a photon that probes the quark. This isn't the most natural space in which to determine if a particle is point-like; indeed, we would rather express the form factor in position space. The momentum space representation and position space representation are related by a Fourier transformation. That is, to determine the form factor in position space, we Fourier transform the expression of the form factor in momentum space.

The form factor in position space $\widetilde{F}_2(x, r)$ is therefore

$$\widetilde{F}_2(x, r) = \int \frac{d^4q}{(2\pi)^4} F_2(x, Q^2) e^{iq \cdot r}, \tag{7.36}$$

where q is the four-momentum of the intermediate photon in DIS and r is the four-vector of position conjugate to q. As earlier, Q^2 is the magnitude of the square of the four-vector q: $q^2 = -Q^2 < 0$. Because q is space-like, we can therefore boost to a frame in which $q_0 = 0$. Then, we can express the momentum q as

$$q = (0, \vec{q}). \tag{7.37}$$

This is called the **Breit frame**. Then, in this Breit frame, the Fourier transform that we have to do is

$$\widetilde{F}_2(x, |\vec{r}|) = \int \frac{d^3q}{(2\pi)^3} F_2(x, Q^2) e^{-i\vec{q} \cdot \vec{r}}, \tag{7.38}$$

where $|\vec{r}|$ is now just the magnitude of the vector of spatial position.

Now, assuming Bjorken scaling, we can evaluate this Fourier transform integral to determine the form factor in position space. With the Bjorken scaling Q^2-independence $F_2(x, Q^2) = F_2(x)$, we have

$$\widetilde{F}_2(x, |\vec{r}|) = \int \frac{d^3q}{(2\pi)^3} F_2(x) e^{-i\vec{q}\cdot\vec{r}}. \tag{7.39}$$

To evaluate this integral, we can express it in spherical coordinates. It then becomes

$$\widetilde{F}_2(x, |\vec{r}|) = \int_0^\infty d|\vec{q}| \, |\vec{q}|^2 \int_{-1}^1 d\cos\theta \int_0^{2\pi} d\phi \frac{F_2(x)}{(2\pi)^3} e^{-i|\vec{q}||\vec{r}|\cos\theta}. \tag{7.40}$$

The integral over ϕ is just 2π and the integral over $\cos\theta$ can be done by relabeling $\cos\theta = u$. Doing these integrals, we find

$$\widetilde{F}_2(x, |\vec{r}|) = \frac{F_2(x)}{2\pi^2} \frac{1}{|\vec{r}|} \int_0^\infty d|\vec{q}| \, |\vec{q}| \sin(|\vec{q}||\vec{r}|). \tag{7.41}$$

This remaining integral can be done by noting that it is related to the definition of the δ-function:

$$\int_0^\infty d|\vec{q}| \, |\vec{q}| \sin(|\vec{q}||\vec{r}|) = -\frac{d}{d|\vec{r}|} \int_0^\infty d|\vec{q}| \, \cos(|\vec{q}||\vec{r}|) = -\frac{\pi}{2} \frac{d}{d|\vec{r}|} \delta(|\vec{r}|). \tag{7.42}$$

This follows from the definition of the δ-function,

$$\int_{-\infty}^\infty du \, \cos(uv) = 2\pi\delta(v). \tag{7.43}$$

The integral over $|\vec{q}|$ only ranges over positive $|\vec{q}|$ and cosine is an even function, so we divide this integral by 2. Further, the radius $|\vec{r}|$ is strictly positive, so we divide by 2 again. That is, the Bjorken-scaling form factor in position space is

$$\widetilde{F}_2(x, |\vec{r}|) = -\frac{1}{4\pi} F_2(x) \frac{1}{|\vec{r}|} \frac{d}{d|\vec{r}|} \delta(|\vec{r}|). \tag{7.44}$$

The derivative of the δ-function seems scary, but it's actually very simple. One can show that

$$\frac{d}{d|\vec{r}|} \delta(|\vec{r}|) = -\frac{\delta(|\vec{r}|)}{|\vec{r}|}. \tag{7.45}$$

Finally, we find that

$$\widetilde{F}_2(x, |\vec{r}|) = F_2(x) \frac{1}{4\pi|\vec{r}|^2} \delta(|\vec{r}|). \tag{7.46}$$

This rather weird functional form for the radial dependence is actually the δ-function expressed in spherical coordinates. Note that it is infinite when $|\vec{r}| = 0$ and zero away from $|\vec{r}| = 0$, and it integrates to 1:

$$\int d|\vec{r}| \, |\vec{r}|^2 \, d\cos\theta \, d\phi \frac{1}{4\pi|\vec{r}|^2} \delta(|\vec{r}|) = 1. \tag{7.47}$$

The statement of Bjorken scaling is that the quark has no spatial extent: it is located exclusively at $\vec{r} = 0$. Another way to say this is that, because the form factor is a δ-function of $|\vec{r}|$, if you are displaced at all from $\vec{r} = 0$, then you don't even know that the quark is there. That is, Bjorken scaling additionally assumes that the quark is a free particle at high energies and does not interact with other particles. If the quark did interact with other particles, then you could know about its presence even if you were displaced from $\vec{r} = 0$. This observation that quarks apparently become approximately free, non-interacting particles at high energies is incredibly intriguing. We will provide an explanation for it in the next chapter. Here, we will attempt to understand more generally the consequences for the form factor, only assuming that quarks are point particles.

The assumption that the form factor for point-particle quarks is independent of Q in momentum space or a δ-function in position space is very restrictive and in no way general. Indeed, we are familiar with many examples of point particles that exhibit extended spatial distributions. For example, the electric potential of a point charge q is, in natural units,

$$V(|\vec{r}|) = \frac{q}{4\pi |\vec{r}|} . \tag{7.48}$$

This clearly has support away from $\vec{r} = 0$, and yet the charge is located exclusively at $\vec{r} = 0$. The influence of the charge is allowed to extend away from $\vec{r} = 0$ because the charge interacts electromagnetically. That is, a point charge is not a free particle, and interacts with the photon which is responsible for the long-distance interactions of charges.

Note, however, that the electric potential cannot be an arbitrary function of distance $|\vec{r}|$. Because the charge is still a point, if you zoom in and get closer and closer to the charge, you must still see the $1/|\vec{r}|$ functional form of the potential. This is essentially a consequence of Gauss's law: because you can never zoom in to actually see the size of the point charge, the electric potential or electric field any finite distance away must always be of this $1/|\vec{r}|$ form. Another way to say this is that the $1/|\vec{r}|$ potential is **scale invariant**: if you scale distances as $|\vec{r}| \to \lambda |\vec{r}|$, the $1/|\vec{r}|$ potential remains unchanged:

$$V(\lambda |\vec{r}|) = \frac{q}{4\pi \lambda |\vec{r}|} = \lambda^{-1} V(|\vec{r}|) , \tag{7.49}$$

for any $\lambda > 0$. So, for the form factor $\widetilde{F}_2(x, |\vec{r}|)$ to represent point-like quarks, it doesn't need to be a δ-function of position, just scale invariant. We'll discuss this in much more detail and precisely define what we mean by scale transformations and scale invariance in Chapter 9. For now, however, this analogy with the electrodynamics of point charges is sufficient for our purposes here.

We can see how this scale transformation works with the form factor which obeys Bjorken scaling. In position space, this form factor scales as

$$\widetilde{F}_2(x, \lambda |\vec{r}|) = F_2(x) \frac{1}{4\pi(\lambda |\vec{r}|)^2} \delta(\lambda |\vec{r}|) = \lambda^{-3} \widetilde{F}_2(x, |\vec{r}|) . \tag{7.50}$$

Indeed, this is scale invariant, as expected and so describes a point-particle quark. A much more general expression of the spatial dependence of the form factor that is scale invariant and so can describe a point particle quark is

$$\widetilde{F}_2(x, |\vec{r}|) = \frac{\epsilon}{4\pi} \frac{r_0^{-\epsilon}}{|\vec{r}|^{3-\epsilon}} F_2(x) , \tag{7.51}$$

for some $\epsilon > 0$ where r_0 is some characteristic length scale. This integrates to 1 on $r \in [0, r_0]$ in three-dimensional spherical coordinates. While this is not the most general form factor possible, it illustrates concretely how Bjorken scaling can be violated. We'll see in a bit that this ϵ controls the strength of the interactions of the quark; as $\epsilon \to 0$, the quark becomes a free particle. Note that indeed this form factor is scale invariant to describe a point particle:

$$\widetilde{F}_2(x, \lambda|\vec{r}|) = \frac{\epsilon}{4\pi} \frac{r_0^{-\epsilon}}{(\lambda|\vec{r}|)^{3-\epsilon}} F_2(x) = \lambda^{-3+\epsilon} \widetilde{F}_2(x, |\vec{r}|). \tag{7.52}$$

That is, the functional dependence of the form factor $\widetilde{F}_2(x, |\vec{r}|)$ with $\vec{r}$ is independent of the magnification with which you look at it. What does this form factor look like in momentum space?

Inverse Fourier transforming back to momentum space, the form factor is

$$F_2(x, Q^2) = \int d^3r \, \widetilde{F}_2(x, |\vec{r}|) e^{i\vec{q}\cdot\vec{r}} = \int d^3r \, \frac{\epsilon}{4\pi} \frac{r_0^{-\epsilon}}{|\vec{r}|^{3-\epsilon}} F_2(x) e^{i\vec{q}\cdot\vec{r}} \tag{7.53}$$

$$= \frac{\epsilon r_0^{-\epsilon}}{4\pi} F_2(x) \int_0^\infty d|\vec{r}| \, |\vec{r}|^2 \int_{-1}^1 d\cos\theta \int_0^{2\pi} d\phi \, |\vec{r}|^{-3+\epsilon} e^{i|\vec{q}||\vec{r}|\cos\theta}$$

$$= \frac{\epsilon}{1-\epsilon} \cos\left(\frac{\pi\epsilon}{2}\right) \Gamma(\epsilon) F_2(x) (r_0^2 Q^2)^{-\epsilon/2}.$$

In this expression, we identified $Q = |\vec{q}|$, the magnitude of the momentum vector. Here, $\Gamma(\epsilon)$ is Euler's gamma function, defined as

$$\Gamma(x) = \int_0^\infty dt \, x^{t-1} e^{-t}. \tag{7.54}$$

This tells us that a scale-invariant spatial distribution corresponds to this ϵ power-law distribution in momentum space. However, as written, this doesn't immediately manifest the Bjorken scaling limit. To see that, we can Taylor expand this result in ϵ. This will justify the earlier claim that ϵ controls the strength of interactions of the quarks. Taylor expanding in ϵ the result of the inverse Fourier transformation from Eq. 7.53, we find

$$F_2(x, Q^2) = \frac{\epsilon}{1-\epsilon} \cos\left(\frac{\pi\epsilon}{2}\right) \Gamma(\epsilon) F_2(x) \sum_{n=0}^\infty \frac{(-\epsilon)^n}{2^n n!} \log^n(r_0^2 Q^2). \tag{7.55}$$

As $\epsilon \to 0$, the prefactor $\frac{\epsilon}{1-\epsilon} \cos\left(\frac{\pi\epsilon}{2}\right) \Gamma(\epsilon) \to 1$ and so the Bjorken scaling limit is the $\epsilon \to 0$ limit:

$$F_2(x, Q)|_{\epsilon\to0} = F_2(x). \tag{7.56}$$

Still consistent with quarks being point particles, Bjorken scaling can be broken by logarithms of energy scale Q^2, but there must be an infinite sum over all powers of logarithms. For $\epsilon > 0$, these logarithms are important, and thus through their interactions, quarks have non-trivial spatial distributions.

We will see in Section 9.2 how this infinite sum over powers of logarithms is done in QCD and what this ϵ actually corresponds to. In the next section, we will address the issue of quark interactions directly. In this exercise of studying Bjorken scaling and its violation, we showed that the $\epsilon \to 0$ limit corresponds to Bjorken scaling in momentum space, Eq. 7.56. In position space, that relationship would correspond to

$$\lim_{\epsilon \to 0} \frac{\epsilon}{4\pi} \frac{r_0^{-\epsilon}}{|\vec{r}|^{3-\epsilon}} = \frac{1}{4\pi |\vec{r}|^2} \delta(|\vec{r}|). \tag{7.57}$$

What does this mean? And what consequence does this have for the radial dependence when $\epsilon > 0$? You'll study that in Exercise 7.1 at the end of this chapter. Before we continue, let's apply crossing symmetry and parton distributions to understand a fundamental process at hadron colliders, called Drell–Yan.

Example 7.1 The **Drell–Yan process** is the inclusive scattering of protons into leptons, like $pp \to e^+ e^- + X$. It is named after Sidney Drell and Tung-Mow Yan for their analysis of it in 1970.[9] It is an extremely important process in collisions at the LHC as it enables a very direct measurement of the parton distributions. What is the differential cross section for Drell–Yan?

Solution

For the Drell–Yan process $pp \to e^+ e^-$, the fundamental interaction is the scattering of individual quarks and anti-quarks within the protons, $q\bar{q} \to e^+ e^-$. Let's draw a picture to see what we're dealing with in Drell–Yan:

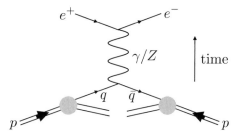

Here, in the initial state, there are two protons that are collided at high energy. At sufficiently high energy (when the de Broglie wavelength of the proton is much smaller than its Compton wavelength), the protons explode apart and their constituents, the quarks or anti-quarks, interact directly. For Drell–Yan, a quark from one proton and an anti-quark from the other proton annihilate into a photon (or Z boson) which then splits into an electron and positron.[10]

The underlying Feynman diagram for this process can be expressed in terms of the Mandelstam variables for the partons, $\hat{s}$, $\hat{t}$, and $\hat{u}$. In this process, we denote the momenta as

---

[9] S. D. Drell and T. M. Yan, "Massive lepton pair production in hadron–hadron collisions at high energies," Phys. Rev. Lett. **25**, 316 (1970), Erratum: [Phys. Rev. Lett. **25**, 902 (1970)].

[10] Interesting fact: The largest Feynman diagram in the world is on the floor of the physics building at the University of Oregon. The Feynman diagram is for the Drell–Yan process. Davison Soper, a professor at the University of Oregon, demonstrated that the Drell–Yan process factorizes, enabling the discussion of this chapter. See J. C. Collins, D. E. Soper and G. F. Sterman, "Transverse momentum distribution in Drell–Yan Pair and W and Z boson production," Nucl. Phys. B **250**, 199 (1985).

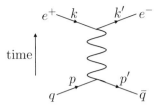

and so

$$\hat{s} = (p + p')^2 = (k + k')^2, \tag{7.58}$$

$$\hat{t} = (p - k)^2 = (p' - k')^2, \tag{7.59}$$

$$\hat{u} = (p - k')^2 = (p' - k)^2. \tag{7.60}$$

The total cross section for $e^+ e^- \to q\bar{q}$ with an intermediate photon in terms of the center-of-mass energy $\hat{s} = E_{\text{cm}}^2$ is

$$\sigma(e^+ e^- \to q\bar{q}) = 4 \frac{\pi\alpha^2}{\hat{s}} \sum_{q \text{ quarks}} Q_q^2. \tag{7.61}$$

This can be crossed into the initial state by accounting for averaging, rather than summing, over color:

$$\sigma(q\bar{q} \to e^+ e^-) = \frac{4}{9} \frac{\pi\alpha^2}{\hat{s}} \sum_{q \text{ quarks}} Q_q^2. \tag{7.62}$$

Averaging over quark colors requires dividing by 9 because there are three possible colors and three possible anti-colors for the initial quark and anti-quark, respectively. Each color–anti-color pair is equally likely and there are nine such pairs. We still need to pull the quarks out of the protons, so let's do that now.

Let's say we pull the quark out of proton 1 with momentum fraction x_1 of proton 1's momentum, while the anti-quark comes from proton 2 with momentum fraction x_2. If we call the momentum of proton 1 P_1 (and P_2 for proton 2) then we define the center-of-mass proton collision energy as

$$s \equiv (P_1 + P_2)^2 = 2P_1 \cdot P_2, \tag{7.63}$$

so that

$$\hat{s} = x_1 x_2 s = 2(x_1 P_1) \cdot (x_2 P_2). \tag{7.64}$$

Here, we assume that we are working at sufficiently high energy so as to ignore the proton mass. The probability distributions of the momentum fractions x_1 and x_2 are defined by the parton distribution functions $f_q(x)$. The cross section differential in both momentum fractions can therefore be written as

$$\frac{d^2\sigma(q\bar{q} \to e^+ e^-)}{dx_1 \, dx_2} = \frac{4}{9} \frac{\pi\alpha^2}{\hat{s}} \sum_{q \text{ quarks}} Q_q^2 f_q(x_1) f_{\bar{q}}(x_2), \tag{7.65}$$

or, written as the cross section for $pp \to e^+e^-$,

$$\frac{d^2\sigma(pp \to e^+e^-)}{dx_1\,dx_2} = \frac{4\,\pi\alpha^2}{9}\frac{1}{s}\sum_{q \text{ quarks}} Q_q^2 \frac{f_q(x_1)}{x_1}\frac{f_{\bar{q}}(x_2)}{x_2}\,. \tag{7.66}$$

Here, q represents all possible quarks (or anti-quarks) that can be pulled out of the proton.

This is interesting, but we can't directly measure the momentum fractions x_1 and x_2. So we want to re-express them in terms of things we can measure. Two useful quantities with which to express the momentum fractions are the invariant mass Q^2 and the rapidity y of the final-state electron–positron pair. The invariant mass is just $\hat{s}$:

$$Q^2 = \hat{s} = x_1 x_2 s\,. \tag{7.67}$$

The rapidity y is defined as

$$y = \frac{1}{2}\log\frac{E+p_z}{E-p_z}\,, \tag{7.68}$$

and can be determined by measuring the sum of the energies and z-component of momenta of the e^+ and e^-. The total energy of the e^+e^- pair is just the sum of the energies of the quark and anti-quark:

$$E = x_1\frac{E_{\text{cm}}}{2} + x_2\frac{E_{\text{cm}}}{2} = \frac{x_1+x_2}{2}E_{\text{cm}}\,. \tag{7.69}$$

Similarly, the p_z of the e^+e^- pair is equal to the p_z of the quark and anti-quark by momentum conservation:

$$p_z = x_1\frac{E_{\text{cm}}}{2} - x_2\frac{E_{\text{cm}}}{2} = \frac{x_1-x_2}{2}E_{\text{cm}}\,. \tag{7.70}$$

Here, we have aligned the momenta of the protons along the $\hat{z}$-axis. Now, from these expressions we can calculate the rapidity:

$$y = \frac{1}{2}\log\frac{E+p_z}{E-p_z} = \frac{1}{2}\log\frac{\frac{x_1+x_2}{2}E_{\text{cm}} + \frac{x_1-x_2}{2}E_{\text{cm}}}{\frac{x_1+x_2}{2}E_{\text{cm}} - \frac{x_1-x_2}{2}E_{\text{cm}}} = \frac{1}{2}\log\frac{x_1}{x_2}\,. \tag{7.71}$$

To write the cross section in terms of Q^2 and y we can solve for x_1 and x_2 as

$$x_1 = \sqrt{\frac{Q^2}{s}}e^y\,, \qquad\qquad x_2 = \sqrt{\frac{Q^2}{s}}e^{-y}\,. \tag{7.72}$$

We also need the Jacobian factor J from the change of variables. To do this, we calculate all derivatives and then take the determinant of the derivative matrix:

$$J = \begin{vmatrix} \frac{\partial x_1}{\partial Q^2} & \frac{\partial x_1}{\partial y} \\ \frac{\partial x_2}{\partial Q^2} & \frac{\partial x_2}{\partial y} \end{vmatrix} = \begin{vmatrix} \frac{e^y}{2\sqrt{sQ^2}} & \sqrt{\frac{Q^2}{s}}e^y \\ \frac{e^{-y}}{2\sqrt{sQ^2}} & -\sqrt{\frac{Q^2}{s}}e^{-y} \end{vmatrix} \tag{7.73}$$

$$= \frac{1}{s}\,.$$

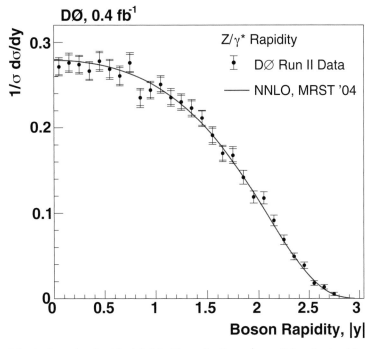

Distribution of the absolute value of rapidity $|y|$ of the Z boson in $p\bar{p} \to e^+e^-$ collisions. Reprinted figure with permission from V. M. Abazov *et al.* [D0 Collaboration], Phys. Rev. D **76**, 012003 (2007). Copyright 2007 by the American Physical Society.

Putting it all together, we predict that the differential cross section is

$$\frac{d^2\sigma(pp \to e^+e^-)}{dQ^2\, dy} = \frac{d^2\sigma(pp \to e^+e^-)}{dx_1\, dx_2} \cdot J \tag{7.74}$$

$$= \frac{4}{9}\frac{\pi\alpha^2}{Q^2 s} \sum_{q\ \text{quarks}} Q_q^2 f_q\left(\sqrt{\frac{Q^2}{s}}e^y\right) f_{\bar{q}}\left(\sqrt{\frac{Q^2}{s}}e^{-y}\right) .$$

This is an incredible result! Rapidity dependence only enters in the parton distributions, and therefore a measurement of the rapidity is very sensitive to the parton distribution functions.

Figure 7.2 shows the rapidity distribution of the final-state e^+e^- pair as measured in the D0 experiment at the Tevatron, located at Fermilab outside of Chicago. The Tevatron accelerated and collided protons on anti-protons and measured the momenta of electron–positron pairs produced in the process $p\bar{p} \to e^+e^- + X$, where X is anything else. Because protons and anti-protons are anti-particles of one another, it is easiest to pull a quark out of the proton and an anti-quark out of the anti-proton. This configuration then means that the quark and anti-quark parton distributions f_q and $f_{\bar{q}}$ that appear in Eq. 7.74 are identical. Additionally, D0 looked for e^+e^- pairs with invariant masses near the Z boson mass, which enables an interpretation of this plot as the rapidity of a Z boson produced in collision. From these data, you will be able to extract the functional form of the parton distribution function $f_q(x)$ in Exercise 7.7.

7.3 Three-Jet Events

In our study thus far of attempting to understand the quarks and their interactions, we have learned a huge amount:

- Particles composed of quarks (hadrons) arrange themselves into irreducible representations of flavor symmetry groups (e.g., isospin).
- There seem to be three "colors" of each flavor of quark, which is necessary to account for the measured value of R, the ratio of the cross section for $e^+e^- \to$ hadrons to that of $e^+e^- \to \mu^+\mu^-$.
- The parton model predicts that quarks are point particles: their description is independent of the energy or wavelength that probes them.
- At high energies, quarks become free particles and the form factor $F_2(x)$ exhibits Bjorken scaling.

While these observations are extremely informative, they also seem to raise more questions:

- What is the force that binds the quarks together in hadrons?
- What exactly is color? What are the transformations that relate the three colors?
- Related to this, if color is conserved, what is the corresponding symmetry, by Noether's theorem?

In this section, we will posit part of a solution to these problems, and in the next chapter we will introduce the theory that answers all of them: quantum chromodynamics.

Box 7.1   **Historical Profile: Mary Gaillard and Sau Lan Wu**

Mary Gaillard is a professor at the University of California, Berkeley, and became that school's first tenured female physics professor in 1982. In the 1970s, she contributed extensively to the physics of heavy quarks, notably predicting the mass of the charm quark prior to its discovery.[11] Additionally, Gaillard worked on elucidating the phenomena of the Higgs boson long before its discovery, as well as on grand unification, the idea that at sufficiently high energies the fundamental forces unify into a single force.[12] In the late 1980s, she led the theoretical physics group at Lawrence Berkeley Laboratory.

Sau Lan Wu was a member of the TASSO experiment at the PETRA accelerator located at DESY, a German particle physics laboratory in Hamburg. Wu was tasked with preparing the analysis for searching

---

[11] M. K. Gaillard and B. W. Lee, "Rare decay modes of the K-mesons in gauge theories," Phys. Rev. D **10**, 897 (1974).

[12] A. J. Buras, J. R. Ellis, M. K. Gaillard and D. V. Nanopoulos, "Aspects of the grand unification of strong, weak and electromagnetic interactions," Nucl. Phys. B **135**, 66 (1978).

for the gluon in these $e^+e^- \rightarrow q\bar{q}g$ events. She wrote a paper describing the identification technique in 1979,[13] and the gluon was discovered shortly thereafter.[14] In addition to gluons, Wu was also a member of the collaborations that discovered the charm quark (in 1974) and the Higgs boson (in 2012). Since 1977, she has been a professor at the University of Wisconsin-Madison.

More information about the time of the discovery of the gluon can be found in a review article by John Ellis: J. Ellis, "The discovery of the gluon," Int. J. Mod. Phys. A **29**, no. 31, 1430072 (2014) [arXiv:1409.4232 [hep-ph]].

7.3.1 The Glue That Binds the Proton

Our goal in this section is quite restricted: we will just attempt to make predictions for what the force carrier is that binds quarks together in hadrons. We call this force carrier the **gluon** and we want a hypothesis for what properties it has that we can test in experiment. If the gluon is responsible for binding quarks, then we might postulate that it is something like the photon. Electromagnetism, through its force carrier the photon, is responsible for binding the proton and the electron into hydrogen. So, it is feasible that the gluon is a spin-1 particle and massless, just like the photon. Let's work with this assumption and see what the predictions are.

The force that the gluon carries cannot be electromagnetism; that is, the gluon is not the photon. Perhaps this is obvious, but it is a very important point. Within the quark model, the argument for this is simple. There exist bound states of quarks for which all quarks have the same electric charge; for example, the Ω^- baryon that is the bound state of three strange quarks. It is not possible for three particles of the same electric charge to form a bound state electromagnetically. Additionally, this means that the leptons, particles like the electron or muon, do not feel the force carried by the gluon.

So, we now have a model for the strong force, the force that binds hadrons together. It is carried by the gluon, which only talks directly to the quarks, and not to the leptons. What prediction does this simple model make?

Let's go back to our good friend e^+e^- collisions, in which we first found evidence for quarks. Can we "see" a gluon in e^+e^- collisions? Because gluons do not talk directly to electrons or positrons, we cannot just produce gluons directly from e^+e^- collisions. Within our assumptions, gluons can be radiated from quarks, just as photons can be radiated from accelerating electric charges. The simplest process in which a gluon g can be produced in e^+e^- collisions is then

$$e^+e^- \rightarrow q\bar{q}g \,. \tag{7.75}$$

---

[13] S. L. Wu and G. Zobernig, "A method of three jet analysis in e^+e^- annihilation," Z. Phys. C **2**, 107 (1979).

[14] R. Brandelik *et al.* [TASSO Collaboration], "Evidence for planar events in e^+e^- annihilation at high energies," Phys. Lett. **86B**, 243 (1979); D. P. Barber *et al.*, "Discovery of three jet events and a test of quantum chromodynamics at PETRA energies," Phys. Rev. Lett. **43**, 830 (1979); C. Berger *et al.* [PLUTO Collaboration], "Evidence for gluon bremsstrahlung in e^+e^- annihilations at high energies," Phys. Lett. **86B**, 418 (1979); W. Bartel *et al.* [JADE Collaboration], "Observation of planar three jet events in e^+e^- annihilation and evidence for gluon bremsstrahlung," Phys. Lett. **91B**, 142 (1980).

We will calculate the Feynman diagrams and the cross section for this process. The prediction of the gluon to be discovered in such a process was first emphasized by John Ellis, Mary Gaillard, and Graham Ross in 1976.[15]

By the rules of Feynman diagrams, we need to sum over all possible diagrams we can draw, consistent with the interactions that we define. For the process $e^+e^- \to q\bar{q}g$, there are two diagrams that we must sum together:

$$\mathcal{M}(e^+e^- \to q\bar{q}g) = \qquad\qquad\qquad\qquad\qquad\qquad\qquad (7.76)$$

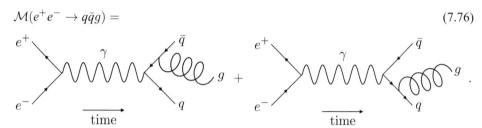

That is, the final-state gluon could have been emitted off of either the final-state quark or anti-quark, and there is no measurement we can perform to distinguish them. Here, we identify the photon and the gluon by their symbols: the photon is a wavy line, while the gluon is a curly line (like a spring).

What do we need to know to evaluate this diagram? We know how to evaluate just the stuff involving the photon and external fermions from our analysis in Chapter 6. However, there are two new things in these diagrams that are unfamiliar from $e^+e^- \to q\bar{q}$ scattering: what the wavefunction of an external gluon is, and what an intermediate fermion (quark) propagator is. First the external gluon wavefunction.

7.3.2 External Gluon Wavefunction

Because we are assuming that the gluon is a spin-1 massless particle, its wavefunction should satisfy the same equation of motion as the photon. Then, the external wavefunction of a gluon will be represented by a polarization four-vector ϵ, with the requirement that $p \cdot \epsilon = 0$, for gluon momentum p. In Section 2.2.3, we constructed the polarization vectors for right- and left-handed circular polarization, for momentum aligned along the $+\hat{z}$-axis. It is easy enough to rotate these polarization vectors to correspond to arbitrary momentum, but here we will present another, more useful, representation of the gluon polarization vector. We won't derive the results that follow, but will provide consistency checks to verify that it has the properties that a polarization vector should have.

The right-handed circularly polarized (right-handed helicity) gluon polarization can be written as

$$\epsilon_R^\mu(p) = -\frac{1}{\sqrt{2}} \frac{u_R^\dagger(r)\sigma^\mu u_R(p)}{u_R^\dagger(r)u_L(p)}, \qquad\qquad (7.77)$$

for a gluon with momentum p. Here, the u and $u^\dagger$ objects are the two-component spinors with the corresponding helicity assignment. r is an arbitrary massless momentum four-vector that represents the gauge freedom of the gluon's vector potential. The

[15] J. R. Ellis, M. K. Gaillard and G. G. Ross, "Search for gluons in e^+e^- annihilation," Nucl. Phys. B **111**, 253 (1976), Erratum: [Nucl. Phys. B **130**, 516 (1977)].

arbitrariness of r will be very important for simplifying calculations. We'll show this property shortly.

First, note that this expression for the polarization vector indeed satisfies the equation of motion. We have

$$p \cdot \epsilon_R(p) = -\frac{1}{\sqrt{2}} \frac{u_R^\dagger(r)(p \cdot \sigma)u_R(p)}{u_R^\dagger(r)u_L(p)} = 0, \tag{7.78}$$

by the Dirac equation for the massless spinor $u_R(p)$: $(p \cdot \sigma)u_R(p) = 0$. Additionally, the polarization vector is zero when dotted into the arbitrary vector r:

$$r \cdot \epsilon_R(p) = -\frac{1}{\sqrt{2}} \frac{u_R^\dagger(r)(r \cdot \sigma)u_R(p)}{u_R^\dagger(r)u_L(p)} = 0, \tag{7.79}$$

by the Dirac equation for the massless spinor $u_R^\dagger(r)$.

For a momentum p, we can explicitly evaluate this polarization vector. For simplicity, let's align p with the $+\hat{z}$-axis, so that

$$u_R(p) = \sqrt{2E}\begin{pmatrix}1\\0\end{pmatrix}, \qquad u_L(p) = \sqrt{2E}\begin{pmatrix}0\\-1\end{pmatrix}, \tag{7.80}$$

where E is the energy of the four-momentum p. For the massless momentum r that points at an angle θ from the $+\hat{z}$-axis and an angle ϕ about the $+\hat{z}$-axis, the spinor is

$$u_R^\dagger(r) = \sqrt{2E_r}\left(e^{i\phi/2}\cos\frac{\theta}{2} \quad e^{-i\phi/2}\sin\frac{\theta}{2}\right). \tag{7.81}$$

Here, E_r is the energy of four-momentum r. Then, the spinor product in the denominator of the polarization vector is

$$u_R^\dagger(r)u_L(p) = \sqrt{2E_r}\left(e^{i\phi/2}\cos\frac{\theta}{2} \quad e^{-i\phi/2}\sin\frac{\theta}{2}\right)\sqrt{2E}\begin{pmatrix}0\\-1\end{pmatrix} = -2\sqrt{E_rE}\,e^{-i\phi/2}\sin\frac{\theta}{2}. \tag{7.82}$$

The spinor product in the numerator of the polarization vector is

$$u_R^\dagger(r)\sigma^\mu u_R(p) = \sqrt{2E_r}\left(e^{i\phi/2}\cos\frac{\theta}{2} \quad e^{-i\phi/2}\sin\frac{\theta}{2}\right)(\mathbb{I}, \sigma_1, \sigma_2, \sigma_3)^\mu\sqrt{2E}\begin{pmatrix}1\\0\end{pmatrix} \tag{7.83}$$

$$= 2\sqrt{E_rE}\left(e^{i\phi/2}\cos\frac{\theta}{2}, e^{-i\phi/2}\sin\frac{\theta}{2}, ie^{-i\phi/2}\sin\frac{\theta}{2}, e^{i\phi/2}\cos\frac{\theta}{2}\right)^\mu.$$

Putting these together, we find that the polarization vector can be expressed as

$$\epsilon_R^\mu(p) = -\frac{1}{\sqrt{2}}\frac{u_R^\dagger(r)\sigma^\mu u_R(p)}{u_R^\dagger(r)u_L(p)} = \frac{1}{\sqrt{2}}\left(e^{i\phi}\cot\frac{\theta}{2}, 1, i, e^{i\phi}\cot\frac{\theta}{2}\right)^\mu. \tag{7.84}$$

This looks different than the polarization vector that we quoted in Section 2.2.3. However, recall that the gauge freedom of the polarization vector ϵ means that we can add an arbitrary contribution proportional to the gluon momentum p without affecting any observable quantities. Then, this expression for the polarization vector can be written as

$$\epsilon_R^\mu(p) = \frac{1}{\sqrt{2}}\left(e^{i\phi}\cot\frac{\theta}{2}, 1, i, e^{i\phi}\cot\frac{\theta}{2}\right)^\mu = \frac{1}{\sqrt{2}}(0, 1, i, 0)^\mu + \frac{e^{i\phi}\cot\frac{\theta}{2}}{\sqrt{2E}}p^\mu, \tag{7.85}$$

where $p^\mu = (E, 0, 0, E)$. Indeed, the contribution to the polarization vector that depends on r is proportional to the gluon momentum p, and so cannot affect any observable. The rightmost term of Eq. 7.85 can be set to 0 for $\theta = \pi$; that is, when the vector r is in the opposite direction of p.

The polarization vector for a left-handed circularly polarized (left-handed helicity) gluon can correspondingly be expressed as

$$\epsilon_L^\mu(p) = \frac{1}{\sqrt{2}} \frac{u_L^\dagger(r)\bar{\sigma}^\mu u_L(p)}{u_L^\dagger(r)u_R(p)}. \tag{7.86}$$

One can verify, just as we did with the right-handed helicity polarization, that this expression satisfies all of the expected properties, though we won't do that here.

7.3.3 Fermion Propagator

Okay, now on to the intermediate quark propagator. The way that we can think about the quark propagator (or any propagator, for that matter) is as the Green's function for the appropriate equation of motion. Let's see what this means in more detail. The quark propagator occurs in the part of the diagram like:

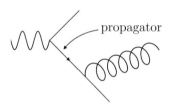

Vertices, where particles interact, correspond to particles at the same spacetime location, just like the nodes in a circuit diagram. So, we think of the propagator as representing a quark produced, say, at the spatial point that corresponds to the vertex with the photon. It is then annihilated at the vertex with the gluon. That is, the propagator for the quark is the solution to the Dirac equation with a source at a single point. Let's call this solution $G(x)$, which depends on the position four-vector x. Then,

$$i(\gamma \cdot \partial)G(x) = i\delta^{(4)}(x), \tag{7.87}$$

for a massless quark propagator. $G(x)$ is indeed a Green's function and the multi-dimensional δ-function $\delta^{(4)}(x)$ is defined as

$$\delta^{(4)}(x) = \delta(x_0)\delta(x_1)\delta(x_2)\delta(x_3), \tag{7.88}$$

where x_0 is the temporal coordinate and x_1, x_2, x_3 are the spatial coordinates. To solve this, we can Fourier transform to the momentum space representation. Fourier transforming Eq. 7.87 yields the equation

$$(\gamma \cdot p)\widetilde{G}(p) = i, \qquad \text{or that} \qquad \widetilde{G}(p) = \frac{i}{\gamma \cdot p}. \tag{7.89}$$

Here, $\widetilde{G}(p)$ is the Fourier transform of $G(x)$:

$$\widetilde{G}(p) = \int d^4x\, G(x) e^{-ip\cdot x}.\tag{7.90}$$

We call $\widetilde{G}(p)$ the propagator for a massless quark.

This is still slightly weird because $\gamma \cdot p$ is a matrix, and we have to invert it. To do this, we can use the anti-commutation relation of the γ-matrices:

$$\frac{1}{\gamma \cdot p} = \frac{\gamma \cdot p}{(\gamma \cdot p)(\gamma \cdot p)} = \frac{\gamma \cdot p}{p_\mu p_\nu \gamma^\mu \gamma^\nu} = \frac{\gamma \cdot p}{\frac{1}{2} p_\mu p_\nu \{\gamma^\mu, \gamma^\nu\}} = \frac{\gamma \cdot p}{p^2}.\tag{7.91}$$

In this expression, we have used the relationship

$$\gamma^\mu \gamma^\nu + \gamma^\nu \gamma^\mu = 2\eta^{\mu\nu},\tag{7.92}$$

from Eq. 2.98. Thus, we can write the propagator for a massless quark as

$$\widetilde{G}(p) = i\frac{\gamma \cdot p}{p^2}.\tag{7.93}$$

For two-component spinors of definite helicity, the corresponding propagators are

$$\widetilde{G}_R(p) = i\frac{\bar{\sigma} \cdot p}{p^2}, \qquad\qquad \widetilde{G}_L(p) = i\frac{\sigma \cdot p}{p^2},\tag{7.94}$$

where R and L denote right- and left-handed helicity, respectively.

7.3.4 The Cross Section for $e^+e^- \rightarrow q\bar{q}g$

Okay, we're now ready to evaluate the Feynman diagrams and calculate the cross section for the process $e^+e^- \rightarrow q\bar{q}g$. This section will be quite dense on mathematical manipulations. If you aren't interested in the steps in getting there, you can just skip to the result of all of this, Eq. 7.113.

As we did in the previous chapter, we will evaluate the Feynman diagrams for different helicity configurations. We'll do this explicitly for one helicity configuration, and we'll just quote the result summed over all helicities. Let's consider the process with helicity assignments

$$e_L^- e_R^+ \rightarrow q_R \bar{q}_L g_L.\tag{7.95}$$

The Feynman diagrams for this process are

$$= v_R^\dagger(p_2)\bar{\sigma}^\mu e\, u_L(p_1)\frac{1}{(p_1 + p_2)^2}u_R^\dagger(k_1)Q_q e\, \sigma_\mu \frac{\bar{\sigma}\cdot(-k_2 - k_3)}{(k_2 + k_3)^2} g\,\sigma \cdot \epsilon_L(k_3) v_L(k_2)\tag{7.96}$$

$$+ v_R^\dagger(p_2)\bar{\sigma}^\mu e\, u_L(p_1)\frac{1}{(p_1 + p_2)^2}u_R^\dagger(k_1) g\,\sigma \cdot \epsilon_L(k_3)\frac{\bar{\sigma}\cdot(k_1 + k_3)}{(k_1 + k_3)^2}Q_q e\, \sigma_\mu v_L(k_2).$$

There are a lot of moving parts in these diagrams and expressions, so let's focus on the first diagram in a bit more detail. Let's identify different parts of the diagram and how they appear in the mathematical evaluation. For the first diagram, we have

$$(7.97)$$

$$\boxed{v_R^\dagger(p_2)\bar\sigma^\mu e\, u_L(p_1)} \underbrace{\frac{1}{(p_1+p_2)^2}} u_R^\dagger(k_1) Q_q e\, \sigma_\mu \frac{\bar\sigma\cdot(-k_2-k_3)}{(k_2+k_3)^2} g\,\sigma\cdot\epsilon_L(k_3) v_L(k_2) .$$

We have matched the boundary shape of a subpart of the diagram with the term in the mathematical expression. The left side of this diagram should be familiar: the part in the solid rectangular boundary is just the electromagnetic interactions of electrons and positrons, while the photon propagator is the part with the solid oval boundary. The right of this diagram is a bit less familiar. The quark–anti-quark pair is produced in the region with the dotted rectangular boundary; this is proportional to the electric charge of the quark, Q_q. The quark propagator is included in this region and the momentum flowing through the propagator is $-k_2 - k_3$. The negative sign means that the direction of momentum (outward) is opposite to the direction of fermion arrow (inward). The gluon is emitted in the rectangular dashed region and the strength of its coupling to quarks is denoted by g. Finally, the external quark and anti-quark spinors are in the dotted and dashed oval regions, respectively. The second diagram can be broken apart in a similar way. Conservation of energy and momentum for this process is

$$p_1 + p_2 = k_1 + k_2 + k_3 . \tag{7.98}$$

The sum of Feynman diagrams in Eq. 7.96 can be simplified by contracting the free Lorentz indices on the Pauli matrices. Contracting the Lorentz indices rearranges the Feynman diagrams to be expressed in terms of spinor inner products, which is an example of a Fierz identity. The matrix element then becomes

$$\mathcal{M}(e_L^- e_R^+ \to q_R \bar q_L g_L) \tag{7.99}$$

$$= -\frac{2Q_q e^2 g}{(p_1+p_2)^2} u_R^\dagger(k_1) u_L(p_1) v_R^\dagger(p_2) \frac{\bar\sigma\cdot(k_2+k_3)}{(k_2+k_3)^2} \sigma\cdot\epsilon_L(k_3) v_L(k_2)$$

$$+ \frac{2Q_q e^2 g}{(p_1+p_2)^2} u_R^\dagger(k_1) \sigma\cdot\epsilon_L(k_3) \frac{\bar\sigma\cdot(k_1+k_3)}{(k_1+k_3)^2} u_L(p_1) v_R^\dagger(p_2) v_L(k_2) .$$

Now, we need to evaluate the remaining σ-matrix structure and the gluon polarization vector. Plugging in the explicit expression for the gluon polarization vector in terms of spinors from Eq. 7.86 into the first term of Eq. 7.99, we have

$$v_R^\dagger(p_2) \frac{\bar\sigma \cdot (k_2 + k_3)}{(k_2 + k_3)^2} \sigma_\mu v_L(k_2) \epsilon_L^\mu(k_3) \tag{7.100}$$

$$= v_R^\dagger(p_2) \frac{\bar\sigma \cdot (k_2 + k_3)}{(k_2 + k_3)^2} \sigma_\mu v_L(k_2) \frac{1}{\sqrt{2}} \frac{u_L^\dagger(r)\bar\sigma^\mu u_L(k_3)}{u_L^\dagger(r)u_R(k_3)}$$

$$= \sqrt{2} v_R^\dagger(p_2) \frac{\bar\sigma \cdot (k_2 + k_3)}{(k_2 + k_3)^2} u_L(k_3) \frac{u_L^\dagger(r)v_L(k_2)}{u_L^\dagger(r)u_R(k_3)}\,.$$

By gauge invariance, we can choose any convenient reference four-vector r. An especially nice choice is $r = k_2$, the momentum of the final-state anti-quark. With this choice, note that

$$u_L^\dagger(k_2)v_L(k_2) = 0\,, \tag{7.101}$$

and so this entire term vanishes!

We have to also evaluate the second term in the matrix element with the choice $r = k_2$ for the gluon polarization reference vector. For the factor involving the polarization vector, we find

$$\epsilon_L^\mu(k_3) u_R^\dagger(k_1) \sigma_\mu \frac{\bar\sigma \cdot (k_1 + k_3)}{(k_1 + k_3)^2} u_L(p_1) = \sqrt{2} \frac{u_R^\dagger(k_1)u_L(k_3)}{u_L^\dagger(k_2)u_R(k_3)} u_L^\dagger(k_2) \frac{\bar\sigma \cdot (k_1 + k_3)}{(k_1 + k_3)^2} u_L(p_1)\,.$$
$$\tag{7.102}$$

The matrix element then nicely simplifies to

$$\mathcal{M}(e_L^- e_R^+ \to q_R \bar{q}_L g_L) \tag{7.103}$$

$$= \frac{\sqrt{2} 2 Q_q e^2 g}{(p_1 + p_2)^2} \frac{u_R^\dagger(k_1)u_L(k_3)}{u_L^\dagger(k_2)u_R(k_3)} u_L^\dagger(k_2) \frac{\bar\sigma \cdot (k_1 + k_3)}{(k_1 + k_3)^2} u_L(p_1) v_R^\dagger(p_2) v_L(k_2)\,.$$

Further simplification can be accomplished by momentum conservation. Note that $k_1 + k_3 = p_1 + p_2 - k_2$, so we can make this replacement in the numerator of the quark propagator (the remaining term with the $\bar\sigma$):

$$\mathcal{M}(e_L^- e_R^+ \to q_R \bar{q}_L g_L) \tag{7.104}$$

$$= \frac{2\sqrt{2} Q_q e^2 g}{(p_1 + p_2)^2} \frac{u_R^\dagger(k_1)u_L(k_3)}{u_L^\dagger(k_2)u_R(k_3)} u_L^\dagger(k_2) \frac{\bar\sigma \cdot (p_1 + p_2 - k_2)}{(k_1 + k_3)^2} u_L(p_1) v_R^\dagger(p_2) v_L(k_2)$$

$$= \frac{2\sqrt{2} Q_q e^2 g}{(p_1 + p_2)^2} \frac{u_R^\dagger(k_1)u_L(k_3)}{u_L^\dagger(k_2)u_R(k_3)} \frac{u_L^\dagger(k_2)u_R(p_2)}{(k_1 + k_3)^2} \left(u_R^\dagger(p_2)u_L(p_1) \right) \left(v_R^\dagger(p_2)v_L(k_2) \right)\,.$$

In going from the second to the third line, we have used the Dirac equation for the spinors $u_L^\dagger(k_2)$ and $u_L(p_1)$:

$$u_L^\dagger(k_2)\bar\sigma \cdot k_2 = 0\,, \qquad\qquad \bar\sigma \cdot p_1 u_L(p_1) = 0\,. \tag{7.105}$$

Additionally, the σ matrix on the second line is expressed as an outer product of spinors on the third line:

$$\bar\sigma \cdot p_2 = u_R(p_2)u_R^\dagger(p_2)\,. \tag{7.106}$$

The final simplification that we can make to the matrix element is to rewrite the factors from the propagators in terms of spinor inner products. The remaining momentum dot products can be expressed as

$$(p_1 + p_2)^2 = 2p_1 \cdot p_2 = \left(u_R^\dagger(p_2)u_L(p_1)\right)\left(u_L^\dagger(p_1)u_R(p_2)\right), \qquad (7.107)$$

$$(k_1 + k_3)^2 = 2k_1 \cdot k_3 = \left(u_R^\dagger(k_1)u_L(k_3)\right)\left(u_L^\dagger(k_3)u_R(k_1)\right).$$

You will prove these identities in Exercise 7.6. With these results, the matrix element finally simplifies to

$$\mathcal{M}(e_L^- e_R^+ \to q_R \bar{q}_L g_L) \qquad (7.108)$$

$$= 2\sqrt{2}Q_q e^2 g \frac{\left(v_R^\dagger(p_2)v_L(k_2)\right)\left(u_L^\dagger(k_2)u_R(p_2)\right)}{\left(u_L^\dagger(p_1)u_R(p_2)\right)\left(u_L^\dagger(k_2)u_R(k_3)\right)\left(u_L^\dagger(k_3)u_R(k_1)\right)},$$

which is just a rational function of spinor inner products. This is a remarkable and remarkably compact expression. To compute the cross section, we just need to calculate the matrix elements with other helicity assignments, square them, and insert them into Fermi's Golden Rule.

To calculate the cross section, we first need to take the absolute square of the matrix element. For the process we've been considering thus far, this is

$$|\mathcal{M}(e_L^- e_R^+ \to q_R \bar{q}_L g_L)|^2 = 4Q_q^2 e^4 g^2 \frac{(p_2 \cdot k_2)^2}{(p_1 \cdot p_2)(k_1 \cdot k_3)(k_3 \cdot k_2)}. \qquad (7.109)$$

This expression follows from Eq. 7.108 and the use of identities such as those introduced in Eq. 7.107. Matrix elements corresponding to different helicity choices can be found by simply permuting indices appropriately. For example, if the helicities of the quark and anti-quark are flipped, we have

$$|\mathcal{M}(e_L^- e_R^+ \to q_L \bar{q}_R g_L)|^2 = 4Q_q^2 e^4 g^2 \frac{(p_2 \cdot k_1)^2}{(p_1 \cdot p_2)(k_1 \cdot k_3)(k_3 \cdot k_2)}. \qquad (7.110)$$

These matrix elements can be rewritten in the three-body phase space coordinates represented by energy fractions x_i. These energy fractions are defined as

$$x_i = \frac{2Q \cdot k_i}{Q^2}, \qquad (7.111)$$

for $i = 1, 2, 3$. These energy fractions satisfy $x_1 + x_2 + x_3 = 2$ and, in the center-of-mass frame, the four-vector $Q = k_1 + k_2 + k_3 = (E_{cm}, 0, 0, 0)$. Using these x_i variables, the squared matrix element for the process we've been studying becomes

$$|\mathcal{M}(e_L^- e_R^+ \to q_R \bar{q}_L g_L)|^2 = \frac{2Q_q^2 e^4 g^2}{E_{cm}^2} \frac{x_2^2(1 + \cos\theta)^2}{(1 - x_1)(1 - x_2)}. \qquad (7.112)$$

Here, $\cos\theta$ is the scattering angle: the angle between the electron–positron beam and the final-state anti-quark. For just studying the dynamics of the gluon, we can integrate over $\cos\theta$.

Note the physics contained in the expression of Eq. 7.112. The squared matrix element diverges when either $x_1 \to 1$ or $x_2 \to 1$. Physically, this corresponds to either $k_2 \cdot k_3 \to 0$ or $k_1 \cdot k_3 \to 0$ from Eq. 7.109. For two massless four-vectors, this can occur either when $k_3 \to 0$ (the energy of the gluon is small) or when $\vec{k}_3 \parallel \vec{k}_1$ or $\vec{k}_2$, called the **collinear limit**. The existence of divergences in the matrix element in the **soft** (= low-energy) and/or collinear limits will have profound physical consequences for observing phenomena of the gluon. We will discuss this in detail in Chapter 9.

We won't provide the details, but by summing over external particle spins and integrating over the scattering angle $\cos\theta$, the cross section differential in the center-of-mass quark and anti-quark energy fractions x_q, $x_{\bar{q}}$ is

$$\frac{d^2\sigma(e^+e^- \to q\bar{q}g)}{dx_q\, dx_{\bar{q}}} = \sigma(e^+e^- \to q\bar{q})\frac{\alpha_s}{2\pi}C_F\frac{x_q^2 + x_{\bar{q}}^2}{(1-x_q)(1-x_{\bar{q}})}. \tag{7.113}$$

Here, $\alpha_s = g^2/(4\pi)$ is called the strong coupling constant and $C_F = 4/3$ is the factor that accounts for the possible different colors of the gluon emitted off of the final-state quarks. The cross section for quark–anti-quark production is

$$\sigma(e^+e^- \to q\bar{q}) = 4\frac{\pi\alpha^2}{E_{\text{cm}}^2}\sum_{q\text{ quarks}} Q_q^2. \tag{7.114}$$

7.3.5 Tests of a Spin-1 Gluon

Our analysis of the process $e^+e^- \to q\bar{q}g$, where the gluon couples to quarks in a similar way to photons, makes a number of concrete predictions. First and foremost, just the existence of such a process is a major prediction. In the same way that the process $e^+e^- \to q\bar{q}$ manifests itself in experiment as dijet events, $e^+e^- \to q\bar{q}g$ manifests itself as three-jet events. An example event display of a three-jet event recorded at the L3 experiment at LEP is shown in Fig. 7.3. Three well-defined, collimated streams of hadrons are observed to have been created from electron–positron collisions, indicative of the production of a gluon in the final state.

The cross section differential in the quark and anti-quark energy fractions in Eq. 7.113 enables us to learn more about this gluon. Just as we inferred the spin of quarks from the distribution of the scattering angle, we are able to infer the spin of the gluon from distributions of the relative angles of the three jets in the final state. In principle, if we were able to determine which jets corresponded to the quark and to the anti-quark, then we could measure their energy fractions and compare directly to Eq. 7.113. However, this isn't possible, not the least reason being that we observe hadrons in our detectors, and not quarks and gluons. So, we have to think a bit more about how to connect Eq. 7.113 to something we actually measure.

What we can do that is sensible is to measure a function of the three-jet energy fractions that treats them all equally. This procedure doesn't require a unique map from a jet back to quark or gluon. For example, we might measure the largest energy fraction of the three jets

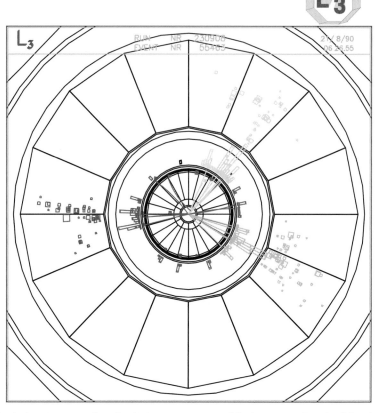

Fig. 7.3   An example of a three-jet event collected at the L3 experiment, one of the detectors on the LEP collider. The three jets are visible as significant energy deposits in the calorimeters to the left, upper right, and lower right of the figure. Credit: L3 Experiment © CERN.

in the final state. This observable is called **thrust** τ,[16] and is defined in terms of the energy fractions x_i as

$$\tau = \max\{x_i\} = \max\{x_q, x_{\bar{q}}, 2 - x_q - x_{\bar{q}}\}. \tag{7.115}$$

On the right, we've just written out the three energy fractions for the quark, the anti-quark, and the gluon x_g, which is constrained by their sum equaling 2. The distribution of thrust τ can be calculated from the double differential cross section of Eq. 7.113 by summing over all phase space regions that result in the same value of τ. This is accomplished by inserting a δ-function into the integral over phase space. In particular, the cross section differential in τ is

$$\frac{d\sigma}{d\tau} = \int_0^1 dx_q \int_0^1 dx_{\bar{q}} \, \Theta(x_q + x_{\bar{q}} - 1) \frac{d^2\sigma}{dx_q \, dx_{\bar{q}}} \, \delta\left(\tau - \max\{x_i\}\right). \tag{7.116}$$

---

[16] E. Farhi, "A QCD test for jets," Phys. Rev. Lett. **39**, 1587 (1977).

Here, $\Theta(x)$ is the Heaviside Θ-function, which equals 1 if $x > 0$ and 0 if $x < 0$. Therefore, to predict the distribution of thrust for any model of the gluon, we just need to know the differential cross section for that model and then do this integral.

This is a very general procedure and enables us to make predictions for any observable defined as a function of the phase space coordinates. All we need to do is replace the appearance of τ and its functional form in the δ-function of Eq. 7.116 with the observable under study. To test the spin of the gluon, a very sensitive set of observables are the energy fractions of the most energetic jet, the second most energetic jet, and the least energetic jet. We call these energy fractions x_1, x_2, and x_3, respectively. Of course, from our discussion above, x_1 is just thrust τ. The theoretical predictions for the distributions of the energy fractions x_1, x_2, and x_3 for a spin-1 gluon can be made and compared to data. This comparison was done at the SLD experiment at SLAC in the 1990s, and the result is illustrated in Fig. 7.4. The data from SLD are compared to three predictions for the spin of the gluon: spin-0, spin-1, and spin-2. The data agree beautifully with the spin-1 hypothesis,

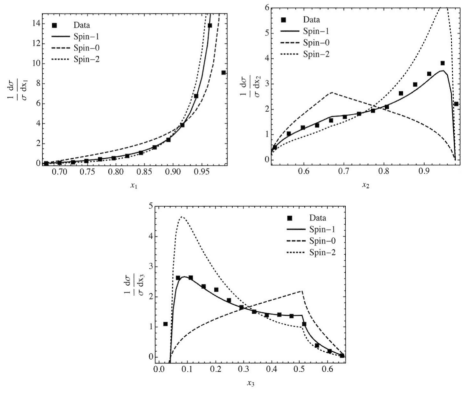

Fig. 7.4 Measurements of the three energy fractions of the jets in $e^+e^- \to$ three-jet events measured at the SLD experiment. The data are compared to three predictions for the spin of the gluon: spin-0, spin-1, and spin-2. The uncertainties in the data points are comparable to or smaller than the squares that denote the data. Reprinted data from table with permission from K. Abe *et al.* [SLD Collaboration], Phys. Rev. D **55**, 2533 (1997). Copyright 1997 by the American Physical Society.

while disagreeing significantly with both spin-0 and spin-2 predictions. Apparently, the gluon is a spin-1 particle!

We'll work through the analytic prediction for thrust in Example 7.2 and you'll study it in more detail in Exercise 7.8. For the predictions of the distributions of x_2 and x_3, the lower-energy jets, the calculation is a bit subtle. For x_2, for example, it is possible that $x_1 = 1$, while x_2 is anywhere in the range of $[0.5, 1]$. However, if $x_1 = 1$, then the cross section diverges, and any resulting prediction is meaningless. Technically, the energy fraction of the second or third hardest jets lacks a property called **infrared and collinear (IRC) safety**, which is the statement that an observable's distribution can be predicted within Feynman diagram perturbation theory. When a particle's energy gets very small (infrared limit) or becomes collinear to another particle, the cross section in general diverges, as mentioned in Section 7.3.4. In these limits, to ensure that we get sensible results from our calculation, the observable that we study must push this divergence to a single point on phase space. For example, for thrust defined as $\max\{x_i\}$, the cross section diverges if one of the energy fractions approaches 1. However, this is also the largest energy fraction possible, and so the divergent low-energy and collinear limits of the cross section are isolated at the point where $\tau = 1$. Away from this point, our prediction from performing the integral of Eq. 7.116 is sensible.

This point of IRC safety was not mentioned in the original paper that did the comparison presented in Fig. 7.4, so it is not known if they were aware of this subtlety. At any rate, to make the predictions for spin-0, spin-1, and spin-2 particles, the results in Fig. 7.4 impose a cut of $x_1 < 0.98$, which avoids this singular region. Unfortunately, the consequence of this is that the cross section that one measures is then an exclusive, rather than an inclusive, cross section. Thus, one should interpret the conclusions of Fig. 7.4 with some care, but a more detailed calculation can also be done and essentially the exact same results are obtained.

Example 7.2 Eq. 7.116 is the abstract expression that enables us to determine the predicted distribution of thrust in $e^+e^- \rightarrow q\bar{q}g$ scattering. Let's calculate this using the explicit expression for the double differential cross section, Eq. 7.113.

Solution

The first thing that we will do in our prediction of thrust is to determine the range of possible values of τ. Thrust is defined as

$$\tau = \max\{x_i\}, \tag{7.117}$$

where, in the center-of-mass frame, $x_i = 2E_i/E_{\text{cm}}$. By momentum conservation, the largest value one of the x_i can take is 1; this limit corresponds to back-to-back jets with equal and opposite momenta. The x_i are restricted to sum to 2 by total energy conservation and so the minimum value of the maximum of the x_i is when all three are equal to $2/3$. Therefore, thrust $\tau \in [2/3, 1]$. These two limits separate out different kinematic configurations of the jets. For $\tau \rightarrow 1$, this is the dijet limit. For $\tau \rightarrow 2/3$, there are three well-defined jets

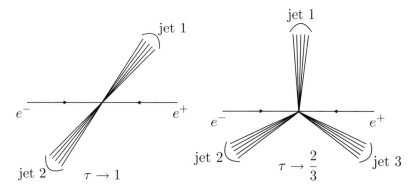

Fig. 7.5 An illustration of the (left) dijet and (right) three-jet Mercedes-Benz configurations of final states in $e^+e^- \rightarrow$ hadrons processes which correspond to different limits of thrust τ.

with equal energy $120°$ degrees from one another, which is colloquially referred to as the **Mercedes-Benz configuration**. These limits are illustrated in Fig. 7.5.

To calculate the distribution of thrust, we need to do the integrals in the expression

$$\frac{d\sigma}{d\tau} = \int_0^1 dx_q \int_0^1 dx_{\bar{q}}\, \Theta(x_q + x_{\bar{q}} - 1)\, \frac{d^2\sigma}{dx_q\, dx_{\bar{q}}}\, \delta\left(\tau - \max\{x_i\}\right). \tag{7.118}$$

The first thing we need to do is to write the δ-function that restricts to the measured value of thrust $\delta(\tau - \max\{x_i\})$ in a useful form.

To do this, let's first assume that x_q is the largest of the x_i. Then, x_q is larger than both $x_{\bar{q}}$ and $x_g = 2 - x_q - x_{\bar{q}}$, which can be enforced by the Θ-functions:

$$\Theta(x_q - x_{\bar{q}})\Theta(x_q - x_g) = \Theta(x_q - x_{\bar{q}})\Theta(2x_q + x_{\bar{q}} - 2). \tag{7.119}$$

Then, restricting x_q to be the largest of the x_i yields a contribution to the δ-function of

$$\delta(\tau - \max\{x_i\}) \supset \Theta(x_q - x_{\bar{q}})\Theta(2x_q + x_{\bar{q}} - 2)\delta(\tau - x_q). \tag{7.120}$$

We can do the same thing assuming that $x_{\bar{q}}$ and x_g are the largest. For $x_{\bar{q}}$, the double differential cross section from Eq. 7.113 is symmetric in $x_q \leftrightarrow x_{\bar{q}}$ and so it will lead to the same value of the integral as when x_q is largest. So, for the configuration when $x_{\bar{q}}$ is largest, we can simply multiply the result of Eq. 7.120 by a factor of 2.

Doing the same thing for when x_g is the largest, and noting that $x_g = 2 - x_q - x_{\bar{q}}$, we find a contribution to the δ-function of

$$\delta(\tau - \max\{x_i\}) \supset \Theta(2 - 2x_q - x_{\bar{q}})\Theta(2 - x_q - 2x_{\bar{q}})\delta(\tau - 2 + x_q + x_{\bar{q}}). \tag{7.121}$$

Combining these three possible configurations, we find

$$\delta(\tau - \max\{x_i\}) = 2\Theta(x_q - x_{\bar{q}})\Theta(2x_q - 2 + x_2)\delta(\tau - x_q) \tag{7.122}$$
$$+ \Theta(2 - 2x_q - x_{\bar{q}})\Theta(2 - x_q - 2x_{\bar{q}})\delta(\tau - 2 + x_q + x_{\bar{q}}).$$

With this expression for the δ-function, we can now do the integral over x_q from Eq. 7.116. That is, we must integrate

$$\int_0^1 dx_q \, \Theta(x_q + x_{\bar{q}} - 1) \frac{x_q^2 + x_{\bar{q}}^2}{(1 - x_q)(1 - x_{\bar{q}})} \left[2\Theta(x_q - x_{\bar{q}})\Theta(2x_q - 2 + x_{\bar{q}})\delta(\tau - x_q) \right.$$
$$\left. + \Theta(2 - 2x_q - x_{\bar{q}})\Theta(2 - x_q - 2x_{\bar{q}})\delta(\tau - 2 + x_q + x_{\bar{q}}) \right] . \qquad (7.123)$$

Focusing first on the δ-function of $\delta(\tau - x_q)$, we have

$$\int_0^1 dx_q \, \Theta(x_q + x_{\bar{q}} - 1) \frac{x_q^2 + x_{\bar{q}}^2}{(1 - x_q)(1 - x_{\bar{q}})} 2\Theta(x_q - x_{\bar{q}})\Theta(2x_q - 2 + x_{\bar{q}})\delta(\tau - x_q)$$
$$= 2 \frac{\tau^2 + x_{\bar{q}}^2}{(1 - \tau)(1 - x_{\bar{q}})} \Theta(\tau - x_{\bar{q}})\Theta(x_{\bar{q}} - 2(1 - \tau))\Theta(1 - \tau) . \qquad (7.124)$$

Now, let's integrate over the other δ-function term:

$$\int_0^1 dx_q \, \Theta(x_q + x_{\bar{q}} - 1) \frac{x_q^2 + x_{\bar{q}}^2}{(1 - x_q)(1 - x_{\bar{q}})}$$
$$\times \Theta(2 - 2x_q - x_{\bar{q}})\Theta(2 - x_q - 2x_{\bar{q}})\delta(\tau - 2 + x_q + x_{\bar{q}})$$
$$= \frac{(2 - x_{\bar{q}} - \tau)^2 + x_{\bar{q}}^2}{(\tau + x_{\bar{q}} - 1)(1 - x_{\bar{q}})} \Theta(\tau - x_{\bar{q}})\Theta(x_{\bar{q}} - 2(1 - \tau))\Theta(1 - \tau) . \qquad (7.125)$$

In this expression, we have made the replacement that $x_q = 2 - x_{\bar{q}} - \tau$ as enforced by the δ-function everywhere.

In this form, we can then combine the results of the two integrals over x_q:

$$\int_0^1 dx_q \, \Theta(x_q + x_{\bar{q}} - 1) \frac{x_q^2 + x_{\bar{q}}^2}{(1 - x_q)(1 - x_{\bar{q}})} \delta(\tau - \max\{x_i\}) \qquad (7.126)$$
$$= \left[2 \frac{\tau^2 + x_{\bar{q}}^2}{(1 - \tau)(1 - x_{\bar{q}})} + \frac{(2 - x_{\bar{q}} - \tau)^2 + x_{\bar{q}}^2}{(\tau + x_{\bar{q}} - 1)(1 - x_{\bar{q}})} \right] \Theta(\tau - x_{\bar{q}})\Theta(x_{\bar{q}} - 2(1 - \tau))\Theta(1 - \tau) .$$

This is written in a nice way so we can straightforwardly integrate this over the anti-quark energy fraction $x_{\bar{q}}$. The integral we need to do is then

$$\int_{2(1-\tau)}^{\tau} dx_{\bar{q}} \left[2 \frac{\tau^2 + x_{\bar{q}}^2}{(1 - \tau)(1 - x_{\bar{q}})} + \frac{(2 - x_{\bar{q}} - \tau)^2 + x_{\bar{q}}^2}{(\tau + x_{\bar{q}} - 1)(1 - x_{\bar{q}})} \right] \Theta(1 - \tau) . \qquad (7.127)$$

In this expression, we have explicitly written the integration bounds as enforced by the Θ-functions. For this integral to be non-zero, the upper bound of the integral must be larger than the lower bound of the integral. This imposes

$$\tau > 2(1 - \tau) \qquad \text{or that} \qquad \tau > 2/3 , \qquad (7.128)$$

which is the exact same lower bound of thrust τ that we argued earlier.

There are a lot of integrals to do in Eq. 7.127, and we no longer have δ-functions around to simplify things. This integral can be done in closed form in numerous ways. Here, we'll present one technique for doing them and then state the result. In Eq. 7.127, we have to do integrals like

$$\int dx \, \frac{1}{1-x} = -\log(1-x) \,. \tag{7.129}$$

We also have integrals like

$$\int dx \, \frac{x}{1-x} \,. \tag{7.130}$$

To evaluate this, we can use integration by parts, but here we use another trick. We can evaluate it by Taylor series expansion. The Taylor series of the integrand is

$$\frac{x}{1-x} = x \sum_{n=0}^{\infty} x^n = \sum_{n=0}^{\infty} x^{n+1} \,. \tag{7.131}$$

Now, integrating this, we find

$$\int dx \, \frac{x}{1-x} = \int dx \sum_{n=0}^{\infty} x^{n+1} = \sum_{n=0}^{\infty} \frac{1}{n+2} x^{n+2} \,. \tag{7.132}$$

To do the infinite sum, make the replacement $n + 2 = m$ and so

$$\int dx \, \frac{x}{1-x} = \sum_{m=2}^{\infty} \frac{1}{m} x^m \,. \tag{7.133}$$

Then, we add and subtract x to this result, which yields

$$\int dx \, \frac{x}{1-x} = \sum_{m=2}^{\infty} \frac{1}{m} x^m + x - x = \sum_{m=1}^{\infty} \frac{1}{m} x^m - x \,. \tag{7.134}$$

Now, we can identify the infinite sum with the Taylor series for the logarithm:

$$\sum_{m=1}^{\infty} \frac{1}{m} x^m = -\log(1-x) \,. \tag{7.135}$$

Therefore, we have calculated

$$\int dx \, \frac{x}{1-x} = -\log(1-x) - x \,. \tag{7.136}$$

For the last general integral that remains, it can be massaged into a simple form. Note that

$$\int dx \, \frac{x^2}{1-x} = \int dx \left[\frac{x}{1-x} + \left(\frac{x^2}{1-x} - \frac{x}{1-x} \right) \right] \tag{7.137}$$

$$= \int dx \left[\frac{x}{1-x} - x \right] = -\log(1-x) - x - \frac{x^2}{2} \,.$$

These are all the integrals that we need.

The second term in the integrand of Eq. 7.127 consists of a denominator with two terms that depend on $x_{\bar{q}}$. To separate them out into simpler denominators for which we can use the results of the integrals that we just calculated, we can partial fraction them into two terms. Doing this yields

$$\frac{1}{(\tau - 1 + x_{\bar{q}})(1 - x_{\bar{q}})} = \frac{1}{\tau}\left[\frac{1}{\tau - 1 + x_{\bar{q}}} + \frac{1}{1 - x_{\bar{q}}}\right]. \tag{7.138}$$

Now, each term can be integrated straightforwardly.

Using these results for the general integrals that remain and the partial fractioning, to get to the result there's a lot of long and tedious algebra which we won't do here. When the dust settles, we find

$$\int_{2(1-\tau)}^{\tau} dx_{\bar{q}} \left[2\frac{\tau^2 + x_{\bar{q}}^2}{(1-\tau)(1-x_{\bar{q}})} + \frac{(2 - x_{\bar{q}} - \tau)^2 + x_{\bar{q}}^2}{(\tau + x_{\bar{q}} - 1)(1 - x_{\bar{q}})}\right]\Theta(1-\tau) \tag{7.139}$$

$$= \left[\frac{6\tau^2 - 6\tau + 4}{\tau(1-\tau)}\log\frac{2\tau - 1}{1 - \tau} - \frac{3(3\tau - 2)(2 - \tau)}{1 - \tau}\right]\Theta(1-\tau)\Theta(\tau - 2/3).$$

To get the differential cross section in thrust τ, we just need to restore the overall multiplicative factors from Eq. 7.113. It is common to denote the $e^+e^- \to q\bar{q}$ cross section by σ_0:

$$\sigma(e^+e^- \to q\bar{q}) \equiv \sigma_0, \tag{7.140}$$

as it is governed by an intermediate photon and has nothing to do with gluons. Putting it all together, the cross section differential in the thrust τ is therefore

$$\frac{1}{\sigma_0}\frac{d\sigma}{d\tau} = \frac{\alpha_s}{2\pi}C_F\left[\frac{6\tau^2 - 6\tau + 4}{\tau(1-\tau)}\log\frac{2\tau - 1}{1 - \tau} - \frac{3(3\tau - 2)(2 - \tau)}{1 - \tau}\right]\Theta(1-\tau)\Theta(\tau - 2/3). \tag{7.141}$$

This expression manifests the IRC safety of thrust. This cross section is finite for $\tau < 1$ and only diverges at the single point where $\tau = 1$. In Chapter 9, we'll interpret what this divergence means physically.

7.4 Spinor Helicity

By the way, the formalism applied in this chapter for calculating matrix elements with two-component spinors is called **spinor helicity** and is widely used for modern calculations involving complicated Feynman diagrams.[17] In spinor helicity, the two-component spinors are denoted as

$$u_R(p) = v_L(p) \equiv |p\rangle, \qquad\qquad u_R^\dagger(p) = v_L^\dagger(p) \equiv [p, \tag{7.142}$$

$$u_L^\dagger(p) = v_R^\dagger(p) \equiv \langle p, \qquad\qquad u_L(p) = v_R(p) \equiv p]. \tag{7.143}$$

---

[17] Now-classic reviews of spinor helicity and related calculational tricks are: M. L. Mangano and S. J. Parke, "Multiparton amplitudes in gauge theories," Phys. Rept. **200**, 301 (1991) [arXiv:hep-th/0509223]; L. J. Dixon, "Calculating scattering amplitudes efficiently," in *QCD and Beyond: Proceedings, Theoretical Advanced Study Institute in Elementary Particle Physics, TASI-95, Boulder, CO, June 4–30, 1995*, World Scientific (1996), pp. 539–582 [arXiv:hep-ph/9601359].

With this notation, note that

$$u_L^\dagger(p)u_R(k) = \langle pk\rangle\,, \quad \text{and} \quad \left(u_L^\dagger(p)u_R(k)\right)^* = [kp]\,. \tag{7.144}$$

Also, $[pk]\langle kp\rangle = 2k\cdot p$.

In this notation, the matrix element that we calculated is written as

$$\mathcal{M}(e_L^- e_R^+ \to q_R \bar{q}_L g_L) = 2\sqrt{2}Q_q e^2 g\frac{\langle k_2 p_2\rangle^2}{\langle p_1 p_2\rangle\langle k_1 k_3\rangle\langle k_3 k_2\rangle}\,, \tag{7.145}$$

which is exceptionally compact. In addition to its compactness, the spinor helicity formalism enables simple determination of the matrix element, from physical requirements. By enforcing the correct limits in which the gluon becomes collinear to one of the external quarks or has low energy, one can prove that Eq. 7.145 is the unique amplitude.

Exercises

7.1 *Plus-Function Expansion.* In the limit that $\epsilon \to 0$, the function

$$f(x) = \frac{\epsilon}{x^{1-\epsilon}} \tag{7.146}$$

has some very strange properties. In Eq. 7.57, we said that it turned into a δ-function as $\epsilon \to 0$. In this exercise, we will study this in more detail and demonstrate how to expand this function as an infinite sum of distributions.

(a) As defined above, the function $f(x)$ integrates to 1 on $x \in [0,1]$:

$$\int_0^1 dx\frac{\epsilon}{x^{1-\epsilon}} = 1\,. \tag{7.147}$$

If we could just Taylor expand the function in ϵ, we would have

$$\frac{\epsilon}{x^{1-\epsilon}} = \frac{\epsilon}{x}e^{\epsilon \log x} = \frac{\epsilon}{x}\sum_{n=0}^{\infty}\frac{(\epsilon \log x)^n}{n!}\,. \tag{7.148}$$

Attempt to integrate this expansion on $x \in [0,1]$ for $\epsilon \to 0$. What do you find? Using this result, argue that the Taylor expansion is unjustified.

(b) If $x > 0$, what is the value of the function $f(x)$ for $\epsilon \to 0$? With this observation, argue that

$$\lim_{\epsilon\to 0}\frac{\epsilon}{x^{1-\epsilon}} = \delta(x)\,. \tag{7.149}$$

(c) As long as $x > 0$, the function $f(x)$ is finite and so the Taylor expansion is well defined. All of the problems with the Taylor expansion in ϵ come from the region where $x = 0$. With this in mind, we can use a form of the Taylor expansion that is modified for $x = 0$. We denote this by the **+-function expansion**

$$\frac{\epsilon}{x^{1-\epsilon}} = \delta(x) + \sum_{n=0}^{\infty}\frac{\epsilon^{n+1}}{n!}\left(\frac{\log^n x}{x}\right)_+\,. \tag{7.150}$$

Argue that every +-function integrates to 0 on $x \in [0, 1]$:

$$\int_0^1 dx \left(\frac{\log^n x}{x} \right)_+ = 0 . \tag{7.151}$$

+-functions are properly distributions, not functions, because they are formally infinite at $x = 0$.

(d) This expansion in +-functions is extremely useful, especially when integrating against an analytic function. An analytic function $g(x)$ with a convergent Taylor series about $x = 0$ can be expressed as

$$g(x) = \sum_{n=0}^{\infty} c_n x^n , \tag{7.152}$$

for some numerical coefficients c_n. Argue that the integration over a +-function with an analytic function $g(x)$ is defined as

$$\int_0^1 dx \left(\frac{\log^n x}{x} \right)_+ g(x) = \int_0^1 dx \, \frac{\log^n x}{x} \left(g(x) - g(0) \right) . \tag{7.153}$$

Note that on the right everything in the integrand consists of regular functions.

Hint: Show that the left and right side of this equation agree order-by-order in the Taylor expansion of $g(x)$.

7.2   *Breit Frame.* In the interpretation of Bjorken scaling in Section 7.2.1, we introduced the Breit frame, in which the photon that mediates the DIS process has zero energy:

$$q_{\text{Breit}} = (0, \vec{q}) . \tag{7.154}$$

In general, any space-like four-vector for which $q^2 < 0$ can be Lorentz transformed into the Breit frame. The Breit frame is also called the **brick wall frame**.

(a) For DIS in the Breit frame, how are the initial and final electron momenta related? Does the term "brick wall" frame make sense?

(b) In the frame in which the proton is at rest, we can express the momenta of the initial and final electrons p_e and $p_{e'}$ and the proton p_p as

$$p_e = E_e(1, 0, 0, 1) , \quad p_{e'} = E_{e'}(1, \sin\theta, 0, \cos\theta) , \quad p_p = (m_p, 0, 0, 0) , \tag{7.155}$$

where E_e and $E_{e'}$ are the initial and final electron energies and m_p is the mass of the proton. What Lorentz boost would you need to perform to get the electron momenta in the Breit frame? For DIS in the Breit frame, what is the momentum of the proton?

7.3   *Form Factor Evolution Equation.* Evolution equations are very useful tools in theoretical particle physics. They enable controlled predictive power as once you know the solution at one point, you can evolve to find the solution at any other point. In this exercise, we will explore an evolution equation for the form factor $F_2(x, Q^2)$, assuming point-particle quarks, but that interact in a scale-invariant manner. This will set the stage for the DGLAP evolution equations which are the complete theory that describes the energy dependence of parton distribution functions and form factors.

(a) Show that the expression for the form factor in Eq. 7.53 satisfies the homogeneous differential equation

$$Q^2 \frac{\partial}{\partial Q^2} F_2(x, Q^2) = -\frac{\epsilon}{2} F_2(x, Q^2) . \tag{7.156}$$

(b) This evolution equation in energy scale Q^2 can be solved in terms of the form factor defined at a specified energy Q_0^2, $F_2(x, Q_0^2)$. Show that the general solution to Eq. 7.156 can be written as

$$F(x, Q^2) = F(x, Q_0^2) e^{-\frac{\epsilon}{2} \log \frac{Q^2}{Q_0^2}} . \tag{7.157}$$

Written in this way, the exponential factor ϵ is called the **anomalous dimension**. From this general expression, determine the form factor $F(x, Q_0^2)$ at energy Q_0^2 from the result of Eq. 7.53.

(c) Assuming that $\epsilon > 0$, what is the solution to Eq. 7.156 as $Q^2 \to \infty$? What about when $Q^2 \to 0$? These limits are present in quantum chromodynamics and their interpretation for high-energy physics will be profound.

7.4 *Infrared and Collinear Safety.* The principle of infrared and collinear safety is of fundamental importance when making theoretical predictions. Observables that are IRC safe have distributions that are finite and can be predicted by calculating Feynman diagrams and integrating over phase space appropriately. In this exercise, we'll identify the IRC safety or unsafety of several standard observables that are used to study hadronic final states at electron–positron colliders.

These observables are expressed in the three-body phase space coordinates x_q and $x_{\bar{q}}$, the energy fractions of the quark and anti-quark. For each of the observables presented below and their functional form on this three-body phase space, determine whether the observable is IRC safe or IRC unsafe. Recall that an IRC-safe observable is one for which the divergences in the double differential cross section of Eq. 7.113 are mapped to a single, isolated value of the observable.

(a) C-parameter:[18]

$$C = 6 \frac{(1 - x_q)(1 - x_{\bar{q}})(x_q + x_{\bar{q}} - 1)}{x_q x_{\bar{q}} (2 - x_q - x_{\bar{q}})} \tag{7.158}$$

(b) Relative squared energy fractions:

$$\frac{3}{2} \left(\frac{x_2^2 + x_3^2}{x_1^2 + x_2^2 + x_3^2} - \frac{1}{2} \right), \qquad \text{where} \qquad x_1 > x_2 > x_3 \tag{7.159}$$

---

[18] G. Parisi, "Super inclusive cross-sections," Phys. Lett. **74B**, 65 (1978); J. F. Donoghue, F. E. Low and S. Y. Pi, "Tensor analysis of hadronic jets in quantum chromodynamics," Phys. Rev. D **20**, 2759 (1979); R. K. Ellis, D. A. Ross and A. E. Terrano, "The perturbative calculation of jet structure in e^+e^- annihilation," Nucl. Phys. B **178**, 421 (1981).

(c) Broadening:[19]

$$B = \sqrt{\frac{x_2}{x_1}}(1 - x_1)^{1/2}(1 - x_2)^{1/2}, \qquad \text{where} \qquad x_1 > x_2 > x_3 \qquad (7.160)$$

(d) Ratio of $1-$ thrust to broadening:

$$\frac{1 - \tau}{B} = \sqrt{\frac{x_1(1 - x_1)}{x_2(1 - x_2)}}, \qquad \text{where} \qquad x_1 > x_2 > x_3. \qquad (7.161)$$

7.5 *Properties of Helicity Spinors.* As introduced in Section 7.4, the spinor helicity notation provides a very compact representation of inner products of two-component Weyl spinors. Because Weyl spinors transform under the fundamental two-dimensional representation of the Lorentz group, any particle of higher spin can be represented by an appropriate product of spinors. In this exercise, we'll study two important properties of helicity spinors: anti-symmetry and the Schouten identity.

(a) Show that the spinor inner product is anti-symmetric:

$$\langle pk \rangle = -\langle kp \rangle, \qquad (7.162)$$

for two massless four-momenta p and k.

(b) Show that the inner spinor products involving four massless four-momenta p, k, q, r satisfy the **Schouten identity**:

$$\langle pk \rangle \langle qr \rangle = \langle pq \rangle \langle kr \rangle + \langle pr \rangle \langle qk \rangle. \qquad (7.163)$$

Note that both of these identities also hold for the square bracket inner product $[\cdot]$ because it is just the complex conjugate of the angle bracket inner product (Eq. 7.144).

7.6 *More Helicity Spinors.* There were a few points in our calculation of the matrix element for $e^+e^- \to q\bar{q}g$ scattering where we left the justification to the exercises. We'll address those here.

(a) In Eq. 7.107, we used a relationship between spinor inner products and four-vector dot products. For arbitrary massless four-vectors p and k, prove the equalities

$$2p \cdot k = \frac{1}{2} \left(u_R^\dagger(p)\sigma^\mu v_L(p) \right) \left(v_R^\dagger(k)\bar{\sigma}_\mu u_L(k) \right) \qquad (7.164)$$

$$= \left(u_R^\dagger(p)u_L(k) \right) \left(v_R^\dagger(k)v_L(p) \right).$$

You'll need to use the expressions for the spinors presented in Eq. 6.23 of Chapter 6. Without loss of generality, you can work in the frame in which the three-momentum $\vec{p}$ is aligned along the $+\hat{z}$-axis.

---

[19] P. E. L. Rakow and B. R. Webber, "Transverse momentum moments of hadron distributions in QCD jets," Nucl. Phys. B **191**, 63 (1981); R. K. Ellis and B. R. Webber, "QCD jet broadening in hadron hadron collisions," in *Physics of the Superconducting Supercollider: Proceedings of the 1986 Summer Study Meeting, June 23 to July 11, 1986, Snowmass, CO*, American Physical Society (1986), pp. 74–76; S. Catani, G. Turnock and B. R. Webber, "Jet broadening measures in e^+e^- annihilation," Phys. Lett. B **295**, 269 (1992).

(b) In Section 7.3.2, we constructed the right-handed helicity polarization vector ϵ_R for the gluon from spinors to be

$$\epsilon_R^\mu(p) = -\frac{1}{\sqrt{2}} \frac{u_R^\dagger(r)\sigma^\mu u_R(p)}{u_R^\dagger(r)u_L(p)}. \tag{7.165}$$

p is the momentum of the gluon and r is an arbitrary massless reference four-vector. We showed explicitly that this expression agrees with our familiar right-handed polarization vector. However, it is enlightening to show that this indeed carries spin-1 from its properties under rotations.

For simplicity, align the three-momentum of the gluon $\vec{p}$ along the $+\hat{z}$-axis and perform a rotation by an angle ϕ about the $+\hat{z}$-axis. You only have to rotate the spinors that depend on the gluon momentum p, and a spinor u is rotated by a matrix $\mathbb{M}$ as

$$u \to \mathbb{M}u. \tag{7.166}$$

For a rotation about the $+\hat{z}$-axis by an angle ϕ, the rotation matrix is

$$\mathbb{M}(\phi) = \exp\left[i\frac{\phi}{2}\begin{pmatrix} 1 & 0 \\ 0 & -1 \end{pmatrix}\right]. \tag{7.167}$$

How does the polarization vector $\epsilon_R(p)$ transform under this rotation? Does this transformation tell you what the spin of the gluon is?

Hint: What is the smallest angle ϕ for which the polarization vector is unchanged by this rotation?

(c) The four-vector r is arbitrary and cannot in any way affect the physical properties of the gluon. Performing a rotation of the spinors that depend on r, show that the polarization vector $\epsilon_R(p)$ is completely unaffected. You can now align the momentum r along the $+\hat{z}$-axis and then rotate the r spinors about the $+\hat{z}$-axis by an angle ϕ. This demonstrates that the vector r does not affect the spin of the gluon at all.

7.7 *The Drell–Yan Process.* In Example 7.1, we analyzed the Drell–Yan process in which lepton pairs are produced from hadron collisions. In Eq. 7.74, we were able to construct a very useful expression for the cross section differential in the invariant mass of the final-state leptons Q^2 and their rapidity y. A measurement of the rapidity distribution of the Z boson from proton–anti-proton collisions at the Tevatron was presented in Fig. 7.2. In this exercise, we will use these data and the expression for the cross section to extract the functional form of the parton distribution function for quarks in the proton, $f_q(x)$. This will be a very simplified procedure; collaborations that do this combine multiple data sets from different experiments. Nevertheless, we will be able to see interesting features of the pdf in this exercise. For ease of reference, we reprint the differential cross section here:

$$\frac{d^2\sigma(p\bar{p} \to e^+e^-)}{dQ^2\,dy} = \frac{4\pi\alpha^2}{9Q^2 s} \sum_{q\text{ quarks}} Q_q^2 f_q\left(\sqrt{\frac{Q^2}{s}}e^y\right) f_{\bar{q}}\left(\sqrt{\frac{Q^2}{s}}e^{-y}\right). \tag{7.168}$$

(a) If the invariant mass of the final-state leptons Q^2 is restricted to be the square of the mass of the Z boson m_Z^2, express the momentum fraction x_2 of the anti-quark in terms of x_1, s, and m_Z.

(b) As a function of x_1, m_Z and s, what is the rapidity y of the Z boson? You should find

$$y = \log \frac{x_1 \sqrt{s}}{m_Z}. \tag{7.169}$$

(c) The Tevatron collided protons and anti-protons at a center-of-mass energy of 1.96 TeV. From the result of part (b), what is the maximum value of rapidity of the Z? Does this agree with the plot of data in Fig. 7.2?

(d) This rapidity distribution can be used to extract the parton distribution functions of the quarks in the proton or anti-proton. An anti-proton consists of two valence anti-up quarks and one valence anti-down quark, as it's just the anti-particle of the proton. Thus, in $p\bar{p} \to Z$ events at the Tevatron, typically the quark comes from the proton and the anti-quark from the anti-proton. We expect that the pdf of a quark in a proton is identical to that of an anti-quark in an anti-proton because of symmetry under charge conjugation.

The power of Eq. 7.168 is that the only dependence on rapidity y appears in the parton distribution functions for the quark and anti-quark in collision. This tells us that there is a direct map from the rapidity distribution in Fig. 7.2 to the parton distribution function $f_q(x)$, regardless of the form of the rest of the cross section.

Interpreting Fig. 7.2 as a probability distribution $p(y)$, determine the functional relationship between the quark pdf $f_q(x)$ and the rapidity distribution $p(y)$. Don't forget the Jacobian factor, and note that the rapidity should be evaluated at the value determined in part (b). You should find

$$xf_q(x) = p(y)\big|_{y=\log \frac{x\sqrt{s}}{m_Z}}. \tag{7.170}$$

(e) To determine the pdf $f_q(x)$, the experiments would take their rapidity data point by point and map them to a pdf point by point in x. For the sake of brevity here, just take ten data points on Fig. 7.2 and map them to values of the pdf $f_q(x)$. From these ten points, make a table of momentum fraction x values versus the pdf value times momentum fraction $xf_q(x)$. Make sure the points are relatively uniform over $0 < |y| < 3$, so that you get good coverage.

(f) Figure 7.6 is a plot from the CTEQ collaboration (Coordinated Theoretical–Experimental Project on QCD; read: "see-teck") of their detailed and precise extraction of parton distributions from various collider data. In this plot, the parton center-of-mass collision energy is $\sqrt{\hat{s}} = 85$ GeV, which is close (enough) to the mass of the Z boson for comparison. While not labeled on the plot, the abscissa is the momentum fraction x, and the ordinate is $xf_q(x)$.

Focus on the pdfs of the up and down quarks (the two pdfs with maxima visible on the plot). Compare their distributions to your extraction from part (e).

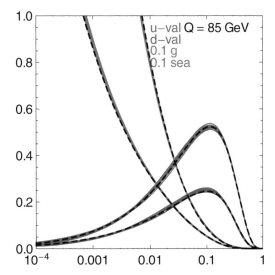

Fig. 7.6   Parton distribution functions from the CTEQ collaboration at a parton collisions energy of $\sqrt{\hat{s}} = 85$ GeV. The abscissa is the parton momentum fraction x and the ordinate is the momentum fraction times the pdf $xf(x)$. From P. Nadolsky *et al.*, "Progress in CTEQ-TEA PDF Analysis," doi:10.3204/DESY-PROC-2012-02/301 [arXiv:1206.3321 [hep-ph]].

How do they compare; e.g., are the peaks in about the same place with the same height?

7.8   *Thrust in Data.* Thrust has been extensively measured in data. A huge collection of plots and tables of measurements from e^+e^- collision experiments are collected at the website `hepdata.cedar.ac.uk/review/shapes/`. This is a very useful tool for data archiving as well as for comparisons of theory predictions to data. In this exercise, we will make plots of distributions from data of thrust and compare to our theoretical prediction from Eq. 7.141.

Navigate to this website and click on one of the links for "THRUST" (they all go to the same place). There are a lot of measurements here! The cool thing about this website is that you can select several measurements and overlay their plots. To see this at work, first, at the top of the page, find the "plot" button. To the right of that button, select the x- and y-axes as both linear ("lin").

From here, we'll select some experimental results to plot on top of one another. We'll look at the measurement of thrust in collisions at a center-of-mass energy of 91.2 GeV corresponding to the mass of the Z boson. A common way to plot thrust is as $1 - \tau$ rather than τ, which is what we will do here. Measurements that can be directly compared are:

- ALEPH (Heister *et al.* ZP C35(2004)457 [R]), Thrust 91.2 GeV
- DELPHI (Abreu *et al.* Zeit.Phys.C73(1997)11 [R]), 1-Thrust 91.2 GeV charged+neutral
- DELPHI (Abreu *et al.* Zeit.Phys.C59(1993)21 [R]), Thrust 91.2 GeV

- OPAL (Abbiendi *et al.* Eur.Phys.J.C40(2005)287 [R]), 1-Thrust 91.2 GeV
- SLD (Abe *et al.* Phys.Rev.D51(1995)962 [R]), Thrust 91.2 GeV.

Select these measurements and then click "plot" at the top of the page. You should then see an overlaid plot of the data from these experiments.

Each set of color points corresponds to a different measurement. These distributions are normalized (integrate to 1), so we'll only be able to compare the shape of our analytic prediction from Eq. 7.141 to these data. Plot this prediction and rescale it by multiplying by an overall constant to make it as close as possible to the data plots. How does the prediction compare to the data? What differences do you notice? How do the endpoints of data and prediction compare? Do both distributions exhibit the same behavior as $1 - \tau \to 0$?

In Chapter 9, we'll discuss the problems with the prediction and the first steps to correct it.

7.9   *Research Problem.* In our analysis of DIS or Drell–Yan, we had to posit the existence of parton distribution functions to describe constituent quarks in the proton. Parton distributions are extremely well tested and enable quantitative results for scattering processes involving hadrons. Nevertheless, pdfs are an approximation for a particular property of the proton wavefunction. What is the limit of this approximation? What is the complete set of scattering processes for which parton distributions are sensible?

8

Quantum Chromodynamics

Models are useful tools for providing a framework in which to think about a problem. For example, the quark model enables a powerful organizing principle for thinking about the zoo of hadrons and their relationships. Nevertheless, models always have limitations. The quark model provides no mechanism for why or how quarks are bound into hadrons in the first place, nor can the quark model explain the phenomena of Bjorken scaling and three-jet production that we observed in Chapter 7. The parton model has similar limitations. While it provides a concrete explanation for Bjorken scaling, the parton model doesn't shed light on why the form factor $F_2(x)$ has the form it does from Fig. 7.1. Neither of these models can claim to be the definitive explanation of the interactions of quarks and gluons.

Models have brought us a long way, but we want a theory of this strong force that binds quarks into hadrons. A scientific theory provides the explanation without assumptions or qualifications that limit its realm of applicability. For developing the theory of the strong force, we will follow our noses guided by principles of quantum mechanics. This theory must conserve probability and this restriction will essentially provide a unique understanding and interpretation for the three colors of quarks we need to explain data. Quarks and gluons are spin-1/2 and spin-1 particles, respectively, and enforcing conservation of angular momentum restricts their possible interactions. As the gluon is a massless, spin-1 particle like the photon, it will have a gauge symmetry that is necessary for it to have only two physical degrees of freedom. Consideration of all of these constraints will lead us to quantum chromodynamics, or QCD.

In this chapter, we construct QCD and then discuss some of its consequences. The formulation of QCD is in many ways similar to that of electromagnetism, studied in Section 2.2.3, but there are small, but profound, differences. Unlike the photon, the gluon is charged under the force that it carries and so interacts with itself. This feature of QCD results in the phenomenon of asymptotic freedom, for which the strength of the interactions between quarks and gluons decreases at higher energies. Asymptotic freedom provides the explanation of Bjorken scaling and the existence of jets, both of which we explore in detail in Chapter 9.

8.1 Color Symmetry

The first thing we need to do to formulate QCD is to understand the consequences of the property of quark color. To keep our discussion simple, let's consider just one massless quark, described by a spinor solution to the Dirac equation, ψ. This quark is allowed to have any of three colors, red, green, or blue, which we will denote by an index i: ψ_i. $i = 1$ is red, $i = 2$ is green, and $i = 3$ is blue. As in our discussion of isospin from Section 3.3, in this colored world we say that the three colors represent a symmetry. Any linear combination of different colored quarks produces an equally valid description of the physics. Also like isospin, this color symmetry has nothing to do with Lorentz transformations; changing a quark's color does not rotate its spin or change its momentum. Such a symmetry that is independent of spacetime transformations is referred to as an **internal symmetry**. A general theorem of quantum field theory called the **Coleman–Mandula theorem** states that the only possible collection of symmetries are spacetime symmetries (rotations, boosts, and the like) and internal symmetries.[1]

We can transform ψ_i as

$$\psi_i \rightarrow U_{ij}\psi_j = U_{i1}\psi_1 + U_{i2}\psi_2 + U_{i3}\psi_3 \,, \tag{8.1}$$

for some numbers U_{ij}, and there must be no change in the result of any possible measurement we could do. We're dealing with quantum mechanics, and so there are constraints on the U_{ij}. As a wavefunction, ψ_i represents the probability amplitude for measuring a quark with color i. Also, we can take ψ_1, ψ_2, and ψ_3 to be an orthonormal color basis. With this set-up, the probability density for color i then transforms as

$$\begin{aligned}
\psi^{\dagger i}\psi_i &\rightarrow (U_{i1}\psi_1 + U_{i2}\psi_2 + U_{i3}\psi_3)^\dagger (U_{i1}\psi_1 + U_{i2}\psi_2 + U_{i3}\psi_3) \\
&= U_{i1}^* U_{1i}\psi^{\dagger 1}\psi_1 + U_{i2}^* U_{2i}\psi^{\dagger 2}\psi_2 + U_{i3}^* U_{3i}\psi^{\dagger 3}\psi_3 \,.
\end{aligned} \tag{8.2}$$

The subscript and superscript placement of color indices denotes color and anti-color, respectively, so that the probability density $\psi^{\dagger i}\psi_i$ has total color 0. The orthogonal assumption enables us to set products of different color quarks to zero, $\psi^{\dagger i}\psi_j = 0$, for $i \neq j$. The normalization assumption means that the product of a quark and an anti-quark is the same, regardless of color:

$$\psi^{\dagger 1}\psi_1 = \psi^{\dagger 2}\psi_2 = \psi^{\dagger 3}\psi_3 \,. \tag{8.3}$$

Then, to preserve probability, we must enforce

$$|U_{i1}|^2 + |U_{i2}|^2 + |U_{i3}|^2 = 1 \,. \tag{8.4}$$

We can use vector notation to represent any linear combination of the three colored quarks. Let's take the spinor ψ with no index to denote an arbitrary linear combination of quark colors:

$$\psi = a_1\psi_1 + a_2\psi_2 + a_3\psi_3 \equiv \begin{pmatrix} a_1 \\ a_2 \\ a_3 \end{pmatrix}, \tag{8.5}$$

---

[1] S. R. Coleman and J. Mandula, "All possible symmetries of the S matrix," Phys. Rev. **159**, 1251 (1967). It was later realized that this could be extended slightly by introduction of a new spacetime symmetry called supersymmetry. See R. Haag, J. T. Lopuszanski and M. Sohnius, "All possible generators of supersymmetries of the S-matrix," Nucl. Phys. B **88**, 257 (1975).

for some complex numbers a_1, a_2, a_3 such that ψ is normalized so

$$(a_1^* \ a_2^* \ a_3^*) \begin{pmatrix} a_1 \\ a_2 \\ a_3 \end{pmatrix} = |a_1|^2 + |a_2|^2 + |a_3|^2 = 1 \,. \tag{8.6}$$

A color rotation can be expressed compactly by action of a matrix $\mathbb{U}$ on the spinor ψ:

$$\psi \to \mathbb{U}\psi \,, \tag{8.7}$$

where the action of $\mathbb{U}$ on a quark of a given color i is defined by Eq. 8.1. Conservation of probability is then just the statement that $\mathbb{U}$ is a unitary matrix: $\mathbb{U}^\dagger \mathbb{U} = \mathbb{I}$. $\mathbb{U}$ is a 3×3 matrix because there are three colors, and so it is an element of the group U(3). We can restrict the determinant of $\mathbb{U}$ to $\det \mathbb{U} = 1$ so that the matrix only implements proper color rotations. Then, rotations of colors into one another are implemented by matrices in the group SU(3). Quarks transform in the three-dimensional fundamental representation of color SU(3).

    A general 3×3 matrix with complex entries has 18 parameters: 9 real and 9 imaginary numbers. The unitarity constraint fixes 9 of the parameters (one relation for each entry of the matrix). Restricting to SU(3) matrices $\mathbb{U}$ with $\det \mathbb{U} = 1$ provides one more constraint. Therefore, an arbitrary 3×3 matrix that is an element of SU(3) has 8 independent parameters. The Lie algebra of the group SU(3), denoted by $\mathfrak{su}(3)$, is therefore an eight-dimensional vector space with basis matrices T^a, where $a = 1, \ldots, 8$. These basis matrices are Hermitian, $(T^a)^\dagger = T^a$, and a general element of SU(3) can be expressed by exponentiating the Lie algebra:

$$\mathbb{U} = e^{i\alpha^a T^a} \,, \tag{8.8}$$

where α^a are eight real constants, indexed by a. This formalism should be familiar from our general analysis in Chapter 3, but applied to SU(3) in particular. In the following example, we will explicitly construct the Lie algebra basis matrices of $\mathfrak{su}(3)$.

Example 8.1 What are the eight matrices T^a that span the Lie algebra $\mathfrak{su}(3)$?

Solution

We're looking for eight 3×3 matrices that are linearly independent. By unitarity of elements of SU(3), the matrices in the Lie algebra are Hermitian: $(T^a)^\dagger = T^a$. Also, a matrix $\mathbb{U}$ in SU(3) has determinant 1. In terms of exponentiation of the Lie algebra, this determinant constraint imposes

$$\det \mathbb{U} = \det e^{i\alpha^a T^a} = \lambda_1 \lambda_2 \lambda_3 = 1 \,. \tag{8.9}$$

On the right, we have identified the determinant with the product of eigenvalues λ_1, λ_2, and λ_3 of the matrix $\mathbb{U}$.

    Now for a bit of trickery. Let's consider taking the logarithm of matrix $\mathbb{U}$. The trace of $\log \mathbb{U}$ is the sum of its eigenvalues. These are just the logarithms of the eigenvalues of $\mathbb{U}$ itself:

$$\text{tr} \log \mathbb{U} = \log \lambda_1 + \log \lambda_2 + \log \lambda_3 \,. \tag{8.10}$$

While we won't prove this, all that's needed to do so is to note that $\mathbb{U}$ is non-singular and that it can be put into diagonal form with an orthogonal matrix formed from its eigenvectors. Because $\det \mathbb{U} = 1$, $\operatorname{tr} \log \mathbb{U} = 0$, which immediately tells us that elements of the Lie algebra are traceless:

$$\operatorname{tr} \log \mathbb{U} = i\alpha^a \operatorname{tr} T^a = 0 \, . \tag{8.11}$$

So, the Lie algebra $\mathfrak{su}(3)$ consists of eight 3×3 matrices T^a that are Hermitian, traceless, linearly independent, and satisfy the commutation relation

$$[T^a, T^b] = i f^{abc} T^c \, . \tag{8.12}$$

The standard basis for $\mathfrak{su}(3)$ are called the **Gell-Mann matrices** and are

$$T^1 = \frac{1}{2}\begin{pmatrix} 0 & 1 & 0 \\ 1 & 0 & 0 \\ 0 & 0 & 0 \end{pmatrix}, \quad T^2 = \frac{1}{2}\begin{pmatrix} 0 & -i & 0 \\ i & 0 & 0 \\ 0 & 0 & 0 \end{pmatrix}, \quad T^3 = \frac{1}{2}\begin{pmatrix} 1 & 0 & 0 \\ 0 & -1 & 0 \\ 0 & 0 & 0 \end{pmatrix},$$

$$T^4 = \frac{1}{2}\begin{pmatrix} 0 & 0 & 1 \\ 0 & 0 & 0 \\ 1 & 0 & 0 \end{pmatrix}, \quad T^5 = \frac{1}{2}\begin{pmatrix} 0 & 0 & -i \\ 0 & 0 & 0 \\ i & 0 & 0 \end{pmatrix}, \quad T^6 = \frac{1}{2}\begin{pmatrix} 0 & 0 & 0 \\ 0 & 0 & 1 \\ 0 & 1 & 0 \end{pmatrix},$$

$$T^7 = \frac{1}{2}\begin{pmatrix} 0 & 0 & 0 \\ 0 & 0 & -i \\ 0 & i & 0 \end{pmatrix}, \quad T^8 = \frac{1}{2\sqrt{3}}\begin{pmatrix} 1 & 0 & 0 \\ 0 & 1 & 0 \\ 0 & 0 & -2 \end{pmatrix} \, . \tag{8.13}$$

You'll study more properties of $\mathfrak{su}(3)$ and the Gell-Mann matrices in Exercise 8.6.

Let's now see how this SU(3) color symmetry works in a Lagrangian for the quark. The Dirac Lagrangian for a massless quark, from which the Dirac equation can be derived, is

$$\mathcal{L} = \bar{\psi} i\gamma \cdot \partial \psi = \bar{\psi}^i i\gamma \cdot \partial \psi_i \, . \tag{8.14}$$

On the right, the sum over colors $i = 1, 2, 3$ is implemented with Einstein summation notation. That is,

$$\bar{\psi}^i i\gamma \cdot \partial \psi_i = \bar{\psi}^1 i\gamma \cdot \partial \psi_1 + \bar{\psi}^2 i\gamma \cdot \partial \psi_2 + \bar{\psi}^3 i\gamma \cdot \partial \psi_3 \, . \tag{8.15}$$

What happens to the Dirac Lagrangian under a color rotation? We know how ψ transforms:

$$\psi \to \mathbb{U}\psi \, , \tag{8.16}$$

and, as its conjugate, $\bar{\psi}$ transforms as

$$\bar{\psi} \to \bar{\psi}\mathbb{U}^\dagger \, . \tag{8.17}$$

Under a color rotation, the Dirac Lagrangian transforms as

$$\mathcal{L} = \bar{\psi} i\gamma \cdot \partial \psi \to \bar{\psi}\mathbb{U}^\dagger i\gamma \cdot \partial \mathbb{U}\psi = \bar{\psi} i\gamma \cdot \partial \psi \, , \tag{8.18}$$

where we have used that $\mathbb{U}$ is just a constant, unitary matrix and so the derivative passes through it. So, the Dirac Lagrangian is invariant under SU(3) color transformations.

However, does this conclusion actually make sense? Is the Dirac Lagrangian of the quark just invariant to color rotations, full stop? Our arguments for properties of color rotations have so far only used restrictions from quantum mechanics, such as conservation of probability. However, there are also restrictions from special relativity, and the implicit assumption that an SU(3) color matrix $\mathbb{U}$ is independent of spacetime position is not the most general nor the preferred structure.

8.2 Non-Abelian Gauge Theory

So far, we've written a color matrix $\mathbb{U}$ as

$$\mathbb{U} = e^{i\alpha^a T^a}, \tag{8.19}$$

for $\mathfrak{su}(3)$ Lie algebra elements T^a and constants α^a, independent of spacetime position. Do we have to make this assumption? No, and actually special relativity would prefer that we do not.

To see what other constraints are imposed by special relativity, let's perform the following thought experiment. Consider two people, call them Emmy and Albert, located at opposite sides of the universe from one another. They are studying color transformations of quarks and each defines a basis for color. Emmy's color basis for quarks is, say, ψ, while Albert's is ψ'. Because there are three colors and probability is conserved, there must exist some unitary matrix $\mathbb{U} \in \mathrm{SU}(3)$ that relates the two bases:

$$\psi = \mathbb{U}\psi'. \tag{8.20}$$

So, in principle, Emmy and Albert could communicate to figure out this unitary transformation and align their color bases.

However, they are very far apart, and can only exchange information about their color bases at the speed of light. Therefore, they cannot instantaneously align their color bases. Of course, they didn't have to be on opposite sides of the universe; they could have been in the same room, or even only separated by the radius of the proton. As long as they are not at exactly the same spacetime point, they can't instantaneously align their color bases. From the perspective of special relativity, it is more natural to use a different color basis at every spacetime point. The different bases will be reconciled and related to one another by an object that implements a color rotation that travels at the speed of light. Spoiling the punchline slightly, this will turn out to be the gluon.

So, instead of considering unitary color transformations that are independent of spacetime position, we consider transformations that are general functions of position:

$$\mathbb{U}(x) = e^{i\alpha^a(x) T^a}, \tag{8.21}$$

where now the coefficients $\alpha^a(x)$ depend on the position four-vector x. We are still allowed to use the same set of basis matrices T^a because we are always describing the group of symmetries for SU(3) color. Spacetime-dependent unitary transformations are referred to as **gauge transformations**. That is, we will now study the SU(3) color gauge theory.

While this change might seem small, it has profound consequences. Let's look at the Dirac Lagrangian again with this spacetime-dependent SU(3) unitary transformation:

$$\bar{\psi} i\gamma \cdot \partial \psi \to \bar{\psi} \mathbb{U}^\dagger i\gamma \cdot \partial \mathbb{U}\psi \,. \tag{8.22}$$

Now the derivative doesn't commute with the matrix $\mathbb{U}$ so we can't just set $\mathbb{U}^\dagger \mathbb{U} = \mathbb{I}$. The action of the spacetime derivative on the matrix $\mathbb{U}$ from Eq. 8.21 is

$$i\gamma \cdot \partial \mathbb{U} = i\left(i(\gamma \cdot \partial \alpha^a(x))T^a \mathbb{U} + \mathbb{U}\gamma \cdot \partial\right) = \mathbb{U}i\gamma \cdot \partial - T^a \mathbb{U}\gamma \cdot \partial \alpha^a(x) \,. \tag{8.23}$$

Using this, the Dirac Lagrangian transforms under a color rotation as

$$\mathcal{L} = \bar{\psi} i\gamma \cdot \partial \psi \to \bar{\psi}\mathbb{U}^\dagger i\gamma \cdot \partial \mathbb{U}\psi = \mathcal{L} - \bar{\psi}T^a \left(\gamma \cdot \partial \alpha^a(x)\right)\psi \,, \tag{8.24}$$

which is no longer invariant! This is just another way of saying that color bases at different spacetime points are in general different. The interpretation of this is the following. The derivative operator ∂_μ slightly displaces the fermion fields that appear in the Dirac Lagrangian, just as the derivative displaces the argument of a function through the Taylor expansion. We saw this when understanding the Lorentz invariance of the Dirac Lagrangian in Section 2.2.2. This implies that the unitary matrix that implements a color rotation of $\bar{\psi}$ is no longer exactly the Hermitian conjugate of the matrix that transforms ψ. The amount by which these two unitary transformations are not Hermitian conjugates is given by the final term in Eq. 8.24. If the coefficients $\alpha^a(x)$ change rapidly with spacetime position, then SU(3) color matrices can significantly differ at neighboring spacetime points.

To restore invariance of the Lagrangian, we need to introduce an object that can rotate color bases at the speed of light. Recall that we want the Lagrangian to be invariant under color rotations because we believe that this is a symmetry of Nature. As a symmetry, by Noether's theorem, this means that color is conserved in interactions. This highly restricts the possible interactions of quarks, which is a good thing for predictivity!

8.2.1 Covariant Derivative

With this in mind, what can we possibly do to construct a Lagrangian that is invariant to a spacetime-dependent color rotation? We can't change the transformation properties of the quarks, but we can introduce a new field which transforms in such a way as to accomplish this invariance. Let's introduce a field A_μ^a that has a transformation law that exactly cancels the leftover term in Eq. 8.24. This field has an index a which corresponds to the color $\mathfrak{su}(3)$ Lie algebra, which is necessary because the residual transformation of the Lagrangian in Eq. 8.24 contains a T^a matrix. This new field also has a Lorentz index μ, which is necessary because the Dirac Lagrangian has γ^μ matrices in it. Then, we will consider the augmented Lagrangian

$$\mathcal{L} = \bar{\psi}\left(i\gamma \cdot \partial + g\gamma \cdot A^a T^a\right)\psi \,, \tag{8.25}$$

where g is called the **coupling** that controls the strength of the interaction between the quark and the field A_μ^a. The Lie algebra index a is summed over. For this Lagrangian to be invariant under a color rotation, the field A_μ^a must transform inhomogeneously:

$$A_\mu^a \to A_\mu^a + \Delta A_\mu^a \,, \tag{8.26}$$

where ΔA_μ^a is the transformation. Let's determine what this must be.

Performing a color rotation on the new Lagrangian of Eq. 8.25, we have

$$\bar{\psi}\left(i\gamma\cdot\partial + g\gamma\cdot A^a T^a\right)\psi \to \bar{\psi}\mathbb{U}^\dagger\left(i\gamma\cdot\partial + g\gamma\cdot A^a T^a + g\gamma\cdot\Delta A^a T^a\right)\mathbb{U}\psi \qquad (8.27)$$
$$= \bar{\psi}\gamma\cdot\partial\psi - \bar{\psi}\gamma\cdot\partial\alpha^a(x)T^a\psi + \bar{\psi}\mathbb{U}^\dagger\left(g\gamma\cdot A^a T^a + g\gamma\cdot\Delta A^a T^a\right)\mathbb{U}\psi.$$

For this to be invariant, we need everything to the right of $\bar{\psi}\gamma\cdot\partial\psi$ to equal $\bar{\psi}g\gamma\cdot A^a T^a\psi$:

$$-\bar{\psi}\gamma\cdot\partial\alpha^a(x)T^a\psi + \bar{\psi}\mathbb{U}^\dagger\left(g\gamma\cdot A^a T^a + g\gamma\cdot\Delta A^a T^a\right)\mathbb{U}\psi = \bar{\psi}g\gamma\cdot A^a T^a\psi, \qquad (8.28)$$

or, by removing the quark and anti-quark fields, that

$$-\partial_\mu\alpha^a(x)T^a + g\mathbb{U}^\dagger A_\mu^a T^a\mathbb{U} + g\mathbb{U}^\dagger\Delta A_\mu^a T^a\mathbb{U} = gA_\mu^a T^a. \qquad (8.29)$$

We can solve for the transformation ΔA_μ^a by multiplying by $\mathbb{U}$ on the left and $\mathbb{U}^\dagger$ on the right, using $\mathbb{U}^\dagger\mathbb{U} = \mathbb{I}$:

$$-\mathbb{U}\left(\partial_\mu\alpha^a(x)\right)T^a\mathbb{U}^\dagger + gA_\mu^a T^a + g\Delta A_\mu^a T^a = g\mathbb{U}A_\mu^a T^a\mathbb{U}^\dagger. \qquad (8.30)$$

Rearranging, we find the transformation to be

$$\Delta A_\mu^a T^a = \mathbb{U}A_\mu^a T^a\mathbb{U}^\dagger + \frac{1}{g}\mathbb{U}\left(\partial_\mu\alpha^a(x)T^a - g\mathbb{U}^\dagger A_\mu^a T^a\mathbb{U}\right)\mathbb{U}^\dagger. \qquad (8.31)$$

An exceptionally nice way to package the field A_μ^a is in a **covariant derivative** D_μ, which is defined as

$$D_\mu \equiv \partial_\mu - igA_\mu^a T^a. \qquad (8.32)$$

Under a spacetime-dependent color rotation, the covariant derivative transforms to

$$D_\mu \to \partial_\mu - igA_\mu^a T^a - ig\Delta A_\mu^a T^a \qquad (8.33)$$
$$= \partial_\mu - igA_\mu^a T^a - ig\left[\mathbb{U}A_\mu^a T^a\mathbb{U}^\dagger + \frac{1}{g}\mathbb{U}\left(\partial_\mu\alpha^a(x)T^a - g\mathbb{U}^\dagger A_\mu^a T^a\mathbb{U}\right)\mathbb{U}^\dagger\right]$$
$$= \partial_\mu - i\mathbb{U}\left(gA_\mu^a + \partial_\mu\alpha^a(x)\right)T^a\mathbb{U}^\dagger.$$

This transformation can be equivalently expressed in the compact form

$$D_\mu \to \mathbb{U}D_\mu\mathbb{U}^\dagger. \qquad (8.34)$$

This simple transformation is why it is called a "covariant" derivative. Then, we can write our new Lagrangian as

$$\mathcal{L} = \bar{\psi}i\gamma\cdot D\psi. \qquad (8.35)$$

This then transforms as

$$\mathcal{L} = \bar{\psi}i\gamma\cdot D\psi \to \bar{\psi}\mathbb{U}^\dagger i\gamma^\mu\left(\mathbb{U}D_\mu\mathbb{U}^\dagger\right)\mathbb{U}\psi = \bar{\psi}i\gamma\cdot D\psi, \qquad (8.36)$$

which is indeed invariant under an SU(3) gauge transformation. As such, we refer to this Lagrangian as SU(3) color gauge invariant.

So, in summary, to construct an SU(3) color-invariant Lagrangian for the quarks, we covariantize the derivative by introducing a new field A_μ^a whose transformation exactly cancels that of the quarks. The covariant derivative D_μ respects the gauge transformation properties of quarks, or any field that carries color. In differential geometry, this field A_μ^a is referred to as the **connection** of the manifold SU(3). This connection enables us, in a well-defined way, to differentiate on this non-linear manifold. Consequently, with a well-defined derivative, we can perform spacetime translation of fields via a Taylor expansion to give them kinetic energy. We'll explore these esoteric mathematical observations in the following section.

8.2.2 Connections and Curvature

The kinetic energy, and therefore the Lagrangian, of the field A_μ^a must be both SU(3) color gauge invariant and Lorentz invariant, just like the Lagrangian for the quarks, Eq. 8.35. From the earlier arguments, to construct the Lagrangian for A_μ^a we must use the covariant derivative, D_μ. One way to do this is to systematically add terms into a potential Lagrangian to eventually make it gauge invariant, a technique called the **Noether procedure**. In some ways, this would be similar to our approach for constructing the scalar field Lagrangian in Section 2.2.1. Here, however, we'll take a different and much simpler approach.

The kinetic energy of A_μ^a, or any object in general, is a measure of the size of the derivative of A_μ^a. The derivative of A_μ^a, in turn, implements a small displacement in spacetime position of the field. If the field A_μ^a changes rapidly (corresponding to large kinetic energy), then the color basis with which the covariant derivative is defined also changes rapidly in spacetime. Similarly, if the field A_μ^a changes slowly (corresponding to small kinetic energy), then the color basis with which the covariant derivative is defined also changes slowly in spacetime. So, if we had a way to measure how quickly the color basis changed in spacetime, this would provide a measure of kinetic energy.

Determining how the color basis changes in spacetime is a subtle procedure, however. For instance, let's consider two nearby spacetime points, x_1 and x_2. We imagine traveling from x_1 to x_2 along a path ℓ like so:

As we travel from x_1 to x_2, we want to measure how much our color basis is rotated. Then, once at point x_2, we can compare before and after. Unfortunately, this comparison isn't well defined in general. We can't in a well-defined way compare quantities at two separate points on a non-linear manifold. As an illustration of this, consider two people, one in Minsk, Belarus, and the other in Johannesburg, South Africa.[2] If you ask both of those people to point "up," they will point something like:

---

[2] Minsk and Johannesburg have almost identical longitudes, and the figure accurately represents their latitudes.

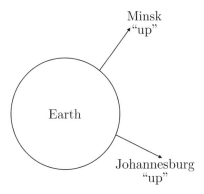

Though the person in Minsk and the person in Johannesburg are doing the exact same thing, these "up" directions are not the same! It's unclear how one can compare them, and actually one can't. What you can do, though, is to compare at the same point.

We can see a way forward to defining kinetic energy of A_μ^a by taking this analogy to the round Earth seriously. At every point in spacetime, there is an SU(3) color manifold which defines the color basis at that point. A schematic illustration of this idea is:

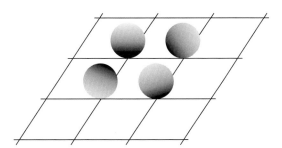

The grid corresponds to spacetime points while the light/dark gray spheres represent the internal color space. Note that the light/dark poles on these spheres are rotated with respect to one another at each point. As we move around in real spacetime, we move in the internal color space because the color basis everywhere in spacetime is in general different. The rate at which color bases change is a measure of kinetic energy of A_μ^a.

These color manifolds have been drawn suggestively: they are non-linear manifolds, in general. A non-linear manifold, simply, is one for which the Pythagorean theorem and Euclid's postulates of planar geometry do not hold. Colloquially, we would say that non-linear manifolds are not flat: they have a non-zero curvature. If the curvature of these manifolds is small, then one must perform a substantial color rotation between neighboring spacetime points to change the color basis; while if the curvature is large, then a small color rotation can produce a large change in color basis. We're getting closer to our goal. If we can determine the curvature of these color manifolds, then we have a kinetic energy.

A powerful way to measure curvature on a manifold is to use the notion of **parallel transport**. Let's introduce parallel transport by going back to our picture of the curvature of the Earth. Parallel transport can be used to prove that the Earth is curved and not flat. The

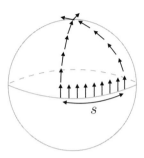

Fig. 8.1 An illustration of parallel transport of two vectors from the Earth's Equator to the North Pole. One vector travels directly to the North Pole, while the other first travels a distance s along the Equator, maintaining its orientation, and then north.

idea is the following. Start somewhere on the Equator of the Earth, say with two vectors that each point north. With each of these vectors, one takes a different route to the North Pole while maintaining the orientation of the vectors. First, we can take the route directly to the North Pole. We can also take the route that first travels a distance s along the Equator, and then goes to the North Pole along a line of longitude. Though different paths to the North Pole are taken, the vectors always point north; they are parallel transported from the Equator to the North Pole. An illustration of this is presented in Fig. 8.1.

Once both vectors reach the North Pole, we can then compare them. For the vector $\vec{v}_1$ along the first path that went directly to the North Pole, at the North Pole it is, say,

$$\vec{v}_1 = (1, 0). \tag{8.37}$$

For the vector $\vec{v}_2$ that first traveled along the Equator, once at the North Pole it is

$$\vec{v}_2 = \left(\cos\frac{s}{r}, \sin\frac{s}{r}\right). \tag{8.38}$$

The radius of Earth is r and therefore the angle θ about which the vector was rotated when traveling along the Equator is $\theta = s/r$. The magnitude of the difference between these vectors is

$$|\vec{v}_1 - \vec{v}_2| = 2\sin\frac{s}{2r}. \tag{8.39}$$

If $s \ll r$, then we can Taylor expand the sine. The curvature of the Earth, κ (the inverse of its radius), is then just the magnitude of the difference of these parallel transported vectors, divided by the difference in distance of the paths taken:

$$\kappa = \frac{1}{r} = \frac{|\vec{v}_1 - \vec{v}_2|}{s}, \tag{8.40}$$

for $s \ll r$.

With this result in mind, we want to perform a similar action for SU(3) color to determine the curvature, and therefore the kinetic energy, of the A_μ^a field. Our approach is the following. Consider parallel transport to spacetime point x_2 from x_1 along two different paths: in the μ direction first and then in the ν direction, and vice-versa. The picture of this is:

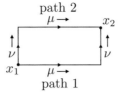

Concretely, we consider the Taylor expansion of quark field ψ about point x_1 to point x_2 in two different ways. Let's call the distance in the μ direction from x_1 to x_2 Δx while the distance in the ν direction is Δy. Then, we have

$$x_2 = x_1 + \Delta x + \Delta y. \tag{8.41}$$

Consider path 1 first. Along this path, going backward from x_2 to x_1, we Taylor expand in Δy first and then in Δx. This produces

$$\psi_{\text{path 1}}(x_1 + \Delta x + \Delta y) \tag{8.42}$$
$$= \psi(x_1 + \Delta x) + \Delta y^\nu D_\nu \psi(x_1 + \Delta x) + \cdots$$
$$= \psi(x_1) + \Delta x^\mu D_\mu \psi(x_1) + \Delta y^\nu D_\nu \psi(x_1) + \Delta x^\mu \Delta y^\nu D_\nu D_\mu \psi(x_1) + \cdots,$$

where terms at higher order in the expansion are in the $\cdots$. In the final line, the derivatives are taken with respect to point x_1. By using covariant derivatives, we have parallel transported the color of the quark from x_1 to x_2. For path 2, we do the Taylor expansions in the opposite order: first expand in Δx and then in Δy. This produces

$$\psi_{\text{path 2}}(x_1 + \Delta x + \Delta y) \tag{8.43}$$
$$= \psi(x_1 + \Delta y) + \Delta x^\mu D_\mu \psi(x_1 + \Delta y) + \cdots$$
$$= \psi(x_1) + \Delta x^\mu D_\mu \psi(x_1) + \Delta y^\nu D_\nu \psi(x_1) + \Delta x^\mu \Delta y^\nu D_\mu D_\nu \psi(x_1) + \cdots.$$

If the internal SU(3) color manifold were linear, then regardless of the path from x_1 to x_2 we would find the exact same expansion. However, as with the example of the Earth's curvature, we can determine the curvature of the internal manifold by taking the difference of the parallel transport of the quark along the two paths:

$$\psi_{\text{path 2}}(x_1 + \Delta x + \Delta y) - \psi_{\text{path 1}}(x_1 + \Delta x + \Delta y) \tag{8.44}$$
$$= \Delta x^\mu \Delta y^\nu \left(D_\mu D_\nu - D_\nu D_\mu \right) \psi(x_1),$$

where we have ignored terms in the expansion at cubic and higher orders in Δx and Δy.

The object in parentheses on the right side of Eq. 8.44 turns out to be independent of the quark ψ, as it should be. This difference of covariant derivatives is strictly a measure of the curvature of the internal SU(3) color space, and is independent of what objects are defined on it. This difference is the commutator of covariant derivatives:

$$D_\mu D_\nu - D_\nu D_\mu = [D_\mu, D_\nu] \tag{8.45}$$
$$= \partial_\mu \partial_\nu - ig \partial_\mu A_\nu^a T^a - ig A_\mu^a T^a \partial_\nu - g^2 A_\mu^a A_\nu^b T^a T^b - \partial_\nu \partial_\mu$$
$$+ ig \partial_\nu A_\mu^a T^a + ig A_\nu^a T^a \partial_\mu + g^2 A_\mu^a A_\nu^b T^b T^a$$
$$= -ig \left(\partial_\mu A_\nu^a T^a - \partial_\nu A_\mu^a T^a - ig A_\mu^a A_\nu^b [T^a, T^b] \right)$$
$$\equiv -ig F_{\mu\nu}^a T^a.$$

Using the commutation relations of the Lie algebra matrices T^a, the object $F^a_{\mu\nu}$ is defined to be

$$F^a_{\mu\nu} = \partial_\mu A^a_\nu - \partial_\nu A^a_\mu + g f^{abc} A^b_\mu A^c_\nu \,, \tag{8.46}$$

where f^{abc} are the structure constants of the $\mathfrak{su}(3)$ Lie algebra. Because SU(3) is a non-Abelian group, the structure constants cannot all be 0.

This commutator is called the **curvature tensor** in geometry and the **Yang–Mills field strength tensor** $F^a_{\mu\nu}$ in physics. It is named after Chen-Ning Yang and Robert Mills, who first constructed the non-Abelian case in the 1950s.[3] $F^a_{\mu\nu}$ is a measure of the curvature of the internal symmetry group SU(3) with connection A^a_μ. If the curvature is small, the connection A^a_μ changes slowly, $F^a_{\mu\nu}$ is small, and thus the radius of curvature is large. This corresponds to a long-wavelength or low-energy excitation produced by A^a_μ. By contrast, if the curvature is high, $F^a_{\mu\nu}$ is large and wavelengths are small. Thus, this corresponds to a high-energy excitation. This motivates $F^a_{\mu\nu}$ as a measure of the momentum of the field A^a_μ. As a momentum, we just need to square it appropriately, and we have the Lagrangian.

Just like the covariant derivative, $F^a_{\mu\nu}$ transforms covariantly under an SU(3) gauge transformation:

$$F^a_{\mu\nu} T^a \to e^{i\alpha^b(x)T^b} F^a_{\mu\nu} T^a e^{-i\alpha^b(x)T^b} = \mathbb{U} F^a_{\mu\nu} T^a \mathbb{U}^\dagger \,. \tag{8.47}$$

$F^a_{\mu\nu}$ has naked Lorentz indices (and so transforms under boosts and rotations) and is not invariant under color rotations. However, this can be made invariant by squaring and taking the trace over color matrices. The quantity

$$\mathrm{tr}\left[F^a_{\mu\nu} T^a F^{\mu\nu\,b} T^b\right] \tag{8.48}$$

is both Lorentz and color invariant. The Lorentz invariance is clear: all Lorentz indices μ and ν are contracted. To see the gauge invariance, note that the trace is cyclic. The SU(3) gauge transformation of the trace is

$$\mathrm{tr}\left[F^a_{\mu\nu} T^a F^{\mu\nu\,b} T^b\right] \to \mathrm{tr}\left[\mathbb{U} F^a_{\mu\nu} T^a \mathbb{U}^\dagger \mathbb{U} F^{\mu\nu\,b} T^b \mathbb{U}^\dagger\right] = \mathrm{tr}\left[F^a_{\mu\nu} T^a \mathbb{U}^\dagger \mathbb{U} F^{\mu\nu\,b} T^b \mathbb{U}^\dagger \mathbb{U}\right] \tag{8.49}$$
$$= \mathrm{tr}\left[F^a_{\mu\nu} T^a F^{\mu\nu\,b} T^b\right] \,.$$

By convention, we can normalize the T^a matrices so that

$$\mathrm{tr}[T^a T^b] = \frac{1}{2}\delta^{ab} \,, \tag{8.50}$$

which is called the **Killing form**.

Finally, the gauge-invariant, Lorentz-invariant description of the theory of SU(3) color symmetry is the Lagrangian

$$\mathcal{L}_{\mathrm{QCD}} = -\frac{1}{4} F^a_{\mu\nu} F^{\mu\nu\,a} + \bar{\psi} i\gamma \cdot D\psi \,. \tag{8.51}$$

---

[3] C. N. Yang and R. L. Mills, "Conservation of isotopic spin and isotopic gauge invariance," Phys. Rev. **96**, 191 (1954).

In this expression, we have already evaluated the Killing form, and so the Lie algebra index a is summed over. The $-1/4$ comes from matching the kinetic energy of the field A_μ^a to its canonical value. The field strength $F_{\mu\nu}^a$ creates gluons, which correspond to the field A_μ^a, the force carrier that talks to the quarks through the covariant derivative. This theory combines every observation we have discussed about the strong force: spin-1/2 point particle quarks, three colors, and a force carrier gluon. As it is the quantum theory of color it is called **quantum chromodynamics**, or QCD. Extremely importantly, connecting to our original motivation, note that there is no mass in the Lagrangian of QCD for the gluon field A_μ^a. The gluon is necessarily massless, which is a requirement for color conservation.[4]

By the way, general relativity can be formulated in the same language as SU(3) color. In this way we think of general relativity as the gauge theory of Lorentz transformations. That is, we allow every point in spacetime to have a different Lorentz transformation. This exactly corresponds to general covariance, or diffeomorphisms.

8.3 Consequences of Quantum Chromodynamics

Superficially, the Lagrangian of QCD in the form of Eq. 8.51 appears similar to the Lagrangian of quantum electrodynamics, QED, presented in Eq. 4.73. The QCD Lagrangian, however, has a Lie algebra index a on the field strength tensor and this makes all the difference. Because the group SU(3) is non-Abelian, the structure constants f^{abc} are not all 0 and so there are terms in Eq. 8.51 that are cubic and quartic in the gluon field A_μ^a. Such terms are responsible for self-interactions of the gluon, which is ultimately responsible for the phenomenon of asymptotic freedom. In this section, we study consequences of the QCD theory, described by the Lagrangian in Eq. 8.51.

8.3.1 Masslessness of the Gluon

As mentioned earlier, the gluon in QCD is massless. If the gluon were massive, then there would be a term in the Lagrangian of the form

$$\mathcal{L} \supset \frac{m^2}{2} A_\mu^a A^{\mu\, a} \,, \tag{8.52}$$

where m is the mass of the gluon as required by the Klein–Gordon equation. However, such a term is forbidden by color conservation as it is not invariant under an SU(3) color rotation. So, the gluon communicates color at the speed of light, just as the photon communicates electromagnetism at the speed of light. You will show that this term is indeed not gauge invariant in Exercise 8.1.

---

[4] If you are interested in the mathematical aspects of this section, they go under the name of "fiber bundles." A good book written for physicists that discusses this and much more of the relation of mathematics to physics is M. Nakahara, *Geometry, Topology and Physics*, Taylor & Francis (2003).

8.3.2 Gluon Degrees of Freedom

Though the gluon field A_μ^a has four components (denoted by the μ index), it actually only has two degrees of freedom. To see this, let's focus on $F_{\mu\nu}^a$, and ignore interactions (set $g = 0$). Then, we have

$$F_{\mu\nu}^a\big|_{g=0} = \partial_\mu A_\nu^a - \partial_\nu A_\mu^a \,. \tag{8.53}$$

Those gluons that are allowed to propagate, i.e., move through space and time, must have a time derivative in the Lagrangian. This ensures that their equation of motion has a time derivative. So, we must have, say, $\mu = 0$ corresponding to the time entry as

$$F_{0\nu}^a\big|_{g=0} = \partial_0 A_\nu^a - \partial_\nu A_0^a \,. \tag{8.54}$$

Naïvely, A_μ^a has four spacetime components, one for each μ index. However, $\mu = 0$ is not allowed to propagate. If $\mu = \nu = 0$, then $F_{00}^a = 0$, and so A_0^a has no time derivative in the Lagrangian. So, A_μ^a has (at most) only three degrees of freedom.

However, it's worse than that. A_μ^a transforms under color rotations as

$$A_\mu^a T^a \to \mathbb{U} A_\mu^a T^a \mathbb{U}^\dagger + \frac{1}{g} \mathbb{U} \left(\partial_\mu \alpha^a(x) \right) T^a \mathbb{U}^\dagger \,. \tag{8.55}$$

We have the freedom or ability to choose $\alpha^a(x)$ as we please. Doing this fixes a gauge, and can be used to remove another degree of freedom of A_μ^a. For instance, we might impose that

$$\partial \cdot A^a = 0 \,. \tag{8.56}$$

This eliminates a degree of freedom along the direction in which A_μ^a is propagating. To enforce this we just require that $\alpha^a(x)$ satisfies

$$\partial^2 \alpha^a(x) = 0 \,. \tag{8.57}$$

This is called Lorenz or Landau gauge.

So, there are only two propagating degrees of freedom of A_μ^a. These correspond to the two helicity configurations of the gluon: left- and right-handed.

8.3.3 Self-Interaction of the Gluon

Another prediction just from the Lagrangian of QCD is that gluons interact with themselves. This is very different than electromagnetism! Recall that the photon part of the electromagnetic Lagrangian is

$$\mathcal{L}_{\text{E\&M}} \supset -\frac{1}{4} F_{\mu\nu} F^{\mu\nu} = -\frac{1}{4} \left(\partial_\mu A_\nu - \partial_\nu A_\mu \right) \left(\partial^\mu A^\nu - \partial^\nu A^\mu \right) , \tag{8.58}$$

where A_μ is the photon field. The equation of motion found from varying the Lagrangian with respect to A_μ is

$$\partial_\mu F^{\mu\nu} = 0 = \partial^2 A^\nu - \partial^\nu \partial \cdot A = 0 \,. \tag{8.59}$$

In Lorenz or Landau gauge, $\partial \cdot A = 0$ and so the equation of motion reduces to the Klein–Gordon equation,

$$\partial^2 A^\nu = 0 . \tag{8.60}$$

That this is a linear differential equation (as are all of Maxwell's equations) has huge consequences for electromagnetic phenomena. One of the most central features of electromagnetism is the principle of superposition. That is, the electric or magnetic fields at any point can be found by just summing individual components of the fields that come from different sources.

In the context of the Klein–Gordon equation, if A_1^μ and A_2^μ are two electromagnetic potentials that each satisfy the Klein–Gordon equation, then so does their sum:

$$\partial^2 (A_1^\mu + A_2^\mu) = 0 . \tag{8.61}$$

Another way to state the property or of linearity/superposition of electromagnetism is in relation to the interaction of the photon. The photon carries no electric charge and so does not interact with itself. A photon traveling through space will pass right by another photon, without so much as a hello. Because photons do not interact with themselves, Maxwell's equations are linear. In a pithy way we might say that photons cannot beget more photons.

This is to be contrasted with the case in QCD. The field strength tensor for the gluon is

$$F_{\mu\nu}^a = \partial_\mu A_\nu^a - \partial_\nu A_\mu^a + g f^{abc} A_\mu^b A_\nu^c . \tag{8.62}$$

The final term, which is quadratic in the gluon fields, is not present in electromagnetism, and is responsible for gluons interacting with themselves.

From the pure gluon component of the QCD Lagrangian, we can vary it with respect to A_μ^a to determine the Euler–Lagrange equations of motion. One finds

$$\frac{\delta \mathcal{L}_{\text{gluon}}}{\delta A_\mu^a} = 0 = \frac{\delta}{\delta A_\mu^a} \left(-\frac{1}{4} F_{\mu\nu}^a F^{\mu\nu\,a} \right) \tag{8.63}$$

$$= \partial^\mu F_{\mu\nu}^a + g f^{abc} A^{\mu\,b} F_{\mu\nu}^c .$$

Expanded out, this is

$$\partial^2 A_\nu^a - \partial_\nu \partial \cdot A^a + g f^{abc} \partial^\mu (A_\mu^b A_\nu^c) + g f^{abc} A^{\mu\,b} \partial_\mu A_\nu^c \tag{8.64}$$
$$- g f^{abc} A^{\mu\,b} \partial_\nu A_\mu^c + g^2 f^{abc} f^{cde} A^{\mu\,b} A_\mu^d A_\nu^e = 0 .$$

This is a highly non-linear differential equation for the gluon potential field A_μ^a. In this expression, the importance of non-linear terms is controlled by the coupling of QCD, g. If g goes to 0, the equations of motion become linear and turn into the same expression as for electromagnetism, Eq. 8.59, but in general g is non-zero.

Because this equation of motion is non-linear, the field A_μ^a does not satisfy the principle of superposition. Two fields $A_{1\,\mu}^a$ and $A_{2\,\mu}^a$, each of which satisfy the equation of motion, cannot be summed into another solution. This makes QCD a hard theory to understand. It is not known how (or even if) the equations of motion can be solved exactly, in general. However, some special exact solutions to the equations of motion do exist. You'll study the instanton solution to the equations of motion in Exercise 8.3.

Another thing to note is that the gluon, unlike the photon, itself carries a charge of the force that it communicates. The gluon field A_μ^a has that SU(3) color index "a," which can take one of eight possible values as there are eight basis matrices of the $\mathfrak{su}(3)$ Lie algebra. We say that gluons carry color, and we might say that

Gluons beget more gluons.

We aren't yet to the bottom of all the weirdness of QCD.

Tests of Gluon Self-Interaction

QCD predicts that the gluon interacts with itself, and this should have experimental consequences that can be observed. One way to test the gluon self-interaction and whether it is as described by QCD is to study properties of the final state in $e^+e^- \to$ four jets events. We've exploited $e^+e^- \to$ two jets and $e^+e^- \to$ three jets previously to determine the spin of quarks and the gluon. However, in each of these processes, either no identified gluon was produced or it was produced just like a photon would be from accelerating charged particles. We can first directly probe the gluon self-interaction with four identified jets in the final state.

We won't present a complete calculation of the Feynman diagrams for four-jet production. Two of the Feynman diagrams for $e^+e^- \to$ four jets production with two gluon jets are

$$\mathcal{M}(e^+e^- \to q\bar{q}gg) \supset \qquad\qquad\qquad\qquad\qquad\qquad\qquad\qquad (8.65)$$

The first diagram would exist if the gluon were just like the photon. Multiple gluons can be emitted from color-charged quarks, just as photons can be emitted from electrically charged electrons. The second diagram has no analogue in electromagnetism and only exists because the gluon couples to itself. In particular, the terms in the QCD Lagrangian that are cubic in the gluon field A_μ^a are responsible for the gluon–gluon–gluon vertex in the second diagram. In observing $e^+e^- \to$ four jets events, that second diagram will affect the differential cross section and produce a prediction that differs from what one would expect if the gluon were just like the photon.

When squared to calculate the cross section, the diagrams pick up factors that count the number of colors that can be shared at each vertex. We've already seen one of these before.

In the calculation of the cross section for $e^+e^- \to q\bar{q}g$ in Section 7.3.4, we included a factor $C_F = 4/3$ which accounted for how gluons that are coupled to quarks could share color. C_F is called the **fundamental Casimir** of SU(3) and appears whenever particles in the fundamental representation of SU(3) color (like quarks) emit gluons. There are more colors of gluon than of quark and so when gluons couple to gluons, there are more colors that can be exchanged. Whenever the process has a diagram like the one on the second line of Eq. 8.65, factors of $C_A = 3$ appear, called the **adjoint Casimir** of SU(3). Thus a consequence of the non-Abelian nature of QCD for gluon self-interaction is its proportionality to C_A. You will calculate both of these Casimir factors in Exercises 8.6 and 8.7.

So if we have a method for measuring C_A we can verify the non-Abelian structure of QCD as well as verifying that the gluon interacts with itself as predicted with an SU(3) gauge symmetry. This test was done at the DELPHI experiment at LEP. DELPHI measured numerous observables that were sensitive to gluon self-interaction such as the angle between the two lowest-energy jets. Most of the time, the two jets with the least energy will be the gluons from Eq. 8.65. In the calculation of the cross section for $e^+e^- \to q\bar{q}g$, we observed that the result had divergences when the gluon had low energy in Eq. 7.113. This holds true for the process $e^+e^- \to q\bar{q}gg$, and the two diagrams in Eq. 8.65 predict different angular distributions for those gluons. This can be used to determine the relative size of the C_F versus C_A contributions to the cross section.

The results of the DELPHI analysis are presented in Fig. 8.2. Here, the two axes are the ratio of the number of quark to gluon colors N_C/N_A and the ratio of Casimirs C_A/C_F. In QCD, there are eight gluon colors and three quark colors and so $N_C/N_A = 3/8$, while, from earlier, $C_A/C_F = 9/4$. The location on this plot for SU(3) QCD is denoted by the shaded circle. Numerous other points on this plot denote where other Lie groups would sit, which would predict different numbers of colors and different Casimirs. In particular, if QCD were just like electromagnetism, it would be located at the point $(0, 1)$, with a gluon that does not interact with itself. The mean value of their data on this plot is denoted by the star, and errors are represented by the ovals. Excellent agreement with the QCD prediction is observed, giving confidence that SU(3) is indeed its gauge group. There are other groups that are also consistent with data, though, so this analysis alone isn't the final word on determination of the gauge group of QCD. We'll see other evidence for the validity of SU(3) QCD in the next section.

8.3.4 The Running Coupling and Asymptotic Freedom

We defined the fine structure constant α as the strength of electromagnetism. α controls the cross section for $e^+e^- \to$ hadrons scattering, as well as setting the size of the electric potential between charged particles. Its value is $\alpha \simeq 1/137$, so corrections to the cross section for $e^+e^- \to \mu^+\mu^-$ that we calculated were small, and so could be ignored. But, this is quantum mechanics, and so everything is only as good as our measurements allow. To actually determine the value of the fine structure constant, we need to describe the measurement we would perform.

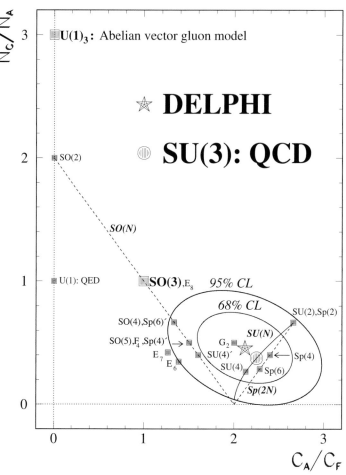

Fig. 8.2   Result of an analysis of $e^+e^- \rightarrow$ four jets events collected at the DELPHI experiment at LEP. Using multiple observables sensitive to the self-interaction of the gluon, DELPHI was able to constrain the ratio of quark to gluon colors N_C/N_A versus the ratio of Casimirs C_A/C_F. Reprinted by permission from Springer Nature: Springer Nature Z. Phys. C, "Measurement of the triple gluon vertex from four-jet events at LEP," P. Abreu *et al.* [DELPHI Collaboration] (1993).

The Running Coupling of Quantum Electrodynamics

One way in principle to measure α is the following. It will happen to be impractical, but provides insight into the physics of what we are sensitive to when measuring α. Let's imagine we have an electron sitting in space. To measure α, we need to know how strongly a photon couples to that electron. So, we could shine a laser onto the electron and observe what happens. Schematically, the set-up is illustrated as:

$$e^- \quad \leftarrow\!\!\text{/}\!\text{WWW}$$
laser light

This can be represented by the Feynman diagrams

This process is called **Compton scattering** after Arthur Compton.[5] In setting the electron out like this, however, some strange things happen.

Because electromagnetic waves propagate through vacuum, the vacuum must be a medium. As a medium, it is a dielectric; it can be polarized in the presence of a charged particle, like an electron. How does the electron polarize the vacuum? It creates an electric field in which virtual particles have a preferred orientation. That is, because of the negative charge of the electron, virtual positrons and electrons orient themselves in this field as:

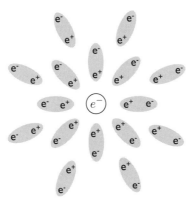

Here, the virtual e^+e^- pairs are denoted by the blobs. These virtual particles screen the electron's electric field so that it appears weaker at larger distances than what you would expect for a point charge. Correspondingly, if you defined the electron's charge from a measurement of its electric field, you would find that the value of the fundamental charge e is smaller when you are further away. Because the fine structure constant is defined in natural units as

$$\alpha = \frac{e^2}{4\pi},$$ (8.66)

it appears that α is smaller when you are further away.

With this insight, let's go back to our experiment of shining light on the electron. If the wavelength of light is long (that is, low energy), then the light is scattered by the virtual particles before it gets close to the electron:

[5] A. H. Compton, "A quantum theory of the scattering of X-rays by light elements," Phys. Rev. **21**, 483 (1923).

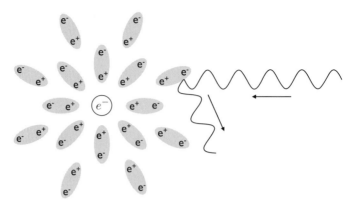

So, a low-energy photon sees a small value of α. By contrast, if the photon is high-energy or short-wavelength, it penetrates further into the cloud:

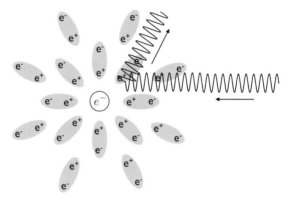

This photon would see a larger value of α! That is, the fine structure constant depends on the energy or wavelength at which it is measured. To denote this, we refer to it as a **running coupling** (its value changes or "runs" with energy) and write $\alpha(Q)$, for energy Q.

This energy dependence can be calculated in quantum field theory. The Feynman diagram which describes the first approximation to the energy dependence of α is

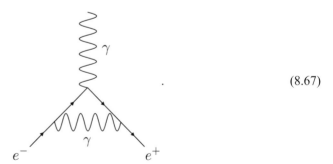

$$(8.67)$$

This diagram represents corrections to the electron–positron–photon vertex, whose value is proportional to the fundamental electric charge e. An external photon comes in from the top of the diagram and interacts with virtual electrons and positrons which then communicate the photon interaction to the real electron and positron at the bottom of the diagram. Note that this diagram has a "loop" in it. By conservation of energy and momentum at every vertex in this diagram, any possible four-momentum can flow around the loop, regardless of the momenta of the external electron, positron, or photon. Consideration of how the value of this diagram changes with the momentum that is flowing in the loop enables a determination of the running coupling $\alpha(Q)$.

The way that this energy dependence is typically expressed is via a quantity called the β-**function** (read: "beta-function") for α. The β-function encodes the derivative of α with energy:

$$\beta(\alpha) \equiv Q\frac{d\alpha}{dQ}. \tag{8.68}$$

The leading-in-α term of the β-function for α calculated from the diagram in Eq. 8.67 is

$$\beta(\alpha) = \frac{2}{3}\frac{\alpha^2}{\pi}. \tag{8.69}$$

We can solve the differential equation for α as a function of energy Q:

$$Q\frac{d\alpha}{dQ} = \frac{2}{3}\frac{\alpha^2}{\pi} \qquad \Longrightarrow \qquad \frac{d\alpha}{\alpha^2} = \frac{2}{3\pi}\frac{dQ}{Q}. \tag{8.70}$$

Integrating both sides, we have

$$\frac{1}{\alpha(Q_0)} - \frac{1}{\alpha(Q)} = \frac{2}{3\pi}\log\frac{Q}{Q_0}, \tag{8.71}$$

or that

$$\alpha(Q) = \frac{\alpha(Q_0)}{1 - \frac{2}{3\pi}\alpha(Q_0)\log\frac{Q}{Q_0}}. \tag{8.72}$$

Here, Q_0 is some reference energy scale (like the Z boson mass, for example) and $\alpha(Q_0)$ is the value of the fine structure constant at that energy. Note that this indeed expresses the expectation for the energy dependence of α: as Q increases, the denominator decreases, and so $\alpha(Q)$ increases. Again, we have a nice picture of this as resulting from polarization of the vacuum.

The Running Coupling of Quantum Chromodynamics

In the QCD Lagrangian, we introduced the coupling g that controls the strength of interaction of the gluon to itself and to quarks. When used in Feynman diagrams and to calculate cross sections, g will be squared, and so we introduce the **strong coupling** α_s:

$$\alpha_s = \frac{g^2}{4\pi}. \tag{8.73}$$

Not surprisingly, the value of α_s, like α, depends on the energy at which it is probed. In QCD, we can calculate appropriate diagrams to determine the β-function of α_s. There are three diagrams to compute now:

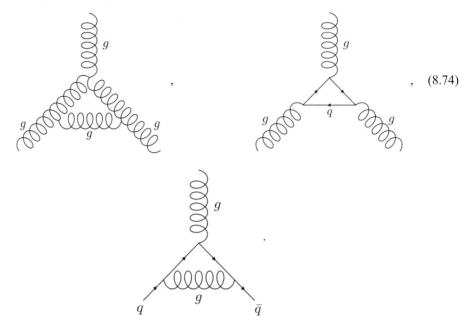

That is, we determine the corrections to the coupling g of quarks to gluons and gluons to gluons from these one-loop diagrams. This calculation is more involved than determining the β-function for α, but the same procedure applies. The β-function for α_s was first calculated by Gerardus 't Hooft, David Politzer, David Gross, and Frank Wilczek.[6] They found

$$Q\frac{d\alpha_s}{dQ} = \beta(\alpha_s) = \frac{\alpha_s^2}{2\pi}\left(-11 + \frac{2}{3}n_f\right).$$ (8.75)

This result is arguably one of the most important in particle physics. Unlike α, there are two contributions to the β-function for α_s. First, the positive $2/3n_f$ term is the contribution from quarks. n_f is the number of active quarks that can contribute to the β-function as virtual particles in the loop of the second diagram in Eq. 8.74. An active quark contributes to the β-function if the energy Q at which the interaction is being probed is larger than twice the quark's mass. The interpretation of this $2/3n_f$ term is exactly like the β-function for α. Pairs of quarks and anti-quarks pop in and out of the vacuum and effectively screen the color charge of a particle. Therefore this contribution works to increase α_s as energy increases.

The -11 term is the contribution from gluons. Because the gluon itself carries color, gluons can pop in and out of the vacuum to affect the color charge of a particle. This is totally different than for α, and has no analogue in classical mechanics. Apparently gluons are responsible for strong anti-screening: gluons work to decrease the size of α_s

[6] H. D. Politzer, "Reliable perturbative results for strong interactions?," Phys. Rev. Lett. **30**, 1346 (1973); D. J. Gross and F. Wilczek, "Ultraviolet behavior of nonabelian gauge theories," Phys. Rev. Lett. **30**, 1343 (1973). See also Box 8.1 for historical context of this result.

Because of the structure of hadrons and the strongly interacting nature of quarks, in the 1960s very few people in the world accepted that quantum field theory could describe these phenomena. Two of the few who were continuing to develop quantum field theory were Martinus Veltman and his student, Gerardus 't Hooft. **Gerardus 't Hooft** made fundamental contributions to theoretical physics in the 1970s, by showing that the theory of the weak interactions was mathematically consistent,[7] developing crucial aspects of what would become string theory,[8] first calculating the quantum effects of topological objects called instantons,[9] and identifying issues with a potential quantum theory of gravity.[10] 't Hooft knew of asymptotic freedom and the value of the β-function in the SU(3) color theory as early as 1972 (it was not yet called QCD). During a conference in Marseille, he wrote the result on a blackboard, but never published it, as he had other projects to finish first. Politzer, Gross, and Wilczek published their results a year later. For more history, see 't Hooft's recount of the time in G. 't Hooft, "When was asymptotic freedom discovered? Or the rehabilitation of quantum field theory," Nucl. Phys. Proc. Suppl. **74**, 413 (1999) [arXiv:hep-th/9808154].

as it is probed at higher and higher energies. The Standard Model has six quarks (up, down, strange, charm, top, bottom), and so n_f is at most 6. Even with all of the Standard Model quarks around,

$$-11 + \frac{2}{3} \cdot 6 = \frac{-33 + 12}{3} = -7, \qquad (8.76)$$

and so the β-function of α_s is negative! Apparently, the strong force QCD gets "weaker" at higher energies.

Solving the β-function equation for $\alpha_s(Q)$, we find

$$\alpha_s(Q) = \frac{\alpha_s(Q_0)}{1 + \frac{\alpha_s(Q_0)}{2\pi}\left(11 - \frac{2}{3}n_f\right)\log\frac{Q}{Q_0}} . \qquad (8.77)$$

As $Q \to \infty$, $\alpha_s(Q) \to 0$. This feature is called **asymptotic freedom**: at asymptotically high energies, quarks interact more and more weakly, becoming like free (= non-interacting) particles. This feature of asymptotic freedom will have profound consequences for the phenomena of QCD at high energies. We'll explore two phenomena that are consequences of asymptotic freedom, parton evolution and jets, in Chapter 9.

Asymptotic freedom and the running of α_s in general can be tested in experiment. Figure 8.3 plots numerous measurements of $\alpha_s(Q)$ as a function of energy scale Q, ranging from $Q = 5$ GeV up to almost $Q = 2$ TeV. Many of these measurements come from the CMS experiment at the LHC in which the value of α_s is extracted from measurements

[7] G. 't Hooft, "Renormalizable Lagrangians for massive Yang–Mills fields," Nucl. Phys. B **35**, 167 (1971).
[8] G. 't Hooft, "A planar diagram theory for strong interactions," Nucl. Phys. B **72**, 461 (1974).
[9] G. 't Hooft, "Computation of the quantum effects due to a four-dimensional pseudoparticle," Phys. Rev. D **14**, 3432 (1976), Erratum: [Phys. Rev. D **18**, 2199 (1978)].
[10] G. 't Hooft and M. J. G. Veltman, "One loop divergencies in the theory of gravitation," Ann. Inst. H. Poincare Phys. Theor. A **20**, 69 (1974).

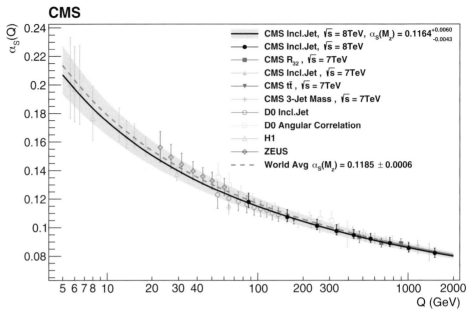

CMS

Fig. 8.3 A collection of various measurements of the strong coupling α_s at energy scales Q ranging from 5 GeV to nearly 2 TeV. The solid line is the predicted running from the QCD β-function with the input value $\alpha_s(m_Z) = 0.1164$. From V. Khachatryan *et al.* [CMS Collaboration], "Measurement and QCD analysis of double-differential inclusive jet cross sections in pp collisions at $\sqrt{s} = 8$ TeV and cross section ratios to 2.76 and 7 TeV," J. High Energy Phys. **1703**, 156 (2017) [arXiv:1609.05331 [hep-ex]].

of cross sections for jet production. Because individual quark and gluon partons actually interact in proton collisions and produce other quarks and gluons that turn into jets in the final state, the value of α_s controls the rate of jet production. By measuring the cross section for jet production as a function of the jet transverse momentum, the energy dependence of α_s can be extracted. Excellent agreement between the data points and the predicted running coupling from the QCD β-function is observed (the solid line). The predicted energy dependence requires an input of the value of α_s at one energy; CMS uses the value at $Q = m_Z$ to be $\alpha_s(m_Z) = 0.1164$, and the uncertainty on that value is represented by the shaded band.

8.3.5 Low-Energy QCD

While asymptotic freedom of QCD leads to the explanation of numerous phenomena at high energies, it provides an intriguing first step to understanding low-energy phenomena of QCD as well. Let's go back to the expression of the running coupling of QCD:

$$\alpha_s(Q) = \frac{\alpha_s(Q_0)}{1 + \frac{\alpha_s(Q_0)}{2\pi}\left(11 - \frac{2}{3}n_f\right)\log\frac{Q}{Q_0}}. \tag{8.78}$$

This was calculated with Feynman diagrams, and the accuracy of Feynman diagrams improves as the value of α_s decreases. Thus the prediction of asymptotic freedom is **robust**: at higher and higher energies, α_s gets smaller, and so we trust the prediction of asymptotic freedom more. However, it seems like we can make another statement, too. Apparently, as the energy decreases, the value of α_s increases. In particular, there is an energy scale at which the coupling diverges. The denominator of Eq. 8.78 is zero at an energy scale Λ for which

$$1 + \frac{\alpha_s(Q_0)}{2\pi} \left(11 - \frac{2}{3}n_f\right) \log \frac{\Lambda}{Q_0} = 0 \,, \tag{8.79}$$

or that

$$\Lambda = Q_0 \exp\left[-\frac{2\pi}{\alpha_s(Q_0)\left(11 - \frac{2}{3}n_f\right)}\right] \,. \tag{8.80}$$

for some initial energy scale Q_0. The energy Λ at which the coupling diverges is called the **Landau pole**, after the Russian physicist Lev Landau.[11] This is also a manifestation of the phenomenon of **dimensional transmutation**: the energy dependence of the dimensionless coupling α_s has introduced a new dimensionful scale Λ.

The location of the Landau pole is extremely suggestive. We have to be very careful interpreting it because the prediction of the running coupling used Feynman diagrams, and when the coupling is very large, Feynman diagrams cease making sense. Nevertheless, let's see what we find. For concreteness, let's set the scale $Q_0 = m_Z$, the mass of the Z boson for which $m_Z = 91.18$ GeV. According to the PDG, the value of α_s at the mass of the Z boson is $\alpha_s(m_Z) = 0.118$. Knowing the answer ahead of time, we will set the number of active quarks at $n_f = 3$, which is the number of quarks which have masses less than Λ. With these parameters, we then find Λ to be

$$\Lambda = m_Z \exp\left[-\frac{2\pi}{\alpha_s(m_Z)\left(11 - \frac{2}{3} \cdot 3\right)}\right] = 246 \text{ MeV} \,. \tag{8.81}$$

Recall that the mass of the pions is about 135 MeV and the masses of the proton and of the neutron are each about 938 MeV. Other hadrons such as kaons, rho mesons, sigma baryons, etc., have masses in this ballpark also. That is, from calculating some Feynman diagrams with quarks and gluons to determine the running coupling, the location of the Landau pole of α_s is comparable to masses of hadrons! Very cool.

So, does this mean that the running coupling predicts the existence of hadrons in QCD? Not quite. Because we cannot trust the accuracy of Feynman diagrams when α_s is large, a Feynman diagram analysis alone isn't sufficient to claim victory of understanding the hadron masses. At any rate, there is still a lot to learn from the Landau pole. Apparently, as one probes QCD at lower energies, quarks and gluons interact more and more strongly. At sufficiently low energies (comparable to the scale of the Landau pole), it takes greater and greater energy to pull quarks apart from one another. That is, at low energies, because

---

[11] L. D. Landau, "On the quantum theory of fields," in *Niels Bohr and the Development of Physics*, Pauli, W. (ed.), Pergamon Press 1955, pp. 52–69.

α_s is getting so large, quarks form states that cannot be separated; that is, they form bound states through a process called **confinement**. These bound states also have zero net color, otherwise they would interact with gluons and get stuck to other colored states. These colorless bound states are hadrons and are a consequence of the strong coupling of QCD at low energies. By the way, this description of low-energy QCD is an example of a **duality** in particle physics. There are two different, yet equivalent descriptions of QCD at low energies. One can either consider strongly interacting quarks and gluons, or consider colorless, and therefore relatively weakly interacting, hadrons and describe the same physics.

However, to actually make predictions for phenomena in low-energy QCD, we have to choose one of these descriptions. Doing so takes us out of our comfort zone of using Feynman diagrams for prediction. Either we describe the physics with quarks and gluons, but α_s is enormous, and so Feynman diagrams are useless; or we can describe the physics with hadrons, but we never directly see hadrons in the Lagrangian of QCD in Eq. 8.51. So, what do we do? One approach is to develop a theory of weakly interacting hadrons, which is called **chiral perturbation theory**. Another approach, which we will discuss in some more detail here, is to keep plowing ahead with the quark-and-gluon description of QCD at low energies. Clearly Feynman diagrams aren't the way to make progress. Instead, one attempts to completely describe all quark and gluon interactions but restricted to discrete points in spacetime. This is known as **lattice QCD**, and was initially developed by Kenneth Wilson in 1974 as an attempt to predict the confinement of quarks.[12]

In lattice QCD, one discretizes spacetime as well as making it finite, with no assumption that α_s is small. On this spacetime lattice, one can then simulate the quark and gluon fields and their interactions as governed by the Lagrangian of QCD. In the limit in which the spacing of the lattice goes to zero and the number of lattice points goes to infinity, one then recovers the continuum description of QCD, as exists in our universe. So, the goal of lattice QCD is to simulate as many spacetime lattice points as possible to get as close as possible to a description of our universe. The challenge is that this is enormously computationally intensive and requires using some of the most powerful supercomputers in the world. The most state-of-the-art lattice QCD calculations use spacetime lattices with about 10^6 sites, corresponding to only about 32 sites in each dimension. Nevertheless, this is an extremely active field that has applications for a broad range of physical phenomena including nuclear physics, astrophysics, collider physics, and purely theoretical questions.[13] Despite these challenges, lattice QCD is able to make precise, quantitative predictions. Figure 8.4 shows the prediction of the spectrum of hadron masses from lattice QCD, just using three measured hadron masses as inputs!

---

[12] K. G. Wilson, "Confinement of quarks," Phys. Rev. D **10**, 2445 (1974).

[13] Some recent reviews of lattice QCD are: T. DeGrand and C. E. Detar, *Lattice Methods for Quantum Chromodynamics*, World Scientific (2006); C. Gattringer and C. B. Lang, "Quantum chromodynamics on the lattice," Lect. Notes Phys. **788**, 1 (2010); Lellouch, L. *et al.* (eds), *Modern Perspectives in Lattice QCD: Quantum Field Theory and High Performance Computing*, Lecture notes of the Les Houches Summer School, August 2009, Vol. 93, Oxford University Press (2011).

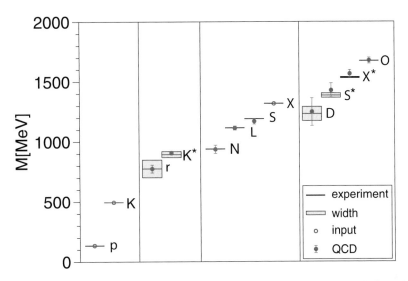

Predicted spectrum of hadron masses from lattice QCD with π, K, and Ξ hadron masses as inputs. The lattice QCD predictions with uncertainties are the filled dots. From S. Durr *et al.*, "Ab-initio determination of light hadron masses," Science **322**, 1224 (2008) [arXiv:0906.3599 [hep-lat]]. Reprinted with permission from AAAS.

Exercises

8.1 *Masslessness of the Gluon.* We hypothesized that the gluon was a massless particle, in analogy with the photon. If the gluon were massive, then there would be a mass term in the Lagrangian of QCD like

$$\mathcal{L}_{\text{QCD}} \supset -\frac{m_g^2}{2} A_\mu^a A^{\mu\,a} = -m_g^2 \operatorname{tr}\left[A_\mu^a T^a A^{\mu\,b} T^b\right],$$

where m_g is the mass of the gluon. Perform an SU(3) color gauge transformation of the gluon field A_μ^a and show that such a mass term is not invariant. What does it transform into? Simplify the expression as much as you can. Are there classes of gauge transformations that do leave the mass term invariant?

8.2 *Bianchi Identity.* Just as with electromagnetism, the field strength tensor of QCD satisfies the Bianchi identity. In QCD, however, the Lie algebra structure is extremely important for consistency of the theory. In QCD, the Bianchi identity is

$$(D_\mu F_{\nu\rho})^a + (D_\rho F_{\mu\nu})^a + (D_\nu F_{\rho\mu})^a = 0. \tag{8.82}$$

(a) Show that the Bianchi identity is only satisfied in QCD if the structure constants satisfy

$$f^{abd}f^{dce} + f^{bcd}f^{dae} + f^{cad}f^{dbe} = 0. \tag{8.83}$$

This relationship is called the Jacobi identity and is a requirement for associativity of a Lie group.

Hint: Use the representation of the field strength tensor as a commutator of covariant derivatives.

(b) Show that, for the commutator $[T^a, T^b]$ for elements T^a and T^b in a Lie algebra, the Jacobi identity is satisfied as

$$[T^a, [T^b, T^c]] + [T^b, [T^c, T^a]] + [T^c, [T^a, T^b]] = 0. \tag{8.84}$$

Prove this relationship by expanding out the commutators.

8.3 *Instantons*. Instantons are solutions to the classical equations of motions of a Yang–Mills theory, like QCD. Their name derives from the property that they are localized in space and time and therefore have similarities to particles. Because of this, they were also historically called "pseudoparticles." They have the property that the field strength tensor that describes them is self-dual: that is, it satisfies the relationship

$$F^a_{\mu\nu} = \frac{i}{2} \epsilon_{\mu\nu\rho\sigma} F^{\rho\sigma a}. \tag{8.85}$$

Here, $\epsilon_{\mu\nu\rho\sigma}$ is the totally anti-symmetric symbol in four dimensions which is equal to 1 for all even permutations of $0, 1, 2, 3$; -1 for odd permutations; and 0 if two indices are repeated. In this exercise, we'll study properties of these self-dual field strength configurations.

(a) In electromagnetism, the self-dual field strength satisfies the relationship

$$F_{\mu\nu} = \frac{i}{2} \epsilon_{\mu\nu\rho\sigma} F^{\rho\sigma}, \tag{8.86}$$

where the form of $F_{\mu\nu}$ is given in Section 2.2.3. What relationship between the electric and magnetic fields $\vec{E}$ and $\vec{B}$ does this self-duality constraint impose?

(b) For a self-dual field strength, its action is

$$S[A^a_\mu] = \int d^4x \ -\frac{1}{4} F^a_{\mu\nu} F^{\mu\nu a} = \int d^4x \ -\frac{i}{8} \epsilon^{\mu\nu\rho\sigma} F^a_{\mu\nu} F^a_{\rho\sigma}. \tag{8.87}$$

Show that the Lagrangian is a total derivative; that is, the action only depends on properties of the instanton at the boundary of spacetime. You should find that the action can be arranged into the form

$$S[A^a_\mu] = -\frac{i}{4} \int d^4x \ \partial_\mu \epsilon^{\mu\nu\rho\sigma} \left(A^a_\nu F^a_{\rho\sigma} - \frac{g}{3} f^{abc} A^a_\nu A^b_\rho A^c_\sigma \right). \tag{8.88}$$

Everything to the right of the partial derivative ∂_μ is called the **Chern–Simons current**.

8.4 *Wilson Lines*. For a gauge theory like SU(3) color, as one moves around in color space even the notion of the identity matrix changes. To compare identity matrices at different spacetime points, we need to parallel transport the identity matrix from the initial point to the final point. On a linear manifold, parallel transport of the identity, 1, just means that the derivative of 1 is 0:

$$\partial_\mu 1 = 0. \tag{8.89}$$

That is, on a linear manifold, the identity is the same everywhere. On a non-linear manifold, we need to use the covariant derivative to parallel transport the identity. This produces the differential equation

$$D_\mu W(x,y) = (\partial_\mu - igA_\mu^a T^a)W(x,y) = 0. \tag{8.90}$$

The parallel transport of the identity $W(x,y)$ is called a **Wilson line** and its arguments x, y are the final (x) and initial (y) points of the parallel transport. In this exercise, we will study properties of Wilson lines in electromagnetism because the interpretation is a bit simpler than in QCD. In electromagnetism, we don't need the Lie algebra matrix, so the Wilson line equation reduces to

$$D_\mu W(x,y) = (\partial_\mu - ieA_\mu)W(x,y) = 0, \tag{8.91}$$

where A_μ is the vector potential. The derivative ∂_μ is taken with respect to x and the vector potential A_μ is a function of final position x.

(a) Show that the Wilson line can be written as

$$W(x,y) = \exp\left[ie\int_y^x ds^\mu A_\mu(s)\right], \tag{8.92}$$

where ds^μ is the four-vector infinitesimal path length along the parallel transport from y to x.

(b) Under a gauge transformation of the vector potential A_μ, how does the Wilson line transform?

(c) A Wilson loop is a Wilson line whose initial and final positions are identical and so it is exclusively defined by the path P of the loop:

$$W_P = \exp\left[ie\oint_P ds^\mu A_\mu\right]. \tag{8.93}$$

The exponential factor can be expanded out in temporal and spatial parts:

$$W_P = \exp\left[ie\oint_P dtA_0 - ie\oint_P d\vec{s}\cdot\vec{A}\right]. \tag{8.94}$$

Assuming that the electric field is 0 so that $A_0 = 0$ and $\vec{A}$ is time-independent, show that the Wilson loop is

$$W_P = \exp\left[-ie\int d\vec{\Omega}\cdot\vec{B}\right], \tag{8.95}$$

where $\vec{B}$ is the magnetic field and $d\vec{\Omega}$ is a differential area element on a surface bounded by the path P.

Hint: You'll need to use Stokes's theorem and the relationships between the electric and magnetic fields and the vector potential from Section 2.2.3.

8.5 $\mathfrak{su}(2)$ *Lie Algebra.* In Chapter 3, we introduced the Lie algebra that defines the group SU(2) to satisfy the commutation relations

$$\left[\frac{1}{2}\sigma_i, \frac{1}{2}\sigma_j\right] = i\epsilon_{ijk}\frac{1}{2}\sigma_k. \tag{8.96}$$

Here, the σ_i are the Pauli spin matrices, with

$$\sigma_1 = \begin{pmatrix} 0 & 1 \\ 1 & 0 \end{pmatrix}, \quad \sigma_2 = \begin{pmatrix} 0 & -i \\ i & 0 \end{pmatrix}, \quad \sigma_3 = \begin{pmatrix} 1 & 0 \\ 0 & -1 \end{pmatrix}, \quad (8.97)$$

and the structure constants are ϵ_{ijk}, the totally anti-symmetric object. The structure constants can be expressed in matrix form and satisfy the Lie algebra themselves. While this is true for any Lie algebra, in this exercise we will demonstrate it for $\mathfrak{su}(2)$.

(a) Let's denote the totally anti-symmetric structure constants as three matrices: $i\epsilon_{1jk}$, $i\epsilon_{2jk}$, and $i\epsilon_{3jk}$. Note that we've multiplied by a factor of the imaginary number i, and j and k denote the row and column of the matrices, respectively. For compactness, let's express the matrices as $T^1 = i\epsilon_{1jk}$, $T^2 = i\epsilon_{2jk}$, and $T^3 = i\epsilon_{3jk}$. Write, in standard matrix form, these three matrices. Are they Hermitian? What other properties do they have?

(b) Now, with these three matrices T^a, compute their commutation relations. You should find that they satisfy a Lie algebra:

$$[T^a, T^b] = if^{abc} T^c. \quad (8.98)$$

What are the structure constants f^{abc} of this Lie algebra? How does it compare to the $\mathfrak{su}(2)$ Lie algebra?

(c) Now, calculate the Killing form of these matrices. That is, take the product of two matrices and then take the trace. You should find

$$\mathrm{tr}\left[T^a T^b\right] = 2\delta^{ab}. \quad (8.99)$$

8.6 *Casimir Invariant.* The Casimir is an object that commutes with every element of the Lie algebra. As discussed in Chapter 3, the Casimir is therefore invariant: no group action can change its value. The Casimir thus quantifies intrinsic properties of the particle such as its spin or total color. In this exercise, we will calculate the Casimir for the fundamental representation of SU(3), which is a measure of the total color of a quark.

(a) For a Lie algebra with basis matrices T^a and a commutation relation

$$[T^a, T^b] = if^{abc} T^c, \quad (8.100)$$

show that the sum of squares of the matrices commutes with all elements of the Lie algebra:

$$[(T^a)^2, T^b] = 0. \quad (8.101)$$

Here,

$$(T^a)^2 = T^1 T^1 + T^2 T^2 + \cdots + T^m T^m, \quad (8.102)$$

for a Lie algebra with m basis matrices. You can assume that the structure constants f^{abc} are completely anti-symmetric: they are identical for even permutations of a, b, c, opposite for odd permutations, and 0 if any of a, b, or c are the same.

(b) For the fundamental representation of SU(N), the Lie algebra consists of $N^2 - 1$ matrices of size $N \times N$. We also typically normalize the Lie algebra according to the Killing form as

$$\text{tr}[T^a T^b] = \frac{1}{2}\delta^{ab}. \tag{8.103}$$

Using this information, show that the Casimir for the fundamental representation of SU(N), which we denote as C_F, is

$$C_F = \frac{N^2 - 1}{2N}. \tag{8.104}$$

Because the Casimir commutes with everything, it must be proportional to the identity matrix, which we typically do not write explicitly.

(c) The fundamental representation of SU(2) has spin 1/2. In Chapter 3 we said that the Casimir for spin 1/2 is

$$\frac{1}{2}\left(1 + \frac{1}{2}\right) = \frac{3}{4}. \tag{8.105}$$

Does the formula of Eq. 8.104 get this value right?

(d) Quarks carry color in the fundamental representation of SU(3). What is the value of C_F for quarks?

(e) Using the Gell-Mann matrices from Eq. 8.13, explicitly calculate the Casimir by squaring and summing matrices. Show that the result agrees with the formula of Eq. 8.104.

8.7 *Adjoint Representation of SU(3).* The adjoint representation of a Lie algebra is the representation in which the size of the basis matrices is equal to the number of basis matrices. For SU(N), the Lie algebra is $N^2 - 1$ dimensional and the adjoint representation consists of matrices of size $(N^2 - 1) \times (N^2 - 1)$. For color SU(3), the adjoint representation is therefore eight-dimensional. This is exactly the number of gluons in QCD, so gluons transform in the adjoint representation of SU(3) color. In this exercise, we will calculate the Casimir of the adjoint representation of SU(N).

(a) For matrices T^a_{adj} in the adjoint representation of SU(N), we typically normalize them according to the Killing form as

$$\text{tr}[T^a_{\text{adj}} T^b_{\text{adj}}] = N\delta^{ab}. \tag{8.106}$$

Using this and the fact that there are $N^2 - 1$ matrices of size $(N^2 - 1) \times (N^2 - 1)$ in the adjoint representation, show that the Casimir of the adjoint C_A is

$$C_A = N. \tag{8.107}$$

(b) Matrices in the adjoint representation can be represented by the structure constants of the Lie algebra. This was explored in Exercise 8.5 above. From the commutation relations

$$[T^a, T^b] = if^{abc}T^c, \tag{8.108}$$

show that the structure constants can be solved for, and one finds

$$f^{abc} = -2i \operatorname{tr}\left[[T^a, T^b]T^c\right] . \tag{8.109}$$

Here, the matrices T^a are in the fundamental representation.

Hint: Multiply by another element of the Lie algebra T^d on both sides of the commutation relation and use the Killing form.

(c) Using the Gell-Mann matrices in Eq. 8.13, calculate the structure constants f^{abc} for the Lie algebra $\mathfrak{su}(3)$.

8.8 *Running Couplings of QED and QCD.* In Section 8.3.4, we introduced the running couplings $\alpha(Q)$, the running fine structure constant, and $\alpha_s(Q)$, the running strong coupling constant. For the fine structure constant, the running coupling (using the one-loop β-function) is

$$\alpha(Q) = \frac{\alpha(Q_0)}{1 - \frac{2}{3\pi}\alpha(Q_0)\log\frac{Q}{Q_0}}, \tag{8.110}$$

while the strong coupling constant is

$$\alpha_s(Q) = \frac{\alpha_s(Q_0)}{1 + \frac{\alpha_s(Q_0)}{2\pi}\left(11 - \frac{2}{3}n_f\right)\log\frac{Q}{Q_0}}. \tag{8.111}$$

Because of the form of the denominator of these expressions, there exists an energy at which the running couplings blow up, corresponding to the Landau pole discussed in Section 8.3.5.

(a) Using the value of the fine structure constant at the electron mass m_e, $\alpha(m_e) = 1/137$, estimate the energy in GeV at which the Landau pole occurs. Provide an estimate of this energy in terms of the total visible energy in the universe.

(b) The value of the fine structure constant increases as energy increases, while the strong coupling decreases with energy. Therefore, there exists an energy Q_{eq} at which $\alpha(Q_{eq}) = \alpha_s(Q_{eq})$. What is this energy, in GeV? Use the value of the strong coupling at the Z boson mass m_Z

$$\alpha_s(m_Z) = 0.118, \tag{8.112}$$

where $m_Z = 91.18$ GeV, and $n_f = 6$. Long before this energy scale, quantum electrodynamics is subsumed into the unified electroweak theory, so this analysis is a bit unreasonable. We'll introduce the electroweak theory in Chapter 11.

(c) The β-function for the fine structure constant in Eq. 8.69 assumes that the electron is the only electrically charged particle in the universe. The Standard Model, however, has six electrically charged quarks and three electrically charged leptons and so we should include the effects of all of these particles. Just as with the calculation of the R ratio in Example 6.2, we also have to include a factor of 3 for each quark to account for their color. Doing this, the β-function for the fine structure constant in the Standard Model is

$$\beta_{\mathrm{SM}}(\alpha) = \frac{2}{3}\frac{\alpha^2}{\pi}\sum_{i \text{ fermions}} Q_i^2, \tag{8.113}$$

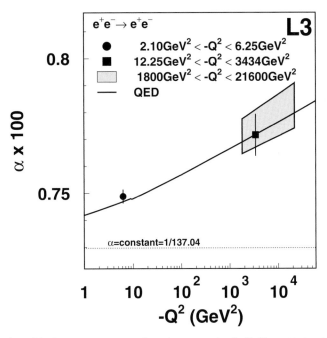

Fig. 8.5   Plot of measured values of the fine structure constant from the cross section for Bhabha scattering at the L3 experiment. From P. Achard *et al.* [L3 Collaboration], "Measurement of the running of the electromagnetic coupling at large momentum-transfer at LEP," Phys. Lett. B **623**, 26 (2005) [arXiv:hep-ex/0507078].

where Q_i is the electric charge of a fermion in units of the elementary charge e and the sum ranges over all charged fermions i of the Standard Model with mass less than the energy scale Q. The modification of our analogy in Section 8.3.4 of vacuum polarization is that there can be particle–anti-particle pairs of any charged fermions of the Standard Model that surround the electron.

Above the top quark mass, what is the squared sum of the electric charges of all of the fermions of the Standard Model? What is the β-function now? Where is the Landau pole for α, in GeV?

(d) Figure 8.5 shows a measurement of the value of the fine structure constant α from the L3 experiment at LEP. The value of α is extracted from the measured cross section for the Bhabha scattering process $e^+e^- \rightarrow e^+e^-$ at a few values of $-t = -Q^2 > 0$. The kink in the prediction of the running coupling around $-Q^2 = 10$ GeV^2 corresponds to the bottom quark now contributing to the β-function (10 GeV $\simeq 2m_b$).

How does the slope of the predicted running fine structure constant in Fig. 8.5 above the location of the bottom quark compare to the β-function in Eq. 8.113? What would the value of $\alpha(Q)$ at $Q \simeq 55$ GeV be if you only accounted for the electron in its running? How does this compare to the data?

8.9   *Research Problem.* Can we ever hope to solve QCD completely and diagonalize its Hamiltonian? What is the best path forward for doing that?

9

Parton Evolution and Jets

Within QCD there is a secret. Of course QCD is Lorentz invariant as it describes the relativistic dynamics and interactions of quarks and gluons. Of course QCD is unitary and probability-conserving as its Lagrangian and Hamiltonian are Hermitian. Of course QCD is gauge invariant because color charge is conserved and the gluon only has two propagating degrees of freedom. These properties were essentially axioms that we used to construct the QCD Lagrangian. A consequence of them, however, is that QCD is approximately scale invariant: systems governed by the QCD Lagrangian look nearly unchanged as you zoom in to shorter and shorter distances.

We've seen glimpses of scale invariance before. Bjorken scaling is the phenomenon in which the structure of quarks appears approximately the same with any photon energy or wavelength probe. In a detector, jets appear to produce particles at every angular scale, down to the resolution of the experiment. Only now are we in the position to understand why these phenomena exist and how their approximate scale invariance is a consequence of asymptotic freedom. At high energies, the strong coupling α_s runs with energy scale or wavelength, but its running is only logarithmic with energy and so is rather slow. So, if the strong coupling at high energies is approximately constant, particles will be produced at nearly the same rate regardless of the energy or distance scale at which we are observing.

In this chapter, we make this consequence of the scale invariance of QCD precise and define what exactly that means. This enables us to straightforwardly derive equations which govern the energy dependence of parton distribution functions. This therefore predicts the phenomenon of Bjorken scaling directly from QCD, but also explains its violation consistently. Jets are a consequence of scale invariance through an arbitrary number of emitted particles over all distance scales. We are able to explicitly sum over any number of particles emitted in a jet, which is also how we compute exclusive cross sections accurately.

9.1 Scale Transformations

We didn't discuss scale transformations and scale invariance in our general analysis of groups in Chapter 3, so let's fix that here. Before studying QCD, let's first consider a much simpler system: the action that describes a massless scalar field:

$$S[\phi] = \int d^4x \, \frac{1}{2} \left(\partial_\mu \phi \right) \left(\partial^\mu \phi \right) . \tag{9.1}$$

This action has many features: it is Lorentz invariant, but it is also invariant under scale transformations. A scale transformation, or **dilation**, is an operation that rescales positions by a factor λ:

$$x^\mu \to \lambda x^\mu \,, \quad \text{for any } \lambda > 0 \,. \tag{9.2}$$

Let's see what this transformation does to the action of Eq. 9.1. We will need to determine how the integration measure d^4x, the derivative ∂_μ, and the field ϕ transform under a dilation.

Let's first consider d^4x. Note that

$$d^4x = dt\, dx\, dy\, dz \,, \tag{9.3}$$

and under a dilation, each individual spacetime coordinate t, x, y, and z is scaled by λ. That is,

$$d^4x = dt\, dx\, dy\, dz \to d(\lambda t)\, d(\lambda x)\, d(\lambda y)\, d(\lambda z) = \lambda^4 d^4x \,. \tag{9.4}$$

Now, on to the derivative. The derivative four-vector ∂_μ in components is

$$\partial_\mu = \left(\frac{\partial}{\partial t}, -\frac{\partial}{\partial x}, -\frac{\partial}{\partial y}, -\frac{\partial}{\partial z} \right)_\mu \,, \tag{9.5}$$

and again, each spacetime coordinate t, x, y, and z is scaled by λ. For example,

$$\frac{\partial}{\partial x} \to \frac{\partial}{\partial(\lambda x)} = \frac{1}{\lambda} \frac{\partial}{\partial x} \,, \tag{9.6}$$

and similarly for all other derivatives. Therefore, $\partial_\mu \to \frac{1}{\lambda}\partial_\mu$ under a dilation.

What about the field, $\phi(x)$? The transformation under dilation can be derived more rigorously within quantum field theory, but here we will just work to justify it. An action S is just the time integral of a Lagrangian, and therefore has dimensions of

$$[S] = [\text{energy}][\text{time}] = [\hbar] \,. \tag{9.7}$$

Because we set $\hbar = 1$ in natural units, the action is dimensionless. We can determine the dimensions of the scalar field ϕ from this requirement and identifying the dimensions of the measure and derivatives. A differential length element dx in natural units has dimensions of inverse energy, while the derivative ∂_μ has units of energy as quantum mechanically it corresponds to the momentum operator. Therefore, for the action to be dimensionless in natural units, the field $\phi(x)$ must have dimensions of energy. That is, $[\phi] = [\text{energy}]$.

We define the scaling of ϕ with λ to be the same as that of the derivative ∂_μ, as they have the same units. Then, $\phi \to \frac{1}{\lambda}\phi$. With this scaling we can then determine how the action scales with λ. Plugging in all scalings that we have identified, under a dilation the action transforms to

$$S[\phi] = \int d^4x\, \frac{1}{2} \left(\partial_\mu \phi \right) \left(\partial^\mu \phi \right) \to \int d^4x \lambda^4 \frac{1}{2} \left(\lambda^{-1} \partial_\mu \lambda^{-1} \phi \right) \left(\lambda^{-1} \partial^\mu \lambda^{-1} \phi \right) \tag{9.8}$$
$$= S[\phi] \,.$$

That is, the action is invariant under dilation. Because the action is invariant under this continuous transformation, there is a conservation law, by Noether's theorem.[1] The action is sufficient to describe everything about a classical or quantum system, and such a system with an action that is invariant under dilations is called **scale invariant**.

Note that scale transformations are not Lorentz transformations. These dilations can be implemented on the spacetime four-vector x by a matrix Λ:

$$\Lambda = \begin{pmatrix} \lambda & 0 & 0 & 0 \\ 0 & \lambda & 0 & 0 \\ 0 & 0 & \lambda & 0 \\ 0 & 0 & 0 & \lambda \end{pmatrix} = \lambda \mathbb{I}, \tag{9.9}$$

where $\lambda > 0$ is on the diagonal, but all other entries are 0. A Lorentz transformation implemented by a matrix Λ satisfies

$$\Lambda^{\mathsf{T}} \eta \Lambda = \eta, \tag{9.10}$$

where η is the spacetime metric matrix. Testing this with the dilation matrix, Eq. 9.9, we find

$$\Lambda^{\mathsf{T}} \eta \Lambda = \lambda \mathbb{I}^{\mathsf{T}} \eta \lambda \mathbb{I} = \lambda^2 \eta \neq \eta, \tag{9.11}$$

for arbitrary $\lambda > 0$. Therefore, in general dilations are not Lorentz transformations.

This is interesting, but is it trivial? Are all relativistic actions scale invariant? The answer is indeed no, because a dilation is not a Lorentz transformation. Let's consider adding a mass to the scalar field. Its action is then

$$S[\phi] = \int d^4x \left[\frac{1}{2} \left(\partial_\mu \phi \right) \left(\partial^\mu \phi \right) - \frac{m^2}{2} \phi^2 \right]. \tag{9.12}$$

This action is still Lorentz invariant because the scalar field ϕ is not affected by a Lorentz transformation. Now, let's apply this dilation to this action. We already identified the kinetic term of the action to be scale invariant, so we only need to study the mass term. Under a dilation, its scaling is

$$S_{m^2}[\phi] = -\frac{m^2}{2} \int d^4x \, \phi^2 \rightarrow -\frac{m^2}{2} \int d^4x \lambda^4 \left(\lambda^{-1} \phi \right)^2 = \lambda^2 S_{m^2}[\phi]. \tag{9.13}$$

Importantly, the mass m is just a number and so is not affected by a dilation. This demonstrates that the theory with a mass is not scale invariant. Again, this isn't a problem, per se, because dilations are not Lorentz transformations.

The physical interpretation of this lack of scale invariance with a mass is related to the scalar field's intrinsic wavelength. For a massless scalar, the only relevant distance scale is its de Broglie wavelength λ_{dB}, where

$$\lambda_{\text{dB}} = \frac{\hbar}{|\vec{p}|}, \tag{9.14}$$

---

[1] The conservation law from scale invariance is a bit non-standard. In a scale-invariant theory, the trace of the stress–energy tensor $T_{\mu\nu}$, T^μ_μ, is zero. The trace of the stress–energy tensor is like a measure of the difference between the total energy and the magnitude of total momentum of the system: $T^\mu_\mu \sim E - |\vec{p}|$. If the system has no explicit masses, then it is scale invariant and $T^\mu_\mu = 0$.

for some momentum $\vec{p}$ and we have inserted explicit factors of $\hbar$. If the scalar is massless, its magnitude of momentum can be arbitrarily large or small; there is no intrinsic momentum or energy scale. Correspondingly, there is no intrinsic length scale because the de Broglie wavelength can in principle be any length from 0 (high energy) to infinite (low energy). A dilation changes the size of the de Broglie wavelength of the scalar, but it always transforms the massless scalar to have a de Broglie wavelength that is a physical value.

By contrast, if the scalar has a mass m, there are now two relevant distance scales. There is of course the de Broglie wavelength, but we can also Lorentz boost to the rest frame of the scalar in which the relevant distance scale is the Compton wavelength λ_C:

$$\lambda_C = \frac{\hbar}{mc}. \tag{9.15}$$

For a massive particle, the Compton wavelength is the absolute largest wavelength that it can have; a massive particle has a minimum energy of its mass. Therefore, under a dilation, we can potentially rescale the wavelength of a massive particle to larger than its Compton wavelength. As this is unphysical, dilations do not in general map physical states of massive particles to other physical states, and therefore such a system is not scale invariant.

One potential way out would seem to be to just restrict to those dilations that mapped physical states to physical states for massive particles. This is possible, but by doing this the action of dilations no longer forms a group. As dilations are just implemented by multiplication by a positive real number, they have an identity (multiplication by 1) and are associative because number multiplication is associative. To form a group, dilations must also be closed. Multiplication by $\lambda_1 > 0$ and $\lambda_2 > 0$ is equivalent to multiplication by $\lambda_3 > 0$, where

$$\lambda_1 \lambda_2 = \lambda_3 > 0. \tag{9.16}$$

For massive particles, we could just agree to only dilate by an amount $0 < \lambda \leq 1$, which would never have the issue of enlarging the Compton wavelength. However, such a restriction then leads to dilations without an inverse, and so dilations do not form a group. For dilations to be a group, if a multiplication by $\lambda > 0$ is in the group, then so too must $\lambda^{-1} > 0$. This enforces dilations to be a rescaling by any positive number. In the example below, we explicitly construct the dilation operator that implements scale transformations. You'll study this operator and the group of dilations further in Exercise 9.1.

Example 9.1 What is the dilation operator $\hat{D}$ whose action implements scale transformations?

Solution

A dilation rescales the position four-vector as

$$x^\mu \to \lambda x^\mu, \tag{9.17}$$

for $\lambda > 0$. The dilation operator $\hat{D}$ is a Hermitian operator which implements an infinitesimal rescaling. In general, we can express $\lambda = e^\epsilon$, for any real number ϵ. Therefore, a rescaling can be expressed as

$$x^\mu \to e^\epsilon x^\mu = e^{i\epsilon\hat{D}} x^\mu . \tag{9.18}$$

On the right, we have inserted the action of the dilation operator $\hat{D}$ into the exponent. In this way, $\hat{D}$ is an element of the Lie algebra of the dilation or scale transformation group. Note the factor of i in the exponent on the right: we want a unitary representation of the scale transformation group.

Assuming that $\epsilon \ll 1$, we can Taylor expand the exponential as

$$x^\mu + \epsilon x^\mu = x^\mu + i\epsilon\hat{D}x^\mu , \tag{9.19}$$

which implies that x^μ is an eigenvector of $\hat{D}$:

$$\hat{D}x^\mu = -ix^\mu . \tag{9.20}$$

A first-order differential operator that accomplishes this is

$$\hat{D} = -ix^\mu \partial_\mu . \tag{9.21}$$

Note here that μ is an internal index and summed over, so we can represent it by any character. The action of this dilation operator on a spacetime position four-vector x^μ is

$$\hat{D}x^\mu = -ix^\nu \partial_\nu x^\mu = -ix^\nu \delta^\mu_\nu = -ix^\mu . \tag{9.22}$$

The object δ^μ_ν is the Kronecker-δ for four spacetime dimensions; $\delta^\nu_\mu = 1$ only if $\mu = \nu$ and is 0 otherwise.

9.1.1 Scale Invariance of QCD

With this insight about the action of dilations, let's now study what happens in QCD. The QCD action with massless quarks is

$$S[A^a_\mu, \psi] = \int d^4x \left[-\frac{1}{4} F^a_{\mu\nu} F^{\mu\nu\,a} + \bar{\psi} i\gamma \cdot D\psi \right] , \tag{9.23}$$

as introduced in Chapter 8. Recall that the field strength tensor $F^a_{\mu\nu}$ in terms of the gluon field A^a_μ is

$$F^a_{\mu\nu} = \partial_\mu A^a_\nu - \partial_\nu A^a_\mu + g f^{abc} A^b_\mu A^c_\nu . \tag{9.24}$$

Using the same principles as we did in analyzing the action of a scalar field ϕ, the gluon field A^a_μ has dimensions of energy, and so will scale like λ^{-1} under a dilation. Similarly, the covariant derivative

$$D_\mu = \partial_\mu - igA^a_\mu T^a \tag{9.25}$$

will also scale like λ^{-1}. This then implies that the quark field ψ has dimensions of

$$[\psi] = [\text{energy}]^{3/2} . \tag{9.26}$$

Then, under a dilation, the quark field ψ scales like $\lambda^{-3/2}$. Note that the γ matrices and the structure constants f^{abc} are just numbers, and so do not transform under a dilation.

With these scalings, the QCD action is scale invariant. Performing a dilation by an amount $\lambda > 0$, the action transforms as

$$S[A^a_\mu, \psi] \rightarrow \int d^4x \lambda^4 \left[-\frac{1}{4} \lambda^{-2} F^a_{\mu\nu} \lambda^{-2} F^{\mu\nu\,a} + \lambda^{-3/2} \bar{\psi} i\gamma \cdot D\lambda^{-1}\psi\lambda^{-3/2} \right] \qquad (9.27)$$
$$= S[A^a_\mu, \psi] \,.$$

Therefore, we expect that QCD is not only Lorentz invariant and gauge invariant, but also scale invariant. We have already seen how Lorentz and gauge invariance highly restrict interactions of particles. We'll discuss what this scale invariance implies shortly.

First, our argument for scale invariance in QCD was a little fast. In this argument, we assumed that the coupling g that appears in the field strength or the covariant derivative is just a number and so is itself scale invariant. That is, we assumed that the coupling g is the same regardless of the wavelength or energy with which we probe it. However, we know this isn't true. The coupling g, or the strong coupling α_s, changes with energy according to the β-function of QCD. With $\alpha_s = g^2/(4\pi)$ the energy dependence of the strong coupling is

$$Q\frac{d\alpha_s}{dQ} = \beta(\alpha_s) \neq 0 \,. \qquad (9.28)$$

Therefore, the non-zero β-function breaks the scale invariance of QCD. This is entirely a quantum phenomenon; that is, the β-function would be 0 if $\hbar = 0$. Even in a system like QCD which is not strictly scale invariant, the manner in which scale invariance is violated is highly restricted. The equation that governs the violation of scale invariance in quantum field theory is called the **Callan–Symanzik equation**.[2]

Nevertheless, there is a sense in which the violation of scale invariance is weak in QCD: the value of the β-function is relatively small at high energies, so we can consider QCD as a scale-invariant theory, with small corrections that violate scale invariance. From Section 8.3.4, the β-function for α_s in QCD is

$$Q\frac{d\alpha_s}{dQ} = \beta(\alpha_s) = \frac{\alpha_s^2}{2\pi}\left(-11 + \frac{2}{3}n_f\right) \,. \qquad (9.29)$$

As a fraction of the value of α_s, the β-function is

$$\frac{\beta(\alpha_s)}{\alpha_s} = \frac{\alpha_s}{2\pi}\left(-11 + \frac{2}{3}n_f\right) \,, \qquad (9.30)$$

which decreases as Q increases. Even at the Z boson mass $Q = m_Z$ where $\alpha_s = 0.118$, the β-function is about 14% the value of α_s. At an energy of $Q = 1$ TeV, which is regularly probed by proton collisions at the LHC, the β-function is less than 10% of the value of α_s. So, to a precision of about 10%, we can say that QCD is scale invariant in the collisions at the LHC.

---

[2] C. G. Callan, Jr., "Broken scale invariance in scalar field theory," Phys. Rev. D **2**, 1541 (1970); K. Symanzik, "Small distance behavior in field theory and power counting," Commun. Math. Phys. **18**, 227 (1970); K. Symanzik, "Small distance behavior analysis and Wilson expansion," Commun. Math. Phys. **23**, 49 (1971).

9.1.2 Fractals and Scale Invariance

A powerful way to think about scale transformations and invariance for the rest of this chapter is the following. Imagine we probe the system with some wavelength λ_1, which can be visualized as viewing the system with some pixel size set by λ_1:

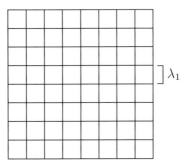

$] \lambda_1$

Scale invariance says that if we probe the system at another wavelength $\lambda_2 < \lambda_1$, then the physics is identical:

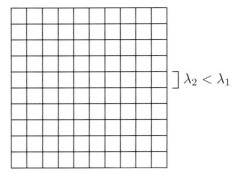

$] \lambda_2 < \lambda_1$

We can freely "zoom in" or "zoom out" of our system and we see the same physical phenomena. For example, if probing at a wavelength λ_1 we observe three particles, then by probing at $\lambda_2 < \lambda_1$, we would also see three particles. This implies that a scale-invariant theory produces an infinite number of particles, because we can zoom in arbitrarily and still see the same number of particles produced at every distance scale. Another way to say this is that a scale-invariant system is one which exhibits structure on all scales.

Mathematical objects that exhibit structure at all scales were studied and defined in detail by Benoit Mandelbrot in the 1960s and 1970s. He called them **fractals**, and there are a huge number of fractal systems in Nature such as Romanesco broccoli, lightning, and tree branches. His seminal paper on fractals is titled "How long is the coast of Britain."[3]

Figure 9.1 shows an example of a fractal called the Mandelbrot set. The Mandelbrot set is defined as the set of all complex numbers c for which the recursive formula

$$z_{n+1} = z_n^2 + c, \tag{9.31}$$

---

[3] B. B. Mandelbrot, "How long is the coast of Britain?" *Science,* **156**, no. 3775, 636 (1967). Mandelbrot's middle initial "B." is recursive: it stands for Benoit B. Mandelbrot.

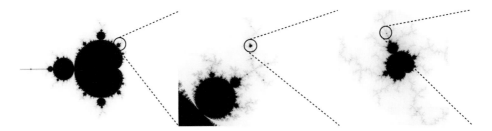

A visualization of the Mandelbrot set fractal. On the left is the whole Mandelbrot set in black, and as one zooms in to the boundary of the Mandelbrot set, arbitrarily detailed structure is resolved.

with $z_0 = 0$, is bounded as $n \rightarrow \infty$. The left panel of Fig. 9.1 shows the whole Mandelbrot set in black, which forms an intricate cardioid shape. One can zoom in anywhere on the boundary of the Mandelbrot set and see finer and finer structure at arbitrarily small scale, as illustrated moving right in the figure. Recursive equations, like that defining the Mandelbrot set, often give rise to fractal structure, and we'll see a similar recursive structure in studying the energy dependence of parton distribution functions in the following section.

9.2 Parton Evolution

In our discussion of deeply inelastic scattering in Chapter 7, the measurement of the form factor $F_2(x)$ (or parton distribution function $f_q(x)$) was used to argue that quarks were non-interacting point particles. If the parton distributions are independent of probing energy scale Q^2, then quarks appear the same regardless of the resolution at which they are probed, a phenomon called Bjorken scaling. But is this actually true? As discussed in Section 7.2.1, even if quarks are point particles, Bjorken scaling can be violated if quarks have non-trivial interactions with one another. Now, with the theory of QCD, we know that they do interact with one another, through the gluon. At best, Bjorken scaling is an approximate feature of QCD. So, can we use QCD to determine the Q^2 dependence of the parton distribution functions and understand the limit in which Bjorken scaling is observed?

9.2.1 Collinear Divergences in QCD

To motivate this discussion, let's remind ourselves about the cross section we calculated for the process $e^+e^- \rightarrow q\bar{q}g$. In Section 7.3.4, we found

$$\frac{d^2\sigma(e^+e^- \rightarrow q\bar{q}g)}{dx_q \, dx_{\bar{q}}} = \sigma(e^+e^- \rightarrow q\bar{q})\frac{\alpha_s}{2\pi}C_F\frac{x_q^2 + x_{\bar{q}}^2}{(1 - x_q)(1 - x_{\bar{q}})}. \tag{9.32}$$

The x_i variables are defined from the final-state momenta p_i as

$$x_i = \frac{2Q_{tot} \cdot p_i}{Q_{tot}^2}, \tag{9.33}$$

for $i = q, \bar{q}, g$, and Q_{tot} is the total system momentum four-vector which in the center-of-mass frame is $Q_{tot} = (E_{cm}, 0, 0, 0)$. Now, in the differential cross section above, let's take the limit in which the quark and gluon become collinear. That is, we will expand in the limit where the angle between the quark and gluon becomes small and we only keep the leading terms. The physical picture we have is

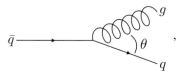

where θ is called the **splitting angle**. Our goal will be to expand Eq. 9.32 in the limit where $\theta \ll 1$. A nice way to express the result of this expansion is in terms of the relative energy fraction of the collinear quark and gluon. We denote the energy fraction of the quark as z, defined as

$$z \equiv \frac{E_q}{E_q + E_g} = \frac{x_q}{x_q + x_g} = \frac{x_q}{2 - x_{\bar{q}}}. \tag{9.34}$$

In the collinear limit, we can set $x_{\bar{q}} = 1$; that is, when the quark and gluon become collinear, then the anti-quark has energy $E_{\bar{q}} = E_{cm}/2$. In this limit, the energy fraction z becomes

$$z = x_q. \tag{9.35}$$

Note also that the energy fraction of the gluon $x_g = 1 - z$.

We also want to express the splitting angle θ between the quark and gluon in terms of the three-body phase space variables x_i. Note that

$$\frac{2p_q \cdot p_g}{Q_{tot}^2} = \frac{2E_q E_g}{E_{cm}^2}(1 - \cos\theta) = \frac{x_q x_g}{2}(1 - \cos\theta). \tag{9.36}$$

Similarly, we can re-express the dot product of the quark and gluon four-vectors $p_q \cdot p_g$ with the total momentum Q_{tot} as

$$\frac{2p_q \cdot p_g}{Q_{tot}^2} = \frac{Q_{tot}^2 - 2Q_{tot} \cdot p_{\bar{q}}}{Q_{tot}^2} = 1 - x_{\bar{q}}. \tag{9.37}$$

Therefore, the cosine of the splitting angle θ is

$$1 - \cos\theta = \frac{2(1 - x_{\bar{q}})}{x_q x_g}. \tag{9.38}$$

In the collinear limit, we can Taylor expand $1 - \cos\theta = \theta^2/2 + \cdots$ at lowest order, and so the splitting angle is

$$\theta^2 = \frac{4(1 - x_{\bar{q}})}{x_q x_g} = \frac{4(1 - x_{\bar{q}})}{x_q(1 - x_q)}. \tag{9.39}$$

On the right, we have used that $x_g = 1 - x_q$ in the collinear limit.

With these expressions for z and θ^2, we will change variables in Eq. 9.32 from x_q and $x_{\bar{q}}$ to z and θ^2. We have

$$z = x_q , \qquad\qquad 4(1 - x_{\bar{q}}) = z(1 - z)\theta^2 , \qquad\qquad (9.40)$$

or that

$$x_{\bar{q}} = 1 - \frac{z(1 - z)}{4}\theta^2 . \qquad\qquad (9.41)$$

The Jacobian of this change of variables is

$$J = \left| \frac{\partial x_{\bar{q}}}{\partial \theta^2} \right| = \frac{z(1 - z)}{4} = \frac{1 - x_{\bar{q}}}{\theta^2} . \qquad\qquad (9.42)$$

Then, in the collinear limit of the quark and gluon the differential cross section of the process $e^+ e^- \to q\bar{q}g$ becomes

$$\frac{d^2\sigma \, (e^+ e^- \to \bar{q}(qg))}{dz \, d\theta^2} = \sigma(e^+ e^- \to q\bar{q})\frac{\alpha_s}{2\pi}C_F\frac{1}{\theta^2}\frac{1 + z^2}{1 - z} . \qquad\qquad (9.43)$$

The parentheses that associate the q and g in the cross section denote that the gluon is becoming collinear to the quark.

This form of the cross section tells us many things. First, the expression diverges in the collinear limit when $\theta^2 \to 0$, and in the low-energy or soft-gluon limit when $1 - z \to 0$. This might be very disconcerting if we attempt to interpret the cross section as a probability. Probabilities can't be larger than 1, so what is going on?

The answer is simply that we can't interpret this cross section as a probability. In the limit when the gluon becomes either collinear to the quark or soft, the configuration of the final state is a degenerate quantum mechanical system. All that our particle detectors measure is the energy and momentum of the final-state particles. In the limit in which the gluon becomes collinear to the quark, for example, we only observe the total energy of the quark–gluon system, which is equivalent to a quark with no emitted gluon. Similarly, if the gluon has zero energy, then the quark's and anti-quark's energies are completely unaffected. There is no measurement exclusively from energy and momentum that we can do to identify an exactly collinear or zero-energy gluon.

Let's visualize more concretely what's going on in the case of collinear gluon emission. We could have just a quark traveling along:

or a quark and one collinear gluon:

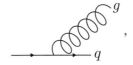

or two collinear gluons:

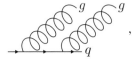

or three, or any number of collinear gluons emitted off of the quark. The important point is that all of these configurations are degenerate: there is no measurement that we can perform to distinguish them. Each configuration produces a cross section that is divergent, but the interpretation of the divergence is physical. Because we can't distinguish these different final states, a divergent cross section tells us that an arbitrary number of collinear gluons can and in general will be emitted. Only by summing over all states with any number of collinear gluon emissions, q, qg, qgg, $qggg$, $\ldots$, do we find a finite result. This is a fundamental result in quantum field theory and is known as the **Kinoshita–Lee–Nauenberg (KLN) theorem**.[4] We'll show how to explicitly do this sum in a second.

Another feature of this collinear limit is the presence of a function of z that governs the distribution of the quark's energy fraction. This function is universal: it is the same for any process in which a quark splits collinearly to a gluon. The argument as to why such a function would be universal is the following. In the limit in which a quark and gluon become collinear, every other particle in the final state is relatively infinitely far away with respect to the splitting angle θ. That is, the quark and gluon only see each other, and all other particles in the final state only see the combined quark–gluon system and can't resolve the quark and gluon individually. As such, this function is called the **universal collinear splitting function** $P_{qg \leftarrow q}(z)$:

$$P_{qg \leftarrow q}(z) = C_F \frac{1 + z^2}{1 - z} . \tag{9.44}$$

With this observation that there are collinear divergences when gluons and quarks get close, we can compute the energy dependence of the parton distribution function, $f_q(x)$.

9.2.2 Energy Dependence of Parton Distributions

Let's work to calculate the energy scale Q^2 dependence of the parton distribution function, $f_q(x)$. Our goal will be to derive a differential equation in Q^2 for $f_q(x)$. To do this, we imagine probing the proton at an energy scale $Q^2 + \delta Q^2$ and then at a slightly lower scale Q^2. Here, δQ^2 is a small energy scale that we will eventually take to 0. Using the analysis of the collinear limit, we will ask what could have happened to the quark in changing the resolution scale from $Q^2 + \delta Q^2$ to Q^2, and taking $\delta Q^2 \to 0$ will produce the desired differential equation. With the differential equation in hand, we will then show what the

---

[4] T. Kinoshita, "Mass singularities of Feynman amplitudes," J. Math. Phys. **3**, 650 (1962); T. D. Lee and M. Nauenberg, "Degenerate systems and mass singularities," Phys. Rev. **133**, B1549 (1964). This was also identified much earlier in quantum electrodynamics: F. Bloch and A. Nordsieck, "Note on the radiation field of the electron," Phys. Rev. **52**, 54 (1937).

Bjorken scaling limit is, and you'll study how Bjorken scaling is broken by a Taylor series of logarithms of Q^2 in Exercise 9.5.

As discussed earlier, there is no such thing as a "bare" quark: all quarks are associated with an arbitrary number of collinear gluons. The number of those collinear gluons that we resolve depends on the energy or wavelength at which we probe the quark. Therefore, the first thing that we need to do is to rewrite the differential probability for a collinear gluon splitting off of a quark in appropriate variables. This differential probability in terms of the energy fraction z of the quark and the splitting angle θ^2 of the gluon is

$$\frac{d^2\sigma(q \to qg)}{dz \, d\theta^2} = \frac{\alpha_s}{2\pi} \frac{1}{\theta^2} P_{qg \leftarrow q}(z) \,, \tag{9.45}$$

which we have just extracted from the collinear limit of the differential cross section for the process $e^+e^- \to q\bar{q}g$, Eq. 9.43. Our goal is to determine the Q^2 dependence of the parton distribution $f_q(x)$, and so we want to write the differential probability in a way that manifests these quantities. First, in $f_q(x)$, x is the energy or momentum fraction of the quark, just like z in the expression above. Q^2, however, is not directly the splitting angle θ^2, so we'll need to change variables to determine the collinear splitting probability differential in energy fraction z and Q^2.

Q^2 is the squared invariant mass of the quark–gluon system. That is, for quark and gluon four-momenta p_q and p_g, respectively,

$$Q^2 = 2p_q \cdot p_g = 2E_q E_g (1 - \cos\theta) \,. \tag{9.46}$$

In the collinear limit, $1 - \cos\theta \to \theta^2/2$ and the quark and gluon energies E_q and E_g can be written as a fraction of their sum total energy E:

$$E_q = zE \,, \qquad\qquad E_g = (1-z)E \,. \tag{9.47}$$

Then, in the collinear limit,

$$Q^2 = z(1-z)E^2\theta^2 \,, \tag{9.48}$$

where E is the total energy of the quark and the gluon. Making the change of variables with the necessary Jacobian, the differential probability in terms of z and Q^2 is then

$$\frac{d^2\sigma(q \to qg)}{dz \, dQ^2} = \frac{\alpha_s}{2\pi} \frac{1}{Q^2} P_{qg \leftarrow q}(z) \,. \tag{9.49}$$

Now, let's consider the parton distribution evaluated at the two scales $Q^2 + \delta Q^2$ and Q^2. If we look at the proton at scale $Q^2 + \delta Q^2$ and find a quark with energy fraction x, the quark is described by the parton distribution function $f_q(x, Q^2 + \delta Q^2)$. We now want to look at the proton at a slightly lower scale of Q^2. There are two possibilities for what can happen. First, nothing can happen; we just see the parton distribution function at scale Q^2: $f_q(x, Q^2)$. Next, there could have been a collinear emission of a gluon at this lower scale that decreased the energy fraction of the quark from some value z down to x. The picture of this is

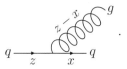

The probability for this to happen is

$$\delta Q^2 \int_0^1 dz \int_0^1 dx'\, f_q(z, Q^2) \frac{d^2\sigma(q \to qg)}{dx'\, dQ^2} \delta(x - zx') \tag{9.50}$$

$$= \frac{\delta Q^2}{Q^2} \frac{\alpha_s}{2\pi} \int_0^1 dz \int_0^1 dx'\, P_{qg\leftarrow q}(x') f_q(z, Q^2) \delta(x - zx')$$

$$= \frac{\delta Q^2}{Q^2} \frac{\alpha_s}{2\pi} \int_x^1 \frac{dz}{z} P_{qg\leftarrow q}\left(\frac{x}{z}\right) f_q(z, Q^2).$$

There are a few moving parts, so let's consider them carefully. In the first line of this equation, we have pulled out of the proton a quark with momentum fraction z at a scale Q^2, hence the factor $f_q(z, Q^2)$. This quark then undergoes a collinear splitting to a quark and a gluon, which is governed by the differential probability from Eq. 9.49. The quark in this splitting takes a relative energy fraction x' in the splitting, corresponding to a fraction $x = zx'$ of the proton's energy. Because this splitting occurred over the energy range δQ^2, we multiply by this differential amount. Finally, we integrate over z and x' because they are unmeasured quantities. On the third line, the factor of $1/z$ is the Jacobian from integrating the δ-function.

Box 9.1   **Historical Profile: Guido Altarelli**

Guido Altarelli was an Italian physicist who earned his Ph.D. from La Sapienza University in Rome in 1963, working with Raoul Gatto. He then went on to work with Gatto's group in Florence, of which his students and young researchers were endearingly nicknamed *gattini* (gatto = cat and gattini = kittens in Italian). After Florence, Altarelli held positions throughout Europe and the US, including at Rockefeller University in New York City, École Normale Superieure in Paris, and back at La Sapienza. Altarelli's work throughout his career spanned the breadth of particle physics, including QCD evolution equations, weak decays of hadrons,[5] neutrino mixing,[6] and more. A particular interest of his was the precision extraction of the value of the strong coupling α_s from experimental data, and he was critical of many of the methods, measurements, and theoretical calculations that are used in the official value provided by the PDG.[7] He was a long-term member of the CERN Theory Division from the mid-1980s until his death in 2015.

---

[5] G. Altarelli, N. Cabibbo, G. Corbo, L. Maiani and G. Martinelli, "Leptonic decay of heavy flavors: A theoretical update," Nucl. Phys. B **208**, 365 (1982).

[6] G. Altarelli and F. Feruglio, "Discrete flavor symmetries and models of neutrino mixing," Rev. Mod. Phys. **82**, 2701 (2010) [arXiv:1002.0211 [hep-ph]].

[7] G. Altarelli, "The QCD running coupling and its measurement," in *Proceedings of the Corfu Summer Institute 2012*, PoS(Corfu2012)002 (2013) [arXiv:1303.6065 [hep-ph]]; G. P. Salam, "The strong coupling: A theoretical perspective" [arXiv:1712.05165 [hep-ph]], and in Forte, S., Levy, A. and Ridolfi, G. (eds), *From My Vast Repertoire: The Legacy of Guido Altarelli*, World Scientific (2018), pp.101–121.

Combining the terms at the scales $Q^2 + \delta Q^2$ and Q^2 we have

$$f_q(x, Q^2 + \delta Q^2) = f_q(x, Q^2) + \frac{\delta Q^2}{Q^2} \frac{\alpha_s}{2\pi} \int_x^1 \frac{dz}{z} P_{qg \leftarrow q}\left(\frac{x}{z}\right) f_q(z, Q^2). \tag{9.51}$$

Then, in the limit as $\delta Q^2 \to 0$, this turns into a differential equation:

$$Q^2 \frac{df_q(x, Q^2)}{dQ^2} = \frac{\alpha_s}{2\pi} \int_x^1 \frac{dz}{z} P_{qg \leftarrow q}\left(\frac{x}{z}\right) f_q(z, Q^2). \tag{9.52}$$

This differential equation determines the energy dependence of the parton distribution function. The running strong coupling, α_s, is evaluated at the scale Q. In this expression, we have only assumed that quarks emit gluons, while a complete analysis requires considering three coupled equations that describe quarks emitting gluons, gluons emitting gluons, and gluons splitting to quarks. These parton evolution equations are among the most important results in QCD and are called the **DGLAP equations** after its discoverers: Yuri Dokshitzer, Vladimir Gribov, Lev Lipatov, Guido Altarelli, and Giorgio Parisi.[8]

The predicted Q^2 evolution of the parton distribution function from the DGLAP equations can be compared to data. Figure 9.2 shows the cross section differential in the exchange energy Q^2 (the abscissa of the plot) and quark momentum fraction x for deeply inelastic scattering that we calculated in Eq. 7.32. The data on this plot come from the ZEUS and H1 experiments at the HERA (H̲adron–E̲lectron R̲ing A̲ccelerator) collider at Deutsches Elektronen-Synchrotron (DESY) in Hamburg, Germany. The measured cross sections at different values of x are multiplied by a power of 2 to separate the curves. The predicted Q^2 dependence is determined from the DGLAP evolution equations and is shown by the solid curves on the plot. Extremely good agreement between the data and prediction over nearly four decades in both Q^2 and x demonstrates that the structure of partons as probed at different energy scales is indeed described by the DGLAP equations. For reference, Bjorken scaling would predict no dependence on Q^2, which would just be flat line on this plot. These data also conclusively demonstrate violation of Bjorken scaling, though the Q^2 dependence is relatively weak.

9.2.3 Physical Interpretation of the DGLAP Equations

The DGLAP evolution equations of the parton distribution function are a prediction directly from the QCD Lagrangian that we constructed in Chapter 8. Therefore, they should subsume the prediction of Bjorken scaling. Indeed, we can see how Bjorken scaling arises as a consequence of asymptotic freedom and the high-energy limit of the DGLAP equations. Bjorken scaling is the property that the quark constituents of the proton were both point particles and non-interacting, and therefore the form factor $F_2(x)$ or the parton distribution $f_q(x)$ was independent of Q^2. The QCD Lagrangian assumes that quarks have

[8] V. N. Gribov and L. N. Lipatov, "Deep inelastic e p scattering in perturbation theory," Sov. J. Nucl. Phys. **15**, 438 (1972) [Yad. Fiz. **15**, 781 (1972)]; Y. L. Dokshitzer, "Calculation of the structure functions for deep inelastic scattering and e^+e^- annihilation by perturbation theory in quantum chromodynamics.," Sov. Phys. JETP **46**, 641 (1977) [Zh. Eksp. Teor. Fiz. **73**, 1216 (1977)]; G. Altarelli and G. Parisi, "Asymptotic freedom in parton language," Nucl. Phys. B **126**, 298 (1977).

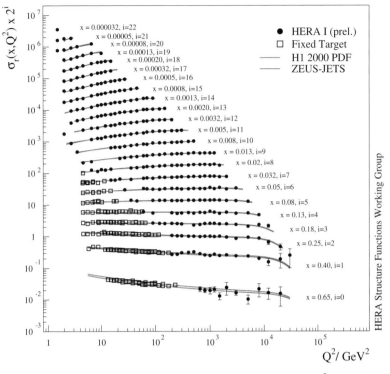

HERA I e$^+$p Neutral Current Scattering - H1 and ZEUS

Fig. 9.2 Plot of the cross section for deeply inelastic scattering differential in the exchanged energy Q^2 and quark momentum fraction x. The data come from the H1 and ZEUS experiments and the predicted Q^2 dependence (solid lines) comes from applying the DGLAP evolution equations. Reprinted from Nucl. Phys. Proc. Suppl. **191**, C. Gwenlan [HERA Combined Structure Functions Working Group], "Combined HERA Deep Inelastic Scattering Data and NLO QCD Fits," 2 (2009), with permission from Elsevier.

no spatial extent, and therefore the derivation of the DGLAP equations used point-particle quarks and gluons. The interactions of quarks and gluons are controlled by the coupling α_s: as $\alpha_s \to 0$, the strength of the interaction between quarks and gluons decreases. Indeed, from Eq. 9.52, if $\alpha_s = 0$, then the parton distribution is independent of Q^2:

$$\frac{df_q(x, Q^2)}{dQ^2} \stackrel{\alpha_s \to 0}{=} 0 . \qquad (9.53)$$

It is not possible to just turn off interactions in QCD, but we can exploit the energy dependence of the coupling, α_s. The value of α_s decreases with increasing energy Q, which is the property of asymptotic freedom. The leading behavior with energy of the running coupling evaluated at scale Q is

$$\alpha_s(Q) = \frac{\alpha_s(Q_0)}{1 + \frac{\alpha_s(Q_0)}{2\pi} \left(11 - \frac{2}{3}n_f\right) \log \frac{Q}{Q_0}} , \qquad (9.54)$$

where Q_0 is some reference energy scale and n_f is the number of relevant quarks. The strength of quark–gluon interaction is decreased by probing the proton at higher and higher energies. Thus, the Bjorken scaling limit in QCD is the high-energy limit: as $Q \to \infty$, we have that

$$\frac{df_q(x, Q^2)}{dQ^2} \stackrel{Q\to\infty}{=} 0 \,. \tag{9.55}$$

This is the reason why Bjorken scaling is observed in high-energy collisions of electrons and protons. At high energies, the coupling α_s is small, and so to first approximation quarks are indeed free particles.

In Section 7.2.1, we discussed how interactions could break Bjorken scaling while maintaining the point-particle property of quarks. The resolution was that a point particle has no spatial extent, and so there must be some nice scaling properties of the parton distributions. This is analogous to the scale invariance of the electric potential of a charged point particle in electromagnetism. Just assuming this scaling property, we showed that Bjorken scaling (independence of Q) could be violated by an infinite Taylor series of logarithms of Q. That is, we can in general express the parton distribution as a Taylor series in $\log Q^2$:

$$f_q(x, Q^2) = \sum_{n=0}^{\infty} c_n(x, Q_0^2) \log^n \frac{Q^2}{Q_0^2} \,, \tag{9.56}$$

where Q_0 is some reference scale and $c_n(x, Q_0^2)$ are coefficients that depend on energy fraction x and are evaluated at the scale Q_0. This property is also a consequence of the DGLAP equations. We can write the DGLAP equations in a more suggestive form:

$$\frac{df_q(x, Q^2)}{d \log Q^2} = \frac{\alpha_s}{2\pi} \int_x^1 \frac{dz}{z} P_{qg\leftarrow q}\left(\frac{x}{z}\right) f_q(z, Q^2) \,. \tag{9.57}$$

Inserting Eq. 9.56 into this expression, one finds a recursion relation between the coefficients $c_n(x, Q_0^2)$ and $c_{n-1}(x, Q_0^2)$ for all $n > 1$. Because there is no upper limit on n in this recursion relation, an infinite number of terms are required, just as our general analysis predicted. As the DGLAP equations sum an infinite number of $\log^n Q^2$ terms, we say that the DGLAP equations **resum** logarithms of Q^2. We'll explore the procedure of resummation and its interpretation in another context in Section 9.3. Also, you'll explicitly construct the recursion relation between the $c_n(x, Q_0^2)$ and $c_{n-1}(x, Q_0^2)$ coefficients in Exercise 9.5.

Connecting to the beginning of this chapter, the DGLAP equations describe scale-invariant gluon emission, up to the effects of the running of α_s. It is straightforward to validate this directly from Eq. 9.52. First consider the left side of the DGLAP equation. Under a dilation, the energy Q scales as

$$Q \to \lambda^{-1} Q \,. \tag{9.58}$$

We will assume that the parton distribution function $f_q(x, Q^2)$ scales simply, to:

$$f_q(x, Q^2) \to \lambda^l f_q(x, Q^2) \,. \tag{9.59}$$

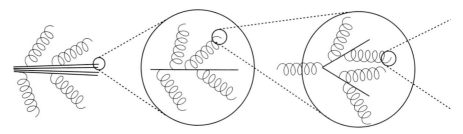

An illustration of particle production over a range of energy or distance scales described by the DGLAP equations. As we zoom in, we see self-similar particle production, just like the structure of a fractal.

Here, we denote the scaling of $f_q(x, Q^2)$ as λ^f. If this assumed dilation of $f_q(x, Q^2)$ is inconsistent, then the scaling of left- and right-hand sides of the DGLAP equation will not agree. The left side of the quark DGLAP equation therefore scales with λ as

$$Q^2 \frac{df_q}{dQ^2} \to (\lambda^{-2}Q^2) \frac{d(\lambda^f f_q)}{d(\lambda^{-2}Q^2)} = \lambda^f Q^2 \frac{df_q(x, Q^2)}{dQ^2} . \tag{9.60}$$

On the right side of the DGLAP equation, there are no explicit scales whatsoever: z and x are dimensionless energy fractions, and so do not scale with λ under a dilation. The only thing that might transform, ignoring the running of α_s, is $f_q(z, Q^2)$. Therefore, the right side of the DGLAP equation scales with λ as

$$\frac{\alpha_s}{2\pi} \int_x^1 \frac{dz}{z} P_{qg \leftarrow q}\left(\frac{x}{z}\right) f_q(z, Q^2) \to \frac{\alpha_s}{2\pi} \int_x^1 \frac{dz}{z} P_{qg \leftarrow q}\left(\frac{x}{z}\right) \left(\lambda^f f_q(z, Q^2)\right) \tag{9.61}$$

$$= \lambda^f \frac{\alpha_s}{2\pi} \int_x^1 \frac{dz}{z} P_{qg \leftarrow q}\left(\frac{x}{z}\right) f_q(z, Q^2) .$$

The left and right sides of the DGLAP equation therefore scale the same under a dilation, and up to running coupling effects, DGLAP describes scale-invariant gluon emission.

Figure 9.3 illustrates the interpretation of this scale-invariant particle production from the DGLAP equations. At some initial scale Q^2, we see some number of quarks and gluons. We can zoom in to any one of the particles and we will see similar particle production at a smaller value of Q^2, described by the same DGLAP equations. We can continue zooming in to see more particle production always described by the DGLAP equations. If we could zoom in indefinitely, we would see particle production and structure at all scales and correspondingly the DGLAP equations would predict a fractal structure of QCD at high energies. The running of α_s and hadron masses prohibit the possibility of zooming in ad infinitum, but nevertheless a fractal-like structure of QCD is exhibited over a wide range of energy scales.

9.3 Jets

This fractal or self-similar feature of high-energy QCD has other experimental consequences. Our topic in this section will be what is perhaps the most shocking observation and prediction of QCD: collimated streams of high-energy particles, called "jets." We've seen jets before, especially in understanding the process $e^+e^- \to$ hadrons, but here we'll work to understand their features quantitatively.

The DGLAP equations say there is no such thing as a "bare" quark. A quark is always associated with an arbitrary number of gluons that are approximately collinear to the direction of the quark. Thus, in an experiment where a quark is produced, as in $e^+e^- \to q\bar{q}$ events, that quark (or anti-quark) will always be associated with an arbitrary number of collinear particles. This collimated stream of particles is a **jet**: it is the manifestation of the approximate scale invariance of QCD, in addition to the relative smallness of α_s at high energies.

In the late 1960s and early 1970s this prediction of jets produced in $e^+e^- \to$ hadrons events was believed by only a few to possibly be a consequence of the strong force.[9] At the time, the strong interactions binding hadrons were thought by many to just be strong; effectively, the coupling of the strong force was thought to be very large. Such a large coupling would not produce jets in experiment: one would instead observe a nearly spherical distribution of particles emanating from the collision point. However, the discovery and interpretation of Bjorken scaling and asymptotic freedom in QCD made jets a robust prediction of high-energy QCD and one that could be compared to data.

What we see most often in a high-energy $e^+e^- \to$ hadrons event is something like what is displayed in Fig. 9.4. This is an event display of e^+e^- collisions at a center-of-mass energy of about 91 GeV (i.e., at the Z boson mass) from the OPAL (an <u>O</u>mni-<u>P</u>urpose <u>A</u>pparatus at <u>L</u>EP) experiment located at LEP at CERN. This is a head-on view of the OPAL detector, with the beams of electrons and positrons going into and out of the page. Two back-to-back, high-energy, collimated streams of particles are clearly visible; these are the jets. Note that because the e^+e^- collision occurs in the center-of-mass frame, there must be at least two jets in the final state, by momentum conservation.

As collimated streams of particles, jets invoke water jets, and one potential source of the etymology of "jet" in particle physics is from the Jet d'Eau, a spectacular 140-meter-high water fountain that sits in Lake Geneva. To my knowledge, the first use of the term "jet" was by George Rochester and Clifford Butler in their analysis of cosmic ray showers in the upper atmosphere that resulted in the discovery of the kaon in 1953.[10] By the way, the Jet d'Eau dates from 1886, long before there was any understanding of jets, let alone special relativity and quantum mechanics!

---

[9] Before the development of QCD, many potential models for particle production were considered. For example, one paper that studied the energy dependence of the particle number in $e^+e^- \to$ hadrons focused on a statistical model that would produce a spherically symmetric distribution of hadrons. Though they didn't study jets, the authors acknowledged:

> The observation of such "jets" in colliding-beam processes would be most spectacular. It is *not* our intention here to study such a possibility further.

Reprinted excerpt with permission from J. D. Bjorken and S. J. Brodsky, Phys. Rev. D **1**, 1416 (1970). Copyright 1970 by the American Physical Society.

[10] G. D. Rochester and C. C. Butler, "The new unstable cosmic-ray particles," Rep. Prog. Phys. **16**, no. 1, 364 (1953).

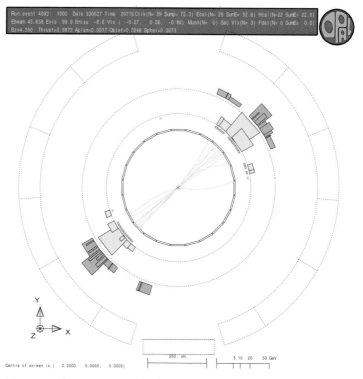

Fig. 9.4   Event display of the process $e^+e^- \rightarrow$ two jets from the OPAL experiment at LEP. The two jets are visible as collimated tracks in the center of the display and as high-energy deposits in the calorimetry. Credit: OPAL experiment © CERN.

9.3.1 All-Orders Predictions: Thrust

In the remainder of this chapter, we'll work to understand a prediction of jets and their structure, which will illustrate more properties of (approximate) scale invariance in QCD. In Example 7.2 of Chapter 7, we calculated the order-α_s prediction for the differential cross section of thrust τ. This result in Eq. 7.141 followed from the calculation of the inclusive cross section to the process $e^+e^- \rightarrow q\bar{q}g$. However, in that differential cross section, we noted that the limit when $\tau \rightarrow 1$ is divergent: this corresponds to the emitted gluon being collinear with either the quark or the anti-quark, or the gluon having low energy. In any of these limits, the matrix element diverges, producing a differential cross section which correspondingly diverges. This is not physical behavior and our prediction breaks down in the limit where $\tau \rightarrow 1$.

Armed with our new understanding of scale invariance of QCD, we can fix this divergence by appropriately including the emission of an arbitrary number of collinear or low-energy gluons in the measurement of thrust. Thus, this analysis will focus on the thrust observable τ in the limit when $\tau \rightarrow 1$. Formally, we will impose the requirement that $1 - \tau$ is parametrically smaller than 1:

$$1 - \tau \ll 1 . \tag{9.62}$$

This requirement restricts the possible regions of phase space in which final-state particles can live and still contribute to τ. That is, by restricting $1 - \tau \ll 1$, we will be calculating an exclusive differential cross section. The value of $1 - \tau$ corresponds to a new energy scale in the final state that is distinct from the center-of-mass collision energy. We need to account for this new energy scale appropriately, and doing so will require consideration of the emission of an arbitrary number of gluons.

So, let's make a prediction of the thrust observable τ in the limit when $\tau \to 1$. We found earlier that $\tau \to 1$ corresponds to the limit in which the gluon in $e^+e^- \to q\bar{q}g$ events become collinear, for example. This is a singular limit, in the sense that the matrix element describing this configuration diverges like $1/\theta^2$, where θ is the splitting angle of the gluon to the closer of the quark or anti-quark:

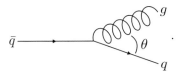

Earlier in this chapter, we argued that this singular limit means that an arbitrary number of collinear gluons will be emitted from the quark and anti-quark. How do these numerous gluons affect the differential cross section of thrust?

In Section 9.2.1, we showed that the differential probability distribution $P_q(z, \theta^2)$ for emission of a gluon at angle θ and energy fraction $1 - z$ off of a quark is

$$P_q(z, \theta^2) = \frac{\alpha_s}{2\pi} C_F \frac{1 + z^2}{1 - z} \frac{1}{\theta^2} dz\, d\theta^2 , \qquad (9.63)$$

which holds in the collinear limit when $\theta \to 0$. The thrust observable τ was defined as a function of the three-body phase space variables x_i in Chapter 7 as

$$\tau = \max\{x_i\} . \qquad (9.64)$$

If the quark and gluon have energy fractions x_q and x_g, respectively, then in the collinear limit, by momentum conservation, the anti-quark has the largest energy fraction $x_{\bar{q}}$. In the limit when the gluon becomes collinear to the quark, we showed earlier in Eq. 9.41 that

$$\tau = x_{\bar{q}} = 1 - \frac{z(1 - z)}{4}\theta^2 , \qquad \text{or that} \qquad 1 - \tau = \frac{z(1 - z)}{4}\theta^2 . \qquad (9.65)$$

As a further simplification, we will work in the soft and collinear limit in which the gluon has both low energy and is collinear to the quark. This corresponds to additionally taking the $z \to 1$ limit, and so the differential probability distribution becomes

$$P_q(z, \theta^2) = \frac{\alpha_s}{\pi} C_F \frac{dz}{1 - z} \frac{d\theta^2}{\theta^2} , \qquad (9.66)$$

and the thrust is

$$1 - \tau = \frac{(1 - z)}{4}\theta^2 . \qquad (9.67)$$

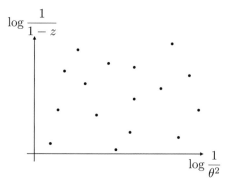

Fig. 9.5 A semi-infinite Lund diagram that illustrates soft and collinear gluon emissions uniformly populating phase space. The limit in which gluons have zero energy is moving vertically to infinity and exactly collinear gluons are found moving horizontally to infinity.

While we only considered the collinear emission of a gluon off of the quark, of course it could also have been emitted off of the anti-quark. The probability for emission of the gluon off of the anti-quark $P_{\bar{q}}(z, \theta^2)$ is exactly the same as that for the quark:

$$P_{\bar{q}}(z, \theta^2) = \frac{\alpha_s}{\pi} C_F \frac{dz}{1-z} \frac{d\theta^2}{\theta^2} . \tag{9.68}$$

Because the gluon could have been emitted collinearly off of either the quark or the anti-quark, the total probability for a soft and collinear gluon in the final state, $P_{\text{soft\&coll}}(z, \theta^2)$, is just the sum of these probabilities:

$$P_{\text{soft\&coll}}(z, \theta^2) = P_q(z, \theta^2) + P_{\bar{q}}(z, \theta^2) = 2\frac{\alpha_s}{\pi} C_F \frac{dz}{1-z} \frac{d\theta^2}{\theta^2} . \tag{9.69}$$

We will use this differential probability for gluon emission in the calculations that follow.

There is a beautiful way to visualize this system and calculate the thrust in the soft and collinear limit. The probability $P_{\text{soft\&coll}}(z, \theta^2)$ governs the splitting probability of an arbitrary number of soft and collinear gluons. Note that it is a flat distribution in $\log 1/(1-z)$ and $\log 1/\theta^2$:

$$P_{\text{soft\&coll}}(z, \theta^2) = 2\frac{\alpha_s}{\pi} C_F \frac{dz}{1-z} \frac{d\theta^2}{\theta^2} = 2\frac{\alpha_s}{\pi} C_F \, d\log\frac{1}{1-z} \, d\log\frac{1}{\theta^2} . \tag{9.70}$$

Therefore, we can imagine gluons uniformly populating the $(\log 1/(1-z), \log 1/\theta^2)$ plane, as illustrated in Fig. 9.5. This plane is called a **Lund diagram**,[11] and each dot corresponds to an emitted gluon with some energy fraction $1 - z$ emitted at an angle θ. To orient you in this plane, the origin corresponds to large-angle ($\theta \sim 1$) and high-energy ($z \sim 0$) gluon emission. Off at infinity are the various divergent limits of the emission probability, and depending on how you approach infinity you are sensitive to a different limit. Moving vertically in this plane corresponds to the soft (low-energy) gluon limit

---

[11] B. Andersson, G. Gustafson, L. Lonnblad and U. Pettersson, "Coherence effects in deep inelastic scattering," Z. Phys. C **43**, 625 (1989).

$(z \to 1)$; horizontally is the collinear-gluon limit $(\theta \to 0)$, and diagonally is some soft and collinear limit. In the following example, we use this picture of gluon emissions in the Lund plane to calculate the differential cross section for thrust in the exclusive limit when $\tau \to 1$.

Example 9.2 What is the differential cross section for thrust τ in the limit in which $\tau \to 1$? To illustrate the physics, we'll just work in the soft and collinear limit as described by a uniform distribution of emissions in the Lund plane. We'll also ignore the running of α_s.

Solution

From Eq. 9.67, the expression for thrust in the soft and collinear limit corresponds to a straight line on a Lund plane:

$$\log \frac{1}{1-\tau} = \log \frac{1}{1-z} + \log \frac{1}{\theta^2} + \log 4 . \tag{9.71}$$

We can draw this line on the Lund diagram:

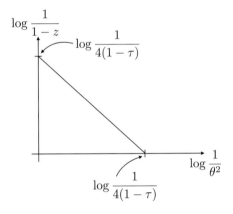

On this diagram, we've denoted the x- and y-intercepts; that is, the value of $\log 1/\theta^2$ or $\log 1/(1-z)$ when the other is 0.

Importantly, note that emissions that land above this line contribute a tiny (negligible) value to thrust, while emissions below the line will increase the value of $1 - \tau$ beyond its measured value. Gluon emissions that land below the line would have a larger energy fraction $1 - z$ or splitting angle θ^2 than that allowed by the measured value of τ. Because gluon emissions are uniformly distributed in this logarithmic space, they are exponentially far apart in "real" space. Therefore, in the soft and collinear limit, emissions above this line contribute an exponentially small amount to the value of thrust. To good approximation, they are indeed negligible. From this picture, let's calculate the cumulative distribution of thrust; i.e., we will calculate the probability that the thrust is no larger than the measured value of $4(1 - \tau)$. We will call this cumulative probability distribution $\Sigma(\tau)$.

Recall that emissions in the Lund diagram are uniformly distributed, but for the measured value of thrust, we must forbid emissions that land below the line:

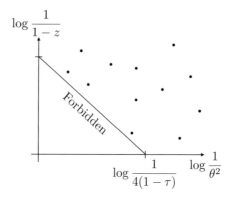

To determine the probability that no gluon emissions land in the triangle, we imagine breaking it up into tiny pieces and forbidding any emission in each of the pieces. The picture of this is:

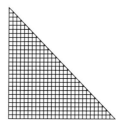

The probability that no gluon is emitted into one of these squares is set by the area of a square:

$$P(\text{no gluon}) = 1 - 2\frac{\alpha_s}{\pi}C_F \cdot (\text{area of square}) . \tag{9.72}$$

For simplicity, let's take the area of each square to be identical. If there are N squares, then the probability that no gluon is emitted into one of the squares is

$$P(\text{no gluon}) = 1 - 2\frac{\alpha_s}{\pi}C_F \cdot \frac{(\text{area of triangle})}{N} . \tag{9.73}$$

To find the total probability that no gluons land anywhere in the triangle, we need to multiply the probabilities of each square together:[12]

$$P(\text{no gluon in triangle}) = \prod_{i=1}^{N}\left[1 - \frac{\alpha_s}{\pi}C_F \cdot (\text{area of square } i)\right] \tag{9.74}$$

$$= \left(1 - 2\frac{\alpha_s}{\pi}C_F \cdot \frac{(\text{area of triangle})}{N}\right)^{N} .$$

---

[12] A rule of thumb to remember with probabilities is how to treat statements with "and" and "or". The probability that independent events A **and** B happen is the product of the individual probabilities: $P(A \wedge B) = P(A)P(B)$. However, the probability that A **or** B happens is the sum of the individual probabilities: $P(A \vee B) = P(A) + P(B)$.

As the number of squares $N \to \infty$, this probability transmogrifies into an exponential:

$$P(\text{no gluon in triangle}) = \exp\left[-2\frac{\alpha_s}{\pi}C_F \cdot (\text{area of triangle})\right]. \tag{9.75}$$

The area of the forbidden triangle is just

$$\text{area of triangle} = \frac{1}{2}\log^2[4(1-\tau)], \tag{9.76}$$

and so the probability of no gluon in the triangle or the cumulative probability of τ in the soft and collinear limits is

$$P(\text{no gluon in triangle}) = \Sigma(\tau) = \exp\left[-\frac{\alpha_s}{\pi}C_F \log^2(4(1-\tau))\right]. \tag{9.77}$$

This result is remarkable, and this exponential factor is called the **Sudakov form factor**.[13] It is responsible for exponentially suppressing the soft and collinear region of phase space (when $\tau \to 1$), which is physical behavior. The Sudakov form factor says that it becomes exponentially unlikely to measure final states with τ very close to 1. If τ is very close to 1, then there must be no gluon emissions at energy scales above that set by the measured value of thrust. In an approximately scale-invariant theory like QCD, we expect emissions at all scales, and so suppressing them over a wide range of energies by restricting $\tau \to 1$ is extremely unlikely.

To calculate the probability distribution for τ, $P(\tau)$, we just take the derivative of the cumulative distribution $\Sigma(\tau)$:

$$P(\tau) = \frac{d\Sigma(\tau)}{d(1-\tau)} = -2\frac{\alpha_s C_F}{\pi}\frac{\log(4(1-\tau))}{1-\tau}\exp\left[-\frac{\alpha_s}{\pi}C_F \log^2(4(1-\tau))\right]. \tag{9.78}$$

Divergences in Feynman diagrams at each order in α_s are turned into exponential suppressions when all orders are included!

Note that in this Sudakov factor we have summed a series on α_s to all powers in α_s. This procedure is called **resummation**, and $P(\tau)$ defined in Eq. 9.78 is called the resummed differential cross section for thrust, τ. We denote this by

$$\frac{1}{\sigma}\frac{d\sigma^{\text{resum}}}{d\tau} = P(\tau) = -2\frac{\alpha_s C_F}{\pi}\frac{\log(4(1-\tau))}{1-\tau}\exp\left[-\frac{\alpha_s}{\pi}C_F \log^2(4(1-\tau))\right]. \tag{9.79}$$

The existence of exponential suppression via a Sudakov form factor is a general feature of exclusive cross sections. This also demonstrates why we said earlier that exclusive cross sections are not well approximated by individual Feynman diagram calculations. Because of multiple scales in the process, exclusive cross sections in general are functions of logarithms of ratios of those scales. These logarithms can become large or even divergent in some phase space regions and invalidate the assumption that corrections to the cross section are controlled by the small value of the coupling. In many (or perhaps most) cases of exclusive cross sections, actually calculating the Sudakov form factor is an extremely

---

[13] V. V. Sudakov, "Vertex parts at very high energies in quantum electrodynamics," Sov. Phys. JETP **3**, 65 (1956) [Zh. Eksp. Teor. Fiz. **30**, 87 (1956)].

non-trivial procedure. This is why, in Section 6.1.5, we stated that exclusive cross sections were much more challenging to work with than inclusive cross sections.

The exponential Sudakov form factor is a manifestation of a semi-classical object in particle physics and quantum field theory. The exponent of the Sudakov form factor is quantum mechanical: to calculate it, we evaluated Feynman diagrams. However, to construct the exponential form, we summed classical probabilities represented by squared matrix elements and cross sections. In its form, this is similar to the WKB approximation for approximating tunneling rates in quantum mechanics.

Resummation is an aspect of a more general procedure in quantum field theory called **renormalization**. The history of renormalization is long and often fraught with misunderstanding. The opponents of renormalization (who included Feynman) thought of it as a terribly ill-defined procedure for "sweep[ing] the infinities under the rug."[14] This is unfortunately a wrong characterization of renormalization. Renormalization is a systematic and well-defined procedure for accounting for and eliminating infinities in a quantum field theory.

The property of renormalizability of a theory is vital for predictivity within that theory. Both QED and QCD are renormalizable theories, and a consequence of renormalizability is that their couplings run and depend on the energy scale at which the theory is probed. Gerardus 't Hooft and Martinus Veltman proved that the entire theory of the Standard Model is renormalizable, and is therefore consistent and predictive.[15] The modern interpretation of renormalization with sound theoretical and physical footing was accomplished by Kenneth Wilson. Using techniques of renormalization, he was able to solve an outstanding problem in condensed matter physics called the Kondo problem.[16] After his work, renormalization became a standard tool for particle physicists and condensed matter physicists alike.

Exercises

9.1   *Dilation Operator.* In Example 9.1, we identified the dilation operator $\hat{D}$ as

$$\hat{D} = -ix^\mu \partial_\mu \, . \tag{9.80}$$

In this exercise, we will study some properties of this operator.

(a) The momentum operator is $\hat{P}_\mu = -i\partial_\mu$. What is the commutator of the dilation and momentum operators,

$$[\hat{D}, \hat{P}_\mu] = ? \tag{9.81}$$

Can the result of this commutator be expressed in terms of familiar operators?

---

[14] R. P. Feynman, *The Character of Physical Law*, © 1967 Richard Feynman, published by the MIT Press.
[15] G. 't Hooft, "Renormalization of massless Yang–Mills fields," Nucl. Phys. B **33**, 173 (1971); G. 't Hooft and M. J. G. Veltman, "Regularization and renormalization of gauge fields," Nucl. Phys. B **44**, 189 (1972).
[16] K. G. Wilson, "The renormalization group: Critical phenomena and the Kondo problem," Rev. Mod. Phys. **47**, 773 (1975).

(b) What are the eigenfunctions of the dilation operator $\hat{D}$? Show that a function f of spacetime position x of the form

$$f(x) = (x \cdot x)^{\Delta/2}, \tag{9.82}$$

for some number Δ, is an eigenfunction. What is its eigenvalue? What does the eigenvalue represent?

9.2 *Expansion of Differential Cross Section for Thrust.* In Example 9.2, we calculated the resummed cross section for thrust to all orders in α_s, but in the soft and collinear limit. In Example 7.2, we calculated the differential cross section of thrust throughout its phase space, but only to leading order in α_s. Do these two calculations yield the same result when the same limits are taken?

(a) From the result of Example 7.2, Eq. 7.141, expand in the limit where $\tau \to 1$. Only keep those terms that diverge most rapidly as $\tau \to 1$. This should correspond to the soft and collinear limit of the differential cross section for thrust.

(b) From the result of Example 9.2, Eq. 9.78, expand to lowest order in α_s.

(c) How do these two limits compare? Do they agree, as expected?

9.3 *Jet Multiplicity.* In Exercise 6.7 from Chapter 6, we studied the exclusive cross section for n-jet production in the ALEPH experiment at LEP. The experiment counted n jets in an event if, for every pair i, j of jets, the following inequality is satisfied:

$$2 \min[E_i^2, E_j^2](1 - \cos\theta_{ij}) > y_{\text{cut}} E_{\text{cm}}^2, \tag{9.83}$$

for some number $y_{\text{cut}} < 1$, where E_i is the energy of jet i and θ_{ij} is the angle between jets i and j. The fractional rates for $n = 2, 3, 4, 5$, and 6 or more jets as a function of y_{cut} measured at ALEPH are reprinted in Fig. 9.6. In this exercise, we will work to calculate these distributions.

(a) When y_{cut} is very small, most of the jets in the final state are likely soft and relatively collinear to the quark or anti-quark off of which gluons are emitted. For a soft and collinear gluon emitted off of a quark, express the phase space restriction

$$2 \min[E_q^2, E_g^2](1 - \cos\theta) > y_{\text{cut}} E_{\text{cm}}^2 \tag{9.84}$$

in terms of z, the relative energy fraction of the quark, and their splitting angle θ. Expand in the limits where $z \to 1$ and $\theta \to 0$. You should find

$$\frac{(1-z)^2 \theta^2}{4} > y_{\text{cut}}. \tag{9.85}$$

(b) For the process $e^+ e^- \to q\bar{q}g$, if Eq. 9.85 is satisfied then ALEPH would identify the final state as having three jets. On the Lund plane, identify the region where emissions must be forbidden for the final state to be identified by as having two jets; that is, the region defined by Eq. 9.85 must be forbidden.

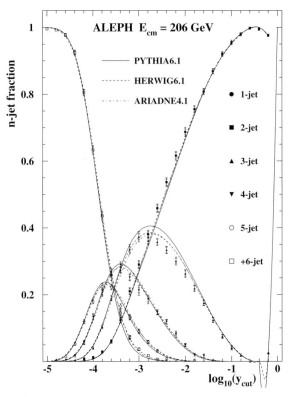

Fig. 9.6 Plot of the *n*-jet fraction in $e^+e^- \rightarrow$ hadrons events from the ALEPH experiment at the Large Electron–Positron Collider as a function of the jet resolution variable y_{cut}. Reprinted by permission from Springer Nature: Springer Nature Eur. Phys. J. C "Studies of QCD at e^+e^- centre-of-mass energies between 91 GeV and 209 GeV," A. Heister *et al.* [ALEPH Collaboration] (2004).

(c) From this forbidden region, calculate the probability that there are no emissions in the forbidden region; that is, the probability state has two identified jets as a function of y_{cut} that the final $P_2(y_{cut})$. You should find

$$P_2(y_{cut}) = \exp\left[-\frac{\alpha_s}{\pi}\frac{C_F}{2}\log^2(4y_{cut})\right] . \qquad (9.86)$$

(d) How does the predicted two-jet rate of Eq. 9.86 compare to the corresponding distribution in Fig. 9.6? From the prediction of Eq. 9.86, below approximately what value of y_{cut} is the cross section for $e^+e^- \rightarrow$ two jets not a good approximation to the cross section for $e^+e^- \rightarrow$ hadrons?

(e) Can you think how you would determine the probability for three, four, five, or any number of jets greater than two? Remember, the total probability to produce any number of jets in the final state must be unity.

9.4 *Properties of the DGLAP Equation.* An important property of the DGLAP equation is that it preserves the normalization of the parton distribution functions. For instance, the proton has a net of two up quarks. That is, the sum of the number of up quarks minus the number of anti-up quarks in the proton is 2. In terms of the parton distributions for up quarks f_u and anti-up quarks $f_{\bar{u}}$, this is

$$\int dx\, (f_u(x, Q^2) - f_{\bar{u}}(x, Q^2)) = 2\,. \tag{9.87}$$

This expression must be independent of energy scale Q for a proton to always be a proton regardless of the energy at which you probe it. Take the derivative with respect to Q^2 of both sides of the expression, and use the DGLAP equation, Eq. 9.52, to simplify it. Show that if the DGLAP equation conserves the normalization, then it must be that

$$\int_0^1 dx\, P_{qg \leftarrow q}(x) = 0\,. \tag{9.88}$$

This is a bit of a weird equation, but it is actually true. The reason that this equation is true is the collinear splitting function $P_{qg \leftarrow q}(x)$ that we have been using is missing a key contribution. We considered only the emission of real gluons, but the quark could have emitted a virtual gluon that was then reabsorbed by the quark. This virtual "loop" contribution to the splitting function is negative, and exactly cancels off the divergence as $x \to 1$ in $P_{qg \leftarrow q}(x)$. Another way to say this is that the collinear splitting function $P_{qg \leftarrow q}(x)$ should actually be regularized in the $x \to 1$ limit. This regularization is accomplished by the +-function expansion described in Exercise 7.1 of Chapter 7. The integral of a +-function is 0, just like the requirement in Eq. 9.88.

9.5 *Resummation of Q^2 with DGLAP.* We can see how the DGLAP equations explicitly resum logarithms of energy scale Q^2 to all orders by expressing the parton distribution function as a series in logarithms:

$$f_q(x, Q^2) = \sum_{n=0}^{\infty} c_n(x, Q_0^2) \log^n \frac{Q^2}{Q_0^2}\,, \tag{9.89}$$

where Q_0 is a reference energy scale. The c_n coefficient functions will be related to one another through the DGLAP equations. In this exercise, we will see how this works.

(a) Insert the expansion of the parton distribution function, Eq. 9.89, into the DGLAP equation, Eq. 9.52. Derive a recursion relation that relates the coefficient functions $c_n(x, Q_0^2)$ and $c_{n-1}(x, Q_0^2)$. Leave the collinear splitting functions implicit in the recursion relation.

Hint: Match powers of logarithms on both sides of the DGLAP equations.

(b) The first coefficient function, $c_0(x, Q_0^2)$, is just the functional form of the parton distribution function evaluated at the energy scale Q_0^2:

$$c_0(x, Q_0^2) = f_q(x, Q_0^2)\,. \tag{9.90}$$

Solve the recursion relation for the coefficient functions $c_n(x, Q_0^2)$ in terms of $f_q(x, Q_0^2)$. You should find

$$c_n(x, Q_0^2) \tag{9.91}$$

$$= \frac{1}{n!} \left(\frac{\alpha_s}{2\pi}\right)^n \underbrace{\int_x^1 \frac{dz_1}{z_1} P_{qg \leftarrow q}\left(\frac{x}{z_1}\right) \cdots \int_{z_{n-1}}^1 \frac{dz_n}{z_n} P_{qg \leftarrow q}\left(\frac{z_{n-1}}{z_n}\right)}_{n \text{ nested integrals}} f(z_n, Q_0^2).$$

(c) A common way to make progress in solving the DGLAP equations is to perform a Mellin transformation. For the parton distribution function $f_q(x, Q^2)$, its Mellin transformation is defined to be

$$\tilde{f}_q(N, Q^2) = \int_0^1 dx\, x^N f(x, Q^2). \tag{9.92}$$

In the DGLAP equation of Eq. 9.52, multiply both sides by x^N and integrate over $x \in [0, 1]$. Show that the Mellin-transformed parton distribution functions satisfy the differential equation

$$Q^2 \frac{d\tilde{f}_q(N, Q^2)}{dQ^2} = \left(\frac{\alpha_s}{2\pi} \int_0^1 dx\, x^N P_{qg \leftarrow q}(x)\right) \tilde{f}_q(N, Q^2). \tag{9.93}$$

(d) Show that the Mellin-transformed parton distribution function that solves the DGLAP equation of Eq. 9.93 is

$$\tilde{f}_q(N, Q^2) = \tilde{f}_q(N, Q_0^2) \exp\left[\left(\frac{\alpha_s}{2\pi} \int_0^1 dx\, x^N P_{qg \leftarrow q}(x)\right) \log \frac{Q^2}{Q_0^2}\right], \tag{9.94}$$

where $\tilde{f}_q(N, Q_0^2)$ is the Mellin-transformed parton distribution function defined at an energy scale Q_0^2.

9.6 *Jet Mass at the LHC.* Jets, the collimated streams of particles from QCD interactions, are created in copious numbers at the Large Hadron Collider. One of the fundamental properties of a jet is its mass; that is, the square of the sum of the four-vectors of the particles of a jet. In this problem, we'll perform some simple calculations to understand these jets.

(a) The simplest jet that can have a non-zero mass (with massless quarks) is one that has two constituent particles. Assume that a jet has just two particles in it. Call the four-momentum of particle 1 p_1 and the four-momentum of particle 2 p_2. Express the squared mass of the jet m_J^2 as

$$m_J^2 = 2p_1 \cdot p_2, \tag{9.95}$$

in terms of the transverse momenta $p_{\perp 1}, p_{\perp 2}$, pseudorapidity η_1, η_2, and azimuthal angles ϕ_1, ϕ_2 of the two particles.

(b) Now, Taylor expand the result to lowest non-zero order in the small-angle limit. To take this limit, assume that $|\eta_1 - \eta_2| \ll 1$ and $|\phi_1 - \phi_2| \ll 1$ and that the pseudorapidity differences are about the same size as the azimuth differences. Express the result in terms of $R = \sqrt{(\eta_1 - \eta_2)^2 + (\phi_1 - \phi_2)^2}$.

(c) In this limit, we can express the jet mass in terms of the relative transverse momentum fraction. Define the jet's transverse momentum to be $p_{\perp J} = p_{\perp 1} + p_{\perp 2}$ and $z = p_{\perp 1}/p_{\perp J}$. Rewrite the squared jet mass in the collinear limit from part (b) in terms of z, $p_{\perp J}$, and R. You should find

$$m_J^2 = z(1-z)p_{\perp J}^2 R^2 . \tag{9.96}$$

(d) With this expression, we will calculate the average jet mass. Using the universal collinear splitting function for gluon emission off of a quark as the differential probability distribution

$$\frac{d^2\sigma}{dz\, d\theta^2} = \frac{\alpha_s}{2\pi} C_F \frac{1}{\theta^2} \frac{1+z^2}{1-z}, \tag{9.97}$$

calculate the average squared jet mass $\langle m_J^2 \rangle$ using the result of part (c).
   The definition of the average squared jet mass is

$$\langle m_J^2 \rangle = \int_0^{R_0^2} d\theta^2 \int_0^1 dz\, \frac{d^2\sigma}{dz\, d\theta^2}\, m_J^2(z, \theta^2) . \tag{9.98}$$

To do this integral, $\theta^2 = R^2$ and $m_J^2(z, \theta^2)$ is the functional form for the squared jet mass from part (c) in terms of energy fraction z and splitting angle θ. The boundaries of the angular integral extend from 0 to an angle called the **jet radius**, which we can denote by R_0. To define the angular region of interest in a jet, jet algorithms are used that identify collimated, energetic particles in a detector. The jet algorithm used in this problem is to just associate two particles as a jet if they are within an angle R_0 of one another, which is referred to as a **k_T-type algorithm** (read: "kay-tee").[17]

(e) The plot in Fig. 9.7 presents data from the ATLAS experiment in which they measured the jet mass on jets identified with a k_T-type algorithm called **Cambridge–Aachen**.[18] They used a jet radius of $R_0 = 1.2$ and the transverse momentum of the jets ranges from $300 < p_{\perp J} < 400$ GeV. Using your expression from part (d), predict the root-mean-square of the jet mass for these data. How does this prediction compare to the plot from ATLAS?

9.7 *Underlying Event at Hadron Colliders.* In this chapter, we studied the probability for collinear gluon emission from a quark. Consider a quark from a proton that is involved in a high-energy collision at the LHC. This quark is traveling along the

---

[17] S. Catani, Y. L. Dokshitzer, M. H. Seymour and B. R. Webber, "Longitudinally invariant K_t clustering algorithms for hadron hadron collisions," Nucl. Phys. B **406**, 187 (1993); S. D. Ellis and D. E. Soper, "Successive combination jet algorithm for hadron collisions," Phys. Rev. D **48**, 3160 (1993) [arXiv:hep-ph/9305266].

[18] Y. L. Dokshitzer, G. D. Leder, S. Moretti and B. R. Webber, "Better jet clustering algorithms," J. High Energy Phys. **9708**, 001 (1997) [arXiv:hep-ph/9707323]; M. Wobisch and T. Wengler, "Hadronization corrections to jet cross-sections in deep inelastic scattering," in Doyle, A.T., Grindhammer, G., Ingleman, G. and Jung, H. (eds), *Monte Carlo Generators for HERA Physics*, Desy (1999), pp. 270–279 [arXiv:hep-ph/9907280].

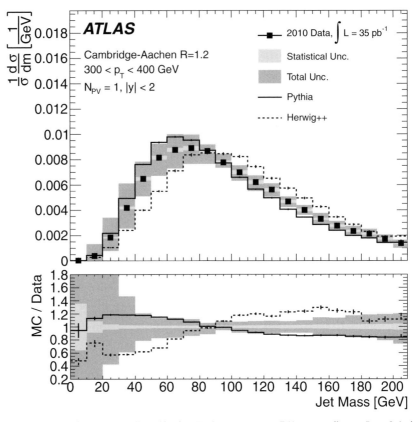

Fig. 9.7 Distribution of the mass of jets in data collected by the ATLAS experiment in 7 TeV proton collisions. From G. Aad *et al.* [ATLAS Collaboration], "Jet mass and substructure of inclusive jets in $\sqrt{s} = 7$ TeV *pp* collisions with the ATLAS experiment," J. High Energy Phys. **1205**, 128 (2012) [arXiv:1203.4606 [hep-ex]].

proton beam, in the $+\hat{z}$-direction. Using the probability distribution for the energy fraction $1 - z$ and the splitting angle θ of a soft gluon off of a quark,

$$P_q(z, \theta^2) = \frac{\alpha_s}{\pi} C_F \frac{dz}{1-z} \frac{d\theta^2}{\theta^2} , \qquad (9.99)$$

you will study the distribution of soft (= low-energy) radiation that permeates every collision event at the LHC. This radiation is called the **underlying event**.

(a) Change variables in the probability distribution written above to $p_\perp$, the transverse momentum of the gluon, and η, the pseudorapidity of the gluon. You will need the energy of the quark before emitting the gluon; call the energy of the quark $x\frac{E_{CM}}{2}$, where $x \in [0, 1]$ is the energy fraction of the quark from the proton and $E_{CM}/2$ is the energy of the proton. Recall that in Eq. 9.99 above, z is the energy fraction of the quark after emitting the gluon.

Hint: Make sure to expand the expression for pseudorapidity, Eq. 5.11, in the small-angle limit.

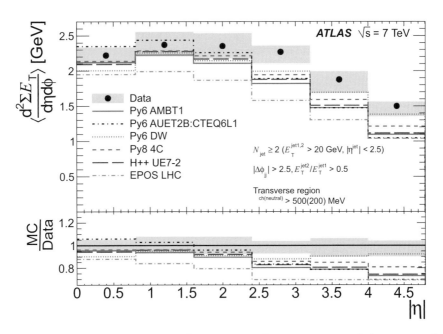

Fig. 9.8 Pseudorapidity distribution of the average particle activity measured as transverse energy which is sensitive to the underlying event in $pp \rightarrow$ two-jet events in 7 TeV collisions at ATLAS. From G. Aad *et al.* [ATLAS Collaboration], "Measurements of the pseudorapidity dependence of the total transverse energy in proton–proton collisions at $\sqrt{s} =$ 7 TeV with ATLAS," J. High Energy Phys. **1211**, 033 (2012) [arXiv:1208.6256 [hep-ex]].

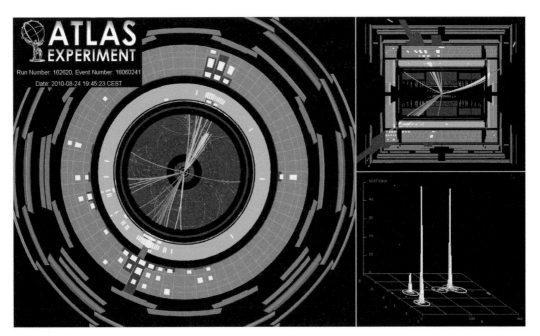

Fig. 9.9 Event display of a $pp \rightarrow$ jets event in 7 TeV collisions at ATLAS. Credit: ATLAS Experiment © 2018 CERN.

(b) Does the distribution of this radiation depend on pseudorapidity? In an experiment like ATLAS or CMS, describe in words what the underlying event would look like in your detector. Can you give a reason as to why it is so useful to use pseudorapidity as a coordinate in a hadron collider?

(c) Figure 9.8 is a plot of the pseudorapidity η distribution of the average transverse energy which is sensitive to the underlying event in the ATLAS experiment. It extends out to $\eta = 4.5$. What angle in degrees above the proton beam does $\eta = 4.5$ correspond to?

(d) Are these data in the plot (approximately) consistent with your result in part (a)? Explain why or why not.

9.8 *Jet Event Display.* Figure 9.9 is an event display of jets produced in proton collisions at the LHC and recorded by ATLAS. How many jets do there appear to be in this event? For each jet, estimate its transverse momentum and determine the pseudorapidity η and azimuth angle ϕ at which the center of the jet is approximately located.

9.9 *Research Problem.* The approximate scale invariance of QCD gave us insight into the structure of jets and the interactions of quarks and gluons at high energies. How far can this insight be pushed to understand QCD? Can we completely understand a theory that is like QCD but exactly scale invariant?

10

Parity Violation

With the prediction and observation of jets, we close our discussion of QCD. In this chapter, as the foundation for the remainder of the book, we are going to turn back the clock to the 1930s and 1950s when another fundamental force was beginning to be identified. The evidence for this new force was very different from that of QCD. While QCD was identified as responsible for hadronic bound states of quarks, the evidence for this other force was almost the antithesis of a bound state. As more particles were discovered in the mid-twentieth century, the ways in which they decayed exhibited some odd and very counterintuitive properties. The most shocking feature of many of these decays was that they violated parity conservation, previously a sacred physical principle. Left- and right-handedness were observed to be treated differently in these decays, something that would never happen with gravity, electromagnetism, or QCD. This new force, called the weak force, completely changed ideas of what a force could be, and required significant theoretical developments to fully understand it.

The violation of parity in nuclear decays is our invitation into studying the weak force. From this observation, we are able to construct a phenomenological model that describes the data on the weak force. This model, called the $V - A$ theory, is extremely precise in its realm of applicability, but is not a satisfactory fundamental theory of Nature. In Chapters 11 through 13, we will keep peeling back the layers of the weak force, discovering surprise after surprise (massive spin-1 force carriers, particle–anti-particle asymmetry, particle oscillations, etc.) along the way. The ultimate prediction of the weak force is the requirement of a spin-0 boson, called the Higgs boson, upon which this whole framework sits.

But it all starts with the seemingly innocuous observation of parity violation.

10.1 Decay of the Neutron

In the 1930s, with the discovery of the neutron, it was shortly thereafter realized that the neutron decays. Free neutrons not bound in a stable atomic nucleus have a lifetime of about 15 minutes and are observed to decay to a proton and an electron:

$$n \to p + e^- . \tag{10.1}$$

This decay is allowed by energy–momentum conservation (the mass of the neutron is 939.6 MeV, the mass of the proton is 938.3 MeV, and the mass of the electron is 511 keV) and by charge conservation, because the neutron is neutral. This model of the decay makes definite predictions: the neutron must be a boson (spin-0 or spin-1) because the proton and electron are both spin-1/2 particles, and in an experiment, the energy of the electron is a unique value in the neutron rest frame. In the neutron rest frame, we would see

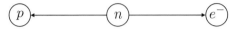

We can determine the energy of the electron via four-momentum conservation. Note that the invariant masses before and after the decay must be equal:

$$p_n^2 = (p_p + p_e)^2 = m_n^2 = m_p^2 + m_e^2 + 2p_p \cdot p_e. \tag{10.2}$$

If we work in the frame in which the neutron is at rest and the proton and electron momenta are along the $\hat{z}$-axis, then their four-momenta are

$$p_p = \left(\sqrt{m_p^2 + p_z^2}, 0, 0, p_z \right), \qquad p_e = \left(\sqrt{m_e^2 + p_z^2}, 0, 0, -p_z \right), \tag{10.3}$$

where p_z is the as-of-yet undetermined z-component of momentum. Solving for p_z using Eq. 10.2 and plugging in the appropriate masses of particles, we expect that the electron in this decay has an energy of $E_e = 1.2$ MeV. Thus, running this experiment over and over, we should repeatedly find electrons with energy of 1.2 MeV.

However, when this experiment was done, that is not what was observed. Instead of the electron always having this energy, it was observed to have a distribution of energies that was always less than 1.2 MeV! Additionally, the prediction that the neutron is a boson is inconsistent with experiments observing radioactive decays of unstable elements. Atoms before and after radioactive decays were observed to still be fermions (half-integer spin), which is inconsistent with the spin-0 hypothesis.

These observations led Wolfgang Pauli,[1] Enrico Fermi, and others to postulate the existence of the **neutrino**, a massless (or, rather, very small-mass) spin-1/2 particle that was produced in the decay of a neutron:

$$n \to p + e^- + \bar{\nu}. \tag{10.4}$$

The (anti-)neutrino $\bar{\nu}$ is electrically neutral (neutrino means "little neutral one" in Italian). Fermi introduced a phenomenological model to describe this decay,[2] but it turned out to be incorrect for subtle reasons. We won't discuss his theory here but we'll explore in detail the properties of the interaction that governs this decay. Our goal for now will be to describe one of the most startling particle physics experiments of the twentieth century, led by Chinese-American physicist Chien-Shiung Wu. To describe it, though, we need some background.

---

[1] W. Pauli, "Dear radioactive ladies and gentlemen," from a letter to Lise Meitner, dated Dec. 1930 [reprinted in Phys. Today **31**, no. 9, 27 (1978)]. Pauli's proposal for the neutrino (which he called at the time the "neutron") was in a letter addressed to the attendees of a conference on radioactivity in Tübingen, Germany. Pauli was unable to attend the conference because his attendance was required at a ball in Zürich, Switzerland.

[2] E. Fermi, "Tentativo di una teoria dell'emissione dei raggi beta," Ric. Sci. **4**, 491 (1933); E. Fermi, "An attempt of a theory of beta radiation. 1," Z. Phys. **88**, 161 (1934).

10.2 Discrete Lorentz Transformations

10.2.1 Parity Transformations

In our discussion of Lorentz transformations in Chapter 2, there were classes of possible transformations that we ignored, mostly for brevity. These were the "large" transformations that corresponded to matrices with negative determinant. One example of such a transformation is **parity**; or, the transformation that inverts spatial axes. The parity operator $\mathbb{P}$, when acting on the position vector $\vec{x}$, negates it:

$$\mathbb{P}\vec{x} = -\vec{x}. \tag{10.5}$$

Parity can be thought of as viewing the physical system in a mirror. Vectors, such as position $\vec{x}$, will be flipped in the mirror. For example, imagine that there is a ball that rolls by to your right with velocity $\vec{v}$:

In this figure, I've denoted your left and right arms. In a mirror this would look like:

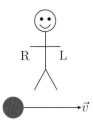

In a mirror, the velocity of the ball has flipped: it is going to the left. Under a parity transformation the velocity vector $\vec{v}$ is negated, just like position:

$$\mathbb{P}\vec{v} = -\vec{v}. \tag{10.6}$$

To see how this follows from Eq. 10.5, the velocity is the time derivative of position:

$$\vec{v} = \frac{d\vec{x}}{dt}, \tag{10.7}$$

and time t does not change under a parity transformation.

Properly, only objects that flip when seen in a mirror (that is, a parity transformation) are vectors. Objects that do not flip under parity are called **pseudovectors**. Because pseudovectors do not flip when viewed in a mirror, one could call them "vampire vectors."[3]

---

[3] This was a colorful name that an undergraduate professor of mine used to describe pseudovectors.

An example of a pseudovector is angular momentum. Consider a spinning wheel that from your perspective is rotating bottom-over-top:

rotation

The direction of angular momentum $\vec{L}$ has also been denoted here. In a mirror, you would see the exact same thing:

rotation

The reason for this is that angular momentum $\vec{L}$ is formed from the cross-product of two (true) vectors, position and momentum. Angular momentum is

$$\vec{L} = \vec{x} \times \vec{p}, \tag{10.8}$$

and under parity

$$\mathbb{P}\vec{x} = -\vec{x}, \qquad\qquad \mathbb{P}\vec{p} = -\vec{p}, \tag{10.9}$$

and so $\mathbb{P}\vec{L} = \vec{L}$. Pseudovectors, such as angular momentum, do not transform under parity. Other examples of pseudovectors are torque and magnetic field. Both are defined from a cross-product of vectors, which is why they do not transform.

Note that two applications of parity returns us to the initial state, and so $\mathbb{P}^2 = 1$. That is, the eigenvalues of the parity operator are $+1$ and -1, and this can potentially be a useful quantum number to classify particles. For example, the members of the pion triplet π^+, π^-, and π^0 are **pseudoscalars**. Scalars, like pure numbers, do not transform under parity. Three apples in a mirror are still three apples. Pseudoscalars, however, do transform under parity. The neutral pion, for example, is negated under the action of parity:

$$\mathbb{P}\pi^0 = -\pi^0. \tag{10.10}$$

Often, though, parity is not so interesting because many physical systems are invariant under it. For example, Maxwell's equations are

$$\vec{\nabla} \cdot \vec{E} = \rho, \qquad\qquad \vec{\nabla} \times \vec{E} = -\frac{\partial \vec{B}}{\partial t}, \tag{10.11}$$

$$\vec{\nabla} \cdot \vec{B} = 0, \qquad\qquad \vec{\nabla} \times \vec{B} = \vec{J} + \frac{\partial \vec{E}}{\partial t},$$

of course in natural units. Maxwell's equations are unchanged under a parity transformation of all of the objects that appear in them:

$$\mathbb{P}\vec{E} = -\vec{E}, \quad \mathbb{P}\vec{\nabla} = -\vec{\nabla}, \quad \mathbb{P}\rho = \rho, \quad \mathbb{P}\vec{B} = \vec{B}, \quad \mathbb{P}\vec{J} = -\vec{J}, \quad \mathbb{P}\frac{\partial}{\partial t} = \frac{\partial}{\partial t}. \quad (10.12)$$

Note that $\vec{E}$, $\vec{\nabla}$, and the current density $\vec{J}$ are vectors, while the charge density ρ is a scalar and the magnetic field $\vec{B}$ is a pseudovector. Because Maxwell's equations are unchanged under parity, we say that electromagnetism is **parity invariant**.

One can also show that the Lagrangian of QCD is invariant to parity transformations. So, with these two examples as our guide, the expectation is that fundamental interactions are invariant to parity.[4] More on this in a second.

Starting in Section 10.3, it will be enlightening to know the parity transformation of a fermion. A more rigorous derivation for the parity transformations of spinors can be developed through construction of unitary operators that act on the individual momentum eigenstates in an appropriate way. While such a derivation would potentially be enlightening, it requires significant technical details, and so we will take a different route. We start from the Dirac equation and act the parity operator on it. One side of the Dirac equation consists of the product of the particle's momentum and the Pauli sigma matrices acting on the spinor, while the other side of the Dirac equation is simply 0. The parity transformation of both sides of the Dirac equation must agree, and because 0 transforms to itself under parity, a solution of the Dirac equation must transform into another solution of the Dirac equation under parity. Let's see how this works in particular, in the first example of this chapter.

Example 10.1 What is the parity transformation of a spinor solution to the Dirac equation?

Solution

We consider the Dirac equation for a right-handed, initial-state, massless fermion $u_R(p)$ with arbitrary momentum p. In spherical coordinates, we can write p as

$$p = E(1, \sin\theta\cos\phi, \sin\theta\sin\phi, \cos\theta), \quad (10.13)$$

where the energy of the fermion is E, θ is the polar angle to the $+\hat{z}$-axis, and ϕ is the azimuthal angle about the $\hat{z}$-axis. From Section 6.1.1, this spinor satisfies the equation

$$E\begin{pmatrix} 1 - \cos\theta & -e^{-i\phi}\sin\theta \\ -e^{i\phi}\sin\theta & 1 + \cos\theta \end{pmatrix} u_R(p) = 0. \quad (10.14)$$

Under parity, the three-momentum $\vec{p}$ is negated, which is equivalent to combined action of the polar angle θ transforming to $\pi - \theta$ and the azimuthal angle transforming to $\phi + \pi$:

$$\mathbb{P}\theta = \pi - \theta, \qquad\qquad \mathbb{P}\phi = \phi + \pi. \quad (10.15)$$

---

[4] Gravity as defined by general relativity is also parity invariant. So, we actually have three examples of fundamental interactions that are parity invariant.

Note that this transformation maintains the requirement that the polar angle $\theta \in [0, \pi]$. Also, this is not just a proper rotation: a proper rotation can only add a fixed angle to another angle, and not negate an arbitrary angle. A parity transformation does not change the energy of the fermion.

We can then apply this parity transformation to the Dirac equation for the spinor $u_R(p)$. We have

$$\mathbb{P}\left[E \begin{pmatrix} 1 - \cos\theta & -e^{-i\phi}\sin\theta \\ -e^{i\phi}\sin\theta & 1 + \cos\theta \end{pmatrix} u_R(p) \right] \tag{10.16}$$

$$= \mathbb{P}\left[E \begin{pmatrix} 1 - \cos\theta & -e^{-i\phi}\sin\theta \\ -e^{i\phi}\sin\theta & 1 + \cos\theta \end{pmatrix} \right] \mathbb{P}[u_R(p)]$$

$$= E \begin{pmatrix} 1 - \cos(\pi - \theta) & -e^{-i(\phi+\pi)}\sin(\pi - \theta) \\ -e^{i(\phi+\pi)}\sin(\pi - \theta) & 1 + \cos(\pi - \theta) \end{pmatrix} \mathbb{P}[u_R(p)]$$

$$= E \begin{pmatrix} 1 + \cos\theta & e^{-i\phi}\sin\theta \\ e^{i\phi}\sin\theta & 1 - \cos\theta \end{pmatrix} \mathbb{P}[u_R(p)] = 0 \,.$$

The expression on the final line is just the Dirac equation for a left-handed spinor, $u_L(p)$! That is, the final line of this transformation is only satisfied if

$$\mathbb{P}[u_R(p)] = u_L(p) \,. \tag{10.17}$$

So, parity transforms a right-handed fermion into a left-handed fermion. Keep this in mind when we study nuclear decays in Section 10.3.

In the same way, the parity transformation of a left-handed, initial-state, massless fermion can be determined by using the transformations of θ and ϕ on its Dirac equation. In this case, we have

$$\mathbb{P}\left[E \begin{pmatrix} 1 + \cos\theta & e^{-i\phi}\sin\theta \\ e^{i\phi}\sin\theta & 1 - \cos\theta \end{pmatrix} u_L(p) \right] \tag{10.18}$$

$$= \mathbb{P}\left[E \begin{pmatrix} 1 + \cos\theta & e^{-i\phi}\sin\theta \\ e^{i\phi}\sin\theta & 1 - \cos\theta \end{pmatrix} \right] \mathbb{P}[u_L(p)]$$

$$= E \begin{pmatrix} 1 + \cos(\pi - \theta) & e^{-i(\phi+\pi)}\sin(\pi - \theta) \\ e^{i(\phi+\pi)}\sin(\pi - \theta) & 1 - \cos(\pi - \theta) \end{pmatrix} \mathbb{P}[u_L(p)]$$

$$= E \begin{pmatrix} 1 - \cos\theta & -e^{-i\phi}\sin\theta \\ -e^{i\phi}\sin\theta & 1 + \cos\theta \end{pmatrix} \mathbb{P}[u_L(p)] = 0 \,.$$

The expression on the final line is only satisfied by a right-handed spinor, or that

$$\mathbb{P}[u_L(p)] = u_R(p) \,. \tag{10.19}$$

Thus, parity transforms a left-handed fermion into a right-handed fermion.

With these transformations of spinors identified, we can then explicitly construct the matrix that implements a parity transformation. We have demonstrated that

$$\mathbb{P}\begin{pmatrix} u_R(p) \\ u_L(p) \end{pmatrix} = \begin{pmatrix} 0 & 1 \\ 1 & 0 \end{pmatrix} \begin{pmatrix} u_R(p) \\ u_L(p) \end{pmatrix} = \begin{pmatrix} u_L(p) \\ u_R(p) \end{pmatrix} \,. \tag{10.20}$$

Note that the trace of this $\mathbb{P}$ matrix is 0 and the determinant is -1. Therefore, the eigenvalues of the matrix that implements parity transformations on spinors are indeed ± 1, in agreement with the general analysis discussed above.

10.2.2 Time Reversal and Charge Conjugation

There are two other discrete Lorentz transformations that can be used to classify elementary particle physics. One of these is **time reversal**, $\mathbb{T}$, which, as its name suggests, flips time:

$$\mathbb{T}t = -t, \tag{10.21}$$

where t is time. Just like parity, the time-reversal operator has eigenvalues of ± 1, as two applications of $\mathbb{T}$ gives the identity operator: $\mathbb{T}^2 = 1$. Unlike for parity, a vector may or may not flip with $\mathbb{T}$. For example, the position vector $\vec{x}$ is unchanged: $\mathbb{T}\vec{x} = \vec{x}$. By contrast, momentum $\vec{p}$ is flipped:

$$\mathbb{T}\vec{p} = -\vec{p}, \quad \text{because it involves velocity,} \quad \vec{p} = m\frac{d\vec{x}}{dt}. \tag{10.22}$$

Angular momentum $\vec{L}$ is negated under a time-reversal, transformation. If time is reversed, then an object is rotating in the opposite direction:

$$\mathbb{T}\left[\vec{L} \circlearrowleft \right]_{\text{rotation}} = \left[\circlearrowright \vec{L}\right]_{\text{rotation}}. \tag{10.23}$$

Maxwell's equations are invariant under $\mathbb{T}$ as well as $\mathbb{P}$. The $\mathbb{T}$ transformations are

$$\mathbb{T}\vec{E} = \vec{E}, \quad \mathbb{T}\vec{\nabla} = \vec{\nabla}, \quad \mathbb{T}\rho = \rho, \quad \mathbb{T}\vec{B} = -\vec{B}, \quad \mathbb{T}\vec{J} = -\vec{J}, \quad \mathbb{T}\frac{\partial}{\partial t} = -\frac{\partial}{\partial t}. \tag{10.24}$$

I'll leave it for you to determine why the magnetic field $\vec{B}$ flips under time reversal, but the electric field does not.

The simple time-reversal operation as defined in Eq. 10.21 has some strange properties in quantum mechanics. Consider the Schrödinger equation for a wavefunction ψ:

$$i\frac{\partial}{\partial t}\psi = \hat{H}\psi. \tag{10.25}$$

$\hat{H}$ is the Hermitian Hamiltonian that describes the energy states of the system of interest. Let's assume that the Hamiltonian is time-independent; then, it is clearly invariant under the action of $\mathbb{T}$:

$$\mathbb{T}\hat{H} = \hat{H}. \tag{10.26}$$

Considering the action of $\mathbb{T}$ on the whole Schrödinger equation, we find

$$\mathbb{T}\left[i\frac{\partial}{\partial t}\psi\right] = \mathbb{T}\left[\hat{H}\psi\right] = \hat{H}\mathbb{T}\left[\psi\right] . \tag{10.27}$$

If we just posit that $\mathbb{T}$ flips the sign of time, then the Schrödinger equation transforms as

$$-i\frac{\partial}{\partial t}\mathbb{T}\left[\psi\right] = \hat{H}\mathbb{T}\left[\psi\right] . \tag{10.28}$$

The solution to this equation, $\mathbb{T}[\psi]$, will necessarily be different than ψ as defined in Eq. 10.25. This is very weird because the Hamiltonian is time-reversal invariant. The statement of time-reversal invariance is that we cannot tell if time is going forward or backward. However, if the time-reversed solution to the Schrödinger equation is not equal to the original solution, then we can tell what the direction of time is.

The way out of this is to have $\mathbb{T}$ turn all factors of i into $-i$ in addition to flipping the sign of time:

$$\mathbb{T}t = -t \qquad \text{and} \qquad \mathbb{T}i = -i . \tag{10.29}$$

This fixes our earlier problem, but at the expense defining $\mathbb{T}$ to be an **anti-unitary operator**. However, this property is vital as it ensures that positive energy states are mapped to positive energy states under the action of $\mathbb{T}$. This quantum mechanical definition of time reversal was introduced by Eugene Wigner. Note that Maxwell's equations have no imaginary numbers, and so the action of this quantum mechanical $\mathbb{T}$ on Maxwell's equations is exactly the same as identified in Eq. 10.24.

You will study the action of this time-reversal operator on spinor solutions to the Dirac equation in Exercise 10.1. With this operator, one can show that the quantum electrodynamics and QCD Lagrangians are both invariant to time-reversal transformations. So, we expect fundamental interactions to be invariant to time reversal.

There is a third discrete transformation that we can perform. It is **charge conjugation**, which acts to flip the sign of all charges. Unlike parity and time reversal, charge conjugation only acts non-trivially when a system has non-zero charges. Positions, momenta, and angular momenta are all invariant under charge conjugation. The charge conjugation operator, $\mathbb{C}$, negates a charge q:

$$\mathbb{C}q = -q . \tag{10.30}$$

Therefore, the charge density ρ and the current density $\vec{J}$ are also negated by charge conjugation. The full charge conjugation transformations for the objects appearing in Maxwell's equations are

$$\mathbb{C}\vec{E} = -\vec{E}, \quad \mathbb{C}\vec{\nabla} = \vec{\nabla}, \quad \mathbb{C}\rho = -\rho, \quad \mathbb{C}\vec{B} = -\vec{B}, \quad \mathbb{C}\vec{J} = -\vec{J}, \quad \mathbb{C}\frac{\partial}{\partial t} = \frac{\partial}{\partial t} . \tag{10.31}$$

Note that the electric field is negated under charge conjugation: if the charge of a particle is flipped, then the direction of the field lines is flipped, too. As with parity and time reversal, Maxwell's equations are charge-conjugation invariant, which you can verify. Additionally, charge conjugation is also a symmetry of QCD.

Table 10.1 Action of C, P, T on spinors			
Operation:	$\mathbb{C}$	$\mathbb{P}$	$\mathbb{T}$
General Action:	$q \to -q$	$\vec{x} \to -\vec{x}$	$t \to -t, i \to -i$
Action on u_R:	$u_R \leftrightarrow v_L$	$u_R \leftrightarrow u_L$	$u_R \leftrightarrow \epsilon u_R$

The action of charge conjugation flips the sign of charges, but does nothing to spin. Therefore, its action on spinor solutions to the Dirac equation is to turn particles into anti-particles without affecting the helicity of that particle. So, a spinor describing a particle solution to the Dirac equation u should turn into a spinor describing an anti-particle solution:

$$\mathbb{C}u = v, \tag{10.32}$$

and vice-versa. What about helicity assignments? Because helicity is unaffected by charge conjugation, the helicity label is flipped; that is, a u_R spinor is transformed to a v_L spinor:

$$\mathbb{C}u_R = v_L. \tag{10.33}$$

In the massless solutions to the Dirac equation that we studied in Section 6.1.1, we found that $u_R = v_L$, so this charge conjugation operation seems to just be the identity operator. However, this shouldn't be surprising: in solving the Dirac equation there, there was no electric charge to be found, and so we should indeed find the exact same solution to the Dirac equation. In Exercise 10.2, we'll consider the Dirac equation in which the spinor is coupled to the photon. In this case, you'll find a non-trivial transformation of the Dirac equation, as expected.

A summary of the $\mathbb{P}$, $\mathbb{T}$, and $\mathbb{C}$ operators and their action on the spinor u_R is presented in Table 10.1. The time-reversal transformation on a spinor is a bit strange as it does not transform a spinor with well-defined helicity into another spinor with well-defined helicity. For a spinor u_R, the transformation is

$$\mathbb{T}u_R = \epsilon u_R = \begin{pmatrix} 0 & 1 \\ -1 & 0 \end{pmatrix} u_R. \tag{10.34}$$

The ϵ object is called the anti-symmetric symbol. The transformations for other spinors are found by relabeling appropriately.

10.2.3 CPT Theorem

With the definitions of $\mathbb{C}$, $\mathbb{P}$, and $\mathbb{T}$, how they transform vectors, and that they are observed to be symmetries of electromagnetism and QCD, one might postulate that they, individually, are symmetries of all possible fundamental interactions. This is reasonable and most physicists believed it until the mid-1950s, for reasons that we will discuss shortly. Actually, all that is proved regarding $\mathbb{C}$, $\mathbb{P}$, and $\mathbb{T}$ is that the application of all of them together is required to be a symmetry, if you have a Hermitian Hamiltonian, and vice-versa.

This result is known as the CPT theorem (or historically as the Lüders–Pauli theorem) and has been proved independently by many people.[5] It is entirely possible and allowed for two of $\mathbb{C}$, $\mathbb{P}$, and $\mathbb{T}$ to be violated while of the action $\mathbb{CPT}$ is preserved. This would be weird, but Nature does not care about our aesthetic tastes.

Before we take what we learned about $\mathbb{P}$, $\mathbb{T}$, and $\mathbb{C}$ transformations, the statement of the CPT theorem deserves re-emphasizing. The combined action of CPT on the Hamiltonian of your system is a symmetry if and only if the Hamiltonian is Hermitian. Equivalently, CPT is a symmetry of your system if and only if the energy eigenvalues are all real. Another way to say this is that the action of CPT on your Hamiltonian H is just complex conjugation:

$$\mathbb{CPT}\, H = H^{\dagger}. \tag{10.35}$$

Of course, the Hamiltonian is Hermitian only if $H = H^{\dagger}$. So, CPT has deep consequences for the Hilbert space of states that exist for the system under consideration.

10.3 Parity Violation in Nuclear Decays

Now, back to where we began this chapter. Neutron decay, or nuclear decay more generally, was an extremely important field of research politically in the 1940s. However, if you just want to harness the energy from nuclear decay, you don't care that much about whether $\mathbb{C}$, $\mathbb{P}$, or $\mathbb{T}$ is violated. In the mid-1950s, Tsung-Dao Lee and Chen-Ning Yang pointed out that the force that governs neutron decay may indeed violate parity.[6] This would be weird and unfamiliar because both electromagnetism and QCD preserve parity. Nevertheless, up until then no experimental result had confirmed parity conservation or violation in neutron decays.

Enter C. S. Wu. With motivation from Lee and Yang's paper, Wu led an experiment to directly test the parity properties of neutron and nuclear decay. Her experiment was deliciously simple. Wu observed the decay of cobalt-60 to nickel-60,

$$^{60}\text{Co} \;\rightarrow\; ^{60}\text{Ni} + e^- + \bar{\nu} + 2\gamma, \tag{10.36}$$

in a magnetic field. The magnetic field was applied to polarize the cobalt nuclear spin along a preferred axis. Then, one can observe the direction in which the electron is emitted: either

---

[5] See J. S. Schwinger, "The theory of quantized fields. 1," Phys. Rev. **82**, 914 (1951); W. Pauli, L. Rosenfeld, and V. Weisskopf, *Niels Bohr and the Development of Physics*, Pergamon Press (1955); G. Lüders, "On the equivalence of invariance under time reversal and under particle–antiparticle conjugation for relativistic field theories," Kong. Dan. Vid. Sel. Mat. Fys. Med. **28**, no. 5, 1 (1954); J. S. Bell, "Time reversal in field theory," Proc. Roy. Soc. Lond. A **231**, 479 (1955); R. Jost, "A remark on the C.T.P. theorem," Helv. Phys. Acta **30**, 409 (1957). A textbook that reviews these results and much more about their consequences for axiomatic construction of quantum field theory is R. F. Streater and A. S. Wightman, *PCT, Spin and Statistics, and All That*, Princeton University Press (2000).

[6] T. D. Lee and C. N. Yang, "Question of parity conservation in weak interactions," Phys. Rev. **104**, 254 (1956). See also L. D. Landau, "On the conservation laws for weak interactions," Nucl. Phys. **3**, 127 (1957); A. Salam, "On parity conservation and neutrino mass," Nuovo Cim. **5**, 299 (1957).

parallel or anti-parallel to the direction of nuclear spin/magnetic field. So, the set-up of the experiment is:

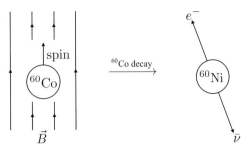

The correlation of the emitted photons with the spin of the ^{60}Co nucleus is used as a control. How does this test parity violation or conservation? Let's see what the predictions are.

If parity is conserved, then let's imagine what the parity-reversed experiment would be. To determine this, we can just parity transform each component of the experiment. First, the magnetic field $\vec{B}$ does not transform under parity, and neither does the nuclear spin. Both are pseudovectors, as they are defined by a cross-product. On the other hand, the momentum of the electron (or anti-neutrino) is a vector, and as such turns into the negative of itself under parity. That is, under a parity transformation, the experimental set-up is unchanged (because the important quantities are pseudovectors), while the final configuration is flipped, because you only care about vectors. So, if parity is conserved, then with equal likelihood you should observe the electron parallel to and anti-parallel to the direction of nuclear spin:

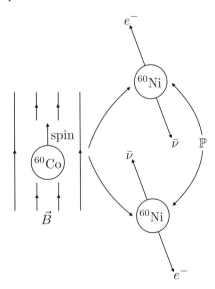

Then, you watch a bunch of ^{60}Co decay, counting the directions of the electrons, and determine how many were parallel and how many were anti-parallel to the nuclear spin. If parity is not conserved, then you will simply see more electrons in one direction than the other.

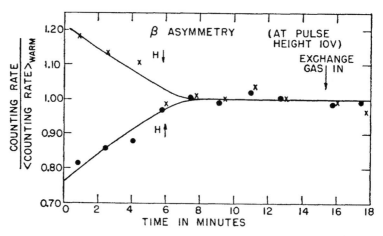

A plot of the relative number of electrons emitted opposite to (x) or in the direction of (●) the nuclear spin of ^{60}Co as a function of time. At late times, nuclear polarization is lost due to thermal effects. At early times, the electron is clearly observed to preferentially be emitted opposite to the direction of nuclear spin. Reprinted figure with permission from C. S. Wu, E. Ambler, R. W. Hayward, D. D. Hoppes and R. P. Hudson, Phys. Rev. **105**, 1413 (1957). Copyright 1957 by the American Physical Society.

In an extremely shocking result, Wu found that more electrons were emitted in the decay opposite to the direction of nuclear spin.[7] The key result from Wu and her team's paper is plotted in Fig. 10.1. This plot shows how the asymmetry in the direction of emission of electrons (historically called "β rays") depends on time. The system was cooled to extremely low temperature so that the thermal excitations of the ^{60}Co nucleus were small and so its spin could be polarized in a magnetic field. As the system warms up, the thermal excitations eventually completely depolarize the ^{60}Co nucleus, and so after about 6 minutes, no preferred electron direction is observed. Before that time, however, the relative number of electrons that are emitted opposite to the spin of the ^{60}Co nucleus (the "x"s on the plot) is different and much larger than the number of electrons emitted in the direction of nuclear spin (the "●"s on the plot). By using the correlation between the two emitted photons and the nuclear spin, Wu estimated that when maximally polarized, at least 70% of the electrons were emitted opposite to the direction of nuclear spin!

When told of this result, Wolfgang Pauli stated that the result was "total nonsense" and that the experiment "must be repeated."[8] It was,[9] and has been many times since, and the same result was observed. Apparently, the force that governs nuclear decays, in contrast

[7] C. S. Wu, E. Ambler, R. W. Hayward, D. D. Hoppes and R. P. Hudson, "Experimental test of parity conservation in beta decay," Phys. Rev. **105**, 1413 (1957).

[8] Copyright 2001. From "Reversal of the parity conservation law in nuclear physics" by R. P. Hudson, in *A Century of Excellence in Measurements, Standards, and Technology*, by Lide, D. R. (ed.). Reproduced by permission of Taylor & Francis Group, LLC, a division of Informa plc.

[9] R. L. Garwin, L. M. Lederman and M. Weinrich, "Observations of the failure of conservation of parity and charge conjugation in meson decays: The magnetic moment of the free muon," Phys. Rev. **105**, 1415 (1957). Actually, this paper appeared immediately after Wu's in the same edition of the Physical Review: these authors had learned about Wu's result at a Friday lunch meeting with T. D. Lee at Columbia University.

to electromagnetism and QCD, violates parity. We now call this force the **weak force**, and we will study the properties and consequences of the weak force in the rest of this book.

10.3.1 Consequences of Parity Violation

Let's dive into Wu's observation of parity violation a bit more. If parity is violated in weak-force interactions, then this means that systems that are related by a parity transformation do not exist with equal probability. Earlier in this chapter, we demonstrated that under a parity transformation, a right-handed electron turns into a left-handed electron, and vice-versa. Thus, because parity is violated in the weak interactions, we expect that the number of right- and left-handed electrons produced in the decay of ^{60}Co will not be equal. Further experimental evidence demonstrated that the electrons produced from the decay of ^{60}Co are actually always left-handed: their direction of momentum is always opposite to that of their spin. That is, not only is parity violated by the weak interactions, but it is violated maximally: the difference between the probability for a left-handed electron versus a right-handed electron to be produced in the decay of ^{60}Co is 100%.

The brilliance of Wu's experiment is that it provides evidence (though somewhat indirectly) for what the helicity of the unobserved anti-neutrino produced from the ^{60}Co decay can be. That is, one can measure the spin of the nucleus before and after the decay, and the spin of the electron. Enforcing conservation of angular momentum on the anti-neutrino then tells you what its spin must be, even though you can't measure it directly. With the electron observed to always be left-handed in these decays, it turns out that conservation of angular momentum enforces that the anti-neutrino must be right-handed, with its spin parallel to its momentum. Every confirmed experiment of the weak interactions has observed left-handed neutrinos and/or right-handed anti-neutrinos. This suggests that only one helicity of neutrinos exists. This is very weird, and unexpected from

---

Box 10.1   **Historical Profile: Chien-Shiung Wu**

Chien-Shiung Wu was a Chinese-American physicist who was a professor at Columbia University when she lead the experiment that discovered parity violation in the weak interactions. Wu had been a student of Ernest Lawrence at UC Berkeley, graduating just before the onset of World War II. Wu's experiment was conducted at the National Bureau of Standards in Washington, D.C., in the days between Christmas 1956 and New Year's 1957. The paper of their results was sent to Physical Review in mid-January and was published shortly thereafter. For their observation that the weak nuclear force could violate parity, T. D. Lee and C. N. Yang were awarded the Nobel Prize in Physics in 1957, only months after Wu's experiment. (That's how shocking the result was!) C. S. Wu, despite leading the experiment that tested parity violation in nuclear decays, was not acknowledged by the Nobel Committee in what is perhaps the grossest oversight in the history of the prize. (Other gross omissions include Lise Meitner, Marietta Blau, and Jocelyn Bell Burnell.) While she did not win a Nobel Prize, Wu was acknowledged with numerous awards later, including the first Wolf Prize in physics. Later in life, Wu was outspoken against gender discrimination, and successfully petitioned for equal pay at Columbia in 1975, 30 years after she first arrived there.

our experience with other spin-1/2 particles like electrons or muons. In particular, we know that from our study of $e^+e^- \to \mu^+\mu^-$ scattering, because electrons and muons can have either left- or right-handed helicity, this leads to the distinctive $1 + \cos^2\theta$ distribution of the differential cross section.

The fundamental process in the ^{60}Co decay to ^{60}Ni is the decay of a neutron to a proton. For a neutron to decay into a proton, one of the constituent down quarks d of the neutron must turn into an up quark u. In terms of the constituent quarks, this decay is

$$d \to u + e^- + \bar{\nu}_e, \tag{10.37}$$

where we have added an "e" subscript to the anti-neutrino, as we call it an electron anti-neutrino. From the discussion of spin in the weak interactions, apparently only left-handed particles know about the weak force. With this observation, the decay of a neutron into a proton can be described by the interaction Lagrangian term:

$$\mathcal{L} \supset \frac{4G_F}{\sqrt{2}} \left(\nu_{eL}^\dagger \bar{\sigma}^\mu e_L \right) \left(d_L^\dagger \bar{\sigma}_\mu u_L \right). \tag{10.38}$$

Here, we've denoted the two-component left-handed spinors of the corresponding fields by their particle name; e_L is the left-handed spinor of the electron, for example. The coefficient G_F is called the **Fermi constant**, and the 4 and $\sqrt{2}$ are there for historical reasons. The theory in which this interaction was constructed is called the **$V - A$ theory** (vector minus axial), developed shortly after Wu's experiment.[10] It is the precursor to the theory of the weak interactions, which is the fundamental theory. In the next section, we will discuss the predictions (and shortcomings) of the $V - A$ theory, and the need for a more fundamental theory that describes all known phenomena and predicts more.

10.4 The $V - A$ Theory

In the previous section, we discussed the weird, or rather, super-weird phenomena of neutron decays. Apparently, unlike electromagnetism, the strong force, and gravity, whatever the force is that mediates neutron decay violates parity. All particles produced in neutron decays are observed to be left-handed: that is, if they are massless or nearly massless (like neutrinos or electrons at high energies), their direction of momentum is always opposite to that of their spin. This observation actually means that parity in nuclear decays is violated maximally: if one performs a parity operation on a nuclear decay, the result of that transformation is a physical configuration that has zero probability to occur.

---

[10] See E. C. G. Sudarshan and R. E. Marshak, "Chirality invariance and the universal Fermi interaction," Phys. Rev. **109**, 1860 (1958); R. P. Feynman and M. Gell-Mann, "Theory of Fermi interaction," Phys. Rev. **109**, 193 (1958); J. J. Sakurai, "Mass reversal and weak interactions," Nuovo Cim. **7**, 649 (1958). The history of who invented the $V - A$ theory is extremely contentious. This is likely a case of the **Matthew effect** (see R. K. Merton, "The Matthew effect in science," Science **159**, no 3810, 56–63 (1968)). Feynman and Gell-Mann were already well known, and happened to publish their paper before Sudarshan (who was a graduate student at the time) and Marshak, though they acknowledge "valuable discussions" with them.

To see what this means, let's look at the left-handed neutrino just traveling through space:

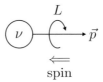

Note that the spin is anti-parallel to the momentum or velocity of the neutrino, so its helicity is

$$h = \hat{p} \cdot \hat{s} = -\frac{1}{2}. \tag{10.39}$$

Under a parity transformation, the spin (= angular momentum) of the neutrino is unchanged, as it is a pseudovector. However, the direction of the neutrino (its momentum) is flipped:

$$\mathbb{P} \left[\begin{array}{c} L \\ \nu \quad \vec{p} \\ \text{spin} \end{array} \right] = \begin{array}{c} R \\ \vec{p} \quad \nu \\ \text{spin} \end{array} \tag{10.40}$$

After this parity transformation, the neutrino's helicity is right-handed: $h = +1/2$. If parity were conserved, then the $V - A$ interaction of Eq. 10.38 would have to include both left- and right-handed particles. If parity were partially violated, then the interaction would have to include both left- and right-handed particles, but with different coefficients. The fact that there are no right-handed particle contributions in this interaction means that parity is **maximally violated**.

This theory is called "vector minus axial" because it is described by one linear combination of vectors and axial vectors (also called pseudovectors). Note that under a parity transformation, this linear combination turns into the orthogonal linear combination:

$$\mathbb{P}(V - A) = -V - A = -(V + A). \tag{10.41}$$

10.4.1 Decay of the Muon

In the rest of this chapter, we study the predictions of the $V - A$ theory. One could calculate the decay of the neutron with this theory, but it is unnecessarily challenging to illustrate the features, as one must account for the non-zero masses of all particles except for the neutrino. So, for simplicity, we study the decay of the muon. The decay of the muon is described by the interaction Lagrangian in the $V - A$ theory:

$$\mathcal{L} = \frac{4G_F}{\sqrt{2}} \left(\nu_{\mu L}^{\dagger} \bar{\sigma}^{\mu} \mu_L \right) \left(e_L^{\dagger} \bar{\sigma}_{\mu} \nu_{eL} \right). \tag{10.42}$$

This governs the decay $\mu^- \to e^- + \bar{\nu}_e + \nu_\mu$. Here, we note that there are two types of neutrinos in this decay. ν_μ is the neutrino associated with the muon, while $\bar{\nu}_e$ is the anti-neutrino associated with the electron. The measurement of the lifetime of the muon is common in upper-level undergraduate laboratory courses.

To start, as we always do, we draw the Feynman diagram that corresponds to the interaction described by the Lagrangian above. The Feynman diagram for the muon decay $\mu^- \to e^- + \bar{\nu}_e + \nu_\mu$ is

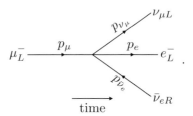

The corresponding matrix element from this Feynman diagram is

$$\mathcal{M}(\mu^- \to e^- \bar{\nu}_e \nu_\mu) = \frac{4G_F}{\sqrt{2}} \left(u_L^\dagger(p_{\nu_\mu}) \bar{\sigma}^\mu u_L(p_\mu) \right) \left(u_L^\dagger(p_e) \bar{\sigma}_\mu v_R(p_{\bar{\nu}_e}) \right) . \tag{10.43}$$

Before we calculate this, there are a few things to note. First, note the spinor assignments: for example, the muon, as the initial-state particle, has a $u_L(p_\mu)$ spinor. By contrast, the electron, as a final-state particle, has a $u_L^\dagger(p_e)$ spinor. The strength of the interaction is controlled by the factor with the Fermi constant, which comes directly from the interaction Lagrangian, Eq. 10.42. The spinors of particles are labeled left-handed and those of anti-particles right-handed, from the form of the $V - A$ theory.

We evaluate the matrix element in the frame in which the muon is at rest; that is, the muon must be massive. For simplicity, we assume that the electron and neutrinos are massless. As the electron is about 200 times less massive than the muon (and neutrinos are further thousands of times less massive than the electron), this is an excellent approximation. Though we pick a particular frame in which to evaluate the matrix element, because it is Lorentz invariant it will have the same value in any frame. In the matrix element, we need to evaluate the spinor product

$$\left(u_L^\dagger(p_{\nu_\mu})_a \bar{\sigma}^\mu_{ab} u_L(p_\mu)_b \right) \left(u_L^\dagger(p_e)_c \bar{\sigma}_{\mu\, cd} v_R(p_{\bar{\nu}_e})_d \right) . \tag{10.44}$$

In this expression, we have written explicit spinor indices to denote the entries of the spinors or elements of the sigma matrices. The indices are repeated, and hence summed over, and range over 1 and 2: $a, b, c, d \in \{1, 2\}$. Apparently, to evaluate this spinor product requires evaluating the four-vector matrix product

$$\bar{\sigma}^\mu_{ab} \bar{\sigma}_{\mu\, cd} . \tag{10.45}$$

What is this?

To evaluate this, we need a nice way to express the sigma matrices. For example, consider the identity matrix, $\mathbb{I}$. What are its elements? Well, it only has non-zero entries if the row and column are equal. We can then express this using the Kronecker-δ symbol:

$$\mathbb{I}_{ab} = \delta_{ab} = \delta_{a1}\delta_{b1} + \delta_{a2}\delta_{b2} . \tag{10.46}$$

Recall that $\delta_{ab} = 1$ if $a = b$, and 0 otherwise. We can use the Kronecker-δ to express the entries of all of the sigma matrices:

$$\mathbb{I}_{ab} = \delta_{a1}\delta_{b1} + \delta_{a2}\delta_{b2}\,, \tag{10.47}$$

$$\sigma_{1\,ab} = \delta_{a1}\delta_{b2} + \delta_{a2}\delta_{b1}\,, \tag{10.48}$$

$$\sigma_{2\,ab} = -i\delta_{a1}\delta_{b2} + i\delta_{a2}\delta_{b1}\,, \tag{10.49}$$

$$\sigma_{3\,ab} = \delta_{a1}\delta_{b1} - \delta_{a2}\delta_{b2}\,. \tag{10.50}$$

Using these expressions, we can evaluate the matrix product $\bar{\sigma}^\mu_{ab}\bar{\sigma}_{\mu\,cd}$.
    We have

$$
\begin{aligned}
\bar{\sigma}^\mu_{ab}\bar{\sigma}_{\mu\,cd} &= (\mathbb{I}, -\sigma_1, -\sigma_2, -\sigma_3)^\mu_{ab}(\mathbb{I}, -\sigma_1, -\sigma_2, -\sigma_3)_{\mu\,cd} \\
&= (\delta_{a1}\delta_{b1} + \delta_{a2}\delta_{b2})(\delta_{c1}\delta_{d1} + \delta_{c2}\delta_{d2}) - (\delta_{a1}\delta_{b2} + \delta_{a2}\delta_{b1})(\delta_{c1}\delta_{d2} + \delta_{c2}\delta_{d1}) \\
&\quad - (-i\delta_{a1}\delta_{b2} + i\delta_{a2}\delta_{b1})(-i\delta_{c1}\delta_{d2} + i\delta_{c2}\delta_{d1}) - (\delta_{a1}\delta_{b1} - \delta_{a2}\delta_{b2})(\delta_{c1}\delta_{d1} - \delta_{c2}\delta_{d2}) \\
&= 2(\delta_{a1}\delta_{c2} - \delta_{a2}\delta_{c1})(\delta_{b1}\delta_{d2} - \delta_{b2}\delta_{d1})\,.
\end{aligned}
\tag{10.51}
$$

The matrix with entries $\epsilon_{ab} \equiv \delta_{a1}\delta_{b2} - \delta_{a2}\delta_{b1}$ is the anti-symmetric symbol. In matrix form, it is

$$\epsilon = \begin{pmatrix} 0 & 1 \\ -1 & 0 \end{pmatrix}\,. \tag{10.52}$$

Therefore, we have shown that $\bar{\sigma}^\mu_{ab}\bar{\sigma}_{\mu\,cd} = 2\epsilon_{ac}\epsilon_{bd}$. This is an example of a Fierz identity. We can use this result to evaluate the spinor products. Writing out all indices, we have

$$
\begin{aligned}
\left(u^\dagger_L(p_{\nu_\mu})_a\bar{\sigma}^\mu_{ab}u_L(p_\mu)_b\right) & \left(u^\dagger_L(p_e)_c\bar{\sigma}_{\mu\,cd}v_R(p_{\bar{\nu}_e})_d\right) \\
&= 2\left(u^\dagger_L(p_{\nu_\mu})_a\epsilon_{ac}u^\dagger_L(p_e)_c\right)\left(u_L(p_\mu)_b\epsilon_{bd}v_R(p_{\bar{\nu}_e})_d\right) \\
&= -2\left(u^\dagger_L(p_{\nu_\mu})_a\epsilon_{ac}u^\dagger_L(p_e)_c\right)\left(v_R(p_{\bar{\nu}_e})_d\epsilon_{db}u_L(p_\mu)_b\right)\,.
\end{aligned}
\tag{10.53}
$$

In the last line, we used that $\epsilon_{bd} = -\epsilon_{db}$.
    Now, we want to figure out what these ϵ anti-symmetric symbols are doing. Let's focus on the matrix product $\epsilon_{ac}u^\dagger_L(p_e)_c$ first. Recall for a momentum vector p at an angle θ with respect to the $\hat{z}$-axis and an angle ϕ about the $\hat{z}$-axis, the spinor is

$$u^\dagger_L(p_e)_c = \sqrt{2E}\left(e^{i\phi/2}\sin\frac{\theta}{2} \quad -e^{-i\phi/2}\cos\frac{\theta}{2}\right)_c\,. \tag{10.54}$$

Let's act on this with the ϵ symbol:

$$
\begin{aligned}
\epsilon_{ac}u^\dagger_L(p)_c &= \begin{pmatrix} 0 & 1 \\ -1 & 0 \end{pmatrix}_{ac}\sqrt{2E}\left(e^{i\phi/2}\sin\frac{\theta}{2} \quad -e^{-i\phi/2}\cos\frac{\theta}{2}\right)_c \\
&= \sqrt{2E}\begin{pmatrix} -e^{-i\phi/2}\cos\frac{\theta}{2} \\ -e^{i\phi/2}\sin\frac{\theta}{2} \end{pmatrix}_a = -u_R(p)_a\,.
\end{aligned}
\tag{10.55}
$$

The ϵ-symbol turns $u_L^\dagger(p)$ into $u_R(p)$! We can do the same thing with $v_R(p_{\bar{\nu}_e})_d \epsilon_{db}$. We have

$$v_R(p)_d \epsilon_{db} = \sqrt{2E} \begin{pmatrix} e^{-i\phi/2} \sin \frac{\theta}{2} \\ -e^{i\phi/2} \cos \frac{\theta}{2} \end{pmatrix}_d \begin{pmatrix} 0 & 1 \\ -1 & 0 \end{pmatrix}_{db} \qquad (10.56)$$

$$= \sqrt{2E} \begin{pmatrix} e^{i\phi/2} \cos \frac{\theta}{2} & e^{-i\phi/2} \sin \frac{\theta}{2} \end{pmatrix}_b = v_L^\dagger(p)_b .$$

Putting this into the spinor products, we have

$$\left(u_L^\dagger(p_{\nu_\mu}) \bar{\sigma}^\mu u_L(p_\mu) \right) \left(u_L^\dagger(p_e) \bar{\sigma}_\mu v_R(p_{\bar{\nu}_e}) \right) = -2 \left(u_L^\dagger(p_{\nu_\mu}) u_R(p_e) \right) \left(v_L^\dagger(p_{\bar{\nu}_e}) u_L(p_\mu) \right) . \qquad (10.57)$$

Then, the matrix element is

$$\mathcal{M}(\mu^- \to e^- \bar{\nu}_e \nu_\mu) = -\frac{8G_F}{\sqrt{2}} \left(u_L^\dagger(p_{\nu_\mu}) u_R(p_e) \right) \left(v_L^\dagger(p_{\bar{\nu}_e}) u_L(p_\mu) \right) . \qquad (10.58)$$

The absolute squared matrix element for muon decay is

$$|\mathcal{M}(\mu^- \to e^- \bar{\nu}_e \nu_\mu)|^2 = 32 G_F^2 \left| u_L^\dagger(p_{\nu_\mu}) u_R(p_e) \right|^2 \left| v_L^\dagger(p_{\bar{\nu}_e}) u_L(p_\mu) \right|^2 . \qquad (10.59)$$

We have evaluated some of these spinor products before. The first squared spinor product is

$$\left| u_L^\dagger(p_{\nu_\mu}) u_R(p_e) \right|^2 = 2 p_{\nu_\mu} \cdot p_e . \qquad (10.60)$$

For the second spinor product we might be tempted to write the same (or the related) expression. However, we have to be a bit careful, because we require that the muon is massive. So, we have to treat the spinor $u_L(p_\mu)$ carefully.

The simplest way to do this is to evaluate it in one frame, and then Lorentz boost to an arbitrary frame. Let's work in the frame where the muon is at rest and the electron anti-neutrino is traveling along the $+\hat{z}$-direction. Then, the spinors are

$$v_L^\dagger(p_{\bar{\nu}_e}) = \sqrt{2E_{\bar{\nu}_e}}(1 \ \ 0), \qquad u_L(p_\mu) = \sqrt{m_\mu}\xi. \qquad (10.61)$$

The expression for $v_L^\dagger$ is familiar, but for $u_L(p_\mu)$ it is probably not. The overall factor of $\sqrt{m_\mu}$ is normalization, just like the $\sqrt{2E}$ for massless spinors. That is, it follows from the trace identity of the muon momentum dotted with the sigma matrices. With $p_\mu = (m_\mu, 0, 0, 0)$, the dot product is

$$\sigma \cdot p_\mu = m_\mu \mathbb{I} = \begin{pmatrix} m_\mu & 0 \\ 0 & m_\mu \end{pmatrix} . \qquad (10.62)$$

In Exercise 6.2 of Chapter 6, you showed that the trace of this dot product matrix is equal to the inner product of the corresponding spinor and its Hermitian conjugate. For the case of the muon at rest, the trace and inner products are then

$$\text{tr}\, \sigma \cdot p_\mu = 2m_\mu = u_L(p_\mu)^\dagger u_L(p_\mu) + u_R(p_\mu)^\dagger u_R(p_\mu) . \qquad (10.63)$$

Note that we sum over left- and right-handed spinor products: for a massive particle helicity is not Lorentz invariant. Taking the normalization of the left- and right-handed spinors to be equal, this then results in the normalization factor of Eq. 10.61.

ξ from Eq. 10.61 is an arbitrary two-component spinor normalized such that $\xi^\dagger \xi = 1$. This represents the probability amplitude of the spin of the muon in its rest frame. To evaluate the spinor product, we need to correspondingly average over the spin of the muon, as its direction of spin is arbitrary and not selected for in our experiment. If the muon is unpolarized, then it has equal probability to be in the spin-up or spin-down state, and so we can take

$$\xi = \frac{1}{\sqrt{2}} \begin{pmatrix} 1 \\ 1 \end{pmatrix}. \tag{10.64}$$

Now, evaluating the spinor product,

$$\left| v_L^\dagger(p_{\bar{\nu}_e}) u_L(p_\mu) \right|^2 = \left| \sqrt{2E_{\bar{\nu}_e}} \begin{pmatrix} 1 & 0 \end{pmatrix} \sqrt{\frac{m_\mu}{2}} \begin{pmatrix} 1 \\ 1 \end{pmatrix} \right|^2 \tag{10.65}$$

$$= m_\mu E_{\bar{\nu}_e}.$$

This is the muon rest frame evaluation of the Lorentz-invariant dot product $p_\mu \cdot p_{\bar{\nu}_e}$.

Therefore, once the dust has settled, the squared matrix element is

$$|\mathcal{M}(\mu^- \to e^- \bar{\nu}_e \nu_\mu)|^2 = 64 G_F^2 (p_{\nu_\mu} \cdot p_e)(p_\mu \cdot p_{\bar{\nu}_e}). \tag{10.66}$$

The final state is described by three-body phase space, and so we use the x_i variables, where

$$x_i = \frac{2Q \cdot p_i}{Q^2}, \text{ for } i = e, \nu_\mu, \bar{\nu}_e. \tag{10.67}$$

Note that for muon decay $Q^2 = m_\mu^2$, and $Q = p_\mu$. Then,

$$p_\mu \cdot p_{\bar{\nu}_e} = Q \cdot p_{\bar{\nu}_e} = m_\mu^2 \frac{x_{\bar{\nu}_e}}{2}, \tag{10.68}$$

$$2p_e \cdot p_{\nu_\mu} = Q^2 - 2Q \cdot p_{\bar{\nu}_e} = m_\mu^2 (1 - x_{\bar{\nu}_e}).$$

The squared matrix element is then written in terms of $x_{\bar{\nu}_e}$ as

$$|\mathcal{M}(\mu^- \to e^- \bar{\nu}_e \nu_\mu)|^2 = 16 G_F^2 m_\mu^4 x_{\bar{\nu}_e} (1 - x_{\bar{\nu}_e}). \tag{10.69}$$

This can then be plugged into Fermi's Golden Rule for decays. The decay rate Γ for three final-state particles is

$$\Gamma = \frac{1}{2E_\mu} \int d\Pi_3 |\mathcal{M}(\mu^- \to e^- \bar{\nu}_e \nu_\mu)|^2. \tag{10.70}$$

While we haven't explicitly discussed Fermi's Golden Rule for decays, the form should be reasonable. Note that there is the squared matrix element which is integrated over three-body phase space Π_3. The overall factor of $1/(2E_\mu)$ is familiar from Fermi's Golden Rule for scattering and comes from consideration of the Compton and de Broglie wavelengths of the decaying muon. Note that unlike the expression for the cross section, there is no relative velocity factor in the decay rate. This is of course because the initial state is a single particle, and not colliding particles. Finally, the decay rate Γ has units of [energy]; that is, it has units of $[\text{time}]^{-1}$. Indeed, Γ^{-1} is the characteristic lifetime of the muon. The probability that the muon has decayed by a time $t = \Gamma^{-1}$ is equal to $e^{-1} = 0.3678\ldots$

Plugging in the appropriate expressions and using the formula for three-body phase space that was derived in Exercise 4.3 in Chapter 4, the decay rate Γ for the muon at rest is

$$\Gamma = \frac{1}{2m_\mu} \frac{m_\mu^2}{128\pi^3} \int_0^1 dx_e \int_0^1 dx_{\bar{\nu}_e} \, \Theta(x_e + x_{\bar{\nu}_e} - 1) \, 16 G_F^2 m_\mu^4 x_{\bar{\nu}_e} (1 - x_{\bar{\nu}_e}) \tag{10.71}$$

$$= \frac{G_F^2 m_\mu^5}{16\pi^3} \int_0^1 dx_e \int_{1-x_e}^1 dx_{\bar{\nu}_e} \, x_{\bar{\nu}_e} (1 - x_{\bar{\nu}_e}) \, .$$

Note that the energy of the muon in its rest frame is $E_\mu = m_\mu$. We can't measure the electron anti-neutrino momentum fraction $x_{\bar{\nu}_e}$, so let's integrate over it:

$$\Gamma = \frac{G_F^2 m_\mu^5}{96\pi^3} \int_0^1 dx_e \, (3x_e^2 - 2x_e^3) \, . \tag{10.72}$$

From here, we can define the decay rate differential in the electron energy fraction x_e:

$$\frac{d\Gamma}{dx_e} = \frac{G_F^2 m_\mu^5}{96\pi^3} (3x_e^2 - 2x_e^3) \, . \tag{10.73}$$

By a change of variables, we can also calculate the decay rate differential in the electron energy E_e in the muon rest frame. With

$$x_e = \frac{2E_e}{m_\mu} \, , \tag{10.74}$$

the decay rate differential in E_e is

$$\frac{d\Gamma}{dE_e} = \frac{d\Gamma}{dx_e} \frac{dx_e}{dE_e} = \frac{G_F^2 m_\mu^2}{12\pi^3} E_e^2 \left(3 - \frac{4E_e}{m_\mu} \right) \, . \tag{10.75}$$

This prediction of the decay rate differential in the energy of the electron can be compared to muon decay data. Figure 10.2 shows muon decay data from the TWIST experiment (the <u>T</u>RIUMF <u>W</u>eak <u>I</u>nteraction <u>S</u>ymmetry <u>T</u>est) located at the Canadian national laboratory, TRIUMF. TWIST measured the energy spectrum of electrons from the decays of muons that were bound in ^{27}Al. Their data are compared to the differential decay rate in Eq. 10.75, using the muon mass of $m_\mu = 105.6$ MeV. There are a few things to note in the comparison of our prediction to the data:

1   The data nicely follow our predicted energy spectrum at intermediate energies, and then diverge for electron energies near the upper endpoint of about 53 MeV ($\simeq m_\mu/2$). If you are considering electron energies near 53 MeV, then you are considering an exclusive process. The electron from muon decay can emit photons and lose energy (through essentially the same process that we studied with quarks and gluons) and this effectively prohibits the electron from ever having an energy of half the muon mass. That is, near the upper endpoint, there is a new relevant energy scale E_{ep} of

$$E_{\text{ep}} \equiv \frac{m_\mu}{2} - E_e \, , \tag{10.76}$$

and $E_{\text{ep}} \ll E_e$ near the endpoint. One can account for this new endpoint energy scale, and doing so results in excellent agreement between theory and data throughout the

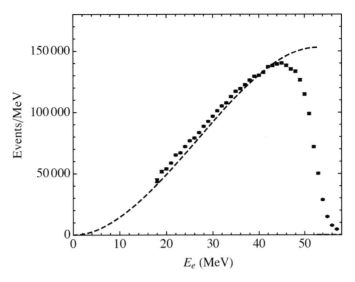

Fig. 10.2   Electron energy spectrum from muon decays recorded at the TWIST experiment. Error bands are included in the data points, and the dashed curve is the prediction from Eq. 10.75. Reprinted data from table with permission from A. Grossheim *et al.* [TWIST Collaboration], Phys. Rev. D **80**, 052012 (2009). Copyright 2009 by the American Physical Society.

spectrum. You'll study features of the effects of photon emission on the measurement of electron energy in Exercise 10.6.

2 At low values of the electron energy E_e, the differential decay rate decreases and actually vanishes as $E_e \to 0$ with the approximation that the electron is massless. This can be understood by considering the spins of the final-state particles with this configuration. If the electron has zero energy, then the two neutrinos must be back to back and their spins aligned:

spin $\Longrightarrow$ spin $\Longrightarrow$

$\nu_\mu \longleftarrow \quad (\mu^-) \quad \longrightarrow \bar{\nu}_e$

This configuration of final-state neutrinos has total angular momentum 1, and so the electron is necessary to ensure that the final state has total spin 1/2 and to conserve angular momentum in this process. However, if the electron is massless and has zero energy, its spin cannot affect the angular momentum of the final state. Therefore, it's impossible for the electron to have zero energy by angular momentum conservation.

For a real electron with a non-zero mass $m_e = 511$ keV, the probability for the electron to have its minimal energy of just the mass is proportional to the ratio of the electron mass to the muon mass, $m_e/m_\mu \simeq 0.005$.

We can also calculate the total decay rate by integrating over x_e. Doing this, we find

$$\Gamma(\mu^- \to e^- \bar{\nu}_e \nu_\mu) = \frac{G_F^2 m_\mu^5}{192\pi^3} \, . \tag{10.77}$$

The Fermi constant is $G_F \simeq 1.17 \times 10^{-5}$ GeV^{-2},[11] and the muon mass is $m_\mu = 105.6$ MeV. Plugging in the numbers, the decay rate is then

$$\Gamma(\mu^- \to e^- \bar{\nu}_e \nu_\mu) = 3.2 \times 10^{-10} \text{ eV}. \tag{10.78}$$

Turning this into a lifetime τ (in seconds) by adding factors of $\hbar$, we have

$$\tau = \frac{\hbar}{\Gamma} \simeq 2.1 \times 10^{-6} \text{ s}. \tag{10.79}$$

The PDG says that the lifetime of the muon is 2.2×10^{-6} s. So, we're very close!

While the $V - A$ theory makes nice predictions like this, theoretically, it leaves a lot to be desired. Most importantly, it provides no explanation for the force that mediates muon decay; that is, there is no force carrier analogous to the photon or gluon. How do we resolve this? More in the next chapter...

Exercises

10.1 *Time Reversal of Spinors.* To determine the transformation of a spinor under time reversal, we start from the Dirac equation, just as we did with parity transformations in Example 10.1. Starting from the Dirac equation for a right-handed spinor ψ_R,

$$i\sigma \cdot \partial \psi_R = 0, \tag{10.80}$$

perform a time-reversal transformation on this equation. Show that the solution to the time-reversed equation, $\mathbb{T}[\psi_R]$, can be expressed as

$$\mathbb{T}[\psi_R] = \epsilon u_R e^{-ip \cdot x}, \tag{10.81}$$

where ϵ is the anti-symmetric symbol.

Hint: Don't forget that one of the Pauli sigma matrices has factors of i in it.

10.2 *Charge Conjugation of Spinors.* We postulated that the action of charge conjugation $\mathbb{C}$ on a right-handed spinor u_R transforms it to v_L:

$$\mathbb{C}[u_R] = v_L. \tag{10.82}$$

When the spinor is not coupled to electromagnetism, this relationship is trivial; however, the Dirac equation for a right-handed spinor ψ_R coupled to the photon field A_μ is

$$i\sigma \cdot D\psi_R = (i\sigma \cdot \partial - e\sigma \cdot A)\psi_R = 0. \tag{10.83}$$

This equation follows from extremizing the QED Lagrangian in Eq. 4.73. Here, e is the electric charge of the field ψ_R.

---

[11] You may rightly ask where this comes from, and one answer is: from measuring the lifetime of the muon. If this seems like a circular argument to you, good; in the next chapter we will provide a fundamental theory for the origin of the value of G_F.

(a) From the charge conjugation transformations of the electric and magnetic fields in Eq. 10.31, show that

$$\mathbb{C}A_\mu = -A_\mu. \tag{10.84}$$

You might want to refer to Section 2.2.3.

(b) Now, charge conjugate the Dirac equation of Eq. 10.83. Show that

$$\mathbb{C}[(i\sigma \cdot \partial - e\sigma \cdot A)\psi_R] = (i\sigma \cdot \partial + e\sigma \cdot A)\mathbb{C}[\psi_R] = 0. \tag{10.85}$$

Does the solution of this equation $\mathbb{C}[\psi_R]$ indeed describe an anti-particle with electric charge $-e$?

10.3 *CPT on Spinors.* Using the transformations established in Table 10.1, act all of the $\mathbb{C}$, $\mathbb{P}$, and $\mathbb{T}$ transformations on the spinor u_R. Show that

$$\mathbb{CPT}[u_R] = -u_R^*. \tag{10.86}$$

This is consistent with the claim from Section 10.2.3 that the combined action of CPT on the Hamiltonian is complex conjugation.

10.4 *C, P, T in Electromagnetism.* We can construct the parity, charge conjugation, and time-reversal transformations of the photon field A_μ from the corresponding transformations of the electric and magnetic fields presented in this chapter. In Exercise 10.2, you've already shown that under charge conjugation, the photon is negated:

$$\mathbb{C}A_\mu = -A_\mu. \tag{10.87}$$

What are some consequences of this, and what about parity and time reversal?

(a) A consequence of charge conjugation negating the photon is a result called Furry's theorem.[12] It states that, because of the properties of charge conjugation, the matrix element of the scattering of an odd number of photons is 0. Can you argue why this is the case? In particular, can you argue that $2 \to 3$ photon scattering vanishes, $\mathcal{M}(\gamma\gamma \to \gamma\gamma\gamma) = 0$?

   Hint: What happens to this matrix element under the action of $\mathbb{C}$? Does that actually do anything?

(b) From the parity transformations of the electric and magnetic fields in Eq. 10.12, determine the parity transformation of A_μ. That is, what is

$$\mathbb{P}A_\mu = ? \tag{10.88}$$

(c) From the time-reversal transformations of the electric and magnetic fields in Eq. 10.24, determine the time-reversal transformation of A_μ. That is, what is

$$\mathbb{T}A_\mu = ? \tag{10.89}$$

(d) What is the combined action of $\mathbb{CPT}$ on the photon field A_μ? Show that

$$\mathbb{CPT}A_\mu = -A_\mu. \tag{10.90}$$

---

[12] W. H. Furry, "A symmetry theorem in the positron theory," *Phys. Rev.* **51**, 125 (1937).

10.5 *Electron Spin in Muon Decay.* In our prediction of the energy distribution of the electron from muon decay, we observed that when the energy of the electron is maximized, the distribution is also maximized; see Eq. 10.75. Draw the configuration of particle momenta and spins when the electron energy is maximized. What is the total angular momentum of the muon neutrino and electron anti-neutrino? How does the spin of the electron compare to the spin of the decaying muon at the endpoint?

10.6 *Endpoint of Electron Energy in Muon Decays.* In comparing our prediction for the electron energy from muon decay to data in Fig. 10.2, we noted that near the upper endpoint the data are rapidly suppressed. Understanding this in detail is quite complicated, but we can make progress on understanding why the electron can never really take away an energy of half the muon mass.

Electrons are electrically charged, and so they can emit photons copiously. These photons take away energy from the electron which we can estimate using the splitting functions that we constructed in Section 9.2. The probability density for an electron to emit a photon at a relative angle $\theta^2 \ll 1$ with an energy fraction $z \ll 1$ is

$$\left| \begin{array}{c} \\ \end{array} \right|^2 = p(z, \theta^2) = \frac{\alpha}{\pi} \frac{1}{z\theta^2} . \tag{10.91}$$

Here, we have just translated the result from Eq. 9.66 to photon emission off of an electron; this involves removing the color factor C_F and replacing the strong coupling α_s by the fine structure constant α. In the diagram, we have labeled θ and z appropriately, to leading order in $z \ll 1$. From this simple expression, we will be able to estimate the average energy that the photon takes from the electron, $\langle zE_e \rangle$.

(a) In our analysis of the observable thrust from gluon emissions off of quarks in Section 9.3.1, we assumed that all particles were massless and so θ could go to 0. When $\theta = 0$, the probability for gluon emission diverged. This followed from the form of the quark propagator, which took the form

$$\frac{1}{p_q + p_g} \propto \frac{1}{p_q \cdot p_g} = \frac{1}{E_q E_g (1 - \cos \theta)} , \tag{10.92}$$

where p_q and p_g are the four-momenta of the quark and gluon, respectively. Assuming that the electron propagator is of the same form, show that when the mass of the electron m_e is included the propagator is

$$\frac{1}{p_e \cdot p_\gamma} = \frac{1}{E_e E_\gamma \left(1 - \sqrt{1 - \frac{m_e^2}{E_e^2}} \cos \theta\right)} . \tag{10.93}$$

Here, E_γ is the energy of the emitted photon. Note that there is no angle θ for which this propagator can diverge.

(b) Assuming that $m_e \ll E_e$ and $\theta \ll 1$, estimate the angle θ_c at which the non-zero electron mass becomes important.

Hint: You'll want to do some Taylor expansions.

(c) This θ_c that you found is effectively the minimal value of the angle θ. (θ can still go to 0, but the propagator is finite in that limit.) With this angular cut $\theta > \theta_c$, determine the probability distribution of the photon energy, $E_\gamma = zE_e$. This is defined to be

$$p(E_\gamma) = \int_0^1 dz \int_{\theta_c^2}^1 d\theta^2 \, p(z, \theta^2) \, \delta(E_\gamma - zE_e) \,. \tag{10.94}$$

You should find that

$$p(E_\gamma) = \frac{\alpha}{\pi} \frac{\log \frac{E_e^2}{m_e^2}}{E_\gamma} \,. \tag{10.95}$$

Don't worry if the argument of your logarithm is just off by a number.

(d) Now, take the mean of this distribution:

$$\langle E_\gamma \rangle = \int dE_\gamma \, p(E_\gamma) \, E_\gamma \,. \tag{10.96}$$

The limits of the integral have been purposefully suppressed. What are the most natural upper and lower bounds on the integral? Evaluate the average value in terms of the electron mass m_e, the electron energy E_e, and the fine structure constant α.

(e) Note that, for intermediate values of the electron energy in Fig. 10.2, the data are at slightly lower values than our prediction. Evaluate this average photon energy $\langle E_\gamma \rangle$ for $\alpha = 1/137$, $E_e = 30$ MeV, and $m_e = 511$ keV. Does the energy that photons take away seem to account for the difference between the distributions, for $20 < E_e \lesssim 40$ MeV?

10.7 *Kinematics of the IceCube Experiment.*[13] The IceCube experiment is an expansive collection of photomultiplier tubes buried over a mile (1.6 kilometers) deep in the ice cap of the South Pole. The goal of IceCube is to observe extremely high-energy neutrinos that were produced in supernovae or other astrophysical phenomena. The way it does it is fascinating. The ice buried deep under the South Pole is extremely clear, except for some dust and ash from mass extinction events in the Earth's history. High-energy neutrinos can pass through the ice and hit a proton in the water molecules of the ice. This interaction typically produces a muon and a neutron. The muon, if it is of high enough energy, can travel through the ice above the speed of light in ice, thereby emitting **Čerenkov radiation**. The Čerenkov radiation is then observed in the photomultiplier tubes, and enables a precise measurement of the energy and direction of the produced muon.

A schematic illustration of the IceCube experiment is shown in Fig. 10.3. In this exercise, we will analyze the kinematics of the process $\bar{\nu}_\mu + p \to \mu^+ + n$. For very high-energy observed anti-muons, we will be able to set a bound on the angle from the anti-muon to the initial anti-neutrino.

---

[13] In this and the next exercise, we are making a simplifying assumption that high-energy neutrino scattering just transforms the proton into a neutron. At sufficiently high energies, however, the proton will dissociate, as in the process of deeply inelastic scattering.

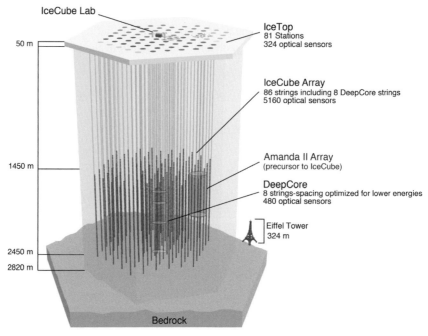

IceCube Lab

50 m

IceTop
81 Stations
324 optical sensors

IceCube Array
86 strings including 8 DeepCore strings
5160 optical sensors

1450 m

Amanda II Array
(precursor to IceCube)

DeepCore
8 strings-spacing optimized for lower energies
480 optical sensors

Eiffel Tower
324 m

2450 m

2820 m

Bedrock

Fig. 10.3 Cartoon of the IceCube Neutrino Detection Experiment located deep in the ice of the Earth's South Pole. The space in which the 5160 photomultiplier tubes that detect Čerenkov radiation are located is approximately 1 km high, with a total volume of about 1 cubic kilometer of ice. Credit: IceCube Collaboration.

(a) The scattering process

$$\bar{\nu}_\mu + p \rightarrow \mu^+ + n \tag{10.97}$$

at IceCube occurs in the frame in which the proton (in the nucleus of a water molecule) is at rest. Assuming that the anti-neutrino's momentum is aligned along the $+\hat{z}$-axis, write the four-momentum of the anti-muon p_μ and the neutron p_n in terms of the anti-neutrino energy $E_{\bar{\nu}}$, the nucleon mass m_N, and the scattering angle θ. You can safely assume that the anti-neutrino and anti-muon are both massless and the proton and neutron have identical masses equal to m_N. The scattering angle is the angle between the original anti-neutrino momentum and the anti-muon's momentum.

You should find:

$$p_\mu = \frac{E_{\bar{\nu}} m_N}{m_N + E_{\bar{\nu}}(1 - \cos\theta)} (1, 0, \sin\theta, \cos\theta), \tag{10.98}$$

$$p_n = \left(E_{\bar{\nu}} + m_N - \frac{E_{\bar{\nu}} m_N}{m_N + E_{\bar{\nu}}(1 - \cos\theta)}, 0, \right.$$

$$\left. - \frac{E_{\bar{\nu}} m_N \sin\theta}{m_N + E_{\bar{\nu}}(1 - \cos\theta)}, E_{\bar{\nu}} - \frac{E_{\bar{\nu}} m_N \cos\theta}{m_N + E_{\bar{\nu}}(1 - \cos\theta)} \right).$$

(b) Figure 10.4 is a plot from the IceCube experiment that shows the deposited energy into the experiment versus the declination angle of the anti-muon from

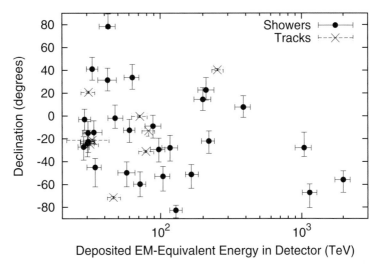

Fig. 10.4   Deposited electromagnetic energy and declination of neutrino events detected by the IceCube experiment between 2010 and 2012. Reprinted figure with permission from M. G. Aartsen *et al.* [IceCube Collaboration], Phys. Rev. Lett. **113**, 101101 (2014). Copyright 2014 by the American Physical Society.

high-energy neutrino scattering. This plot shows the 37 highest-energy events recorded by IceCube. There are three fantastically high-energy events observed above 1000 TeV (= 1 peta-electron volt) of deposited energy. The highest-energy event is affectionately called "Big Bird," while the second and third highest-energy events are called "Bert" and "Ernie." Actually, all of the events on this plot are named after Muppets.

If the Big Bird anti-muon deposited 2 PeV of energy in IceCube, then what is the corresponding largest and smallest energy that the initial anti-neutrino could have had?

(c) What are the corresponding maximum and minimum scattering angles in degrees between the initial anti-neutrino and the measured Big Bird anti-muon? Call the maximum angle $\theta_{\max}$ and the minimum angle $\theta_{\min}$. For simplicity, set the masses of the proton and neutron to $m_N = 1$ GeV. With any possible anti-neutrino energy, is the momentum of the anti-muon close to the direction of the anti-neutrino? You can safely assume that $m_N \ll E_\mu$, so Taylor expanding would likely help.

This property of the scattering angle from part (c) is extremely important for determining the astrophysical source of the high-energy neutrinos observed in IceCube. In 2016, scientists on the Fermi Gamma Ray Space Telescope found evidence that the neutrino that was responsible for the Big Bird muon was created in a **blazar**, an enormously energetic radiation source believed to be generated by a supermassive black hole at the center of a galaxy.[14]

---

[14] M. Kadler *et al.*, "Coincidence of a high-fluence blazar outburst with a PeV-energy neutrino event," Nature Phys. **12**, no. 8, 807 (2016) [arXiv:1602.02012 [astro-ph.HE]].

10.8 *High-Energy Neutrino Cross Sections.* The scattering of a high-energy anti-neutrino off of a proton in the nucleus of a water molecule through the process

$$\bar{\nu}_\mu + p \to \mu^+ + n \qquad (10.99)$$

can be described by the following interaction Lagrangian in the $V - A$ theory:

$$\mathcal{L} = \frac{4G_F}{\sqrt{2}} \left(\mu_L^\dagger \bar{\sigma}^\mu \nu_{\mu L} \right) \left(p_L^\dagger \bar{\sigma}_\mu n_L \right) . \qquad (10.100)$$

The corresponding Feynman diagram and matrix element is

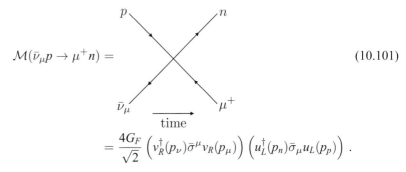

$$\mathcal{M}(\bar{\nu}_\mu p \to \mu^+ n) = \qquad (10.101)$$

$$= \frac{4G_F}{\sqrt{2}} \left(v_R^\dagger(p_\nu) \bar{\sigma}^\mu v_R(p_\mu) \right) \left(u_L^\dagger(p_n) \bar{\sigma}_\mu u_L(p_p) \right) .$$

Here, p_ν, p_μ, p_p, and p_n are the momenta of the anti-neutrino, the anti-muon, the proton, and the neutron, respectively. In this exercise, we will use this matrix element to compute the scattering cross section for this process to occur in the IceCube experiment.

(a) Calculate the squared matrix element $|\mathcal{M}(\bar{\nu}_\mu p \to \mu^+ n)|^2$ in terms of the external particles' four-momenta. You can safely assume that the anti-neutrino and anti-muon are both massless and the proton and neutron have identical masses equal to m_N.

(b) The frame in which the anti-neutrino scattering takes place in IceCube is where the proton is at rest. In this frame, show that the squared matrix element can be written in terms of the anti-neutrino energy E_ν, the anti-muon energy E_μ, and the scattering angle θ as

$$|\mathcal{M}(\bar{\nu}_\mu p \to \mu^+ n)|^2 = 32G_F^2 E_\nu E_\mu m_N (m_N - E_\mu(1 - \cos\theta)). \qquad (10.102)$$

The scattering angle θ is the angle between the initial anti-neutrino momentum and the anti-muon momentum.

Hint: You'll want to exploit momentum conservation, $p_\nu + p_p = p_\mu + p_n$.

(c) To calculate the cross section, we then need to put this matrix element into Fermi's Golden Rule. Before that, though, we need to evaluate two-body phase space in this weird frame. Previously, we had evaluated it in the center-of-mass frame. Because phase space is Lorentz invariant, we could just perform an appropriate Lorentz transformation to get to this frame. That's not what we will do here, and instead let's just recalculate phase space in this frame. In the frame in which the proton is at rest, show that the two-body phase space can be written as

$$\int \frac{d^3 p_\mu}{(2\pi)^3} \frac{d^3 p_n}{(2\pi)^3} \frac{1}{4 E_\mu E_n} (2\pi)^4 \delta^{(4)} (p_\nu + p_p - p_\mu - p_n) \qquad (10.103)$$

$$= \int \frac{dE_\mu \, d\cos\theta}{8\pi} \frac{E_\mu}{E_\nu + m_N - E_\mu} \delta(E_\nu + m_N - E_\mu - E_n) .$$

To express the final integrals, we have written the integral over the anti-muon momentum in spherical coordinates. Here, E_n is the energy of the neutron.

(d) Because IceCube only measures the energy of the anti-muon, we want to use the δ-function to integrate over the scattering angle $\cos\theta$. Using the result for the muon energy from part (a) of Exercise 10.7, show that the energy δ-function can be written for $\cos\theta$ as

$$\delta(E_\nu + m_N - E_\mu - E_n) \qquad (10.104)$$

$$= \frac{m_N}{E_\mu^2} \delta \left(\cos\theta - \frac{m_N E_\mu + E_\mu E_\nu - m_N E_\nu}{E_\mu E_\nu} \right) .$$

Hint: You'll need a Jacobian from the change of variables in the δ-function.

(e) Combining the results for the matrix element and the two-body phase space, show that the cross section σ for the process $\bar{\nu}_\mu + p \to \mu^+ + n$ can be written as

$$\sigma(\bar{\nu}_\mu p \to \mu^+ n) = \frac{m_N^2 G_F^2}{\pi E_\nu} \int dE_\mu \frac{E_\mu}{E_\nu + m_N - E_\mu} . \qquad (10.105)$$

Hint: Don't forget to enforce the δ-function for $\cos\theta$, and remember what frame we are in for calculating the prefactors in Fermi's Golden Rule.

(f) By removing the integral over E_μ, Eq. 10.105 defines the cross section for this scattering differential in the muon energy, E_μ. To get the inclusive cross section, we need to integrate over E_μ, which requires knowing the lower and upper bounds of the anti-muon energy. Using the result for the muon energy from part (a) of Exercise 10.7, determine the minimum and maximum muon energies, $E_\mu^{\min}$ and $E_\mu^{\max}$. With these bounds, perform the integral, and show that the total cross section is

$$\sigma(\bar{\nu}_\mu p \to \mu^+ n) \qquad (10.106)$$

$$= \frac{m_N^2 G_F^2}{\pi} \left[\frac{E_\nu + m_N}{E_\nu} \log \left(1 + \frac{2 E_\nu^2}{m_N(2 E_\nu + m_N)} \right) - \frac{2 E_\nu}{2 E_\nu + m_N} \right] .$$

Hint: What are the minimum and maximum scattering angles and how are they related to the minimum and maximum anti-muon energies?

(g) IceCube ran for three years to find three events which each deposited more than a PeV of energy into the Antarctic ice. Using the cross section in Eq. 10.106, approximately how many PeV neutrinos passed right through IceCube during that time? The total volume of the IceCube detector is about 1 km^3, and the density of ice is about 1000 kg $\cdot$ m^{-3}.

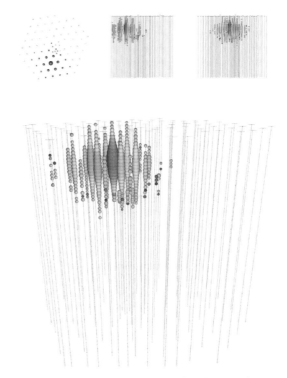

Fig. 10.5   Display of photomultiplier tube response in the IceCube experiment from the Big Bird neutrino event that deposited 2000 TeV of energy. Reprinted figure with permission from M. G. Aartsen *et al.* [IceCube Collaboration], Phys. Rev. Lett. **113**, 101101 (2014). Copyright 2014 by the American Physical Society.

Figure 10.5 is a display of the photomultiplier tube response in the Big Bird neutrino event. The strings on this figure correspond to the strings of photomultiplier tubes, while the bubble region represents the detection of Čerenkov radiation. Larger bubbles correspond to higher energy Čerenkov light detected. I guess if you squint, it kind of looks like the head of a Muppet, hence the naming scheme.

10.9   *Research Problem.* In this chapter, we stated the CPT theorem, which enforces that a Hamiltonian (or Lagrangian) is only Hermitian if it is invariant under the combined action of $\mathbb{C}$, $\mathbb{P}$, and $\mathbb{T}$. The requirement of Hermitivity of the Hamiltonian means that time evolution is unitary. Is the converse true? That is, is invariance under the combined action of $\mathbb{C}$, $\mathbb{P}$, and $\mathbb{T}$ sufficient to prove that a theory is unitary, with or without a well-defined Hamiltonian?

11

The Mass Scales of the Weak Force

The $V - A$ theory, for all its success, leaves much to be desired. It is severely lacking from the perspective of QCD or electromagnetism. In the $V - A$ theory, we just postulate a four-fermion interaction, whose strength is controlled by the Fermi constant, G_F. In QCD and electromagnetism, interactions of four fermions are mediated by spin-1 bosons: either the gluon (in QCD) or the photon (in electromagnetism). If we are to understand this weak force at a fundamental level, we want it to have a force carrier, that, at most, communicates the force at the speed of light. Where is this force carrier for the weak force?

This issue is just the tip of the iceberg. Just asking the question about what these force carriers are demands that we confront serious problems that never arose with QCD or electromagnetism. We will see that the structure of the $V - A$ theory requires the force carriers to be massive! This would seem to be at odds with gauge invariance as a guiding principle, but there's an out: we are able to maintain gauge invariance at the cost of introducing new fields into our theory. This procedure for giving mass to spin-1 force carriers is called the Higgs mechanism, and its construction will be the focus of this chapter. First, though, let's enumerate the litany of problems with the $V - A$ theory.

11.1 Problems with the $V - A$ Theory

In the $V-A$ theory, the interaction of, say, electrons, muons, and their neutrinos is governed by the interaction Lagrangian

$$\mathcal{L} = \frac{4G_F}{\sqrt{2}} \left(\nu_{\mu L}^\dagger \bar{\sigma}^\mu \mu_L \right) \left(e_L^\dagger \bar{\sigma}_\mu \nu_{e L} \right) . \tag{11.1}$$

Recall that the dimensionality of the Fermi constant is GeV^{-2}. With this observation, we can use dimensional analysis to estimate the rate for electron–muon collisions that produce neutrinos. The cross section for the process $e^- + \mu^+ \to \nu_e + \bar{\nu}_\mu$ must have dimensions of $[\text{energy}]^{-2}$ (because it is an area). The Feynman diagram for this scattering in the $V - A$ theory is proportional to G_F:

$$\mathcal{M}(e^- \mu^+ \to \nu_e \bar{\nu}_\mu) =$$ 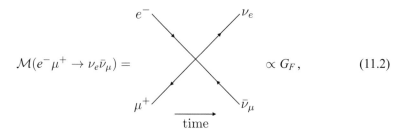 $$\propto G_F, \qquad (11.2)$$

and so the squared matrix element is proportional to G_F^2. At energies well above the muon mass, the rest of the dimensions in the cross section must be made up by factors of the center-of-mass collision energy E_{cm}, as that is the only other energy scale around. Therefore, in the high-energy limit, the cross section must scale like

$$\sigma(e^- \mu^+ \to \nu_e \bar{\nu}_\mu) \sim E_{\text{cm}}^2 G_F^2. \qquad (11.3)$$

This is a bit weird: the cross section for electron–muon scattering diverges as the center-of-mass energy gets large? This doesn't make physical sense. At higher energies, the electron and muon have smaller de Broglie wavelengths and so it should be less likely for their wavefunctions to overlap and therefore collide. This physical picture is consistent with our calculation of electron–muon scattering via a photon. Recall that the cross section for the process $e^+ e^- \to \mu^+ \mu^-$ is

$$\sigma(e^+ e^- \to \mu^+ \mu^-) = \frac{4\pi \alpha^2}{3 E_{\text{cm}}^2}. \qquad (11.4)$$

As expected, this vanishes as $E_{\text{cm}} \to \infty$.

Okay, perhaps this is weird, but let's stomach it and soldier on. Let's just assume that indeed there is some force carrier that is responsible for the interactions in the $V - A$ theory. In analogy with electromagnetism and QCD, we would expect this force carrier to have spin-1 (like the photon and gluon) and perhaps, if it is like the gluon, carry some of the charge which it communicates. Somehow this spin-1 force carrier needs to be responsible for the particular value of G_F. More importantly, it needs to introduce the appropriate dimensions of G_F. Very oddly, this means that the force carrier must be massive!

We can interpret G_F as some sensitivity to a mass scale, according to its dimensionality. That is, we can write

$$G_F = \frac{1}{m_F^2}, \qquad (11.5)$$

for some mass m_F, that we might call the "Fermi mass" (an instance of the Matthew effect). With $G_F = 1.17 \times 10^{-5} \text{ GeV}^{-2}$, the Fermi mass would be

$$m_F = \frac{1}{\sqrt{G_F}} \simeq 292 \text{ GeV}. \qquad (11.6)$$

Somehow the force carrier of the $V - A$ theory needs to introduce a mass scale of 292 GeV. This is problematic. When we discussed QCD, we argued that the gluon must be massless because a mass term in the Lagrangian was inconsistent with gauge invariance. If it is inconsistent with gauge invariance, then Noether says that the corresponding charge is not conserved. Yet, this seems to be what we need to do in the $V - A$ theory to introduce a force carrier that is spin-1 and has a mass at the scale of the Fermi mass. As Winnie-the-Pooh might say, "Oh bother."

11.2 Spontaneous Symmetry Breaking

How do we ever hope to get out of this conundrum? In the 1930s and 1960s, many people including Ernst Stueckelberg,[1] Yoichiro Nambu,[2] Julian Schwinger,[3] Philip Anderson,[4] François Englert, Robert Brout,[5] Peter Higgs,[6] Gerry Guralnik, Carl Hagen, Tom Kibble,[7] Alexander Migdal, Alexander Polyakov,[8] and Gerardus 't Hooft[9] provided insight and solutions to all of these seemingly insurmountable problems. It's typically called the **Higgs mechanism** and predicts the existence of the Higgs boson, which was discovered at the LHC in 2012. Only Higgs and Englert won the Nobel Prize for this work in 2013.

In this section, we discuss the intuition for the Higgs mechanism, starting with the situation in quantum mechanics. This provides insight in the case of superconductivity in quantum field theory. Then, with this experience under our belts, we identify properties of the weak force carriers and use the Higgs mechanism to construct the complete theory, which is called the unified electroweak force.

11.2.1 Quantum Mechanics Analogy

Harmonic Oscillator

To motivate the Higgs mechanism, let's go back to quantum mechanics and think about what our goals are. We claim victory of "solving" a quantum system when we diagonalize the Hamiltonian; that is, we want to find the eigenenergies of a quantum system. With these eigenenergies, we can then fully classify the system and calculate the time evolution of an arbitrary state. Perhaps the simplest interesting quantum system is the harmonic oscillator. This is a quantum system in which a particle of mass m is placed in a quadratic potential with spring constant k:

$$V(x) = \frac{k}{2}x^2, \tag{11.7}$$

---

[1] E. C. G. Stueckelberg, "Interaction forces in electrodynamics and in the field theory of nuclear forces," Helv. Phys. Acta **11**, 299 (1938).

[2] Y. Nambu, "Quasiparticles and gauge invariance in the theory of superconductivity," Phys. Rev. **117**, 648 (1960); Y. Nambu and G. Jona-Lasinio, "Dynamical model of elementary particles based on an analogy with superconductivity. 1," Phys. Rev. **122**, 345 (1961).

[3] J. S. Schwinger, "Gauge invariance and mass," Phys. Rev. **125**, 397 (1962); J. S. Schwinger, "Gauge invariance and mass. 2," Phys. Rev. **128**, 2425 (1962).

[4] P. W. Anderson, "Plasmons, gauge invariance, and mass," Phys. Rev. **130**, 439 (1963).

[5] F. Englert and R. Brout, "Broken symmetry and the mass of gauge vector mesons," Phys. Rev. Lett. **13**, 321 (1964).

[6] P. W. Higgs, "Broken symmetries and the masses of gauge bosons," Phys. Rev. Lett. **13**, 508 (1964).

[7] G. S. Guralnik, C. R. Hagen and T. W. B. Kibble, "Global conservation laws and massless particles," Phys. Rev. Lett. **13**, 585 (1964).

[8] A. A. Migdal and A. M. Polyakov, "Spontaneous breakdown of strong interaction symmetry and the absence of massless particles," Sov. Phys. JETP **24**, 91 (1967) [Zh. Eksp. Teor. Fiz. **51**, 135 (1966)].

[9] G. 't Hooft, "Renormalizable Lagrangians for massive Yang–Mills fields," Nucl. Phys. B **35**, 167 (1971).

that looks like:

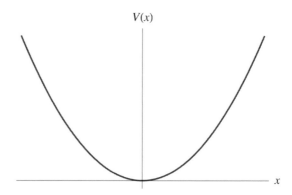

Importantly, note that the minimum of the potential is at $x = 0$. Therefore, to study this system, we can start with the wavefunction localized around $x = 0$, and then perturb it to identify the energy levels. The way in which you may be familiar with doing this is by use of the raising and lowering ladder operators. Any way you do it, you find that the eigenenergies are

$$E_n = \hbar\omega \left(n + \frac{1}{2} \right),$$ (11.8)

where ω is the characteristic frequency of the harmonic oscillator, $\omega = \sqrt{k/m}$.

Double-Well Potential

Okay, that was easy; let's make the system a bit more challenging. Let's consider the system with the potential

$$V(x) = \frac{V_0}{x_0^4} (x^2 - x_0^2)^2 .$$ (11.9)

Here, V_0 is the value of the potential when $x = 0$, and x_0 is some characteristic length or distance of the system. The potential looks like:

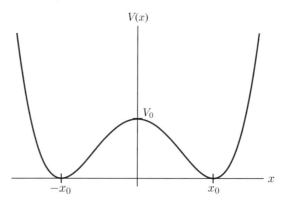

This is called a "double-well" potential for the obvious reasons. Note that, like the harmonic oscillator, it is symmetric in $x \to -x$: $V(x) = V(-x)$, and unlike the harmonic oscillator, has minima at $x = \pm x_0$, away from $x = 0$. Now, to work to diagonalize the Hamiltonian, it doesn't make any sense to consider a wavefunction localized about $x = 0$ and then perturb. Any wavefunction that is localized about $x = 0$ (or any ball placed at or near $x = 0$) will just roll down the potential, and land in one of the two potential wells located at $x = x_0$ or $x = -x_0$. That is, the true ground state is described by a wavefunction localized at either $x = x_0$ or $x = -x_0$, and not about $x = 0$.

There's actually something weird about this, too. Unlike the harmonic oscillator, this potential has two, degenerate (equal energy) minima. Expanding about either of these minima is an equally valid description of the low-energy eigenvalues. Which one we pick or is picked for us is random. Note, however, that while the potential $V(x)$ is symmetric for $x \to -x$, if we pick, say, the minimum located at $x = x_0$ to expand about, our description of the system is no longer symmetric in $x \to -x$. This is a manifestation of a phenomenon called **spontaneous symmetry breaking**: our complete quantum system (in this case the potential $V(x)$) has a symmetry for $x \to -x$. However, our description of the ground state does not, because we happen to choose the minimum at $x = x_0$ about which to expand. By making this random choice, we have "spontaneously" broken the $x \to -x$ symmetry.

Let's see mathematically what this expansion about $x = x_0$ means. To expand about $x = x_0$, we will write $x = x_0 + \delta x$, for some fluctuation position δx. With this expansion, the potential becomes

$$V(x) = \frac{V_0}{x_0^4} \left((x_0 + \delta x)^2 - x_0^2 \right)^2 = \frac{V_0}{x_0^4} \left(2x_0\delta x + (\delta x)^2 \right)^2 . \tag{11.10}$$

This expression for the potential no longer obviously has the $x \to -x$ symmetry. In particle physics, we typically say that the symmetry is no longer **manifest**. However, it is still there; we just have to translate the $x \to -x$ symmetry to the fluctuation δx. For $\delta x \to -\delta x - 2x_0$, the potential becomes

$$V(x) \to \frac{V_0}{x_0^4} \left(2x_0(-\delta x - 2x_0) + (-\delta x - 2x_0)^2 \right)^2 = \frac{V_0}{x_0^4} \left(2x_0\delta x + (\delta x)^2 \right)^2 , \tag{11.11}$$

which corresponds to expanding about the minimum at $x = -x_0$.

One more thing to note about this potential is its expansion for small δx. In the limit where $\delta x \ll x_0$, we have

$$V(x) = \frac{V_0}{x_0^4} \left(2x_0\delta x + (\delta x)^2 \right)^2 = 4\frac{V_0}{x_0^2} (\delta x)^2 + \cdots , \tag{11.12}$$

where we ignore terms at higher orders in $\delta x / x_0$. This is just the harmonic oscillator potential with

$$k = 8\frac{V_0}{x_0^2} . \tag{11.13}$$

The ground state energy in this minimum is then

$$E_0 = \frac{\hbar\omega}{2} = \frac{\hbar}{2} \frac{2\sqrt{2}}{x_0} \sqrt{\frac{V_0}{m}} = \frac{\sqrt{2}\hbar}{x_0} \sqrt{\frac{V_0}{m}} . \tag{11.14}$$

Note that this is non-zero.

"Mexican Hat" Potential

Okay, enough in one dimension, let's move to two-dimensional quantum systems. Let's take this double-well potential in one dimension and just rotate it about the vertical axis. In two dimensions this is

$$V(x,y) = \frac{V_0}{r_0^4}\left[(x^2 + y^2) - r_0^2\right]^2.$$ (11.15)

This looks like a sombrero and so is colloquially called the **Mexican hat potential**:

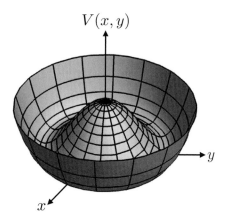

Again, like the double-well potential, the minimum of this potential is not located at the origin $x = y = 0$, but displaced from the origin, where $x^2 + y^2 = r_0^2$. So, just like the double-well potential, we should expand the potential about the minimum. To do this, note that the potential is radially symmetric, so we can re-express coordinates x and y in terms of r, the distance from the origin, and ϕ, the angle about the origin. That is,

$$x = r\cos\phi, \qquad\qquad y = r\sin\phi.$$ (11.16)

In these coordinates, the potential is

$$V(r,\phi) = \frac{V_0}{r_0^4}(r^2 - r_0^2)^2,$$ (11.17)

with no dependence on ϕ.

In rewriting the potential in terms of r and ϕ, we had to choose a value of ϕ. Any choice of $\phi \in [0, 2\pi)$ is equally valid in which to calculate energies, but we picked a particular value. This is like what we did for the double-well potential: we had to pick (arbitrarily) one of the wells in which to expand. However, in this case, the angle coordinate is continuous: by choosing some ϕ, we spontaneously break the continuous rotational symmetry of the system. This has extremely cool consequences. Let's consider the Schrödinger equation for the wavefunction in the r, ϕ coordinates, $\psi(r, \phi)$. The Schrödinger equation for this system is

$$-\frac{\hbar^2}{2m}\left(\frac{1}{r}\frac{\partial}{\partial r}\left(r\frac{\partial}{\partial r}\right) + \frac{1}{r^2}\frac{\partial^2}{\partial\phi^2}\right)\psi + \frac{V_0}{r_0^4}(r^2 - r_0^2)^2\psi = E\psi.$$ (11.18)

This can be solved by separation of variables, where $\psi(r,\phi) = \psi_r(r)\psi_\phi(\phi)$. Doing this, we find the two differential equations

$$-\frac{\hbar^2}{2m}\frac{1}{r}\frac{\partial}{\partial r}\left(r\frac{\partial}{\partial r}\right)\psi_r + \frac{V_0}{r_0^4}(r^2 - r_0^2)^2\psi_r = \left(E - \frac{\alpha}{r^2}\right)\psi_r, \qquad (11.19)$$

$$-\frac{\hbar^2}{2m}\frac{\partial^2}{\partial\phi^2}\psi_\phi = \alpha\psi_\phi, \qquad (11.20)$$

where α is the separation-of-variables constant. Now, we can expand about the minimum of the potential and write $r = r_0 + \delta r$. Only keeping the lowest-order terms in this expansion, the Schrödinger equations become

$$-\frac{\hbar^2}{2m}\frac{\partial^2}{\partial(\delta r)^2}\psi_r + \frac{4V_0}{r_0^2}(\delta r)^2\psi_r = \left(E - \frac{\alpha}{r_0^2}\right)\psi_r, \qquad (11.21)$$

$$-\frac{\hbar^2}{2m}\frac{1}{r_0^2}\frac{\partial^2}{\partial\phi^2}\psi_\phi = \frac{\alpha}{r_0^2}\psi_\phi. \qquad (11.22)$$

With this expansion, the Schrödinger equation for the radial wavefunction ψ_r just turns into a harmonic potential with an effective spring constant of

$$k_{\text{eff}} = 8\frac{V_0}{r_0^2}, \qquad (11.23)$$

which is similar to what we found with the double-well potential. This then implies that the effective energy eigenvalues of the radial wavefunction are just those of the harmonic oscillator, at least for the low-lying states for which our approximation that r only has small fluctuations about r_0 is a good approximation. In particular, the effective radial ground state energy E_r^{gs} is

$$E_r^{\text{gs}} = \frac{\sqrt{2}\hbar}{r_0}\sqrt{\frac{V_0}{m}} = E - \frac{\alpha}{r_0^2}. \qquad (11.24)$$

The total energy of the system E when it is in the radial ground state is then

$$E = \frac{\sqrt{2}\hbar}{r_0}\sqrt{\frac{V_0}{m}} + \frac{\alpha}{r_0^2}. \qquad (11.25)$$

Note that this energy is at least $E_r^{\text{gs}} > 0$ and the precise value depends on the size of α.

The equation for ψ_ϕ is much more interesting than that for ψ_r. In these expressions, α is some constant from separation of variables. In the ψ_ϕ equation, α/r_0^2 has the interpretation as the energy of the state corresponding to ψ_ϕ. We have the boundary condition that $\psi_\phi(\phi) = \psi_\phi(\phi + 2\pi)$, and so the solutions to Eq. 11.22 can be written as

$$\psi_\phi(\phi) = \frac{1}{\sqrt{\pi}}\sin(n\phi + \varphi), \qquad \text{for } n = 1, 2, 3, \ldots, \quad \alpha = \frac{n^2\hbar^2}{2m}. \qquad (11.26)$$

Here, φ is the angle about which we expand the wavefunction (that is, what we define to be the origin). This angle can be anything, $\varphi \in [0, 2\pi)$, and the value we choose does not affect any observable quantities. For $n > 0$, these wavefunctions correspond to an energy

$$E_\phi = \frac{n^2\hbar^2}{2mr_0^2}, \qquad (11.27)$$

that is non-zero; it takes finite energy to excite the system to these states. These energy states are identical to those for the infinite square well of width πr_0. However, because the domain of the angle $\phi \in [0, 2\pi)$ is compact, there is a normalizable solution for $\alpha = 0$. It is the true ground state:

$$\psi_{\phi_0}(\phi) = \frac{1}{\sqrt{2\pi}} \,. \tag{11.28}$$

This state has zero energy, and as such is "always" there. No injection of energy into the system is needed to get to this state.

Note the distinction between this system and the double-well potential. For the double well, we also had to spontaneously break the $x \to -x$ symmetry, but this is not continuous. In that case, to access any state required a positive energy injection into the system. In the next section, we will do the same exercise in quantum field theory. States in quantum mechanics correspond to particles in quantum field theory. That is, the zero-energy angular state we found in the quantum mechanical example will correspond to a particle that can have zero energy. If there is a particle that can have zero energy, then it must be massless; otherwise its energy will be at least the mass. The existence of massless particles corresponding to spontaneously broken continuous symmetries is known as **Goldstone's theorem**, and we'll see how it works and how we can exploit it.

11.2.2 Goldstone's Theorem for the Mexican Hat Potential

All of this was background to prepare you for the situation in quantum field theory, and how this can be exploited to provide mass to spin-1 bosons. The quantum field theory analogy to the quantum mechanical two-dimensional radial potential is a complex spin-0 scalar field $\phi(x)$, with the Lagrangian

$$\mathcal{L} = (\partial_\mu \phi)(\partial^\mu \phi^*) - \lambda(|\phi|^2 - v^2)^2 \,. \tag{11.29}$$

Here, the scalar potential is $V(\phi) = \lambda(|\phi|^2 - v^2)^2$. The value v is the minimum of the potential; that is, when $|\phi| = v$, the potential is 0. v is called the **vacuum expectation value** or vev, because it is the state that the field ϕ will assume in the vacuum, with no injection of energy. We denote this by $\langle |\phi| \rangle = v$, meaning that the average absolute value of ϕ in the vacuum is v. λ is the **quartic coupling**, and should be positive so that energy is always non-negative.

This Lagrangian as it stands is pretty weird. Let's expand out the potential

$$\mathcal{L} = (\partial_\mu \phi)(\partial^\mu \phi^*) + 2\lambda v^2 |\phi|^2 - \lambda|\phi|^4 - \lambda v^4 \,. \tag{11.30}$$

Don't worry about the $|\phi|^4$ or λv^4 terms for now; just focus on the $|\phi|^2$ term and the kinetic term. Ignoring these terms in the Lagrangian means that we are expanding about the point where $\phi = 0$. Varying the Lagrangian with respect to ϕ^* yields the Klein–Gordon equation,

$$(\partial_\mu \partial^\mu - 2\lambda v^2)\phi = 0 \,, \tag{11.31}$$

which has the corresponding energy–momentum relation in special relativity of

$$E^2 - |\vec{p}|^2 = -2\lambda v^2 < 0. \tag{11.32}$$

Huh? A negative mass squared? This is not good! As we argued in quantum mechanics, this just means that you're expanding about the wrong point. We don't want to expand about $\phi = 0$, but rather about $|\phi| = v$.

To do this, let's write

$$\phi(x) = \left(v + \frac{r(x)}{\sqrt{2}}\right) e^{i\frac{\theta(x)}{\sqrt{2}v}}, \tag{11.33}$$

for two real scalar fields $r(x)$ and $\theta(x)$. Note that the conjugate field is

$$\phi^*(x) = \left(v + \frac{r(x)}{\sqrt{2}}\right) e^{-i\frac{\theta(x)}{\sqrt{2}v}}. \tag{11.34}$$

Plugging these into the Lagrangian and simplifying, we find

$$\mathcal{L} = \frac{1}{2}(\partial_\mu r)(\partial^\mu r) + \frac{1}{2}(\partial_\mu \theta)(\partial^\mu \theta) + \frac{r}{\sqrt{2}v}(\partial_\mu \theta)(\partial^\mu \theta) \tag{11.35}$$

$$+ \frac{r^2}{4v^2}(\partial_\mu \theta)(\partial^\mu \theta) - \frac{\lambda}{4}(2\sqrt{2}vr + r^2)^2.$$

Now, the potential is purely a function of the field r, with no θ field. Expanding in the small-r-field limit, where $r \ll v$, this Lagrangian reduces to

$$\mathcal{L} = \frac{1}{2}(\partial_\mu r)(\partial^\mu r) - 2\lambda v^2 r^2 + \frac{1}{2}(\partial_\mu \theta)(\partial^\mu \theta) + \cdots. \tag{11.36}$$

That is, the field $r(x)$ has a mass of $m_r = 2\sqrt{\lambda}v > 0$, while the field θ is massless. This is a manifestation of Goldstone's theorem.[10]

Goldstone's theorem states that when a continuous symmetry is spontaneously broken, there exist zero-energy states in the spectrum of the Hamiltonian. That the field $\theta(x)$ is massless is a consequence of the fact that we had to spontaneously choose a value for θ, which broke the rotational symmetry. This is a continuous symmetry, and so we find that the field $\theta(x)$ is massless. That is, there is no minimum energy required to excite particles from the θ field. This is exactly like what we found in the quantum mechanical example. In this system, we would call the θ field the **Goldstone boson**.

11.2.3 Higgs Mechanism in Superconductivity

Now for the main event. Let's see how we can use this property of spontaneous symmetry breaking to give a spin-1 boson a mass. In the example below, we just consider the simple case of giving the field ϕ an electric charge; therefore, we will show how to give the photon a mass and maintain gauge invariance. In the next section, we will be able to apply this insight to the W and Z bosons, the force carriers of the weak force.

---

[10] J. Goldstone, "Field theories with superconductor solutions," Nuovo Cim. **19**, 154 (1961).

Example 11.1 How does the Higgs mechanism give a mass to the photon?

Solution

The first thing we need to do is to couple the scalar field ϕ to the photon. To do this, we just replace the partial derivative by the covariant derivative:

$$\partial_\mu \to D_\mu = \partial_\mu - ieA_\mu \,, \tag{11.37}$$

where e is the electric charge and A_μ is the photon field. The Lagrangian then becomes

$$\mathcal{L} = -\frac{1}{4}F_{\mu\nu}F^{\mu\nu} + (D_\mu\phi)(D^\mu\phi)^* - \lambda(|\phi| - v^2)^2 \,. \tag{11.38}$$

Recall that the field strength for electromagnetism is

$$F_{\mu\nu} = \partial_\mu A_\nu - \partial_\nu A_\mu \,. \tag{11.39}$$

This Lagrangian is invariant under the U(1) gauge transformations

$$\phi \to e^{i\alpha(x)}\phi \,, \qquad \phi \to e^{-i\alpha(x)}\phi^* \,, \qquad A_\mu \to A_\mu + \frac{1}{e}\partial_\mu\alpha(x) \,. \tag{11.40}$$

Here, $\alpha(x)$ is an arbitrary function of the spacetime coordinate x.

Now, as we did in the case when the scalar field was uncharged, we need to expand about the minimum of the potential to ensure that all fields have non-negative mass-squared. Again, we write

$$\phi(x) = \left(v + \frac{h(x)}{\sqrt{2}}\right) e^{i\frac{\theta(x)}{\sqrt{2}v}} \,. \tag{11.41}$$

Unlike in the previous case, we can simplify the expression of $\phi(x)$ by exploiting gauge transformations. The field $\theta(x)$, the Goldstone boson, can be eliminated at the expense of picking a gauge; that is, choosing a particular $\alpha(x)$. If we choose

$$\alpha(x) = -\frac{\theta(x)}{\sqrt{2}v} \,, \tag{11.42}$$

then by performing a gauge transformation we have

$$\phi(x) = \left(v + \frac{h(x)}{\sqrt{2}}\right) e^{i\frac{\theta(x)}{\sqrt{2}v}} \to \left(v + \frac{h(x)}{\sqrt{2}}\right) e^{i\frac{\theta(x)}{\sqrt{2}v}} e^{-i\frac{\theta(x)}{\sqrt{2}v}} \tag{11.43}$$

$$= v + \frac{h(x)}{\sqrt{2}} \,.$$

We have eliminated the Goldstone boson! By choosing and fixing a gauge, we seem to have broken the symmetries of electromagnetism. However, this breaking was just spontaneous, which means that the symmetries of electromagnetism are still there, just not manifest. Thus, charge is still conserved, just not manifestly so when expanding in the ground state of the potential well.

In this gauge, let's now evaluate the Lagrangian. We have

$$\phi(x) = \phi^*(x) = v + \frac{h(x)}{\sqrt{2}} \,, \tag{11.44}$$

and so

$$\mathcal{L} = -\frac{1}{4}F_{\mu\nu}F^{\mu\nu} + \left[(\partial_\mu - ieA_\mu)\left(v + \frac{h(x)}{\sqrt{2}}\right)\right]\left[(\partial^\mu + ieA^\mu)\left(v + \frac{h(x)}{\sqrt{2}}\right)\right] \quad (11.45)$$
$$- \lambda\left(\left(v + \frac{h(x)}{\sqrt{2}}\right)^2 - v^2\right)^2$$
$$= -\frac{1}{4}F_{\mu\nu}F^{\mu\nu} + \frac{1}{2}(\partial_\mu h)(\partial^\mu h) + e^2v^2A_\mu A^\mu + 2e^2vhA_\mu A^\mu - \frac{\lambda}{4}(2\sqrt{2}vh + h^2)^2.$$

This Lagrangian is insane. We call the field $h(x)$ the Higgs field. Expanding its interactions to quadratic order, we have

$$\mathcal{L}_h = \frac{1}{2}(\partial_\mu h)(\partial^\mu h) - 2\lambda v^2 h^2 + \cdots, \quad (11.46)$$

and so the Higgs field has a mass of $m_h = 2v\sqrt{\lambda}$. The weirder part of this Lagrangian is that which involves the photon. Expanding the photon part to quadratic order, we find

$$\mathcal{L}_A = -\frac{1}{4}F_{\mu\nu}F^{\mu\nu} + e^2v^2A_\mu A^\mu + \cdots. \quad (11.47)$$

The term $e^2v^2A_\mu A^\mu$ is a mass for the photon! What?!? The process of spontaneous symmetry breaking effectively has successfully given a mass to the photon. This mass is $m_A = \sqrt{2}ev$, and so is controlled by the electric charge e and the vev.

Note also that the Lagrangian of Eq. 11.45 describes a photon with three degrees of freedom; that is, there are three components of A_μ that propagate. This is different from the case of a massless photon that only has left- and right-handed helicity. Recall that for a massless photon, the potential degree of freedom in the direction of momentum was removed by a gauge transformation. In the case at hand, however, in expanding the scalar field in its lowest energy state, we have spontaneously broken the gauge symmetry and therefore cannot use it to eliminate the third degree of freedom. This third degree of freedom was formerly the Goldstone field, $\theta(x)$. Because the photon acquired mass through the existence of the Goldstone field $\theta(x)$, we say that the photon "ate the Goldstone boson and became fat," that is, massive.

The process of spin-1 gauge bosons acquiring a mass through spontaneous symmetry breaking with a scalar field is called the Higgs mechanism. This particular application describes superconductivity in the **Bardeen–Cooper–Schrieffer (BCS) theory**.[11] The vacuum expectation value of the scalar field $\langle|\phi|\rangle$ is called the **order parameter** of the symmetry breaking. If $\langle|\phi|\rangle = 0$, the symmetry is preserved, and the photon is massless. If $\langle|\phi|\rangle = v \neq 0$, the gauge symmetry is spontaneously broken, and the photon is massive. The Lagrangian with the massive photon and Higgs boson field is still gauge invariant and so electric charge is still conserved. However, in the expansion about the vev this gauge symmetry is not manifest, and so the individual states of the system will not appear to conserve charge (i.e., the photon will have a non-zero mass).

---

[11] J. Bardeen, L. N. Cooper and J. R. Schrieffer, "Theory of superconductivity," Phys. Rev. **108**, 1175 (1957).

So, we know how to give a spin-1 boson a mass that is consistent with gauge invariance of the system. We will use this knowledge to provide the carriers of the weak force with masses that can account for all of the subtleties we identified with the $V - A$ theory.

11.3 Electroweak Unification

In the previous section, we introduced spontaneous symmetry breaking as a feature of the ground state of a system to break a symmetry of the theory. Applying spontaneous symmetry breaking to a gauge theory, coupling a scalar whose vacuum breaks the symmetry is the Higgs mechanism. A gauge boson can acquire a mass if it is coupled to a scalar field that acquires a vacuum expectation value, or vev.

We argued that, if the weak force were to have spin-1 boson force carriers, then those force carriers are necessarily massive. This is a consequence of the fact that the Fermi constant G_F that appears in the $V - A$ theory is dimensionful. Massive force carriers are unfamiliar in our experience with electromagnetism or QCD, but with the Higgs mechanism, we are equipped to understand how this could work. In this section, we will use the Higgs mechanism to understand what the weak force carriers are.

11.3.1 Properties of the Weak Force Carriers

To begin, we need to enumerate what the properties of the force carrier(s) of the weak force are and are not. Recall that decay of the neutron corresponds to the process

$$n \to p + e^- + \bar{\nu}_e, \tag{11.48}$$

or, in terms of the fundamental quarks,

$$d \to u + e^- + \bar{\nu}_e. \tag{11.49}$$

The interaction that describes this decay in the $V - A$ model is

$$\mathcal{L}_{\text{int}} \supset \frac{4G_F}{\sqrt{2}} \left(\nu_{eL}^\dagger \bar{\sigma}^\mu e_L \right) \left(d_L^\dagger \bar{\sigma}_\mu u_L \right). \tag{11.50}$$

This interaction tells us many things. First, it tells us that the force carrier must be colorless: electrons and neutrinos have no color, and the two factors in the interaction are just multiplied. So, the color of the up and down quarks does not flow outside of their interaction. Next, the force carrier must have electric charge. The sum of the charges of the neutrino and electron is -1 (in terms of the fundamental charge e) and the charge of the up and anti-down quark is $+1$. The total charge of this interaction is 0, which ensures that it is electromagnetically gauge invariant and conserves charge. Therefore, whatever particle is mediating this interaction must have electric charge ± 1. We also already know that this mediating particle must be spin-1 and have a mass comparable to $1/\sqrt{G_F}$. For future reference, we will refer to this particle (until we give it a proper name) as the **charged current**.

But wait, there's more. Let's go back to our analysis of e^+e^- scattering, say $e^+e^- \to \mu^+\mu^-$. At low energies, say $E_{\rm cm} \simeq 1$ GeV, this scattering is extremely well described by the interaction mediated by electromagnetism. The masslessness of the photon means that the only energy scale that the cross section can depend on is the center-of-mass collision energy:

$$\sigma(e^+e^- \to \mu^+\mu^-) \propto \frac{1}{E_{\rm cm}^2} \,. \tag{11.51}$$

However, we know that as we increase the center-of-mass collision energy this $1/E_{\rm cm}^2$ dependence changes, and we observe a peak, or a resonance, in the cross section, as a function of $E_{\rm cm}$. That is, we see something like that shown in Fig. 11.1, where there is a resonance around 91 GeV. This bump in the cross section clearly cannot be ascribed to the photon and electromagnetism, because there is nothing special about 91 GeV in Maxwell's equations. Thus, we call it a new particle, the Z boson. However, for now, we will refer to it as the **neutral current**, as it must be neutral (like the photon) and is massive.

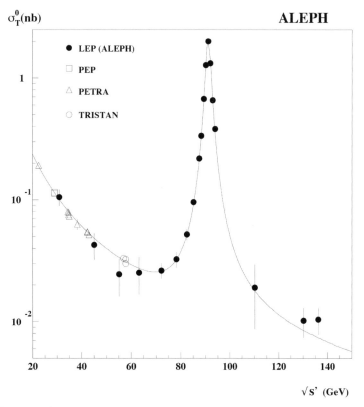

Fig. 11.1   Plot of the $e^+e^- \to \mu^+\mu^-$ cross section in nanobarns as a function of center-of-mass collision energy $\sqrt{s}$. The solid black points are data collected from the ALEPH experiment at LEP. Reprinted from Phys. Lett. B **399**, R. Barate et al. [ALEPH Collaboration], "Study of the muon pair production at center-of-mass energies from 20 GeV to 136 GeV with the ALEPH detector," 329 (1997), with permission from Elsevier.

Additionally, the fact that both the photon and the neutral current mediate the scattering $e^+e^- \to \mu^+\mu^-$ suggests that they are related somehow. Because the photon and the neutral current are intermediate particles in the $e^+e^- \to \mu^+\mu^-$ collisions, we must sum their contribution to the cross section at the matrix element or Feynman diagram level. There is, in principle, no measurement we can do to distinguish what intermediate particle mediated the interaction. Apparently, the photon and neutral current mix into one another quantum mechanically. To do this, they must share (some) quantum numbers, and if they do, are somehow intimately related. More on this in a second.

There's another thing we can look for in an e^+e^- collider and that is the process $e^+e^- \to \nu_e\bar{\nu}_e$. In our detector, because it is incredibly hard to detect neutrinos, we would see electrons and positrons colliding and producing nothing. Experimentally, seeing "nothing" is extremely hard to calibrate and determine uncertainties, so the standard procedure is to let there be an additional final-state particle. In e^+e^- collisions, it is most natural to consider this additional particle to be a photon; thus, we will look for the process $e^+e^- \to \nu_e\bar{\nu}_e\gamma$ in data. What we would actually see in our experiment, however, is just $e^+e^- \to \gamma$, as the neutrinos are unobservable. As the initial state has no net momentum, we would observe that the final state would appear to not conserve momentum. The missing momentum or energy that is needed to conserve momentum of the final state is just missing transverse momentum (MET), and attributed to the neutrinos. That is, to identify neutrino production in e^+e^- collisions, we look for $e^+e^- \to \gamma + $ MET events in our detector.

The fundamental $e^+e^- \to \nu_e\bar{\nu}_e$ scattering can be mediated by the charged current, and the photon can be produced as radiation from the electron or positron. The Lagrangian that governs the interaction of the electron with the neutrino is

$$\mathcal{L}_{\text{int}} \supset \frac{4G_F}{\sqrt{2}} \left(\nu_{eL}^\dagger \bar{\sigma}^\mu e_L \right) \left(e_L^\dagger \bar{\sigma}_\mu \nu_{eL} \right), \tag{11.52}$$

for example. That is, the charged current turns electrons into neutrinos. This interaction cannot be mediated by electromagnetism, because the neutrino is electrically neutral. However, it may be mediated by the neutral current, if the neutral current couples to neutrinos. Just as we did with $e^+e^- \to \mu^+\mu^-$, we can measure the cross section for the observed process of $e^+e^- \to \gamma + $MET as a function of the center-of-mass collision energy, E_{cm}. A plot of this is shown in Fig. 11.2, and one sees that around $E_{\text{cm}} = 91$ GeV or so the cross section changes significantly. This then suggests that the neutral current mediates the process $e^+e^- \to \nu_e\bar{\nu}_e$ and so is somehow related to the charged current.

So, these observations have told us that four bosons – the photon, the neutral current, the $+1$ charged current, and the -1 charged current– are related to one another. Apparently, three of these bosons (the neutral and charged currents) are massive, while the photon is massless. So, we need to develop a theory in which these bosons are related and three of them get mass via the Higgs mechanism. This fundamental theory is called the **unified electroweak force**, as it describes electromagnetism and the weak force in a unified manner.

Let's follow our noses to see if we can construct a sensible theory of this unified electroweak force. Because it has worked in the past (for electromagnetism and QCD), let's

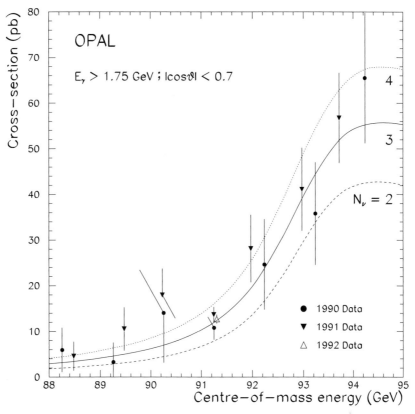

Fig. 11.2 Plot of the $e^+e^- \rightarrow \gamma + $ MET cross section in picobarns as a function of center-of-mass collision energy. The data were collected from the OPAL experiment at LEP. Reprinted by permission from Springer Nature: Springer Nature Z. Phys. C "Measurement of single photon production in e^+e^- collisions near the Z0 resonance," R. Akers *et al.* [OPAL Collaboration] (1995).

try to figure out what the gauge symmetry of this theory is. To do this requires identifying the relationship between the gauge group and the number of force-carrying bosons. In our discussion of QCD, we identified the SU(3) color symmetry. We argued that SU(3) has eight basis matrices T^a, for $a = 1, \ldots, 8$. We determined this by considering how many free parameters there were in an SU(3) matrix. Again, a general 3×3 complex matrix $\mathbb{U}$ has 18 real parameters. Enforcing the unitary constraint $\mathbb{U}^\dagger \mathbb{U} = \mathbb{I}$ fixes nine parameters. Further enforcing the unit determinant is one more constraint. Therefore, there are eight basis matrices of SU(3). For the group SU(N), there are in general $N^2 - 1$ basis matrices, by extension of this argument.

So, for the group SU(N), there are in general $N^2 - 1$ force-carrying bosons. To describe the electroweak force, we need four. $N^2 - 1$ for integer N cannot be 4, so the approach that we used for QCD must be generalized. Note, however, that if we do not impose the unit determinant constraint, then we have the group U(N), which has a total of N^2 basis matrices. $2^2 = 4$ and so the simplest group that has four gauge bosons is U(2), the group of unitary 2×2 matrices. This group can be equivalently expressed as SU(2)⊗U(1)≃U(2), where $\otimes$ means the direct product of the groups SU(2) and U(1). So, this is what we will work with as the electroweak gauge group. We call the SU(2) part **weak isospin** and the U(1) part **hypercharge**. Let's now see how the weak isospin and hypercharge combine to produce the charged and neutral currents, and the photon. The theory that we will develop in the rest of this chapter was first described by Sheldon Glashow, Steven Weinberg, and Abdus Salam.[12]

11.3.2 Spontaneous Breaking of Electroweak Symmetry

Our first goal will be to spontaneously break this gauge symmetry down to just electromagnetism using the Higgs mechanism. For compactness, we denote SU(2) weak isospin by $SU(2)_W$ and the U(1) hypercharge by $U(1)_Y$. We will call the gauge bosons of the weak isospin $SU(2)_W$ A_μ^1, A_μ^2, and A_μ^3, while the gauge boson of hypercharge $U(1)_Y$ will be B_μ. For the Higgs mechanism, we need to introduce the scalar field that can do the symmetry breaking. This field must couple to all of the electroweak gauge bosons, and so must carry weak isospin and hypercharge. The simplest possibility is to have the field $\vec{\phi}$ be in the fundamental representation of $SU(2)_W$, which is a two-component vector:

$$\vec{\phi} = \begin{pmatrix} \phi^+ \\ \phi_0 \end{pmatrix}, \tag{11.53}$$

where ϕ^+ and ϕ_0 are complex scalar fields and the $+$ and 0 notation will become clear in a bit. This field therefore has four degrees of freedom because a complex number has a real and an imaginary part. Such a field is also called a doublet of $SU(2)_W$. It transforms under an $SU(2)_W \otimes U(1)_Y$ gauge transformation as

$$\vec{\phi} = \begin{pmatrix} \phi^+ \\ \phi_0 \end{pmatrix} \rightarrow e^{i\vec{\alpha}\cdot\frac{\vec{\sigma}}{2}} e^{i\frac{\beta}{2}} \begin{pmatrix} \phi^+ \\ \phi_0 \end{pmatrix}, \tag{11.54}$$

where $\vec{\sigma} = (\sigma_1, \sigma_2, \sigma_3)$ are the Pauli sigma matrices (the basis matrices of the fundamental representation of SU(2)), and $\vec{\alpha}$ and β are parameters that determine the gauge transformation. The hypercharge Y of this boson is defined to be $Y = 1$.

Just as we did with QCD, we need to introduce a covariant derivative that enables gauge-invariant interactions between this scalar field and the electroweak gauge bosons A_μ^a and B_μ. This covariant derivative is

$$D_\mu = \partial_\mu - ig_W \frac{\sigma_a}{2} A_\mu^a - ig_Y \frac{1}{2} B_\mu . \tag{11.55}$$

In this expression, g_W and g_Y are the coupling constants of the weak isospin and the hypercharge, respectively. The index $a = 1, 2, 3$ ranges over the three basis matrices of SU(2). Note that the factors of $1/2$ in the covariant derivative come from the identification of the scalar in the fundamental representation of $SU(2)_W$ and the assignment of its

---

[12] S. L. Glashow, "Partial symmetries of weak interactions," Nucl. Phys. **22**, 579 (1961); S. Weinberg, "A model of leptons," Phys. Rev. Lett. **19**, 1264 (1967); A. Salam, "Weak and electromagnetic interactions," in *Elementary particle Theory: Proceedings of the 8th Nobel Symposium, Lerum, Sweden, 1968*, John Wiley & Sons and Almqvist and Wiksell (1968), pp. 367–377.

hypercharge of $Y = 1$. It then follows that the gauge-invariant Lagrangian for the interactions of the scalar with the weak isospin and hypercharge is

$$\mathcal{L} \supset \left(D_\mu \begin{pmatrix} \phi^+ \\ \phi_0 \end{pmatrix} \right)^\dagger \left(D^\mu \begin{pmatrix} \phi^+ \\ \phi_0 \end{pmatrix} \right). \tag{11.56}$$

Now that we know the interactions of the scalar field with the electroweak bosons, we can work to spontaneously break the symmetry. To do this, we need to give the field a vev, which can be accomplished by writing down a gauge-invariant potential. With the vev by convention $\langle |\vec{\phi}| \rangle = v/\sqrt{2}$, the potential is

$$V(\vec{\phi}) = \lambda \left(|\vec{\phi}|^2 - \frac{v^2}{2} \right)^2, \tag{11.57}$$

where λ is the quartic coupling. Note that $|\vec{\phi}|^2$ is gauge invariant. This is

$$|\vec{\phi}|^2 = \left((\phi^+)^* \ (\phi_0)^* \right) \begin{pmatrix} \phi^+ \\ \phi_0 \end{pmatrix} = |\phi^+|^2 + |\phi_0|^2. \tag{11.58}$$

A gauge transformation can rotate the field components ϕ^+ and ϕ_0 into one another, but cannot change the magnitude.

Our next step is to expand the field $\vec{\phi}$ about the vev. To do this, we express the scalar field as

$$\vec{\phi}(x) = \begin{pmatrix} \phi^+ \\ \phi_0 \end{pmatrix} = \begin{pmatrix} \frac{r_1(x)}{\sqrt{2}} e^{i\frac{\theta_1(x)}{\sqrt{2}v}} \\ \left(\frac{v}{\sqrt{2}} + \frac{r_2(x)}{\sqrt{2}} \right) e^{i\frac{\theta_2(x)}{\sqrt{2}v}} \end{pmatrix}, \tag{11.59}$$

where $r_1(x)$, $r_2(x)$, $\theta_1(x)$, and $\theta_2(x)$ are real scalar fields, with each having only one degree of freedom. The average value of each of these fields is 0, and so indeed the vev of the field $\vec{\phi}$ is

$$\langle |\vec{\phi}| \rangle = \left\langle \left[\frac{r_1(x)^2}{2} + \left(\frac{v}{\sqrt{2}} + \frac{r_2(x)}{\sqrt{2}} \right)^2 \right]^{1/2} \right\rangle = \frac{v}{\sqrt{2}}, \tag{11.60}$$

as required from the form of the potential, Eq. 11.57. To describe the unified electroweak theory, we need to give three force-carrying bosons a mass (the neutral and charge currents). Thus, we need to eliminate three of the fields in $\vec{\phi}$ with an appropriate $SU(2)_W \otimes U(1)_Y$ gauge transformation. As there are three basis matrices of $SU(2)_W$ (and therefore three gauge transformation parameters α_1, α_2, and α_3), we will just use $SU(2)_W$ to remove these three scalar fields. In general, though, one could use any three linear combinations of the $\vec{\alpha}$ and β parameters.

An arbitrary $SU(2)$ matrix $\mathbb{U}$ can be written with one rotation angle ψ and two phases χ_1 and χ_2:

$$\mathbb{U} = \begin{pmatrix} e^{i\chi_1} \cos\psi & -e^{i\chi_2} \sin\psi \\ e^{-i\chi_2} \sin\psi & e^{-i\chi_1} \cos\psi \end{pmatrix}. \tag{11.61}$$

It is straightforward to verify that this matrix does indeed satisfy the unitarity requirement and has determinant 1. In Exercise 11.4, you will relate the parameters in this form of an $SU(2)$ matrix to the gauge transformation parameters α_1, α_2, and α_3 in the form of

Eq. 11.54. This matrix rotates away three of the four real scalar fields in $\vec{\phi}$. That is, by acting with this matrix, we want to produce the form

$$\mathbb{U}\vec{\phi} = \begin{pmatrix} e^{i\chi_1}\cos\psi & -e^{i\chi_2}\sin\psi \\ e^{-i\chi_2}\sin\psi & e^{-i\chi_1}\cos\psi \end{pmatrix} \begin{pmatrix} \frac{r_1(x)}{\sqrt{2}}e^{i\frac{\theta_1(x)}{\sqrt{2}v}} \\ \left(\frac{v}{\sqrt{2}} + \frac{r_2(x)}{\sqrt{2}}\right)e^{i\frac{\theta_2(x)}{\sqrt{2}v}} \end{pmatrix} = \begin{pmatrix} 0 \\ \frac{v}{\sqrt{2}} + \frac{H(x)}{\sqrt{2}} \end{pmatrix}.$$

(11.62)

In this final form, the real scalar field that remains, $H(x)$, is called the **Higgs boson**.

Let's see how this works. Performing the matrix multiplication, we have

$$\begin{pmatrix} e^{i\chi_1}\cos\psi & -e^{i\chi_2}\sin\psi \\ e^{-i\chi_2}\sin\psi & e^{-i\chi_1}\cos\psi \end{pmatrix} \begin{pmatrix} \frac{r_1(x)}{\sqrt{2}}e^{i\frac{\theta_1(x)}{\sqrt{2}v}} \\ \left(\frac{v}{\sqrt{2}} + \frac{r_2(x)}{\sqrt{2}}\right)e^{i\frac{\theta_2(x)}{\sqrt{2}v}} \end{pmatrix}$$

(11.63)

$$= \begin{pmatrix} \frac{r_1(x)}{\sqrt{2}}e^{i\left(\frac{\theta_1(x)}{\sqrt{2}v}+\chi_1\right)}\cos\psi - \left(\frac{v}{\sqrt{2}} + \frac{r_2(x)}{\sqrt{2}}\right)e^{i\left(\frac{\theta_2(x)}{\sqrt{2}v}+\chi_2\right)}\sin\psi \\ \frac{r_1(x)}{\sqrt{2}}e^{i\left(\frac{\theta_1(x)}{\sqrt{2}v}-\chi_2\right)}\sin\psi + \left(\frac{v}{\sqrt{2}} + \frac{r_2(x)}{\sqrt{2}}\right)e^{i\left(\frac{\theta_2(x)}{\sqrt{2}v}-\chi_1\right)}\cos\psi \end{pmatrix}.$$

From the form of Eq. 11.62, we want to eliminate all phases from the second entry of the vector $\vec{\phi}$. This then forces us to set

$$\frac{\theta_1(x)}{\sqrt{2}v} = \chi_2, \qquad\qquad \frac{\theta_2(x)}{\sqrt{2}v} = \chi_1.$$

(11.64)

Inserting this into the first entry, we then have the requirement that

$$\frac{r_1(x)}{\sqrt{2}}e^{i\left(\frac{\theta_1(x)}{\sqrt{2}v}+\frac{\theta_2(x)}{\sqrt{2}v}\right)}\cos\psi - \left(\frac{v}{\sqrt{2}} + \frac{r_2(x)}{\sqrt{2}}\right)e^{i\left(\frac{\theta_2(x)}{\sqrt{2}v}+\frac{\theta_1(x)}{\sqrt{2}v}\right)}\sin\psi = 0,$$

(11.65)

or, removing the overall phase factor, that

$$r_1(x)\cos\psi - (v + r_2(x))\sin\psi = 0.$$

(11.66)

The ψ that solves this equation is

$$\cos\psi = \frac{v + r_2(x)}{\sqrt{(v + r_2(x))^2 + r_1(x)^2}}, \qquad \sin\psi = \frac{r_1(x)}{\sqrt{(v + r_2(x))^2 + r_1(x)^2}}.$$

(11.67)

With these results, we can then evaluate the second entry of $\vec{\phi}$, and equate it to the expression in Eq. 11.62. We then find

$$\sqrt{(v + r_2(x))^2 + r_1(x)^2} = v + H(x).$$

(11.68)

In terms of the original fields $r_1(x)$ and $r_2(x)$, the Higgs boson $H(x)$ is therefore

$$H(x) = \sqrt{(v + r_2(x))^2 + r_1(x)^2} - v.$$

(11.69)

The Higgs boson field $H(x)$ describes the small deviations of the $\vec{\phi}$ field from its vev. With the interpretation that $H(x)$ is relatively small, we can Taylor expand the square-root in $r_1(x)$ and $r_2(x)$:

$$H(x) \simeq r_2(x) + \mathcal{O}(r_1^2, r_2^2).$$

(11.70)

While this relationship is interesting, from now on we will just use the expression for $\vec{\phi}$ from Eq. 11.62, knowing that we can in general perform an SU(2)$_W$ gauge transformation to reduce to this form. Importantly, once the dust has settled, we have used up three gauge transformation parameters (α_1, α_2, and α_3) to do this. Thus, we have successfully spontaneously broken the SU(2)$_W\otimes$U(1)$_Y$ symmetry and we should find that three of the four force-carrying bosons of this theory have a mass. Let's do that now.

11.3.3 The Broken Weak Theory

With this form of the scalar $\vec{\phi}$ that spontaneously breaks the SU(2)$_W\otimes$U(1)$_Y$ symmetry, we can re-insert it into its Lagrangian kinetic term. We have

$$
\mathcal{L} \supset \left(D_\mu \begin{pmatrix} \phi^+ \\ \phi_0 \end{pmatrix} \right)^\dagger \left(D^\mu \begin{pmatrix} \phi^+ \\ \phi_0 \end{pmatrix} \right) \tag{11.71}
$$

$$
= \begin{pmatrix} 0 & \dfrac{v}{\sqrt{2}} + \dfrac{H(x)}{\sqrt{2}} \end{pmatrix} \left(\overleftarrow{\partial}_\mu + ig_W \frac{\sigma_a}{2} A^a_\mu + i\frac{g_Y}{2} B_\mu \right)
$$

$$
\times \left(\partial^\mu - ig_W \frac{\sigma_b}{2} A^{\mu\,b} - i\frac{g_Y}{2} B^\mu \right) \begin{pmatrix} 0 \\ \dfrac{v}{\sqrt{2}} + \dfrac{H(x)}{\sqrt{2}} \end{pmatrix} .
$$

The notation $\overleftarrow{\partial}_\mu$ means that the partial derivative acts to the left, on the Higgs field. Our focus for now is on determining the masses of the force-carrying bosons A^a_μ and B_μ. To do this, we set the Higgs field $H(x)$ to zero, and therefore the partial derivatives vanish. Then, the Lagrangian becomes

$$
\mathcal{L}_{m^2} = - \begin{pmatrix} 0 & \dfrac{v}{\sqrt{2}} \end{pmatrix} \left(ig_W \frac{\sigma_a}{2} A^a_\mu + i\frac{g_Y}{2} B_\mu \right) \left(ig_W \frac{\sigma_b}{2} A^{\mu\,b} + i\frac{g_Y}{2} B^\mu \right) \begin{pmatrix} 0 \\ \dfrac{v}{\sqrt{2}} \end{pmatrix} . \tag{11.72}
$$

Now, we need to evaluate the matrix multiplication. Using the explicit expressions for the sigma matrices, the matrices of gauge bosons can be written in components as

$$
g_W \frac{\sigma_a}{2} A^{\mu\,a} + \frac{g_Y}{2} B^\mu = \frac{g_W}{2} \left(\sigma_1 A^{\mu\,1} + \sigma_2 A^{\mu\,2} + \sigma_3 A^{\mu\,3} \right) + \frac{g_Y}{2} \mathbb{I} B^\mu \tag{11.73}
$$

$$
= \begin{pmatrix} \frac{g_W}{2} A^{\mu\,3} + \frac{g_Y}{2} B^\mu & \frac{g_W}{2} \left(A^{\mu\,1} - iA^{\mu\,2} \right) \\ \frac{g_W}{2} \left(A^{\mu\,1} + iA^{\mu\,2} \right) & -\frac{g_W}{2} A^{\mu\,3} + \frac{g_Y}{2} B^\mu \end{pmatrix} .
$$

Here, the term with B_μ is proportional to the 2×2 identity matrix, $\mathbb{I}$. Performing the matrix multiplication, we have

$$
\begin{pmatrix} \frac{g_W}{2} A^{\mu\,3} + \frac{g_Y}{2} B^\mu & \frac{g_W}{2} \left(A^{\mu\,1} - iA^{\mu\,2} \right) \\ \frac{g_W}{2} \left(A^{\mu\,1} + iA^{\mu\,2} \right) & -\frac{g_W}{2} A^{\mu\,3} + \frac{g_Y}{2} B^\mu \end{pmatrix} \begin{pmatrix} 0 \\ \dfrac{v}{\sqrt{2}} \end{pmatrix} \tag{11.74}
$$

$$
= \begin{pmatrix} \frac{v}{\sqrt{2}} \frac{g_W}{2} \left(A^{\mu\,1} - iA^{\mu\,2} \right) \\ \frac{v}{\sqrt{2}} \left(-\frac{g_W}{2} A^{\mu\,3} + \frac{g_Y}{2} B^\mu \right) \end{pmatrix} .
$$

We then just need to take the dot product of this vector with its Hermitian conjugate to determine the mass terms in the Lagrangian. We then find

$$
\mathcal{L}_{m^2} = \begin{pmatrix} \frac{v}{\sqrt{2}} \frac{g_W}{2} \left(A^{\mu\,1} - iA^{\mu\,2} \right) \\ \frac{v}{\sqrt{2}} \left(-\frac{g_W}{2} A^{\mu\,3} + \frac{g_Y}{2} B^{\mu} \right) \end{pmatrix}^{\dagger} \begin{pmatrix} \frac{v}{\sqrt{2}} \frac{g_W}{2} \left(A^{\mu\,1} - iA^{\mu\,2} \right) \\ \frac{v}{\sqrt{2}} \left(-\frac{g_W}{2} A^{\mu\,3} + \frac{g_Y}{2} B^{\mu} \right) \end{pmatrix} \tag{11.75}
$$

$$
= \frac{v^2}{2} \frac{g_W^2}{4} \left((A_{\mu}^1)^2 + (A_{\mu}^2)^2 \right) + \frac{v^2}{2} \left(-\frac{g_W}{2} A_{\mu}^3 + \frac{g_Y}{2} B_{\mu} \right)^2 .
$$

Awesome! This then tells us which of the force-carrying bosons has a mass. We call the linear combinations of the weak isospin fields

$$
\frac{A_{\mu}^1 - iA_{\mu}^2}{\sqrt{2}} \equiv W_{\mu}^+ , \qquad\qquad \frac{A_{\mu}^1 + iA_{\mu}^2}{\sqrt{2}} \equiv W_{\mu}^- \tag{11.76}
$$

the positively and negatively charged **W bosons**. They correspond to the charged current that we had identified in the $V - A$ theory. Their mass term in the Lagrangian is

$$
\mathcal{L}_{m^2} \supset \left(\frac{g_W v}{2} \right)^2 W_{\mu}^+ W^{\mu\,-} , \tag{11.77}
$$

with a mass of $m_W = \frac{g_W v}{2}$. The other linear combination of weak isospin and hypercharge gauge fields that gets a non-zero mass is

$$
\frac{g_W}{\sqrt{g_W^2 + g_Y^2}} A_{\mu}^3 - \frac{g_Y}{\sqrt{g_W^2 + g_Y^2}} B_{\mu} \equiv Z_{\mu} , \tag{11.78}
$$

which is called the **Z boson**. It corresponds to the neutral current that we identified as a resonance in $e^+ e^-$ scattering. Its mass term is

$$
\mathcal{L}_{m^2} \supset \frac{1}{2} \frac{v^2}{4} (g_W^2 + g_Y^2) Z_{\mu} Z^{\mu} , \tag{11.79}
$$

with a mass of $m_Z = \frac{v \sqrt{g_W^2 + g_Y^2}}{2}$. The linear combination of fields that does not get a mass via the Higgs mechanism is

$$
\frac{g_Y}{\sqrt{g_W^2 + g_Y^2}} A_{\mu}^3 + \frac{g_W}{\sqrt{g_W^2 + g_Y^2}} B_{\mu} \equiv A_{\mu} , \tag{11.80}
$$

which is called the photon of electromagnetism. Extremely importantly, note that because the photon is still massless, it has a corresponding manifest gauge symmetry, even in this broken theory. This in turn implies that electric charge is (still!) conserved when expanding about the ground state of the potential for the scalar $\vec{\phi}$.

This theory with W, Z, and photon bosons is called the **broken theory**, as the W and Z bosons are massive. Interactions between particles of the Standard Model with these electroweak bosons are still implemented via the covariant derivative (just as in QCD or electromagnetism). However, it is useful to express the covariant derivative in terms of the W, Z, and photons by just inverting their linear combinations. The weak isospin and hypercharge gauge bosons are thus

$$A_\mu^1 = \frac{W_\mu^+ + W_\mu^-}{\sqrt{2}}, \qquad A_\mu^2 = \frac{iW_\mu^+ - iW_\mu^-}{\sqrt{2}}, \tag{11.81}$$

$$A_\mu^3 = \frac{g_W}{\sqrt{g_W^2 + g_Y^2}} Z_\mu + \frac{g_Y}{\sqrt{g_W^2 + g_Y^2}} A_\mu,$$

$$B_\mu = -\frac{g_Y}{\sqrt{g_W^2 + g_Y^2}} Z_\mu + \frac{g_W}{\sqrt{g_W^2 + g_Y^2}} A_\mu.$$

Inserting these expressions into Eq. 11.55, the covariant derivative acting on the scalar doublet $\vec{\phi}$ is then

$$D_\mu = \partial_\mu - i\frac{g_W}{\sqrt{2}} \frac{\sigma_1 + i\sigma_2}{2} W_\mu^+ - i\frac{g_W}{\sqrt{2}} \frac{\sigma_1 - i\sigma_2}{2} W_\mu^- \tag{11.82}$$

$$- i\left(\frac{\sigma_3}{2} \frac{g_W^2}{\sqrt{g_W^2 + g_Y^2}} - \frac{\mathbb{I}}{2} \frac{g_Y^2}{\sqrt{g_W^2 + g_Y^2}}\right) Z_\mu - i\left(\frac{\sigma_3 + \mathbb{I}}{2}\right) \frac{g_W g_Y}{\sqrt{g_W^2 + g_Y^2}} A_\mu.$$

If A_μ is to be the photon, then the coefficient in the covariant derivative must be the electric charge of the particle to which it couples. The fundamental unit of electric charge e is thus

$$e \equiv \frac{g_W g_Y}{\sqrt{g_W^2 + g_Y^2}}. \tag{11.83}$$

The matrix in front of this expression for e in Eq. 11.82 is therefore the charge of $\vec{\phi}$ in units of e. Note that this matrix is

$$\frac{\sigma_3 + \mathbb{I}}{2} = \begin{pmatrix} 1 & 0 \\ 0 & 0 \end{pmatrix}. \tag{11.84}$$

Then, acting this covariant derivative on the doublet $\vec{\phi}$, we find

$$D_\mu \vec{\phi} \supset \left(\partial_\mu - i\left(\frac{\sigma_3 + \mathbb{I}}{2}\right) e A_\mu\right) \begin{pmatrix} \phi^+ \\ \phi^0 \end{pmatrix} = \begin{pmatrix} (\partial_\mu - ieA_\mu)\phi^+ \\ \partial_\mu \phi_0 \end{pmatrix}. \tag{11.85}$$

That is, the complex scalar field ϕ^+ has electric charge $+e$, while the field ϕ_0 is electrically neutral. This justifies the $+$ and 0 notation. For general doublets of SU(2)$_W$ (which include the electron and the electron neutrino, and the up and down quarks), the coupling to the photon is determined by the value of the hypercharge Y for that doublet. For a general doublet, then, the covariant derivative is

$$D_\mu = \partial_\mu - i\frac{g_W}{\sqrt{2}} \frac{\sigma_1 + i\sigma_2}{2} W_\mu^+ - i\frac{g_W}{\sqrt{2}} \frac{\sigma_1 - i\sigma_2}{2} W_\mu^- \tag{11.86}$$

$$- i\left(\frac{\sigma_3}{2} \frac{g_W^2}{\sqrt{g_W^2 + g_Y^2}} - Y\frac{\mathbb{I}}{2} \frac{g_Y^2}{\sqrt{g_W^2 + g_Y^2}}\right) Z_\mu - i\left(\frac{\sigma_3 + Y\mathbb{I}}{2}\right) \frac{g_W g_Y}{\sqrt{g_W^2 + g_Y^2}} A_\mu.$$

The electric charge of the particle in the upper component of an $SU(2)_W$ doublet is $Q_{upper} = \frac{1}{2} + \frac{Y}{2}$, while the charge of the lower component particle is $Q_{lower} = -\frac{1}{2} + \frac{Y}{2}$. We will study many examples of $SU(2)_W$ doublets in Chapter 12.

11.3.4 Four Predictions of the Broken Weak Theory

This electroweak theory and its spontaneously broken gauge symmetry makes a huge number of predictions. In the rest of this chapter, we will review just four of them, and discuss more in later chapters. The power of the electroweak theory is that the interactions of the electroweak force-carrying bosons are defined by three parameters: the couplings g_W and g_Y, and the vev, v. The fact that we will be able to consistently make predictions of four measurable quantities from these three parameters is a detailed test of the theory.

Mass of the Z Boson, m_Z

Numerous times throughout this book we have already seen that the e^+e^- collision cross section exhibits a resonance near a center-of-mass collision energy of 91 GeV. We attribute this resonance to a new particle, the Z boson, which is electrically neutral. According to the PDG, the mass of the Z boson is $m_Z = 91.18$ GeV and the electroweak theory prediction for m_Z from Eq. 11.79 is

$$m_Z = \frac{v\sqrt{g_W^2 + g_Y^2}}{2} = 91.18 \text{ GeV}. \tag{11.87}$$

Value of the Electric Charge, e

The electroweak theory predicts the strength with which the photon couples to electrically charged particles. Within quantum electrodynamics, this is just the fundamental electric charge, e, which is related to the fine structure constant α as

$$\alpha = \frac{e^2}{4\pi} = 0.007297, \tag{11.88}$$

where we have used the value of α from the PDG (truncated to four significant figures). From Eq. 11.83, the electroweak theory predicts that the fundamental electric charge e is

$$e = \frac{g_W g_Y}{\sqrt{g_W^2 + g_Y^2}} = 0.303. \tag{11.89}$$

Mass of the W Boson, m_W

Determining the mass of the W boson is a harder task than for the Z boson. Because the W boson carries electric charge, the process $e^+e^- \rightarrow W^+$, for example, in which electrons and positrons are collided and create a W boson, which subsequently decays, does not exist. So, we have to change our strategy. While we cannot create W bosons singly in

e^+e^- collisions, we can produce pairs of them, if the center-of-mass collision energy E_{cm} is large enough. That is, the process

$$e^+e^- \to W^+W^- \tag{11.90}$$

is completely allowed; for example, the electric charge of the initial and final state is 0. Now, however, when scanning over E_{cm} we won't see a beautiful resonance in the cross section which makes the existence of a particle very obviously known (as we saw for the Z boson). So, what do we see?

Let's work to calculate the cross section for WW production in electron–positron collisions, $\sigma(e^+e^- \to W^+W^-)$. Actually, all we will do here is estimate this cross section; we won't explicitly calculate Feynman diagrams, but this will nevertheless provide us with a concrete prediction. To calculate the cross section, Fermi's Golden Rule tells us that

$$\sigma(e^+e^- \to W^+W^-) = \frac{1}{2E_{cm}^2} \int d\Pi_2 \, |\mathcal{M}(e^+e^- \to W^+W^-)|^2 . \tag{11.91}$$

Here, we have implicitly used many of the results we derived in Chapters 4 and 6. Π_2 is two-body phase space (corresponding to the momentum configurations of the two W bosons) and $\mathcal{M}(e^+e^- \to W^+W^-)$ is the matrix element for this process, that we could calculate with Feynman diagrams. Using the simplification from Example 4.2 of Chapter 4, we are able to express the integral over two-body phase space as a single integral over the scattering angle, θ:

$$\sigma(e^+e^- \to W^+W^-) = \frac{1}{2E_{cm}^2} \frac{2|\vec{p}|}{E_{cm}} \int \frac{d\cos\theta}{16\pi} |\mathcal{M}(e^+e^- \to W^+W^-)|^2 . \tag{11.92}$$

Recall that the scattering angle θ is the angle that, say, the W^+ boson makes with the initial colliding electron. By momentum conservation, the W^+ and W^- bosons have equal and opposite momenta, whose magnitude is given by $|\vec{p}|$.

At this point, to honestly evaluate the cross section, we would need to compute the scattering angle θ dependence in the matrix element $\mathcal{M}(e^+e^- \to W^+W^-)$. However, for our purposes, it will turn out that we don't care about the overall normalization of the cross section. With that in mind, the squared matrix element $|\mathcal{M}(e^+e^- \to W^+W^-)|^2$ is just some function of the scattering angle, θ. Thus, its integral will just be some number. Let's call this number c_{int}, where

$$c_{int} = \int \frac{d\cos\theta}{16\pi} |\mathcal{M}(e^+e^- \to W^+W^-)|^2 . \tag{11.93}$$

c_{int} might have dependence on masses or energy scales, but let's ignore that for now. For on-shell W bosons, their magnitude of momentum $|\vec{p}|$ is just the quadrature difference of their energy E_W and mass m_W:

$$|\vec{p}| = \sqrt{E_W^2 - m_W^2} . \tag{11.94}$$

For center-of-mass collision energy E_{cm}, E_W is just half of E_{cm}. Therefore,

$$|\vec{p}| = \sqrt{\frac{E_{cm}^2}{4} - m_W^2} . \tag{11.95}$$

Using these results, we can write the cross section as

$$\sigma(e^+e^- \to W^+W^-) = \frac{1}{2E_{\text{cm}}^2} \frac{2\sqrt{\frac{E_{\text{cm}}^2}{4} - m_W^2}}{E_{\text{cm}}} c_{\text{int}} \propto \frac{\sqrt{E_{\text{cm}}^2 - 4m_W^2}}{E_{\text{cm}}^3}. \tag{11.96}$$

This provides a concrete prediction for the cross section $\sigma(e^+e^- \to W^+W^-)$ as a function of the center-of-mass energy E_{cm}. For $E_{\text{cm}} < 2m_W$, the cross section is zero, because there isn't enough energy to create two W bosons in the final state. For $E_{\text{cm}} > 2m_W$, the shape of the cross section is determined by Eq. 11.96. The value of E_{cm} at which the cross section is first non-zero is called the **threshold energy**. This cross section can be measured in data and the value of the threshold energy determined, which consequently provides a measurement of the mass of the W boson. Figure 11.3 shows a plot of combined results from all four experiments at LEP (ALEPH, DELPHI, L3, and OPAL) of the $e^+e^- \to W^+W^-$ cross section as a function of E_{cm}. On this plot, we have also plotted the E_{cm} dependence of the cross section as determined by Eq. 11.96 (and multiplied by an appropriate normalization constant) with the W boson mass set to its PDG value,

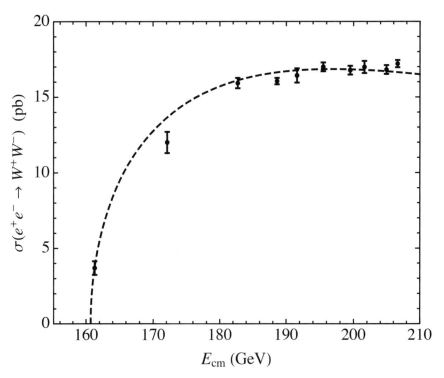

Fig. 11.3 A plot of the cross section in picobarns as a function of the center-of-mass energy E_{cm} for the process $e^+e^- \to W^+W^-$ collected at the four experiments at LEP. The dashed curve is the prediction from Eq. 11.96 with $m_W = 80.38$ GeV. The data are from a table and are reprinted from Phys. Rept. **532**, S. Schael et al. [ALEPH and DELPHI and L3 and OPAL and LEP Electroweak Collaborations], "Electroweak measurements in electron positron collisions at W-boson-pair energies at LEP," 119 (2013), with permission from Elsevier.

$m_W = 80.38$ GeV. The threshold energy is clearly evident, and the prediction of Eq. 11.96 agrees remarkably well with the data. The agreement can be improved by a more detailed analysis, but the fact that this simple calculation agrees so well with data is impressive. This can then be used with the relation derived above to fix another function of the parameters of the electroweak theory:

$$m_W = \frac{g_W v}{2} = 80.38 \text{ GeV} . \tag{11.97}$$

Fermi Constant, G_F

The final prediction we make within the electroweak theory is of the Fermi constant, G_F. In the previous chapter, we calculated the lifetime of the muon in the $V - A$ theory, using the accepted value of G_F. Perhaps a better way to go about it, especially with our current goal, is to measure the lifetime of the muon, and use that measurement to determine the value of G_F. With this in mind, we would like to calculate the lifetime of the muon within the electroweak theory, and then we can identify the relationship between the parameters g_W, g_Y, and v and the value of G_F. The first thing we need to do, though, is to understand how W bosons (that is, the charged current) couple to fermions, like the muon and electron.

As with all gauge bosons, the W bosons couple to fermions through the covariant derivative, so we should see how this works. The first thing we need to do is to determine what the appropriate $SU(2)_W$ doublet is with which the W boson interacts. From the discussion immediately after Eq. 11.86, the upper and lower components of an $SU(2)_W$ doublet have electric charge that differs by one unit. Additionally, they must have the same value of spin, because $SU(2)_W$ is an internal symmetry (i.e., it has nothing to do with the Lorentz group). A natural doublet then is the left-handed electron e_L and its associated neutrino, ν_{eL}. This is called the electron doublet and is written in vector form as

$$\begin{pmatrix} \nu_{eL} \\ e_L \end{pmatrix} .$$

Then, by setting the hypercharge Y of this doublet to be $Y = -1$, we correctly fix the electric charge of the neutrino to be $Q_\nu = 0$, and for the electron $Q_e = -1$ (in units of the fundamental charge, e).

Then, the Lagrangian for this left-handed electron doublet is

$$\mathcal{L} = \begin{pmatrix} \nu_{eL} \\ e_L \end{pmatrix}^\dagger i\bar{\sigma} \cdot D \begin{pmatrix} \nu_{eL} \\ e_L \end{pmatrix} , \tag{11.98}$$

where D_μ is the electroweak covariant derivative, Eq. 11.86, with $Y = -1$. In this expression, the matrix $\bar{\sigma}^\mu$ is the usual Pauli matrix four-vector that acts on the individual spinors e_L and ν_{eL}. As we are interested in their coupling to the W bosons, let's just focus on the W boson contribution to the covariant derivative. We have

$$\mathcal{L} \supset \begin{pmatrix} \nu_{eL} \\ e_L \end{pmatrix}^\dagger i\bar{\sigma}^\mu \left(-i\frac{g_W}{\sqrt{2}}\frac{\sigma_1 + i\sigma_2}{2}W_\mu^+ - i\frac{g_W}{\sqrt{2}}\frac{\sigma_1 - i\sigma_2}{2}W_\mu^- \right) \begin{pmatrix} \nu_{eL} \\ e_L \end{pmatrix} . \tag{11.99}$$

Again, one has to be somewhat careful here with notation: the σ_1 and σ_2 matrices that multiply the W bosons act on the electron doublet as a vector. The matrix with the W bosons is then

$$-i\frac{g_W}{\sqrt{2}}\frac{\sigma_1 + i\sigma_2}{2}W_\mu^+ - i\frac{g_W}{\sqrt{2}}\frac{\sigma_1 - i\sigma_2}{2}W_\mu^- = -i\frac{g_W}{\sqrt{2}}\begin{pmatrix} 0 & W^+ \\ W^- & 0 \end{pmatrix}, \qquad (11.100)$$

using the explicit expressions for the sigma matrices. Performing the matrix multiplication, the interaction of the electron doublet and the W bosons is then

$$\mathcal{L} \supset \frac{g_W}{\sqrt{2}}\nu_{eL}^\dagger\bar{\sigma}\cdot W^+ e_L + \frac{g_W}{\sqrt{2}}e_L^\dagger\bar{\sigma}\cdot W^- \nu_{eL}. \qquad (11.101)$$

The $\bar{\sigma}$ matrices that remain act on the two components of the left-handed spinors. Note that, importantly, both of these terms individually have zero net electric charge. Thus, a W boson turns an electron into an electron neutrino with strength $g_W/\sqrt{2}$.

While we have just considered the left-handed electron doublet, essentially the same analysis goes through for the left-handed muon doublet. The muon, and its associated muon neutrino, couple to the W boson in a similar way as do the electron and its neutrino:

$$\mathcal{L} \supset \frac{g_W}{\sqrt{2}}\nu_{\mu L}^\dagger\bar{\sigma}\cdot W^+ \mu_L + \frac{g_W}{\sqrt{2}}\mu_L^\dagger\bar{\sigma}\cdot W^- \nu_{\mu L}. \qquad (11.102)$$

A prediction of the electroweak force is **lepton universality**: all left-handed lepton doublets (electron, muon, or tau) couple to W bosons with the exact same strength. We'll use these results in the following.

Now that we have identified that the W boson can couple muons, electrons, and their neutrinos, we can work to calculate the muon decay rate in the electroweak theory. The Feynman diagram for this decay, with an intermediate W boson, is

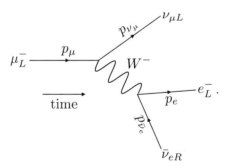

Note that the intermediate, virtual W boson is negatively charged: the charge of the muon flows through the W boson to the final-state electron. To evaluate this Feynman diagram, we know the wavefunctions of the external fermions, and from Eqs. 11.101 and 11.102 we know what the vertices with the W boson are. The remaining thing we need in order to calculate this Feynman diagram is the propagator of the intermediate W boson, $\backslash\!\backslash\!\backslash\!\backslash$. In Section 4.3.2, we argued that the propagator of the massless photon was proportional to $1/q^2$, where q is the momentum that flows through the photon. The W boson, however, is massive, and so its propagator will be different. So, what is it?

First, if the W boson were massless, then its propagator would be the same as that of the photon. That is,

$$\text{(diagram)} \overset{m_W \to 0}{\propto} \frac{1}{q^2}. \tag{11.103}$$

On the other hand, if the momentum flowing through the W boson goes to 0, then it still has a characteristic Compton wavelength, as determined by its non-zero mass, m_W. That is, we expect that

$$\text{(diagram)} \overset{q \to 0}{\propto} \frac{1}{m_W^2}. \tag{11.104}$$

So, in general, we expect that the W boson's propagator has the form

$$\text{(diagram)} \propto \frac{1}{q^2 + \zeta m_W^2}, \tag{11.105}$$

for some proportionality constant ζ.

To determine ζ, let's consider the case when q^2 is close to m_W^2. In that case, because $q^2 > 0$, we can go to a frame in which $q^2 = E^2$, where E is the energy of the intermediate W boson. Further, let's write

$$E = m_W + \delta E, \tag{11.106}$$

for some small energy δE, where $|\delta E| \ll m_W$. With this notation, the propagator is then

$$\frac{1}{q^2 + \zeta m_W^2} = \frac{1}{(m_W + \delta E)^2 + \zeta m_W^2} = \frac{1}{(1 + \zeta)m_W^2 + 2\delta E m_W + \delta E^2}. \tag{11.107}$$

As $\delta E \to 0$, the invariant mass of the momentum flowing through the propagator approaches m_W, and we expect that the W boson can then go on-shell. Another way to say this is from the energy–time uncertainty principle. As $\delta E \to 0$, the amount of time for which the intermediate W boson can exist approaches infinity. So, in the $\delta E \to 0$ limit, the propagator should diverge. This occurs if $\zeta = -1$, and so we finally have

$$\text{(diagram)} \propto \frac{1}{q^2 - m_W^2} \tag{11.108}$$

for the W boson propagator.

Okay, now we have everything we need. Let's evaluate the Feynman diagram for muon decay in the electroweak theory. Getting straight to the answer, we find

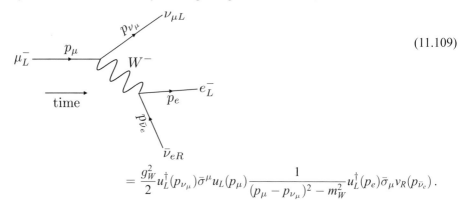

$$= \frac{g_W^2}{2} u_L^\dagger(p_{\nu_\mu}) \bar{\sigma}^\mu u_L(p_\mu) \frac{1}{(p_\mu - p_{\nu_\mu})^2 - m_W^2} u_L^\dagger(p_e) \bar{\sigma}_\mu v_R(p_{\bar{\nu}_e}) .$$

(11.109)

Note that the momentum flowing through the intermediate W boson propagator is $q = p_\mu - p_{\nu_\mu}$. We would like to compare this result to the corresponding calculation in the $V - A$ theory, Eq. 10.43. However, there's still some simplification we can do yet. Let's evaluate the $(p_\mu - p_{\nu_\mu})^2$ that appears in the propagator. We have

$$(p_\mu - p_{\nu_\mu})^2 = m_\mu^2 - 2 p_\mu \cdot p_{\nu_\mu} = m_\mu^2 (1 - x_{\nu_\mu}) ,$$

(11.110)

where x_{ν_μ} is the corresponding three-body phase space variable for the muon neutrino. This invariant mass is maximized when $x_{\nu_\mu} = 0$, but even in that case, we still have that $m_W^2 \gg m_\mu^2$. Actually, the ratio of the squared W boson mass to the squared muon mass is almost a factor of 10^6! So, to extremely good approximation, we can just ignore the mass of the muon in the W boson propagator. With this approximation, the Feynman diagram then becomes

$$\frac{g_W^2}{2} u_L^\dagger(p_{\nu_\mu}) \bar{\sigma}^\mu u_L(p_\mu) \frac{1}{(p_\mu - p_{\nu_\mu})^2 - m_W^2} u_L^\dagger(p_e) \bar{\sigma}_\mu v_R(p_{\bar{\nu}_e})$$

(11.111)

$$\xrightarrow{m_\mu \ll m_W} -\frac{g_W^2}{2 m_W^2} \left(u_L^\dagger(p_{\nu_\mu}) \bar{\sigma}^\mu u_L(p_\mu) \right) \left(u_L^\dagger(p_e) \bar{\sigma}_\mu v_R(p_{\bar{\nu}_e}) \right) .$$

Up to an irrelevant overall minus sign (which doesn't affect the decay rate), this has the same form as the prediction in the $V - A$ theory. Therefore, if these are to agree, we must have the relation

$$\frac{4 G_F}{\sqrt{2}} = \frac{g_W^2}{2 m_W^2} .$$

(11.112)

We say that the $V - A$ theory is a **low-energy effective theory** of the electroweak theory. That is, when the energy or momentum that flows through W boson propagators is small compared to the W boson mass, the predictions of the electroweak theory reduce to those of the $V - A$ theory. So, finally, we have a prediction of the Fermi constant within the electroweak theory:

$$G_F = \frac{\sqrt{2} g_W^2}{8 m_W^2} = \frac{1}{\sqrt{2} v^2} = 1.166 \times 10^{-5} \text{ GeV}^{-2} .$$

(11.113)

Historical Profile: Penguin Diagrams

As we saw in this section, *W* bosons can turn one type of particle into another (for example, an electron into a neutrino). This is a very important observation, and in the 1970s several groups exploited this property to understand the decays of hadrons.[13] In particular, a type of Feynman diagram called a **penguin diagram** was coined that was responsible for these decays. An example of a penguin diagram (which if you squint, looks like a penguin) is:[14]

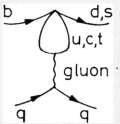

As John Ellis recounted to Mikhail Shifman, the origin of the term "penguin" goes as follows:[15]

> That summer [1977], there was a student at CERN, Melissa Franklin, who is now an experimentalist at Harvard. One evening, she, I and Serge [Rudaz] went to a pub, and she and I started a game of darts. We made a bet that if I lost I had to put the word penguin into my next paper. She actually left the darts game before the end, and was replaced by Serge, who beat me. Nevertheless, I felt obligated to carry out the conditions of the bet.
>
> For some time, it was not clear to me how to get the word into this *b* quark paper that we were writing at the time. Then, one evening, after working at CERN, I stopped on my way back to my apartment to visit some friends living in Meyrin where I smoked some illegal substance. Later, when I got back to my apartment and continued working on our paper, I had a sudden flash that the famous diagrams look like penguins. So we put the name into our paper, and the rest, as they say, is history.

Franklin and two of her fellow students from graduate school, Patricia Burchat at Stanford and Frances Hellman at UC Berkeley, were coincidentally all chairs of their respective physics departments. A fourth fellow student, Mary James, is a professor at Reed College.

Summary

With these four predictions, we can then test the ability of the electroweak theory to consistently describe them. From the first three predictions, the *Z* boson mass, the electric

[13] J. R. Ellis, M. K. Gaillard, D. V. Nanopoulos and S. Rudaz, "The phenomenology of the next left-handed quarks," Nucl. Phys. B **131**, 285 (1977), Erratum: [Nucl. Phys. B **132**, 541 (1978)]. This was studied behind the Iron Curtain a few years earlier. See A. I. Vainshtein, V. I. Zakharov and M. A. Shifman, "A possible mechanism for the $\Delta T = 1/2$ rule in nonleptonic decays of strange particles," JETP Lett. **22**, 55 (1975) [Pisma Zh. Eksp. Teor. Fiz. **22**, 123 (1975)].

[14] Reprinted from Nucl. Phys. B **131**, J. R. Ellis, M. K. Gaillard, D. V. Nanopoulos and S. Rudaz, "The phenomenology of the next left-handed quarks," 285 (1977) Erratum: [Nucl. Phys. B **132**, 541 (1978)], with permission from Elsevier.

[15] Reprinted with permission from "ITEP Lectures on Particle Physics and Field Theory," editor M. A. Shifman, Copyright 1999 World Scientific.

charge, and the W boson mass, we can solve for the three parameters of the electroweak theory:

$$g_Y = \frac{m_Z}{m_W} e = 0.344 \,, \tag{11.114}$$

$$g_W = \frac{m_Z}{\sqrt{m_Z^2 - m_W^2}} e = 0.642 \,, \tag{11.115}$$

$$v = 2 \frac{m_W}{e m_Z} \sqrt{m_Z^2 - m_W^2} = 250.5 \text{ GeV} \,. \tag{11.116}$$

In these expressions, we use the values of m_Z, m_W, and e from the PDG. At this point, this is not a test of the theory: we use three predictions to determine three parameters. However, from these parameters, we can then predict the value of the Fermi constant, G_F. From the value of the vev determined above, we predict G_F to be

$$G_F = \frac{1}{\sqrt{2} v^2} = \frac{1}{\sqrt{2}} (250.5)^{-2} = 1.127 \times 10^{-5} \text{ GeV}^{-2} \,. \tag{11.117}$$

This is within about 3% of the accepted value from the PDG of $G_F = 1.166 \times 10^{-5} \text{ GeV}^{-2}$! This is an extremely non-trivial result and gives us confidence that the electroweak theory is indeed the correct description of these phenomena. The disagreement between the prediction and the measured value of G_F (or any of the other parameters) can be reduced by correctly accounting for additional quantum mechanical effects in the electroweak theory.

By the way, the mixing between the A_μ^3 and B_μ bosons to produce the Z_μ and A_μ bosons is typically characterized by the sine of the **weak mixing angle**:

$$\sin^2 \theta_W \equiv \frac{g_Y^2}{g_W^2 + g_Y^2} = 0.23122 \pm 4 \times 10^{-5} \,. \tag{11.118}$$

Example 11.2 The top quark is the most massive particle of the Standard Model, and its decay is mediated by the W boson. Almost exclusively, the top quark decays to a bottom quark and a W boson:

$$t \to b + W^+ \,. \tag{11.119}$$

Because top quarks and bottom quarks are coupled to one another through a W boson, they form a doublet of $SU(2)_W$, just like the electron and electron neutrino. In the next chapter, we will discuss quark and lepton doublets in much more detail, but all we need here is this observation. In this example, we will study the top quark decay and compare to data.

Solution

The W boson is itself an unstable particle, and so decays to lighter particles. The W boson can be identified through its decay to a charged lepton (such as an electron or a muon) and a neutrino: $W^+ \to l^+ + \nu$. So, the full decay of a top quark as observed in an experiment would be:

$$t \to b + l^+ + \nu \,. \tag{11.120}$$

The Feynman diagram for this decay is basically identical to that for muon decay; we just need to relabel the particles:

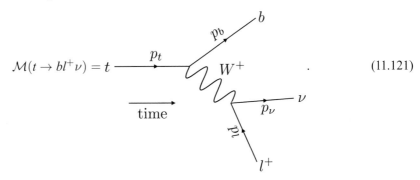

$$\mathcal{M}(t \rightarrow bl^+\nu) = \qquad\qquad\qquad\qquad\qquad\qquad\qquad \tag{11.121}$$

Correspondingly, the matrix element for this decay is essentially just the relabeling of that for the muon in Eq. 11.109:

$$\mathcal{M}(t \rightarrow bl^+\nu) = \frac{g_W^2}{2} \frac{\left(u_L^\dagger(p_b)\bar{\sigma}^\mu u_L(p_t)\right)\left(u_L^\dagger(p_\nu)\bar{\sigma}_\mu v_R(p_l)\right)}{(p_t - p_b)^2 - m_W^2 + im_W\Gamma_W}. \tag{11.122}$$

In the denominator of the propagator, we have included the width of the W boson, Γ_W. This term wasn't relevant for the muon decay, because the momentum flowing through the propagator was so much smaller than the mass of the W boson itself. However, the top quark has a larger mass than the W boson, so it is possible for the W boson to be produced on-shell. The finite decay width of the W boson regulates the divergence of the propagator as the W boson goes on-shell, effectively forcing the W boson to exist only for a finite time.

Further, we will make the **narrow width approximation**, in which we assume that the decay width of the W boson is much smaller than its mass:

$$\Gamma_W \ll m_W. \tag{11.123}$$

This is a good approximation, as from the PDG, the mass of the W boson is $m_W = 80.379$ GeV while its width is only $\Gamma_W = 2.085$ GeV. In this approximation, the matrix element has most of its support when the W boson is on-shell, corresponding to

$$p_W^2 = (p_t - p_b)^2 = m_W^2. \tag{11.124}$$

We then simplify the matrix element with this approximation:

$$\mathcal{M}(t \rightarrow bl^+\nu) = \frac{-ig_W^2}{2m_W\Gamma_W}\left(u_L^\dagger(p_b)\bar{\sigma}^\mu u_L(p_t)\right)\left(u_L^\dagger(p_\nu)\bar{\sigma}_\mu v_R(p_l)\right). \tag{11.125}$$

Now, we just need to evaluate some spinor products.

The spinor products that remain are effectively the same as those for muon decay studied in Section 10.4.1, so we'll just translate those results to decay of the top quark. We have the Fierz identity:

$$\left(u_L^\dagger(p_b)\bar{\sigma}^\mu u_L(p_t)\right)\left(u_L^\dagger(p_\nu)\bar{\sigma}_\mu v_R(p_l)\right) = -2\left(u_L^\dagger(p_b)u_R(p_\nu)\right)\left(v_L^\dagger(p_l)u_L(p_t)\right). \tag{11.126}$$

The squared matrix element is thus

$$|\mathcal{M}(t \to bl^+\nu)|^2 = \frac{g_W^4}{m_W^2 \Gamma_W^2} |u_L^\dagger(p_b) u_R(p_\nu)|^2 |v_L^\dagger(p_l) u_L(p_t)|^2 . \tag{11.127}$$

The mass of the top quark is about 175 GeV, while the mass of the bottom quark is about 4 GeV. The leptons are less massive yet, so, to very good approximation, we can assume that only the top quark has a mass. The squared spinor inner products are then

$$|u_L^\dagger(p_b) u_R(p_\nu)|^2 = 2p_b \cdot p_\nu , \qquad\qquad |v_L^\dagger(p_l) u_L(p_t)|^2 = p_t \cdot p_l . \tag{11.128}$$

The squared matrix element is therefore

$$|\mathcal{M}(t \to bl^+\nu)|^2 = \frac{2g_W^4}{m_W^2 \Gamma_W^2} (p_b \cdot p_\nu)(p_t \cdot p_l) . \tag{11.129}$$

To calculate the decay rate of the top quark Γ_t, we need to put this expression into Fermi's Golden Rule for decays. As developed in the calculation of the muon decay, it is most convenient to express the matrix element in terms of the three-body phase space variables x_i, where

$$x_i = \frac{2p_t \cdot p_i}{m_t^2} , \tag{11.130}$$

for $i = b, l, \nu$. Note that as $p_t = p_b + p_l + p_\nu$, the sum of the x_i variables is 2. So, there are actually only two independent x_i variables. However, the situation is even simpler than that. Because we used the narrow width approximation, we have set

$$m_W^2 = (p_l + p_\nu)^2 = 2p_l \cdot p_\nu = m_t^2 - 2p_t \cdot p_b = m_t^2(1 - x_b) . \tag{11.131}$$

Therefore,

$$x_b = 1 - \frac{m_W^2}{m_t^2} . \tag{11.132}$$

So, there is only one independent phase space variable.

In our experiment, it is essentially impossible to observe neutrinos, so we can't directly measure x_ν. It is therefore natural to take the invariant mass of the visible lepton and bottom quark m_{lb} as the one phase space variable. We define m_{lb} to be

$$m_{lb} = \sqrt{2p_b \cdot p_l} = m_t \sqrt{1 - x_\nu} . \tag{11.133}$$

Not surprisingly, because the total invariant mass of the final state is constrained to be equal to the mass of the top quark, the neutrino energy fraction controls the invariant mass of the lepton and bottom quark. We can now insert all of this into Fermi's Golden Rule. Translating Eq. 10.71 to the top quark decay, we have

$$\Gamma_t = \frac{1}{2m_t} \frac{m_t^2}{128\pi^3} \int_0^1 dx_b \int_0^1 dx_l \, \Theta(x_b + x_l - 1) \, |\mathcal{M}(t \to bl^+\nu)|^2 \tag{11.134}$$

$$= \frac{1}{512\pi^3} \frac{g_W^4 m_t^5}{m_W^2 \Gamma_W^2} \int_0^1 dx_b \int_0^1 dx_l \, \Theta(x_b + x_l - 1)(1 - x_l) x_l \, \delta\left(x_b - \left(1 - \frac{m_W^2}{m_t^2}\right)\right) .$$

In this expression, we have included a δ-function that enforces the intermediate W boson to be on-shell, as specified by Eq. 11.132. Using this, we can integrate over x_b:

$$\Gamma_t = \frac{1}{512\pi^3} \frac{g_W^4 m_t^5}{m_W^2 \Gamma_W^2} \int_{\frac{m_W^2}{m_t^2}}^1 dx_l \, (1 - x_l) x_l \,.$$

Integrating this over x_l is straightforward, but we will instead construct the decay rate differential in m_{lb}. This requires re-expressing m_{lb} in terms of x_l. We have

$$m_{lb} = m_t \sqrt{1 - x_\nu} = m_t \sqrt{x_b + x_l - 1} = m_t \sqrt{x_l - \frac{m_W^2}{m_t^2}} \,. \tag{11.135}$$

So, the differential decay rate is found by including a δ-function that enforces this relationship:

$$\frac{d\Gamma_t}{dm_{lb}} = \frac{1}{512\pi^3} \frac{g_W^4 m_t^5}{m_W^2 \Gamma_W^2} \int_{\frac{m_W^2}{m_t^2}}^1 dx_l \, (1 - x_l) x_l \, \delta\left(m_{lb} - m_t \sqrt{x_l - \frac{m_W^2}{m_t^2}}\right) \,. \tag{11.136}$$

By a change of variables, we can re-express the δ-function to be linear in x_l:

$$\delta\left(m_{lb} - m_t \sqrt{x_l - \frac{m_W^2}{m_t^2}}\right) = \frac{2m_{lb}}{m_t^2} \delta\left(x_l - \frac{m_{lb}^2 + m_W^2}{m_t^2}\right) \,. \tag{11.137}$$

This follows from the δ-function identity

$$\delta(y - f(x)) = \frac{1}{\left|\frac{df}{dx}\right|_{x=f^{-1}(y)}} \delta(x - f^{-1}(y)) \,. \tag{11.138}$$

Finally, we can integrate using the δ-function and find that the differential decay rate is

$$\frac{d\Gamma_t}{dm_{lb}} = \frac{1}{256\pi^3} \frac{g_W^4 m_t^3 m_{lb}}{m_W^2 \Gamma_W^2} \int_{\frac{m_W^2}{m_t^2}}^1 dx_l \, (1 - x_l) x_l \, \delta\left(x_l - \frac{m_{lb}^2 + m_W^2}{m_t^2}\right) \tag{11.139}$$

$$= \frac{1}{256\pi^3} \frac{g_W^4 m_t^3}{m_W^2 \Gamma_W^2} m_{lb} \frac{m_W^2 + m_{lb}^2}{m_t^2} \left(1 - \frac{m_W^2 + m_{lb}^2}{m_t^2}\right) \Theta\left(\sqrt{m_t^2 - m_W^2} - m_{lb}\right) \,.$$

The distribution of m_{lb} from top quark decays can be measured in data and compared to this prediction of the weak force. In Exercise 2.9 of Chapter 2, we studied just the endpoint of this distribution by enforcing relativistic energy and momentum conservation. Now, we actually have the whole distribution in hand and so our comparison with data is more meaningful. A plot of the measurement of m_{lb} from top quarks produced in the CMS experiment is (re-)presented in Fig. 11.4. This plot (and others) were used by CMS to measure the mass of the top quark, by comparing with distributions like that which we derived in Eq. 11.139. Here, we note three features of our prediction and the data, and in Exercise 11.8, you will test the left-handed coupling nature of the weak force with these data. In this comparison, we will just use the approximate measured values of the top and W boson masses of $m_t = 173$ GeV and $m_W = 80$ GeV, respectively.

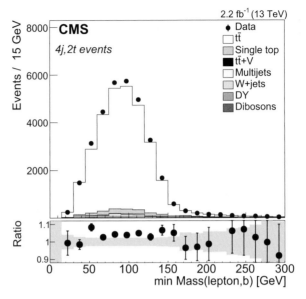

Fig. 11.4   Distribution of the invariant mass of the bottom quark and the charged lepton from the decay of a top quark produced in proton collisions at the CMS experiment. From A. M. Sirunyan *et al.* [CMS Collaboration], "Measurement of the *tt̄* production cross section using events with one lepton and at least one jet in pp collisions at $\sqrt{s} = 13$ TeV," J. High Energy Phys. **1709**, 051 (2017) [arXiv:1701.06228 [hep-ex]].

1 First, note that, as $m_{lb} \to 0$, both our prediction and the data go to 0. That is, there is zero probability for the momenta of the bottom quark and charged lepton to be exactly collinear. However, such a zero probability is called a **Jacobian zero**, because it does not arise from physical restrictions. To see this, we can make a change of variables to a form of the differential decay rate that does not vanish as $m_{lb} \to 0$ (hence the term "Jacobian zero"). The decay rate differential in m_{lb}^2, rather than in m_{lb}, is

$$\frac{d\Gamma_t}{dm_{lb}^2} = \frac{1}{512\pi^3} \frac{g_W^4 m_t^3}{m_W^2 \Gamma_W^2} \frac{m_W^2 + m_{lb}^2}{m_t^2} \left(1 - \frac{m_W^2 + m_{lb}^2}{m_t^2}\right) \Theta\left(m_t^2 - m_W^2 - m_{lb}^2\right).$$

This is non-zero as $m_{lb} \to 0$, as expected. m_{lb}^2 is linear in the three-body phase space variables x_i, so in some sense it's the more natural variable than m_{lb}.

2 The endpoint of this distribution is enforced by the Θ-function, at which

$$m_{lb} = \sqrt{m_t^2 - m_W^2} \simeq 153 \text{ GeV}. \tag{11.140}$$

At this point, the differential decay rate is 0, for physical reasons. At the endpoint, the bottom quark and neutrino momenta are collinear. Additionally, all final-state fermions must be left-handed (or right-handed anti-particles), but this then requires the top quark to have a total spin 3/2! This configuration is:

This isn't possible as it does not conserve angular momentum, so the probability must be 0. An endpoint of 153 GeV is also well represented in the data.

3 Additionally, this distribution has a peak, which we can find by differentiating with respect to m_{lb} and setting it equal to 0. Doing this, we find that the peak is located at

$$m_{lb}^{\text{peak}} = \sqrt{\frac{3m_t^2 - 6m_W^2 + \sqrt{9m_t^4 - 16m_t^2 m_W + 16m_W^4}}{10}} \simeq 113 \text{ GeV}. \qquad (11.141)$$

This peak location is very close to that observed in data, where it appears that the maximum value of the distribution occurs in the bin with masses in $m_{lb} \in [90, 105]$ GeV.

Exercises

11.1 *Maxwell with a Massive Photon.* How badly does electromagnetism break if the photon has a non-zero mass? As we discussed in this chapter, if the photon mass does not come from the Higgs mechanism, then the electromagnetic Lagrangian is not gauge invariant. You may want to consult Section 2.2.3 for electromagnetism with 0 photon mass.

(a) Starting from the massive photon Lagrangian coupled to a current J_μ,

$$\mathcal{L}_{m \neq 0} = -\frac{1}{4} F_{\mu\nu} F^{\mu\nu} - J_\mu A^\mu + \frac{m^2}{2} A_\mu A^\mu, \qquad (11.142)$$

derive the corresponding Maxwell's equations. Which of Maxwell's equations are modified by a non-zero photon mass? Which are unchanged?

(b) Perform a gauge transformation of the action from this Lagrangian. What differential equation must the current J_μ satisfy now? Is this a conservation law? That is, is electric charge conserved with a non-zero photon mass?

11.2 *Scalar Higgs.* Explain why, in a Lorentz-invariant universe, the Higgs particle must be a spin-0 boson.

11.3 *Charge 0 Higgs.* Why do we take the vev to be in the neutral, charge-0 component of the scalar doublet $\vec{\phi}$? Correspondingly, in our universe, why must the Higgs boson be electrically neutral?

11.4 *Forms of SU(2) Matrices.* We can equivalently express an element $\mathbb{U} \in SU(2)$ as

$$\mathbb{U} = \begin{pmatrix} e^{i\chi_1} \cos\psi & -e^{i\chi_2} \sin\psi \\ e^{-i\chi_2} \sin\psi & e^{-i\chi_1} \cos\psi \end{pmatrix} = e^{i\vec{\alpha} \cdot \frac{\vec{\sigma}}{2}}, \qquad (11.143)$$

for real numbers ψ, χ_1, χ_2 and $\alpha_1, \alpha_2, \alpha_3$. The σ-matrices are the usual Pauli matrices. Express $\alpha_1, \alpha_2,$ and α_3 in terms of $\psi, \chi_1,$ and χ_2.

Hint: To do this, Taylor expand both forms of the matrix $\mathbb{U}$ to linear order in the parameters.

11.5 *Unification of Couplings.* The Standard Model consists of three gauge groups that collectively describe the strong and unified electroweak forces. As we discussed in Chapter 8 for QCD and electromagnetism, the couplings of the electroweak theory, g_Y and g_W, also have energy dependence; that is, they are also running couplings. At energies above the top quark mass, the three β-functions of the couplings of the Standard Model that determine their dependence on energy scale Q are

$$Q\frac{d\alpha_1}{dQ} = \frac{41}{10}\frac{\alpha_1^2}{2\pi}, \qquad Q\frac{d\alpha_2}{dQ} = -\frac{19}{6}\frac{\alpha_2^2}{2\pi}, \qquad Q\frac{d\alpha_3}{dQ} = -7\frac{\alpha_3^2}{2\pi}. \qquad (11.144)$$

Here, we use the standard notation that $\alpha_3 = \alpha_s$ and

$$\alpha_1 \equiv \frac{g_Y^2}{4\pi}, \qquad \alpha_2 \equiv \frac{g_W^2}{4\pi}. \qquad (11.145)$$

Note that two of these couplings, α_2 and α_3, are asymptotically free: their β-functions are negative. α_1, on the other hand, has a positive β-function. This will have interesting consequences when these couplings are compared at high energies.

(a) Solve the β-function differential equations for the couplings in terms of their value when $Q = m_Z$, the mass of the Z boson.

Hint: You might find the discussion in Section 8.3.4 helpful.

(b) At $Q = m_Z$, the values of the couplings are

$$\alpha_1(m_Z) = 9.49 \times 10^{-3}, \qquad \alpha_2(m_Z) = 3.16 \times 10^{-2}, \qquad \alpha_3(m_Z) = 0.118. \qquad (11.146)$$

With these coupling values at $Q = m_Z$ and the values of the β-functions, there exist energies at which each pair of couplings have the same value. Find the three energies Q_{12}, Q_{13}, and Q_{23} at which

$$\alpha_1(Q_{12}) = \alpha_2(Q_{12}), \qquad \alpha_1(Q_{13}) = \alpha_3(Q_{13}), \qquad \alpha_2(Q_{23}) = \alpha_3(Q_{23}). \qquad (11.147)$$

How do these three energies compare? Does this hint at a deeper underlying structure, waiting to be discovered? Remember, the mass of the Z boson is about $m_Z = 91$ GeV.

11.6 *When V − A and When Electroweak?* In Exercise 10.7 of Chapter 10, we studied the IceCube experiment, which detects high-energy neutrinos through their interactions with protons or neutrons in the nuclei of water molecules. In the electroweak theory, a Feynman diagram that represents such a scattering would be

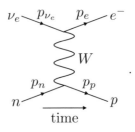

Therefore, the squared momentum that flows through the W boson propagator is just the Mandelstam variable t:

$$t = (p_\nu - p_e)^2. \qquad (11.148)$$

When $|t| \ll m_W^2$, this scattering process is well described by the $V - A$ theory, while when $|t| \gtrsim m_W^2$, the full electroweak theory is necessary. To what neutrino energies do these limits correspond? Throughout this problem, you can assume that the neutrino and electron are massless, and that the proton and neutron are massive and have that the same mass, m_N.

(a) What is the minimum possible value of $|t|$?
(b) In Exercise 10.7, you derived the energy of the anti-muon in the process $\bar{\nu}_\mu + p \to \mu^+ + n$. The same expression can be used for the electron energy here, namely,

$$E_e = \frac{E_\nu m_N}{m_N + E_\nu(1 - \cos\theta)}, \qquad (11.149)$$

where E_ν is the energy of the initial neutrino and θ is the scattering angle. This expression is the energy of the electron in the lab frame in which the neutron is at rest. The scattering angle is the angle between the electron and neutrino momenta. From this expression, determine the scattering angle at which $|t| = m_W^2$. Show that

$$\cos\theta = 1 - \frac{m_W^2 m_N}{2m_N E_\nu^2 - m_W^2 E_\nu}. \qquad (11.150)$$

(c) At what neutrino energy E_ν would you begin to need to account for the full electroweak theory? IceCube has observed neutrino scattering events in which the scattered lepton has over 1 PeV ($= 10^6$ GeV) of energy. In these events do they need to account for the electroweak theory?

11.7 *Charged Current DIS.* In Chapter 7, we studied deeply inelastic scattering, or DIS, in which high-energy electrons were scattered off of protons. At sufficiently high energies, the electron interacted with quarks inside the protons through the electromagnetic force and their properties could be inferred from the scattering angle and energy of the electron. DIS via electromagnetism isn't the only way to probe the inside of the proton; we can also study quarks through the weak force. We still collide quarks off of protons, but now a neutrino is produced in the final state, so that a W boson is exchanged. A pseudo-Feynman diagram of such a process is:

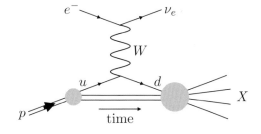

This represents the electron emitting a W boson interacting with an up quark in the proton, turning it into a down quark in the final state. This process is called charged current DIS, as the interaction is mediated by the W boson (the charged current). In this exercise, we will study this charged current DIS and compare a calculation to data.

(a) From this diagram, calculate the cross section for this process differential in Q^2, the invariant mass of the W boson propagator, and x, the momentum fraction of the up quark in the proton. You should find

$$\frac{d^2\sigma(e^- p \to \nu_e + X)}{dx\, dQ^2} = \frac{g_W^4}{4\pi} \frac{1}{(Q^2 + m_W^2)^2} f_u(x) \,. \tag{11.151}$$

Your calculation should parallel what we did in Section 7.2. $f_u(x)$ is the up quark's parton distribution function. Don't forget that the W boson couples exclusively to left-handed fermions.

(b) This process was studied by the H1 detector at HERA, located at DESY. The differential cross section in x at several values of Q^2 as measured at H1 is shown in Fig. 11.5. Do these data seem to exhibit Bjorken scaling?

(c) Using the results presented in Fig. 11.5 and the differential cross section in Eq. 11.151, estimate the mass of the W boson, m_W.

   Hint: It might be helpful to express Q^2 as a fraction of m_W^2: $Q^2 = \chi m_W^2$, for some number χ.

11.8 *Left-Handed Coupling of the W Boson?* In Example 11.2, we calculated the distribution of the invariant mass of the bottom quark and charged lepton from the decay of a top quark. This calculation was done in the context of the weak force, with a W boson that couples exclusively to left-handed particles. We found good agreement between our calculation and data from CMS presented in Fig. 11.4. However, is this actually evidence that the W boson only couples to left-handed particles?

In this exercise, we test this hypothesis by calculating the top quark decay rate with a hypothetical W boson that couples to left- and right-handed particles with equal strength. That is, this W boson is like a massive version of the photon. We then compare this prediction to data and ask whether the data are better described by the electroweak W boson or by this hypothetical W boson.

(a) Unlike in the electroweak theory, there are now multiple spin configurations of the top quark and its decay products that we must consider. We still assume that the neutrino is exclusively a left-handed particle, so there are two matrix elements we must consider:

$$\mathcal{M}(t_L \to b_L l_R^+ \nu_L) = \frac{g_W^2}{2} \frac{\left(u_L^\dagger(p_b)\bar{\sigma}^\mu u_L(p_t)\right)\left(u_L^\dagger(p_\nu)\bar{\sigma}_\mu v_R(p_l)\right)}{(p_t - p_b)^2 - m_W^2 + i m_W \Gamma_W} \,, \tag{11.152}$$

$$\mathcal{M}(t_R \to b_R l_R^+ \nu_L) = \frac{g_W^2}{2} \frac{\left(u_R^\dagger(p_b)\sigma^\mu u_R(p_t)\right)\left(u_L^\dagger(p_\nu)\bar{\sigma}_\mu v_R(p_l)\right)}{(p_t - p_b)^2 - m_W^2 + i m_W \Gamma_W} \,.$$

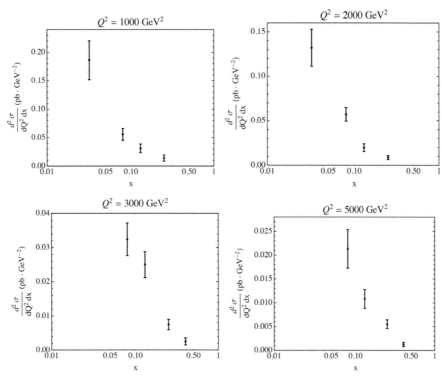

Fig. 11.5 Plots of the cross section differential in momentum fraction x for charged current DIS at several values of Q^2. The double differential cross section is measured in units of pb $\cdot$ GeV^{-2} and the error bars account for statistical and systematic uncertainties. The data come from C. Adloff *et al.* [H1 Collaboration], "Measurement of neutral and charged current cross-sections in electron–proton collisions at high Q^2," Eur. Phys. J. C **19**, 269 (2001) [arXiv:hep-ex/0012052].

We're familiar with how to evaluate the first matrix element, $\mathcal{M}(t_L \rightarrow b_L l_R^+ \nu_L)$, via Fierz identities. The second matrix element, however, contains the matrix product $\sigma^\mu \bar{\sigma}_\mu$, which we haven't manipulated with Fierz identities yet. Prove that

$$\sigma_{ab}^\mu \bar{\sigma}_{\mu\, cd} = 2\delta_{ad}\delta_{bc} \,. \tag{11.153}$$

In this expression, the indices a, b, c, and d define the rows and columns of the Pauli matrices.

Hint: You might want to consult the calculation of the muon decay in Section 10.4.1.

(b) As we did in Example 11.2, we work in the narrow width approximation, setting $(p_t - p_b)^2 = m_W^2$. With this approximation and the result from part (a), square the matrix elements and evaluate all spinor products. For the matrix element of the decay of the right-handed top quark, you should find

$$|\mathcal{M}(t_R \rightarrow b_R l_R^+ \nu_L)|^2 = \frac{2g_W^4}{m_W^2 \Gamma_W^2}(p_b \cdot p_l)(p_t \cdot p_\nu) \,. \tag{11.154}$$

(c) In the calculation of the decay rate Γ_t, we need to average over initial top quark spins. That is, the squared matrix element that we insert into Fermi's Golden Rule is

$$\frac{1}{2}\left(|\mathcal{M}(t_L \to b_L l_R^+ \nu_L)|^2 + |\mathcal{M}(t_R \to b_R l_R^+ \nu_L)|^2\right).\qquad (11.155)$$

Using this average, compute the decay rate differential in the bottom–quark–charged–lepton invariant mass m_{lb}:

$$\frac{d\Gamma_t}{dm_{lb}}.\qquad (11.156)$$

This is then a concrete prediction for this distribution with the hypothesis that the W boson decays to left- and right-handed particles equally.

Hint: It will probably be helpful to follow the steps of the calculation in Example 11.2.

(d) Compare this prediction to the data presented in Fig. 11.4. In particular, is the decay rate 0 at the endpoint of $m_{lb} \simeq 153$ GeV? Explain why or why not by possible spin configurations at the endpoint.

(e) Where is the peak of this distribution? How does it compare to the electroweak prediction in Eq. 11.141?

(f) So, which model for the interactions of the W boson more accurately describes the data, the purely left-handed couplings or equal left- and right-handed couplings?

11.9 *Research Problem.* By postulating the scalar potential in Eq. 11.57, we are able to spontaneously break the $SU(2)_W \otimes U(1)_Y$ symmetry to just electromagnetism. Where does this scalar potential come from?

We successfully resolved the outstanding and exceptionally confusing issue of giving the spin-1 force carriers of the weak force masses. This was accomplished by the Higgs mechanism, where a spin-0 particle, the Higgs, acquires a vacuum expectation value which spontaneously breaks the gauge symmetry and gives the force carrier gauge bosons a mass. We demonstrated that the simplest theory for the unified electroweak theory is an $SU(2)_W \otimes U(1)_Y$ gauge theory in which its parameters are overconstrained by experimental measurements. The fact that there are consistent values for the electroweak theory parameters is a huge success, and strong evidence for it as the correct description of the weak force.

Fermions, for the most part, were not involved in the discussion of the previous chapter. We can just add masses for fermions with no issue; the Dirac equation happily describes both massive and massless fermions. We don't need a Higgs mechanism for fermions. However, the weak force does something that we never encountered with electromagnetism or QCD. The weak force not only treats left- and right-handed particles differently, it also associates different types of fermions. For example, the left-handed electron and electron neutrino combine to form one electroweak doublet. This fact of the weak force then requires some gymnastics to be able to include fermion masses consistent with the $SU(2)_W \otimes U(1)_Y$ gauge symmetry. In particular, because the weak force requires us to be careful with fermion masses, we will find that through interacting with the weak force, fermions can transform into other fermions with different masses!

This is a pretty crazy claim, and we'll need to build intuition for how such a thing is possible. To set the stage for this chapter, let's remind ourselves about how we define particles through the irreps under which they transform and what properties we measure in a detector. This will be of central importance for how we model and understand the properties of fermion interactions with the weak force. In our particle detectors, we are especially sensitive to the energy and momentum of particles, their electric charge, and, well, that's (typically) about it. So, we must identify particles based on their energy, momentum, and electric charge. Identifying particles by their charge is what it is: we see, for example, positive or negative charge based on the direction of bending in a magnetic field. Energy and momentum are a bit more subtle, and thinking about this carefully will lead us down the rabbit hole of fermion mixing.

12.1 Flavor Mixing in the Weak Interactions

Particles can have, in general, any energy or momentum, so these aren't intrinsic properties of a particle. Another way to say this is that the Hermitian operators of energy and momentum, $\hat{H}$ and $\hat{\mathbf{p}}$, have a continuous eigenvalue spectrum on the Hilbert space of free particles. In terms of classifying particles, this property isn't useful, because of the continuous infinity of energy and momentum states. However, from $\hat{H}$ and $\hat{\mathbf{p}}$, we can form an operator that does yield useful information about the particle. The four-momentum squared operator

$$\hat{P}_\mu \hat{P}^\mu \equiv \hat{H}\hat{H} - \hat{\mathbf{p}} \cdot \hat{\mathbf{p}} \tag{12.1}$$

has a discrete spectrum of eigenvalues that correspond to the squared masses of particles. For example, when acting on the electron wavefunction $|e\rangle$, $\hat{P}_\mu \hat{P}^\mu$ returns

$$\hat{P}_\mu \hat{P}^\mu |e\rangle = (E^2 - |\vec{p}|^2)|e\rangle = m_e^2 |e\rangle \,. \tag{12.2}$$

So, by measuring the energy and momentum of a particle, we explicitly collapse its wavefunction into an eigenstate of the squared-momentum operator; that is, into a state of definite mass.

This point, that our experiments identify eigenstates of mass, was not discussed when we introduced QCD, nor is it necessary to discuss for electromagnetism. A free particle traveling through space with a definite mass always has a definite mass. It is possible that interactions affect this, but this does not happen in electromagnetism or QCD. An electron in space that emits a photon must still be an electron; its mass did not change. Similarly, an up quark that emits a gluon is still an up quark. It cannot turn into a down quark or a strange quark, or any other quark for that matter. Another way to say this is the following. In addition to determining a particle based on its mass, we could also determine a particle through its interactions with the force-carrying gauge bosons. We refer to this property of a particle as its **flavor**. The flavor of a particle can be defined by its eigenvalue of a flavor operator $\hat{F}$, which identifies the particle based on its particular set of interactions. In the case of an electron, the action of the flavor operator which is sensitive to electromagnetism would be something like

$$\hat{F}|e\rangle = (\text{electron})|e\rangle \,. \tag{12.3}$$

That is, in this case, the flavor operator identifies that the "electron" can emit photons with a probability that is proportional to the fundamental unit of charge, e. Also, the particle before and after the emission of the photon is identical.

Now, as hinted at above, these two definitions of what an "electron" is (from its mass or from its interactions) are mutually compatible. The particle with mass $m_e = 511$ keV is identical to the particle with electron flavor. In terms of the squared-mass and flavor operators, this means that collapsing to the mass eigenstate with $m_e = 511$ keV is equivalent to collapsing to the eigenstate of electron flavor. Thus, the squared-mass and flavor operators commute:

$$\hat{F}\hat{P}_\mu\hat{P}^\mu|e\rangle = m_e^2\hat{F}|e\rangle = m_e^2(\text{electron})|e\rangle \tag{12.4}$$
$$= \hat{P}_\mu\hat{P}^\mu\hat{F}|e\rangle = (\text{electron})\hat{P}_\mu\hat{P}^\mu|e\rangle = m_e^2(\text{electron})|e\rangle \,.$$

Restricting ourselves to defining flavor according to electromagnetic or QCD interactions, the squared-mass and flavor operators commute for every particle of the Standard Model, for the reason given above. Thus, we have that

$$\hat{P}_\mu\hat{P}^\mu\hat{F}_{\text{EM,QCD}} - \hat{F}_{\text{EM,QCD}}\hat{P}_\mu\hat{P}^\mu = [\hat{P}_\mu\hat{P}^\mu, \hat{F}_{\text{EM,QCD}}] = 0 \,. \tag{12.5}$$

We have explicitly denoted that the flavor operator is only sensitive to electromagnetic and QCD interactions.

This property that these two operators commute means that they can be simultaneously diagonalized in the space of fermionic particles of the Standard Model. That is, the eigenstates of $\hat{P}_\mu\hat{P}^\mu$ and $\hat{F}_{\text{EM,QCD}}$ are identical, and so one can perform the same transformation on these two operators to bring them into a form with only non-zero entries on the diagonal (when expressed as matrices). If we denote the unitary matrix of these orthonormal eigenstates by $\mathbb{V}$, then the transformation that diagonalizes these two operators is

$$(\hat{P}_\mu\hat{P}^\mu)^{\text{diag}} = \mathbb{V}^\dagger\hat{P}_\mu\hat{P}^\mu\mathbb{V}, \qquad (\hat{F}_{\text{EM,QCD}})^{\text{diag}} = \mathbb{V}^\dagger\hat{F}_{\text{EM,QCD}}\mathbb{V}. \tag{12.6}$$

The superscript diag denotes the operator with only diagonal entries. This transformation preserves the commutator of the operators:

$$[(\hat{P}_\mu\hat{P}^\mu)^{\text{diag}}, (\hat{F}_{\text{EM,QCD}})^{\text{diag}}] = [\mathbb{V}^\dagger\hat{P}_\mu\hat{P}^\mu\mathbb{V}, \mathbb{V}^\dagger\hat{F}_{\text{EM,QCD}}\mathbb{V}] \tag{12.7}$$
$$= \mathbb{V}^\dagger[\hat{P}_\mu\hat{P}^\mu, \hat{F}_{\text{EM,QCD}}]\mathbb{V} = 0 \,.$$

This property means that an eigenstate or particle as identified by its mass is equivalent to a particle identified by its type (or its interactions with forces). That is, the particle with mass of 511 keV is the electron flavor, the particle with mass 106 MeV is the muon flavor, etc. In this way we say that electromagnetic and QCD interactions are **flavor-diagonal**: we could simultaneously diagonalize the squared-momentum operator and the flavor operator, if all interactions were only electromagnetism and QCD. This property is analogous to the central-potential problem in quantum mechanics. States can be classified by their energy and z-component of angular momentum, because $\hat{H}$ and $\hat{L}_z$ commute. In quantum mechanics, we can't also label states by $\hat{L}_x$, for example, as $[\hat{L}_x, \hat{L}_z] \neq 0$.

But, just like the situation with quantum mechanical angular momentum, it didn't have to be that the squared-mass operator and the flavor operator commute. It is entirely possible that interactions with the force carrier change the flavor or type of fermion and hence change its mass. In that case then the flavor operator and the squared-momentum operator do not necessarily commute, and in general we are forced to pick a basis in which to define our quantum system. Either states have definite mass but weird interactions with the force, or they have simple interactions with the force but ill-defined masses. Because we measure energy and momentum in our experiments, we pick the first choice (for now anyway). However, this choice is a convention and convenient. If there is a force that is responsible for breaking a flavor symmetry such that

$$[\hat{P}_\mu \hat{P}^\mu, \hat{F}] \neq 0, \tag{12.8}$$

a measurement of the mass and flavor of a particle are incompatible quantum mechanically, in general. In this case, there is no consistent transformation we can perform to bring both the squared-mass and flavor operators into diagonal form.

   If I had told you this a few chapters ago, you might have scoffed because what terribly stupid interaction would do that? However, Nature doesn't care about our sense of aesthetics, and we now know of a force that does just that. The weak force, as mediated by the W boson, turns up quarks into down quarks, or electrons into neutrinos, thereby definitely mixing fermion mass eigenstates. Thankfully, weak interactions still do not mix quarks into leptons, so we can study the effects of the non-flavor diagonal weak interactions for leptons and quarks separately. The story with the leptons is much simpler than for quarks, so let's start with them.

12.2 The Weak Interactions in the Quark Sector: CP Violation

12.2.1 Weak Interactions of Charged Leptons

We need to identify the (squared)-mass matrix for the leptons and their flavor matrix, as defined by their interactions with the W boson. Starting with the mass matrix, in the Standard Model the masses of fermions arise in the same way as the masses of the W and Z bosons: from coupling with the scalar Higgs field, $\vec{\phi}$, and expanding about the vev. So, to determine the masses of the leptons, we just need to construct those $\mathrm{SU}(2)_W \otimes \mathrm{U}(1)_Y$ gauge-invariant lepton–Higgs interactions. These interactions in the Lagrangian of the Standard Model are

$$\mathcal{L} \supset -y_e \left(\nu_{e_L}^\dagger \ \ e_L^\dagger \right) \begin{pmatrix} \phi^+ \\ \phi_0 \end{pmatrix} e_R - y_\mu \left(\nu_{\mu_L}^\dagger \ \ \mu_L^\dagger \right) \begin{pmatrix} \phi^+ \\ \phi_0 \end{pmatrix} \mu_R \tag{12.9}$$

$$- y_\tau \left(\nu_{\tau_L}^\dagger \ \ \tau_L^\dagger \right) \begin{pmatrix} \phi^+ \\ \phi_0 \end{pmatrix} \tau_R + \text{h.c.}$$

The three explicit terms correspond to interactions of the electron, muon, and tau leptons with the Higgs field, and their associated neutrinos. These terms are indeed gauge and Lorentz invariant. Lorentz invariance follows because the scalar field $\vec{\phi}$ has zero spin, while the spins of the right- and left-handed fermion fields are anti-aligned. The coefficients y_e, y_μ, and y_τ are called the lepton **Yukawa couplings**. The symbol h.c. denotes the Hermitian conjugate of all explicitly listed terms, to ensure that the Lagrangian or Hamiltonian is Hermitian, as required by the CPT theorem. For compactness, we will only do the following analysis for the terms that are explicit; for the h.c. terms, one just has to appropriately conjugate.

Electroweak Gauge Transformations

The gauge invariance of these terms is a bit non-trivial, so we'll go through that carefully. Let's just focus on the electron term for simplicity; the muon and tau terms will have the identical transformations. First, we have already discussed the $SU(2)_W \otimes U(1)_Y$ gauge transformation of the scalar field $\vec{\phi}$. This was constructed in Eq. 11.54 in Section 11.3.2. Under a gauge transformation, we have that

$$\vec{\phi} = \begin{pmatrix} \phi^+ \\ \phi_0 \end{pmatrix} \rightarrow e^{i\vec{\alpha}\cdot\frac{\vec{\sigma}}{2}} e^{i\frac{\beta}{2}} \begin{pmatrix} \phi^+ \\ \phi_0 \end{pmatrix}. \tag{12.10}$$

Recall that the scalar field $\vec{\phi}$ is an $SU(2)_W$ doublet and has hypercharge $Y = 1$.

We then need to determine the gauge transformations of the left-handed electron doublet and the right-handed electron field. Let's first work on the doublet. As it's a doublet, its $SU(2)_W$ transformation is identical to the scalar $\vec{\phi}$:

$$\begin{pmatrix} \nu_{e_L} \\ e_L \end{pmatrix} \xrightarrow{SU(2)_W} e^{i\vec{\alpha}\cdot\frac{\vec{\sigma}}{2}} \begin{pmatrix} \nu_{e_L} \\ e_L \end{pmatrix}. \tag{12.11}$$

To determine the $U(1)_Y$ gauge transformation, we need to determine the hypercharge of this doublet. This can be found from the electric charge assignments. Recall that the electric charge of the lower component of the doublet is $Q_{\text{lower}} = -\frac{1}{2} + \frac{Y}{2}$, and if this is to be the electron, then $Y = -1$. Then, the electric charge of the neutrino is $Q_{\text{upper}} = \frac{1}{2} + \frac{Y}{2} = 0$, as required. So, the $U(1)_Y$ gauge transformation of the electron doublet has the opposite sign as for the scalar $\vec{\phi}$:

$$\begin{pmatrix} \nu_{e_L} \\ e_L \end{pmatrix} \xrightarrow{U(1)_Y} e^{-i\frac{\beta}{2}} \begin{pmatrix} \nu_{e_L} \\ e_L \end{pmatrix}. \tag{12.12}$$

Now for the right-handed electron. This is an $SU(2)_W$ singlet; that is, it transforms into itself under an $SU(2)_W$ gauge transformation:

$$e_R \xrightarrow{SU(2)_W} e_R. \tag{12.13}$$

On the other hand, it does have electric charge, but this will be entirely determined by its hypercharge Y. Because it is an $SU(2)_W$ singlet, its electric charge is just $Q = \frac{Y}{2}$. Therefore, for this to correspond to an electron, it must have hypercharge $Y = -2$. Its corresponding $U(1)_Y$ gauge transformation is then

$$e_R \xrightarrow{U(1)_Y} e^{-i\beta} e_R. \tag{12.14}$$

Putting it all together, the full electroweak gauge transformation of the electron–Higgs coupling is

$$- y_e \begin{pmatrix} \nu_{e_L}^\dagger & e_L^\dagger \end{pmatrix} \begin{pmatrix} \phi^+ \\ \phi_0 \end{pmatrix} e_R \qquad (12.15)$$

$$\xrightarrow{\mathrm{SU(2)}_W \otimes \mathrm{U(1)}_Y} -y_e \left[e^{i\vec{\alpha}\cdot\frac{\vec{\sigma}}{2}} e^{-i\frac{\beta}{2}} \begin{pmatrix} \nu_{e_L} \\ e_L \end{pmatrix} \right]^\dagger \left[e^{i\vec{\alpha}\cdot\frac{\vec{\sigma}}{2}} e^{i\frac{\beta}{2}} \begin{pmatrix} \phi^+ \\ \phi_0 \end{pmatrix} \right] \left[e^{-i\beta} e_R \right]$$

$$= -y_e \begin{pmatrix} \nu_{e_L}^\dagger & e_L^\dagger \end{pmatrix} \begin{pmatrix} \phi^+ \\ \phi_0 \end{pmatrix} e_R \,.$$

So, indeed as promised, these interaction terms are gauge invariant.

Lepton Mass Basis

Now, with these interactions established, we set the scalar $\vec{\phi}$ to its vev, and ignore the Higgs boson for the purposes of determining lepton masses. Therefore, setting $\phi^+ = 0$ and $\phi_0 = \frac{v}{\sqrt{2}}$, we have

$$\mathcal{L} \supset -y_e \begin{pmatrix} \nu_{e_L}^\dagger & e_L^\dagger \end{pmatrix} \begin{pmatrix} 0 \\ \frac{v}{\sqrt{2}} \end{pmatrix} e_R - y_\mu \begin{pmatrix} \nu_{\mu_L}^\dagger & \mu_L^\dagger \end{pmatrix} \begin{pmatrix} 0 \\ \frac{v}{\sqrt{2}} \end{pmatrix} \mu_R \qquad (12.16)$$

$$- y_\tau \begin{pmatrix} \nu_{\tau_L}^\dagger & \tau_L^\dagger \end{pmatrix} \begin{pmatrix} 0 \\ \frac{v}{\sqrt{2}} \end{pmatrix} \tau_R + \mathrm{h.c.}$$

$$= -\frac{y_e v}{\sqrt{2}} e_L^\dagger e_R - \frac{y_\mu v}{\sqrt{2}} \mu_L^\dagger \mu_R - \frac{y_\tau v}{\sqrt{2}} \tau_L^\dagger \tau_R + \mathrm{h.c.}$$

Thus, we immediately see that the masses of the charged leptons are proportional both to the Yukawa couplings (the strength of the interaction with the Higgs boson) and to the vev, v. Extremely importantly, in the Standard Model, the neutrinos are massless. (We will find later that this isn't true in Nature, however.) These mass terms have been expressed in the **mass basis**, in which there is no mixing between fields corresponding to different leptons. That is, electron fields are multiplied by electron fields, muon fields by muon fields and tau fields by tau fields. At this stage, no generality has been lost by working in the mass basis.

It is also useful to study the gauge transformations that leave these mass terms invariant. From our discussion in the previous chapter, after expanding about the vev of $\vec{\phi}$, the only electroweak gauge boson that remains massless is the photon. Thus, the broken theory is only manifestly electromagnetically gauge invariant. How this works is a bit non-trivial, so we will go through it in some detail. As earlier, we just consider the gauge transformations of the electron; the muon and tau lepton properties are identical.

First, as it is an $\mathrm{SU(2)}_W$ singlet, the electroweak gauge transformation of the right-handed electron e_R is just

$$e_R \to e^{-i\beta} e_R \,, \qquad (12.17)$$

as discussed above. Note that this is just a U(1) gauge transformation; in particular, it is the right-handed electron's $U(1)_{\text{EM}}$ gauge transformation with electric charge $Q = -1$. The general electromagnetic gauge transformation for a particle with charge Q is

$$e^{iQ\beta} \in U(1)_{\text{EM}}. \tag{12.18}$$

Therefore, for these mass terms to be electromagnetically gauge invariant, the gauge transformation of the left-handed electron e_L must be identical to that of e_R; that is, the left- and right-handed electrons have the same electric charge.

As there are no neutrinos around, to determine the electromagnetic gauge transformation of e_L we must restrict to those $SU(2)_W \otimes U(1)_Y$ transformations that transform e_L proportionally to itself. Thus, we ignore those $SU(2)_W$ transformations that have contributions from σ_1 and σ_2: these matrices only have off-diagonal entries, and so turn the electron into the neutrino, and vice-versa. These gauge transformations correspond to the W bosons, as through their interactions electrons turn into neutrinos. Thus, only the σ_3 term in the $SU(2)_W$ transformation can contribute. Restricting to this case, the action of the remaining gauge transformations of e_L is

$$e_L \to e^{-\frac{i}{2}(\alpha_3+\beta)}e_L = e^{-\frac{i}{2}(\alpha_3-\beta)}e^{-i\beta}e_L. \tag{12.19}$$

On the right, we have expressed this gauge transformation in an equivalent but suggestive way. For this to be identical to the e_R transformation, we must further restrict to the case with $\alpha_3 = \beta$. This restriction then eliminates the factor that contributes when $\alpha_3 \neq \beta$; this transformation corresponds to the Z boson. Therefore, the only gauge transformation of the left-handed electron that remains as a symmetry of the broken theory lepton mass terms is

$$e_L \to e^{-i\beta}e_L. \tag{12.20}$$

This is just the gauge transformation of the $U(1)_{\text{EM}}$ theory, and indeed leaves Eq. 12.16 invariant.

From this form of the Lagrangian, it is then useful to rewrite it to identify the mass matrix of the leptons. We can form the left- and right-handed lepton vectors

$$\begin{pmatrix} e_L^\dagger & \mu_L^\dagger & \tau_L^\dagger \end{pmatrix}, \qquad\qquad \begin{pmatrix} e_R \\ \mu_R \\ \tau_R \end{pmatrix}.$$

With these, the Lagrangian can be written as

$$\mathcal{L} \supset - \begin{pmatrix} e_L^\dagger & \mu_L^\dagger & \tau_L^\dagger \end{pmatrix} \begin{pmatrix} \frac{y_e v}{\sqrt{2}} & & \\ & \frac{y_\mu v}{\sqrt{2}} & \\ & & \frac{y_\tau v}{\sqrt{2}} \end{pmatrix} \begin{pmatrix} e_R \\ \mu_R \\ \tau_R \end{pmatrix} + \text{h.c.} \tag{12.21}$$

The mass matrix for the charged leptons is then manifest and diagonal. The mass matrix for the neutrinos is just the 0 matrix, which is also diagonal. This feature of the Standard Model will be vital for the lepton mass and flavor matrices to commute, as we will see shortly.

Table 12.1 Lepton Electroweak Gauge Transformations

	SU(2)$_W$		U(1)$_Y$		U(1)$_{EM}$
$\begin{pmatrix} \nu_{e_L} \\ e_L \end{pmatrix}$	$e^{i\vec{\alpha}\cdot\frac{\vec{\sigma}}{2}}$ doublet	$e^{-i\frac{\beta}{2}}$	$Y=-1$	$\begin{matrix} 1 \\ e^{-i\beta} \end{matrix}$	$\begin{matrix} Q=0 \\ Q=-1 \end{matrix}$
e_R	1 singlet	$e^{-i\beta}$	$Y=-2$	$e^{-i\beta}$	$Q=-1$

The electroweak gauge transformations and irreps for leptons are collected in Table 12.1. We have written this explicitly for electrons, but the same transformations exist for muons and taus.

Lepton Flavor Basis

With the mass matrices established, let's now construct the flavor matrix for the leptons, as defined by their interactions with the W boson. From the discussion in Section 11.3.4, the interactions between the leptons and the W boson are determined by the SU(2)$_W$ doublet's covariant derivative, Eq. 11.86. First, we write these interactions in the **flavor basis**; that is, in the basis in which the W boson turns electrons into electron neutrinos, muons into muon neutrinos, and taus into tau neutrinos. We will denote the fields in this flavor basis with a subscript F for flavor. The flavor-basis lepton-W boson interactions are then

$$\mathcal{L} \supset \frac{g_W}{\sqrt{2}} \nu^\dagger_{eL,F} \bar{\sigma} \cdot W^+ e_{L,F} + \frac{g_W}{\sqrt{2}} \nu^\dagger_{\mu L,F} \bar{\sigma} \cdot W^+ \mu_{L,F} \tag{12.22}$$

$$+ \frac{g_W}{\sqrt{2}} \nu^\dagger_{\tau L,F} \bar{\sigma} \cdot W^+ \tau_{L,F} + \text{h.c.}$$

In this basis, we can equivalently write this using the lepton vectors we constructed earlier:

$$\mathcal{L} \supset \frac{g_W}{\sqrt{2}} \begin{pmatrix} \nu^\dagger_{eL,F} & \nu^\dagger_{\mu L,F} & \nu^\dagger_{\tau L,F} \end{pmatrix} \bar{\sigma} \cdot W^+ \begin{pmatrix} e_{L,F} \\ \mu_{L,F} \\ \tau_{L,F} \end{pmatrix} + \text{h.c.} \tag{12.23}$$

Both the flavor and mass bases are complete and are therefore related to one another by a unitary transformation. We can introduce two 3×3 unitary matrices that implement this transformation for the charged leptons and neutrinos, respectively. So, we define these unitary transformations via

$$\begin{pmatrix} e_{L,F} \\ \mu_{L,F} \\ \tau_{L,F} \end{pmatrix} = \mathbb{U}_\ell \begin{pmatrix} e_L \\ \mu_L \\ \tau_L \end{pmatrix}, \qquad \begin{pmatrix} \nu_{eL,F} \\ \nu_{\mu L,F} \\ \nu_{\tau L,F} \end{pmatrix} = \mathbb{U}_\nu \begin{pmatrix} \nu_{eL} \\ \nu_{\mu L} \\ \nu_{\tau L} \end{pmatrix}, \tag{12.24}$$

for two unitary matrices $\mathbb{U}_\ell$ and $\mathbb{U}_\nu$. The mass-basis leptons are denoted by a lack of F subscript. With this change of basis established, we can express the interactions in the mass basis:

$$\mathcal{L} \supset \frac{g_W}{\sqrt{2}} \begin{pmatrix} \nu^\dagger_{eL} & \nu^\dagger_{\mu L} & \nu^\dagger_{\tau L} \end{pmatrix} \mathbb{U}^\dagger_\nu \left(\bar{\sigma} \cdot W^+ \right) \mathbb{U}_\ell \begin{pmatrix} e_L \\ \mu_L \\ \tau_L \end{pmatrix} + \text{h.c.} \tag{12.25}$$

Though this is not likely to cause confusion, remember that the $\bar{\sigma}$ matrix that multiplies the W boson acts on the components of the lepton spinors.

From Eq. 12.25, we identify the flavor matrix $\hat{F}$ as the product of the two unitary matrices:

$$\hat{F} = \mathbb{U}_\nu^\dagger \mathbb{U}_\ell. \tag{12.26}$$

Naïvely, we would say that the mass and flavor bases are not the same and the mass and flavor matrices do not commute because $\hat{F}$ is not diagonal. However, we can exploit the property that the neutrinos are massless. As they are massless, their mass matrix is the 0 matrix, which is unchanged with any unitary transformation. Thus, we can, with impunity, pick the matrix $\mathbb{U}_\nu^\dagger$ to be any unitary matrix! So, let's pick the matrix $\mathbb{U}_\nu^\dagger = \mathbb{U}_\ell^\dagger$. With this choice, the flavor matrix is then exceedingly simple:

$$\hat{F} = \mathbb{U}_\ell^\dagger \mathbb{U}_\ell = \mathbb{I}, \tag{12.27}$$

because $\mathbb{U}_\ell$ is a unitary matrix. Therefore, the flavor matrix is the identity matrix, which is of course diagonal. This proves that the lepton mass and flavor matrices can be simultaneously diagonalized and therefore have the same eigenstates. For completeness, the interactions with the W boson are then

$$\mathcal{L} \supset \frac{g_W}{\sqrt{2}} \begin{pmatrix} \nu_{eL}^\dagger & \nu_{\mu L}^\dagger & \nu_{\tau L}^\dagger \end{pmatrix} \bar{\sigma} \cdot W^+ \begin{pmatrix} e_L \\ \mu_L \\ \tau_L \end{pmatrix} + \text{h.c.} \tag{12.28}$$

While we have focused on the interactions with the W bosons, this argument can be extended to all gauge interactions of the leptons. So, precisely because neutrinos are massless in the Standard Model (we can rotate them into one another without affecting the structure of masses), we can simultaneously diagonalize the squared-momentum operator (= mass matrix) and the flavor operator (= gauge interaction matrix) for the leptons. We'll have to correct the masslessness of neutrinos later in this chapter, but for now, it is an extremely good approximation.

12.2.2 Weak Interactions of Quarks

With the leptons under our belts, let's attack the masses and weak interactions of the quarks. The challenge now, unlike the case with leptons, is that all six quarks are massive. This requires the addition of some unfamiliar terms in the Lagrangian to accomplish this, so we'll go through this carefully.

Electroweak Gauge Transformations

As it is simplest, we first construct the gauge transformations for the down-type quarks (the down, strange, and bottom quarks). The weak interactions are responsible for up-type quarks turning into down-type quarks, and so we naturally have SU(2)$_W$ doublets that are formed from one up-type quark and one down-type quark. So, electroweak gauge-invariant interactions between quarks and the scalar field ϕ are

$$\mathcal{L} \supset -y_d \begin{pmatrix} u_L^\dagger & d_L^\dagger \end{pmatrix} \begin{pmatrix} \phi^+ \\ \phi_0 \end{pmatrix} d_R - y_s \begin{pmatrix} c_L^\dagger & s_L^\dagger \end{pmatrix} \begin{pmatrix} \phi^+ \\ \phi_0 \end{pmatrix} s_R \tag{12.29}$$

$$- y_b \begin{pmatrix} t_L^\dagger & b_L^\dagger \end{pmatrix} \begin{pmatrix} \phi^+ \\ \phi_0 \end{pmatrix} b_R + \text{h.c.}$$

Recall that with the electric charge assignments from the quark model, the up-type quarks have charge $Q_u = 2/3$, while the down-type quarks have electric charge $Q_d = -1/3$. Thus, as required, the upper and lower components of the $SU(2)_W$ left-handed quark doublets have electric charge that differs by 1 unit. Also, note that, when expanding about the vev, these terms only give a mass to the down-type quarks.

The gauge invariance of these terms is easy to verify. Let's just consider the first term with the up and down quarks, u and d. We already know the gauge transformation for the scalar field $\vec{\phi}$, so we won't write that down again here. To determine the transformation of the left-handed doublet, we need to know its hypercharge Y. Again, as the electric charge of the up quark, for instance, is 2/3, the hypercharge of the left-handed quark doublet is

$$Y = 2Q_u - 1 = 1/3 . \tag{12.30}$$

It then follows that the electroweak gauge transformation of the left-handed doublet is

$$\begin{pmatrix} u_L \\ d_L \end{pmatrix} \xrightarrow{SU(2)_W \otimes U(1)_Y} e^{i\vec{\alpha}\cdot\frac{\vec{\sigma}}{2}} e^{i\frac{\beta}{6}} \begin{pmatrix} u_L \\ d_L \end{pmatrix} . \tag{12.31}$$

The right-handed down quark d_R is an $SU(2)_W$ singlet, and so its $U(1)_Y$ gauge transformation is just determined by its electric charge. As its electric charge is $Q_d = -1/3$, its hypercharge is $Y = -2/3$ and so its electroweak gauge transformation is

$$d_R \xrightarrow{SU(2)_W \otimes U(1)_Y} e^{-i\frac{\beta}{3}} d_R . \tag{12.32}$$

Then, the electroweak gauge transformation of the whole term in the Lagrangian is

$$- y_d \begin{pmatrix} u_L^\dagger & d_L^\dagger \end{pmatrix} \begin{pmatrix} \phi^+ \\ \phi_0 \end{pmatrix} d_R \tag{12.33}$$

$$\xrightarrow{SU(2)_W \otimes U(1)_Y} -y_d \left[e^{i\vec{\alpha}\cdot\frac{\vec{\sigma}}{2}} e^{i\frac{\beta}{6}} \begin{pmatrix} u_L \\ d_L \end{pmatrix} \right]^\dagger \left[e^{i\vec{\alpha}\cdot\frac{\vec{\sigma}}{2}} e^{i\frac{\beta}{2}} \begin{pmatrix} \phi^+ \\ \phi_0 \end{pmatrix} \right] \left[e^{-i\frac{\beta}{3}} d_R \right]$$

$$= -y_d \begin{pmatrix} u_L^\dagger & d_L^\dagger \end{pmatrix} \begin{pmatrix} \phi^+ \\ \phi_0 \end{pmatrix} d_R ,$$

and so is gauge invariant, as claimed. It's also important to note that this term is invariant under $SU(3)$ color transformations of QCD, as quarks and anti-quarks have opposite color charge.

Now, let's construct the interactions of the up-type quarks (the up, charm, and top quarks) which will give them a mass. This will be a bit subtle, but we have gauge invariance as our guide. A mass for the up quark is generated by a term in the Lagrangian like

$$\mathcal{L} \supset -y_u \begin{pmatrix} u_L^\dagger & d_L^\dagger \end{pmatrix} \begin{pmatrix} A \\ B \end{pmatrix} u_R + \text{h.c.} \tag{12.34}$$

Here, we have denoted the entries of a to-be-determined $SU(2)_W$ doublet as A and B. Such a term must be gauge invariant, and this requirement will tell us everything we need to know. With an electric charge of 2/3, the hypercharge of the right-handed up quark u_R is $Y = 4/3$ and its electroweak gauge transformation is then

$$u_R \xrightarrow{SU(2)_W \otimes U(1)_Y} e^{i\frac{2\beta}{3}} u_R \,. \tag{12.35}$$

The net electroweak gauge transformation of the quark doublet and the right-handed up quark is therefore

$$\begin{pmatrix} u_L^\dagger & d_L^\dagger \end{pmatrix} u_R \xrightarrow{SU(2)_W \otimes U(1)_Y} \left[e^{i\vec{\alpha} \cdot \frac{\vec{\sigma}}{2}} e^{i\frac{\beta}{6}} \begin{pmatrix} u_L \\ d_L \end{pmatrix} \right]^\dagger \left[e^{i\frac{2\beta}{3}} u_R \right] \tag{12.36}$$

$$= \begin{pmatrix} u_L^\dagger & d_L^\dagger \end{pmatrix} e^{-i\vec{\alpha} \cdot \frac{\vec{\sigma}}{2}} e^{i\frac{\beta}{2}} u_R \,.$$

For the term in the Lagrangian to be gauge invariant, it then must transform to cancel this. That is,

$$\begin{pmatrix} A \\ B \end{pmatrix} \xrightarrow{SU(2)_W \otimes U(1)_Y} e^{i\vec{\alpha} \cdot \frac{\vec{\sigma}}{2}} e^{-i\frac{\beta}{2}} \begin{pmatrix} A \\ B \end{pmatrix} \,. \tag{12.37}$$

This transformation is very nearly the same as for the scalar doublet $\vec{\phi}$! However, apparently, this has the opposite value for the hypercharge. This doublet has hypercharge $Y = -1$, and so the electric charge of the upper component is $Q_{\text{upper}} = \frac{1}{2} + \frac{Y}{2} = 0$, while for the lower component it is $Q_{\text{lower}} = -\frac{1}{2} + \frac{Y}{2} = -1$. The electric charge of the field ϕ_0 is 0, so this works as the upper component A of this doublet. The electric charge of the field ϕ^+, however, is $+1$, and so it cannot be the lower component B. However, its complex conjugate $(\phi^+)^* \equiv \phi^-$ does have electric charge -1, by the discussion of charge conjugation from Section 10.2.2. So, we will set the lower component $B = \phi^-$, and this results in gauge-invariant quark–scalar interactions. Putting it all together, all possible couplings of quarks with the scalar field $\vec{\phi}$ are

$$\mathcal{L} \supset -y_d \begin{pmatrix} u_L^\dagger & d_L^\dagger \end{pmatrix} \begin{pmatrix} \phi^+ \\ \phi_0 \end{pmatrix} d_R - y_s \begin{pmatrix} c_L^\dagger & s_L^\dagger \end{pmatrix} \begin{pmatrix} \phi^+ \\ \phi_0 \end{pmatrix} s_R - y_b \begin{pmatrix} t_L^\dagger & b_L^\dagger \end{pmatrix} \begin{pmatrix} \phi^+ \\ \phi_0 \end{pmatrix} b_R$$

$$- y_u \begin{pmatrix} u_L^\dagger & d_L^\dagger \end{pmatrix} \begin{pmatrix} \phi_0 \\ \phi^- \end{pmatrix} u_R - y_c \begin{pmatrix} c_L^\dagger & s_L^\dagger \end{pmatrix} \begin{pmatrix} \phi_0 \\ \phi^- \end{pmatrix} c_R - y_t \begin{pmatrix} t_L^\dagger & b_L^\dagger \end{pmatrix} \begin{pmatrix} \phi_0 \\ \phi^- \end{pmatrix} t_R$$

$$+ \text{h.c.} \tag{12.38}$$

The electroweak gauge transformations of the quarks are collected in Table 12.2.

Table 12.2 Quark Electroweak Gauge Transformations				
	SU(2)$_W$		U(1)$_Y$	U(1)$_{EM}$
$\begin{pmatrix} u_L \\ d_L \end{pmatrix}$	$e^{i\vec{\alpha}\cdot\frac{\vec{\sigma}}{2}}$ doublet	$e^{i\frac{\beta}{6}}$	$Y = 1/3$	$\begin{matrix} e^{i\frac{2\beta}{3}} & Q = 2/3 \\ e^{-i\frac{\beta}{3}} & Q = -1/3 \end{matrix}$
u_R	1 singlet	$e^{i\frac{2\beta}{3}}$	$Y = 4/3$	$e^{i\frac{2\beta}{3}} \quad Q = 2/3$
d_R	1 singlet	$e^{i\frac{\beta}{3}}$	$Y = -2/3$	$e^{-i\frac{\beta}{3}} \quad Q = -1/3$

Quark Mass Basis

With this complete expression, we can then expand about the vev of $\vec{\phi}$ to determine the masses of the quarks. Setting $\phi^+ = \phi^- = 0$ and $\phi_0 = \frac{v}{\sqrt{2}}$, the mass terms for the quarks are then

$$\mathcal{L} \supset -\frac{y_d v}{\sqrt{2}}d_L^\dagger d_R - \frac{y_s v}{\sqrt{2}}s_L^\dagger s_R - \frac{y_b v}{\sqrt{2}}b_L^\dagger b_R - \frac{y_u v}{\sqrt{2}}u_L^\dagger u_R - \frac{y_c v}{\sqrt{2}}c_L^\dagger c_R - \frac{y_t v}{\sqrt{2}}t_L^\dagger t_R \qquad (12.39)$$
$$+ \text{h.c.}$$

As with the leptons, it is convenient to express these mass terms with matrix multiplication of vectors of the up-type and down-type quarks. We then have

$$\mathcal{L} \supset -\begin{pmatrix} d_L^\dagger & s_L^\dagger & b_L^\dagger \end{pmatrix} \begin{pmatrix} \frac{y_d v}{\sqrt{2}} & & \\ & \frac{y_s v}{\sqrt{2}} & \\ & & \frac{y_b v}{\sqrt{2}} \end{pmatrix} \begin{pmatrix} d_R \\ s_R \\ b_R \end{pmatrix} \qquad (12.40)$$

$$- \begin{pmatrix} u_L^\dagger & c_L^\dagger & t_L^\dagger \end{pmatrix} \begin{pmatrix} \frac{y_u v}{\sqrt{2}} & & \\ & \frac{y_c v}{\sqrt{2}} & \\ & & \frac{y_t v}{\sqrt{2}} \end{pmatrix} \begin{pmatrix} u_R \\ c_R \\ t_R \end{pmatrix} + \text{h.c.}$$

In this mass basis, the mass matrices for the down- and up-type quarks are then diagonal and manifest. What do the flavor matrices look like? Let's turn to the interactions with the W boson now.

Quark Flavor Basis

Our approach with quarks will be the same as for the leptons. We first express the interactions with the W boson in the flavor basis, as denoted by an F subscript:

$$\mathcal{L} \supset \frac{g_W}{\sqrt{2}}u_{L,F}^\dagger \bar{\sigma} \cdot W^+ d_{L,F} + \frac{g_W}{\sqrt{2}}c_{L,F}^\dagger \bar{\sigma} \cdot W^+ s_{L,F} \qquad (12.41)$$
$$+ \frac{g_W}{\sqrt{2}}t_{L,F}^\dagger \bar{\sigma} \cdot W^+ b_{L,F} + \text{h.c.}$$

This can be written more compactly with a vector of up- or down-type quarks:

$$\mathcal{L} \supset \frac{g_W}{\sqrt{2}} \begin{pmatrix} u_{L,F}^\dagger & c_{L,F}^\dagger & t_{L,F}^\dagger \end{pmatrix} \bar{\sigma} \cdot W^+ \begin{pmatrix} d_{L,F} \\ s_{L,F} \\ b_{L,F} \end{pmatrix} + \text{h.c.} \qquad (12.42)$$

The mass and flavor bases for both up- and down-type quarks can be related by a unitary transformation, which we define as

$$\begin{pmatrix} u_{L,F} \\ c_{L,F} \\ t_{L,F} \end{pmatrix} = \mathbb{U}_u \begin{pmatrix} u_L \\ c_L \\ t_L \end{pmatrix}, \qquad \begin{pmatrix} d_{L,F} \\ s_{L,F} \\ b_{L,F} \end{pmatrix} = \mathbb{U}_d \begin{pmatrix} d_L \\ s_L \\ b_L \end{pmatrix}, \qquad (12.43)$$

for two unitary matrices $\mathbb{U}_u$ and $\mathbb{U}_d$.

So far, this is all familiar, and likely slightly boring. The excitement happens when we insert these expressions into the Lagrangian. Expressing the interactions with the W boson in the mass basis, we find

$$\mathcal{L} \supset \frac{g_W}{\sqrt{2}} \begin{pmatrix} u_L^\dagger & c_L^\dagger & t_L^\dagger \end{pmatrix} \mathbb{U}_u^\dagger \bar{\sigma} \cdot W^+ \mathbb{U}_d \begin{pmatrix} d_L \\ s_L \\ b_L \end{pmatrix} + \text{h.c.} \qquad (12.44)$$

For the leptons, the product of unitary matrices in the W boson interactions was benign; the neutrinos were massless, and so their mass basis could be rotated arbitrarily and not affect anything. By contrast, all of the quarks are massive, and so we do not have the freedom to choose $\mathbb{U}_d = \mathbb{U}_u$. Such a constraint will necessarily mix the mass basis of the down-type quarks and rotate the mass matrix away from diagonal. Therefore, there is no way in general to ensure that the mass and flavor bases for the quarks are simultaneously diagonal.

So, apparently, for the quarks the squared-momentum operator does not commute with the flavor operator:

$$[\hat{P}_\mu \hat{P}^\mu, \hat{F}] \neq 0. \qquad (12.45)$$

We call the quark flavor-mixing matrix

$$\mathbb{U}_u^\dagger \mathbb{U}_d \equiv V_{\text{CKM}} \qquad (12.46)$$

the **Cabibbo–Kobayashi–Maskawa (CKM) matrix**.[1]

12.2.3 CP Violation of the Weak Interactions

Two-Generation Case

But wait, there's more weirdness to come. It is not enough that the weak interactions mix quark flavors. Hidden in the CKM matrix is a dark and dirty secret: particles and anti-particles do not have the same interactions with the W bosons. Let's see what this means. Let's first consider the simpler case of just two generations of quarks. In this simpler case, the interactions with the W boson are then

$$\mathcal{L}_W^{2G} = \frac{g_W}{\sqrt{2}} \begin{pmatrix} u_L^\dagger & c_L^\dagger \end{pmatrix} V_{\text{CKM}} \bar{\sigma} \cdot W^+ \begin{pmatrix} d_L \\ s_L \end{pmatrix} + \frac{g_W}{\sqrt{2}} \begin{pmatrix} d_L^\dagger & s_L^\dagger \end{pmatrix} V_{\text{CKM}}^\dagger \bar{\sigma} \cdot W^- \begin{pmatrix} u_L \\ c_L \end{pmatrix}. \qquad (12.47)$$

---

[1] N. Cabibbo, "Unitary symmetry and leptonic decays," Phys. Rev. Lett. **10**, 531 (1963); M. Kobayashi and T. Maskawa, "CP violation in the renormalizable theory of weak interaction," Prog. Theor. Phys. **49**, 652 (1973).

The CKM matrix is unitary, as it is the product of two unitary matrices. A 2×2 unitary matrix can be parametrized by one angle and three phases as

$$V_{\text{CKM}} = \begin{pmatrix} e^{i\alpha}\cos\theta_C & -e^{i(\alpha+\gamma)}\sin\theta_C \\ e^{i\beta}\sin\theta_C & e^{i(\beta+\gamma)}\cos\theta_C \end{pmatrix}, \tag{12.48}$$

for an angle θ_C and phases α, β, γ. It is easy to verify that this is a unitary matrix.

This CKM matrix has three non-zero phases in general, which means that $V_{\text{CKM}} \neq V_{\text{CKM}}^*$. As such, up quarks and anti-up quarks would interact with W bosons differently because the complex numbers in Eq. 12.47 are not the same for particles and anti-particles. However, there's a catch. This matrix has three phases and yet mixes four fermions (u_L, d_L, s_L, c_L). So, we can perform yet another change to the fields to remove these phases completely. Recall that the overall phase of a wavefunction is unphysical; it does not affect probabilities. We are able to multiply fermions by an appropriate phase to completely remove the complex numbers from the CKM matrix.

Let's see how this works. Multiplying out the W boson interactions, we find

$$\frac{\sqrt{2}}{g_W}\mathcal{L}_W^{2G} = e^{i\alpha}\cos\theta_C u_L^\dagger \bar{\sigma} \cdot W^+ d_L + e^{i\beta}\sin\theta_C c_L^\dagger \bar{\sigma} \cdot W^+ d_L \tag{12.49}$$

$$- e^{i(\alpha+\gamma)}\sin\theta_C u_L^\dagger \bar{\sigma} \cdot W^+ s_L + e^{i(\beta+\gamma)}\cos\theta_C c_L^\dagger \bar{\sigma} \cdot W^+ s_L + \text{h.c.}$$

If we then rescale the fields as

$$u_L^\dagger \to u_L^\dagger, \qquad d_L \to e^{-i\alpha}d_L, \qquad c_L^\dagger \to e^{i(\alpha-\beta)}c_L^\dagger, \qquad s_L \to e^{-i(\alpha+\gamma)}s_L, \tag{12.50}$$

all phases are removed! With these phase choices, the Lagrangian becomes

$$\mathcal{L}_W^{2G} = \frac{g_W}{\sqrt{2}}\begin{pmatrix} u_L^\dagger & c_L^\dagger \end{pmatrix}\begin{pmatrix} \cos\theta_C & -\sin\theta_C \\ \sin\theta_C & \cos\theta_C \end{pmatrix}\bar{\sigma} \cdot W^+ \begin{pmatrix} d_L \\ s_L \end{pmatrix} + \text{h.c.} \tag{12.51}$$

That is, by appropriate change of phase to the fields, the two-generation CKM matrix is real. The entries of the matrix are defined by the **Cabibbo angle**, θ_C. From the PDG, the measured value of the Cabibbo angle in the Standard Model is

$$\sin\theta_C = 0.22452 \pm 0.00044. \tag{12.52}$$

Note, importantly, that we could remove the three phases because we had four fields: there is an overall phase of the Lagrangian that must be 0 for it to be real. With this adjustment, particles and anti-particles have the identical couplings to the W boson. All is well, for now.

Example 12.1 This two-generation mixing structure has profound consequences for what processes are allowed in the Standard Model, and even resulted in the prediction of extra quarks. In the late 1960s, three quarks had been identified: up, down, and strange quarks. At the same time, the theory of the weak interaction was being formulated, and it was realized that the weak force could mediate the decay of the strange quark to a down quark. The Feynman diagram that describes the process $s \to d\nu_e\bar{\nu}_e$ is

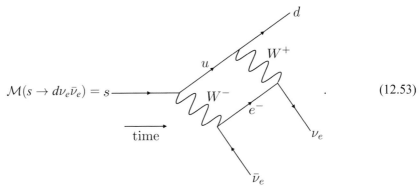

$$\mathcal{M}(s \to d\nu_e \bar{\nu}_e) = s \qquad\qquad\qquad . \qquad (12.53)$$

This decay is allowed kinematically because the mass of the strange quark is $m_s \simeq 95$ MeV, the mass of the down quark is $m_d \simeq 4.5$ MeV, and the neutrinos have very small masses. With only three quarks and mixing between them, this Feynman diagram is non-zero, and so there would be a non-zero rate of strange quark decay to down quarks. Such a process that transforms one quark into another of the same electric charge is called a **flavor-changing neutral current**, or FCNC. However, this decay was never observed.

The resolution of this discrepancy between the three-quark prediction and lack of observation was identified by Sheldon Glashow, John Iliopoulos, and Luciano Maiani in what is now called the **GIM mechanism**.[2] So, what's going on?

Solution

With only three quarks, u, d, and s, we would predict a non-zero decay rate for this process. Glashow, Iliopoulos, and Maiani's insight was to predict the existence of another, new quark whose contribution to the decay rate could exactly cancel that from the Feynman diagram shown above. This is now called the charm quark.

With a full two generations of quarks, they mix according to the Cabibbo matrix from Eq. 12.51. We can then determine the relative importance of the two Feynman diagrams that contribute to strange quark decays, with either an intermediate up or charm quark. The diagram for the intermediate up quark is proportional to

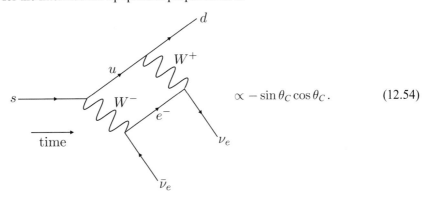

$$\propto -\sin\theta_C \cos\theta_C . \qquad (12.54)$$

[2] S. L. Glashow, J. Iliopoulos and L. Maiani, "Weak interactions with lepton–hadron symmetry," Phys. Rev. D **2**, 1285 (1970).

This follows from the coupling of the strange quark to the up quark (proportional to $-\sin\theta_C$) and the up quark to the down quark (proportional to $\cos\theta_C$). By contrast, the Feynman diagram with an intermediate charm quark is

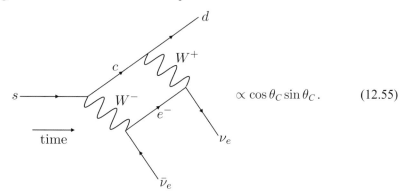

$$\propto \cos\theta_C \sin\theta_C. \qquad (12.55)$$

Now, the coupling of the strange quark to the charm quark is proportional to $\cos\theta_C$, while the coupling of the charm quark to the down quark is proportional to $\sin\theta_C$. Therefore, these two Feynman diagrams have opposite signs. When summed together to calculate the matrix element, they (largely) cancel!

The GIM mechanism is a beautiful resolution to the problem of FCNCs in the Standard Model. As long as the quarks of the Standard Model come in complete generations and mix according to a unitary matrix, FCNCs are highly suppressed. The GIM prediction of the charm quark was made in 1970 and the charm quark was discovered in 1974.[3]

Before we consider the three-generation Standard Model quarks, let's make precise the statement that particles and anti-particles in the two-generation case interact with the W boson identically. That is, our Lagrangian in Eq. 12.51 apparently exhibits a symmetry: we can exchange particles and anti-particles and the Lagrangian is unchanged. Concretely, let's just focus on the coupling of up and down quarks to the W boson. Those terms in the Lagrangian of Eq. 12.51 are

$$\mathcal{L} \supset \frac{g_W \cos\theta_C}{\sqrt{2}} u_L^\dagger \bar{\sigma} \cdot W^+ d_L + \frac{g_W \cos\theta_C}{\sqrt{2}} d_L^\dagger \bar{\sigma} \cdot W^- u_L. \qquad (12.56)$$

This can be equivalently written as

$$\mathcal{L} \supset \frac{g_W \cos\theta_C}{\sqrt{2}} u_L^\dagger \bar{\sigma} \cdot W^+ d_L + \frac{g_W \cos\theta_C}{\sqrt{2}} (u_L^\dagger \bar{\sigma} \cdot W^+ d_L)^*. \qquad (12.57)$$

So, this is invariant under the transformation

$$u_L^\dagger \bar{\sigma} \cdot W^+ d_L \leftrightarrow (u_L^\dagger \bar{\sigma} \cdot W^+ d_L)^*. \qquad (12.58)$$

Note that overall constants (couplings and mixing angles) are unchanged. In general, this is not equivalent to complete Hermitian conjugation of the first term of the Lagrangian in

Eq. 12.56. This is a discrete transformation, and two applications of complex conjugation turn u_L back into itself, for example. This suggests that the operation of exchanging particles and anti-particles is related to the parity, charge conjugation, or time-reversal transformations that we identified in Section 10.2. Recall that two applications of each of these discrete operations is just the identity operator. So, let's see how we can construct a particle–anti-particle transformation from C, P, and T.

The first term of Eq. 12.56 couples a right-handed anti-up quark to a left-handed down quark through a positively charged W boson. (The total charge of this term is therefore 0.) In the second term, however, all charge assignments have been negated, so this implies that the action of charge conjugation $\mathbb{C}$ is necessary. Additionally, all spin assignments are the same in the first and second terms of Eq. 12.56. Charge conjugation doesn't flip the spin, and so just the action of $\mathbb{C}$ would couple a right-handed up quark to a left-handed anti-down quark through the W boson. This never happens in the weak theory, so we need to flip the spins as well. To ensure that the up and down quarks are both appropriately left-handed, we need to act with the parity operator $\mathbb{P}$. Then, the combined action of $\mathbb{C}$ and $\mathbb{P}$ enacts the change we want:

$$\mathbb{C}\mathbb{P}(u_L^\dagger \bar{\sigma} \cdot W^+ d_L) = d_L^\dagger \bar{\sigma} \cdot W^- u_L. \tag{12.59}$$

Again, we emphasize that the mixing angles in the Lagrangian are unchanged by the action of C and P because they are just pure numbers, and not fields.

Now, with this particle–anti-particle transformation identified, we can act on the full Lagrangian. The two-generation quark Lagrangian of Eq. 12.51 thus transforms to itself under the action of CP:

$$\mathbb{C}\mathbb{P}[\mathcal{L}_W^{2G}] = \mathcal{L}_W^{2G}. \tag{12.60}$$

We then say that the two-generation quark Lagrangian is CP invariant. So far, interactions of quarks with the W bosons just violate parity, but preserve CP. By the CPT theorem, this also means that the two-generation Lagrangian must be invariant to time-reversal transformations. The story will be different with three generations of quarks, however.

Three-Generation Case

From earlier, the three quark generations' interactions with the W boson are

$$\mathcal{L}_W^{3G} = \frac{g_W}{\sqrt{2}} \begin{pmatrix} u_L^\dagger & c_L^\dagger & t_L^\dagger \end{pmatrix} V_{\text{CKM}} \bar{\sigma} \cdot W^+ \begin{pmatrix} d_L \\ s_L \\ b_L \end{pmatrix} \tag{12.61}$$

$$+ \frac{g_W}{\sqrt{2}} \begin{pmatrix} d_L^\dagger & s_L^\dagger & b_L^\dagger \end{pmatrix} V_{\text{CKM}}^\dagger \bar{\sigma} \cdot W^- \begin{pmatrix} u_L \\ c_L \\ t_L \end{pmatrix}.$$

The CKM matrix is now a general 3×3 unitary matrix, and as such is defined by three angles (the Euler angles) and six phases. One of these Euler angles can be identified with the Cabibbo angle, θ_C, and the two other angles represent mixings of other quarks. With these angles and phases, the three-generation CKM matrix can be expressed as

$$V_{\text{CKM}} = \begin{pmatrix} 1 & 0 & 0 \\ 0 & e^{i\delta}\cos\theta_{23} & e^{i(\delta+\zeta)}\sin\theta_{23} \\ 0 & -e^{i\epsilon}\sin\theta_{23} & e^{i(\epsilon+\zeta)}\cos\theta_{23} \end{pmatrix} \begin{pmatrix} \cos\theta_{13} & 0 & \sin\theta_{13} \\ 0 & 1 & 0 \\ -\sin\theta_{13} & 0 & \cos\theta_{13} \end{pmatrix}$$

$$\times \begin{pmatrix} e^{i\alpha}\cos\theta_C & -e^{i(\alpha+\gamma)}\sin\theta_C & 0 \\ e^{i\beta}\sin\theta_C & e^{i(\beta+\gamma)}\cos\theta_C & 0 \\ 0 & 0 & 1 \end{pmatrix}, \tag{12.62}$$

for angles θ_C, θ_{23}, and θ_{13} and phases $\alpha, \beta, \gamma, \delta, \epsilon, \zeta$.

Just as in the two-generation case, many of these phases are unphysical and can be removed by judicious choices for the phases of the quarks. As there are six quarks, we can remove five phases; one overall phase cannot be removed because the Lagrangian must be real (its phase is 0). We won't explicitly write those phase choices here, but once this has been done, the three-generation CKM matrix can then be expressed with three angles and a single phase:[4]

$$V_{\text{CKM}} = \begin{pmatrix} 1 & 0 & 0 \\ 0 & \cos\theta_{23} & \sin\theta_{23} \\ 0 & -\sin\theta_{23} & \cos\theta_{23} \end{pmatrix} \begin{pmatrix} \cos\theta_{13} & 0 & e^{-i\delta}\sin\theta_{13} \\ 0 & 1 & 0 \\ -e^{i\delta}\sin\theta_{13} & 0 & \cos\theta_{13} \end{pmatrix}$$

$$\times \begin{pmatrix} \cos\theta_C & \sin\theta_C & 0 \\ -\sin\theta_C & \cos\theta_C & 0 \\ 0 & 0 & 1 \end{pmatrix}. \tag{12.63}$$

Unlike the two-generation CKM matrix, the three-generation CKM matrix is irreducibly a complex matrix. Thus, in our universe, the CKM matrix is not equal to its complex conjugate:

$$V_{\text{CKM}}^* \neq V_{\text{CKM}}. \tag{12.64}$$

Because of this fact, quarks and anti-quarks in general couple to the W boson differently. As we developed above, the operation of turning quarks into anti-quarks is accomplished by a charge conjugation and a parity transformation. Now, acting on the Lagrangian with the combined operation of $\mathbb{CP}$ is not a symmetry:

$$\mathbb{CP}\left[\mathcal{L}_W^{3G}\right] = \frac{g_W}{\sqrt{2}} \begin{pmatrix} u_L^\dagger & c_L^\dagger & t_L^\dagger \end{pmatrix} V_{\text{CKM}}^\dagger \bar{\sigma} \cdot W^+ \begin{pmatrix} d_L \\ s_L \\ b_L \end{pmatrix} \tag{12.65}$$

$$+ \frac{g_W}{\sqrt{2}} \begin{pmatrix} d_L^\dagger & s_L^\dagger & b_L^\dagger \end{pmatrix} V_{\text{CKM}} \bar{\sigma} \cdot W^- \begin{pmatrix} u_L \\ c_L \\ t_L \end{pmatrix}$$

$$\neq \mathcal{L}_W^{3G}.$$

---

[4] L. L. Chau and W. Y. Keung, "Comments on the parametrization of the Kobayashi–Maskawa Matrix," Phys. Rev. Lett. **53**, 1802 (1984).

Box 12.1
Historical Profile: Helen Quinn

While the three-generation CKM matrix is a source of CP violation in the Standard Model, it is not the only source. Earlier, we noted that QCD conserves CP; this is mostly true. Any Feynman diagram that you write down with QCD interactions will preserve CP. However, there are possible sources of CP violation that would never be observed in Feynman diagrams, and come from global, topological properties of QCD, called instantons. One would in general expect instantons in QCD to break CP, but this has never been observed. This lack of CP violation in QCD is called the strong CP problem and the most promising solution was provided by Robert Peccei and Helen Quinn in 1977. They postulated the existence of a new particle, called an axion, that is responsible for preserving CP in QCD.[5] **Helen Quinn** is an Australian-American physicist who spent most of her career at (then called) Stanford Linear Accelerator Center (SLAC). In addition to her work on the strong CP problem, Quinn has made fundamental contributions to understanding unification of the forces of the Standard Model[6] and the physics of bottom quarks; specifically, ways of observing CP violation in the decays of hadrons that contain bottom quarks.[7] Quinn is also an author on what is perhaps the only particle physics paper that is presented as an imagined dialogue,[8] in a similar spirit to Galileo's *Dialogo sopra i due massimi sistemi del mondo*.

So, not only do the weak interactions violate parity P, because there are three generations of quarks, the weak interactions through the CKM matrix also violate the combined action of CP! As the combined action of CPT must be conserved by the Hermitivity of the Hamiltonian, we equivalently say that the CKM matrix violates time reversal T. Recall from Chapter 6 that in a Feynman diagram particles are turned into anti-particles by flipping the direction of time. The value of a Feynman diagram that involves quarks coupling to W bosons is different if the arrow of time is flipped.

By the way, it is possible that, for instance, θ_{13} is zero. As such, there would be no complex phase in the CKM matrix and therefore no CP violation. The best measured values of the angles and phase of the CKM matrix, however, are significantly different from zero. From the PDG, the value of the angles and complex phase in the CKM matrix are

$$\sin\theta_C = 0.2245\,, \qquad\qquad \sin\theta_{23} = 0.042\,, \qquad (12.66)$$
$$\sin\theta_{13} = 0.0035\,, \qquad\qquad \delta = 1.2\,.$$

Uncertainties on these values are at the level of the last quoted digit. None of these values is consistent with 0, and so the quark sector of the Standard Model does indeed violate CP.

---

[5] R. D. Peccei and H. R. Quinn, "CP conservation in the presence of instantons," Phys. Rev. Lett. **38**, 1440 (1977); R. D. Peccei and H. R. Quinn, "Constraints imposed by CP conservation in the presence of instantons," Phys. Rev. D **16**, 1791 (1977).

[6] H. Georgi, H. R. Quinn and S. Weinberg, "Hierarchy of interactions in unified gauge theories," Phys. Rev. Lett. **33**, 451 (1974).

[7] For example, A. E. Snyder and H. R. Quinn, "Measuring CP asymmetry in $B \rightarrow \rho\pi$ decays without ambiguities," Phys. Rev. D **48**, 2139 (1993).

[8] A. De Rujula, H. Georgi, S. L. Glashow and H. R. Quinn, "Fact and fancy in neutrino physics," Rev. Mod. Phys. **46**, 391 (1974).

12.2.4 Fermion Masses in the Standard Model and Tests of Unitarity

With the gauge-invariant interactions of the fermions of the Standard Model established, we then potentially have a mechanism for those fermions to acquire masses. As the masses of the fermions (leptons and quarks) are due to their coupling to the scalar field $\vec{\phi}$, we might be tempted to conclude that the masses of the fermions, as for the W and Z bosons, is a consequence of the Higgs mechanism. While this is somewhat of a common understanding, it is unfortunately not true. As we have argued extensively, the W and Z bosons, as spin-1 particles, cannot have a mass that is consistent with gauge invariance. The only way for these particles to consistently have a mass is via spontaneous symmetry breaking: the full theory is gauge invariant, but the ground state of the system breaks the manifest gauge invariance. The Higgs mechanism is required for the W and Z bosons to have masses.

Such a requirement, by contrast, does not exist for fermions. It is allowable for fermions to have a mass. When we introduced the Dirac equation in Chapter 2, the discussion was exclusively guided by Lorentz invariance, and a mass is perfectly compatible with that. A spin-1/2 particle has two degrees of freedom (spin-up and spin-down) for any mass. The number of degrees of freedom of a spin-1 particle, however, is two if it is massless (left- and right-handed helicity), but three if it is massive. Gauge invariance for massless spin-1 particles means that there are only two degrees of freedom. Massive spin-1 particles must gain a degree of freedom in some manner, and in the Standard Model, this is accomplished by the Higgs mechanism.

So where does this incorrect understanding come from? Because of the left-handed nature of the $SU(2)_W$ weak interactions, the only way within the Standard Model that a gauge-invariant mass arises is from fermions coupling to the scalar field $\vec{\phi}$. Fermion masses are then proportional to the vev v, and the proportionality constant for each fermion is called its Yukawa coupling. An honest mechanism for fermion masses would provide a theory of the Yukawa couplings, which in the Standard Model are just parameters. This is an especially challenging and subtle problem, as the masses of (electrically charged) fermions in the Standard Model range over about six orders of magnitude, from the electron to the top quark. Explaining the hierarchy of fermion masses is called the **flavor problem**, and no satisfactory solution has yet been proposed.

Unitarity of the CKM Matrix

Whether or not we have a solution to the flavor problem, the structure of quark masses and interactions with the W boson make definite predictions that we can test. The three generations of quarks mix because of their interactions with the W boson. Conservation of probability and the CPT theorem require that the CKM mixing matrix V_{CKM} is a unitary matrix such that

$$V_{\text{CKM}}^{\dagger} V_{\text{CKM}} = \mathbb{I}. \tag{12.67}$$

This is a concrete prediction that can be tested in data. Before we discuss the tests and validation that it is indeed unitary, let's discuss the consequences if it were observed to not be unitary.

If the CKM matrix were not unitary, one interpretation would be that probability is not conserved in the Standard Model. This would then seem to call into question the whole of quantum mechanics, which would be an extraordinary claim. A much more pedestrian possibility is simply that the Standard Model, as formulated, is incomplete. The CKM matrix only describes the mixing of three generations of quarks. If only those three generations mix, then the CKM matrix must be unitary, as we understand quantum mechanics. So, if it is observed that the CKM matrix is not unitary, then a possibility is that there is a fourth as-of-yet unobserved generation of quarks. A 3×3 matrix cannot describe generic unitary mixing of four generations of quarks. A measurement of non-unitarity would then be evidence for new quark flavors, which could be verified in other ways. Discovering new particles is always exciting, and the CKM matrix potentially points where to look for them.

We can express the 3×3 CKM matrix in the following form:

$$V_{\text{CKM}} = \begin{pmatrix} V_{ud} & V_{us} & V_{ub} \\ V_{cd} & V_{cs} & V_{cb} \\ V_{td} & V_{ts} & V_{tb} \end{pmatrix}. \tag{12.68}$$

Here, the entries are denoted by the amplitudes for quarks to be coupled through the W boson; for instance, V_{ud} is the amplitude for a u quark to turn into a d quark through emission of a W boson. Demanding that this matrix be unitary then enforces relationships between the entries of the matrix. The constraint from unitarity that is most commonly studied is

$$V_{ud} V_{ub}^* + V_{cd} V_{cb}^* + V_{td} V_{tb}^* = 0. \tag{12.69}$$

This can be interpreted as follows. Each of the three terms in this sum is a complex number and as such can be represented as a two-component vector in the complex plane. The sum of vectors equaling the 0 vector means that, when placed head to tail, the vectors close on themselves. Because there are three vectors, they thus form a triangle in the complex plane. It is therefore useful to measure each of them and see to what degree this so-called **unitarity triangle** closes.

The typical representation of this triangle equation is from dividing by the best-known term, $V_{cd} V_{cb}^*$. Doing this, we have

$$1 + \frac{V_{ud} V_{ub}^*}{V_{cd} V_{cb}^*} + \frac{V_{td} V_{tb}^*}{V_{cd} V_{cb}^*} = 0, \tag{12.70}$$

which is still a triangle, but has one side fixed to be length 1. It is standard to denote the term

$$-\frac{V_{ud} V_{ub}^*}{V_{cd} V_{cb}^*} = \bar{\rho} + i\bar{\eta}, \tag{12.71}$$

where $\bar{\rho}$ and $\bar{\eta}$ can be chosen to be positive real numbers. The sides and angles in the unitarity triangle can be determined from numerous data, such as decay rates of hadrons and quark masses calculated in lattice QCD. The most precise determination of the unitarity triangle from the CKMFitter group as of 2016 is presented in Fig. 12.1.

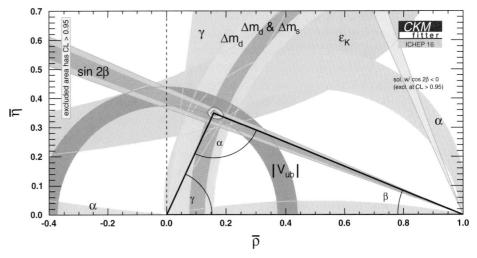

Fig. 12.1 Plot of the unitarity triangle from the CKMFitter group. Various constraints on the sides and angles of the triangle are represented by the shaded regions, and the consistent location of the vertex off of the real axis is hashed. From CKMfitter Group (J. Charles *et al.*), Eur. Phys. J. C41, 1-131 (2005) [hep-ph/0406184]; updated results and plots available at: `http://ckmfitter.in2p3.fr`.

The unitarity triangle is observed to close to extremely high precision, which therefore constrains possible extra generations of quarks which may interact through the W boson.

The area of the unitarity triangle is a parametrization-independent measure of the CP violation of the CKM matrix. It is known as the **Jarlskog invariant**, after the Swedish physicist Cecilia Jarlskog.[9] Note that only if there is an imaginary part (that is, $\bar{\eta} \neq 0$) is this triangle non-degenerate and so has a non-zero area. The current measured value of the Jarlskog invariant J is

$$J = 3.099^{+0.052}_{-0.063} \times 10^{-5}. \tag{12.72}$$

The superscripts and subscripts indicate the upper and lower uncertainties on the determination of J. This is inconsistent with 0 to many standard deviations, and so CP is definitely violated in the quark sector.

12.2.5 CP Violation and the Early Universe: Sakharov Conditions

One of the most striking features of our universe is the fact that there is overwhelmingly more matter than anti-matter. Hydrogen, the most abundant element in the universe, is composed of protons and electrons; to the best of our knowledge, anti-protons and positrons only exist as products of high-energy particle collisions, and not naturally as the composition of stars. The explanation of this matter–anti-matter asymmetry is especially

---

[9] C. Jarlskog, "Commutator of the quark mass matrices in the Standard Electroweak Model and a measure of maximal CP violation," Phys. Rev. Lett. **55**, 1039 (1985).

intriguing because the Standard Model does not seem to prefer matter over anti-matter in any way. We saw in the previous section that CP violation meant that quarks and anti-quarks (that is, matter and anti-matter) interact with W bosons differently. This does not, however, correspond to a particular preference, and cannot account for the matter–anti-matter asymmetry alone.

In the 1960s, this problem was becoming sharper because of a number of recent experimental results. As discussed a couple of chapters ago, C. S. Wu had discovered parity violation in the weak interactions in the late 1950s, CP violation in kaon decays had recently been observed,[10] and the cosmic microwave background had just been identified.[11] Motivated by these results, Andrei Sakharov formulated criteria for matter domination of the universe, which are now referred to as the **Sakharov conditions**.[12] Sakharov identified three criteria of the early universe which together are sufficient to produce an asymmetry in the number of baryons versus anti-baryons, or, correspondingly, an asymmetry in the number of quarks versus anti-quarks.

The three Sakharov conditions are:

1 The early universe must have processes that violate baryon-number conservation. That is, there must be a mechanism to change the number of baryons in the universe.
2 Both of C and CP must be violated. By the CPT theorem, CP violation implies that the early universe must not exhibit time-reversal symmetry.
3 The interactions must occur out of thermal equilibrium. In thermal equilibrium, the amounts of matter and anti-matter are dictated by the Boltzmann distribution, and would be identical as they would have the same energy distribution.

It turns out that all of these conditions can be met in the Standard Model. The second of the Sakharov conditions, C and CP violation, we have studied in detail. The weak interactions violate parity, and the CKM matrix violates CP, so therefore C must also be violated.

The first and third Sakharov conditions are more subtle, and can't easily be understood from properties of the Lagrangian of the Standard Model. For the first condition, it turns out that the vacuum of the Standard Model consists of an infinity of states with exactly the same energy, though separated by a potential energy barrier. Traversing or tunneling between different vacua are accomplished by objects called **sphalerons**. Unlike quantum tunneling, sphalerons can be thought of as thermal tunneling: at a non-zero temperature, there is a possibility for a thermal fluctuation to excite a system over a potential barrier. To release the acquired thermal energy, sphalerons emit particles of the Standard Model in such a way that the number of baryons (or quarks) is not conserved.

The third Sakharov condition can be accomplished through the electroweak phase transition. We observe in our universe today that the ground state of the electroweak theory

[10] J. H. Christenson, J. W. Cronin, V. L. Fitch and R. Turlay, "Evidence for the 2π decay of the K_2^0 meson," Phys. Rev. Lett. **13**, 138 (1964).
[11] A. A. Penzias and R. W. Wilson, "A measurement of excess antenna temperature at 4080 Mc/s," Astrophys. J. **142**, 419 (1965).
[12] A. D. Sakharov, "Violation of CP invariance, C asymmetry, and baryon asymmetry of the universe," Pisma Zh. Eksp. Teor. Fiz. **5**, 32 (1967) [JETP Lett. **5**, 24 (1967)] [Sov. Phys. Usp. **34**, 392 (1991)] [Usp. Fiz. Nauk **161**, 61 (1991)].

does not manifest all of the gauge symmetries. However, in the early universe, when the energy density and temperature were very high, then the particles of the Standard Model would not correspond to small fluctuations about the ground state. The electroweak theory would manifest all of its gauge symmetries. However, as the universe expanded and cooled, regions of the universe where the Standard Model settled to its ground state could form and, as they would be preferred energetically, expand and eventually fill the universe. The regions are called **bubbles**, in analogy with bubbles of steam that form in boiling water. The boiling of water is definitely a non-equilibrium process (liquid water is turning into steam), and this electroweak phase transition from a symmetric state to a spontaneously broken state is also out of thermal equilibrium.

Unfortunately, the observed matter–anti-matter asymmetry is much too large to be explained exclusively by the properties of the Standard Model. In particular, the CP violation of the CKM matrix (as parametrized by the Jarlskog invariant), while non-zero, is extremely small. So, it remains an open problem to provide a satisfactory explanation for the source of all of the Sakharov conditions and why we are composed exclusively of matter.

While his best-known scientific achievements were conditions for matter domination, Sakharov was also a leader of the nuclear bomb program in the USSR in the late 1940s and 1950s. Later, Sakharov was an outspoken critic of nuclear proliferation, human rights advocate, and Soviet dissident. This led to numerous conflicts with the Soviet government, including a forced exile in the city of Gorky, but he was ultimately a force for disintegration of the USSR.

12.3 The Weak Interactions in the Lepton Sector: Neutrino Mixing

In the previous section, we discussed the exceptionally weird property of the weak interactions of mixing different mass eigenstates of quarks. We found that the mass operator (from the measurement of energy and momentum) could not be simultaneously diagonalized with the flavor operator (defined by the interactions of the quarks with W bosons). Additionally, because there are three generations of quarks in the Standard Model, this mixing of quarks introduces CP violation: quarks and anti-quarks interact differently with the W boson.

The story with leptons, on the other hand, was actually quite simple. Unlike quarks, leptons do not mix into one another under the weak interactions. This argument requires neutrinos to be massless; in that case, we are allowed to arbitrarily rotate the neutrinos into one another to diagonalize both the mass matrix and the weak-interaction flavor matrix. The mass matrix for neutrinos in the Standard Model is the 0 matrix, and multiplying by any unitary matrix keeps the masses 0. So, this is the story in the Standard Model; is it true? That is, can we test this prediction?

For simplicity in most of the rest of this chapter, we assume that there are just two different neutrinos. This captures the majority of the interesting physics, and the extension to the case of three neutrinos (that is, three generations of leptons) is straightforward. The prediction in the Standard Model is that neutrinos are massless, and therefore the leptons do not mix under interactions with the weak force. If the leptons do not mix, then the type of neutrino that we somehow produce (in radioactive decay, for example) will be the type of neutrino that we always observe. Let's see if this is the case.

To do this, let's consider the concrete example of production of neutrinos at a nuclear reactor; say, the one at Reed College. The heart of a nuclear reaction, whether it be nuclear fission or fusion, is the decay of the neutron:

$$n \to p + e^- + \bar{\nu}_e. \tag{12.73}$$

By energy, momentum, angular momentum, and charge conservation, the only possible decay of the neutron is to a proton, electron, and electron anti-neutrino. Note, importantly, that this interaction occurs via the weak force, specifically through an intermediate W boson, and so the decay products are eigenstates of flavor (eigenstates of their interaction with the weak force). This is what enables us to state with certainty that the anti-neutrino is electron-type.

To see if the electron anti-neutrino mixes with other neutrino flavors, let's put a neutrino detector some distance away from the reactor; say, somewhere in eastern Oregon. How do we measure a neutrino? Essentially exactly opposite to how it was created. In Exercise 10.7 of Chapter 10 we discussed how this is done in the IceCube experiment. A neutrino that was traveling along strikes a proton in a target and produces a lepton and a neutron. Because this interaction of the neutrino and proton proceeds via the weak interaction, the flavor of the lepton produced must be the same as that of the initial neutrino (because the W boson couples to eigenstates of lepton flavor). So, we have an experiment that can test whether the neutrinos (and therefore leptons) mix under the weak interactions. We produce electron anti-neutrinos in the Reed reactor, let them travel to eastern Oregon, and then see what flavor of charged lepton is produced in the neutrino–proton scattering. Identifying the charged lepton, especially if it is an electron or muon, is pretty easy, so this experiment isn't too challenging, other than the fact that neutrinos interact very, very, very weakly with other matter!

The phenomenon of neutrino mixing is referred to as **neutrino oscillation**, and the physical description of this phenomenon is quite subtle. We discuss in detail the assumptions that are necessary for neutrino oscillation to exist and to correspondingly be observed in experiment.[13]

12.3.1 Neutrino Oscillations

Though there are three generations of neutrinos, as mentioned above, we consider the simpler case of just two generations. The extension to three generations is straightforward and does not add any interesting new physics. So, we identify the neutrinos of the flavor basis to be electron- and muon-type, while the two neutrinos of the mass basis will be

---

[13] This discussion follows closely the analysis presented in A. G. Cohen, S. L. Glashow and Z. Ligeti, "Disentangling neutrino oscillations," Phys. Lett. B **678**, 191 (2009) [arXiv:0810.4602 [hep-ph]].

simply neutrino 1 and neutrino 2. In the Standard Model, with massless neutrinos, of course the flavor and mass bases are identical, but we will not make that restricting assumption here. As the flavor basis, the electron and muon neutrinos have simple and well-defined interactions with the W boson, while the mass basis neutrinos 1 and 2 have well-defined masses m_1 and m_2, respectively. Let's see what the consequences are for the phenomenology of neutrinos with these two, in general different, bases.

Decay Product Entanglement

To begin, as we are considering (anti-)neutrinos produced from neutron decay, we need to understand neutron decay. The neutron decay reaction presented in Eq. 12.73 produces a proton, an electron, and an electron anti-neutrino. The quantum state of these products after neutron decay is special: the momentum, energy, and angular momentum of the particles are all correlated with one another, as required by the appropriate conservation laws. Such a state with strong correlations between particles is referred as an **entangled state**. For the decay of the neutron, we might express such an entangled state of its decay products as

$$|pe^- \bar{\nu}_e\rangle . \tag{12.74}$$

For our purposes here, we define an entangled state as one for which the quantum numbers of multiple particles are constrained by a collective conservation law. For example, in the rest frame of the neutron, the collective energy of the proton, electron, and anti-neutrino must be the mass of the neutron and their collective momentum must be 0. Additionally, the spins of the decay products must add to a total spin of 1/2, which is the spin of the neutron.

We have previously encountered entangled states in our discussion of isospin in Chapter 3, though we didn't use that term there. In Example 3.2, we considered the decomposition of the entangled state of two nucleons into irreducible representations of SU(2) isospin. Such a state was denoted as $|NN\rangle$ and the nucleons were entangled because of their collective isospin. The two irreducible representations, the isospin singlet and triplet, are distinguished by their total isospin. For example, consider the anti-symmetric singlet state:

$$\text{singlet} = \frac{1}{\sqrt{2}}(|pn\rangle - |np\rangle) . \tag{12.75}$$

The total isospin of this state is 0 and so any measurement that we make of this state must correspond to 0 isospin. Consider what the wavefunction tells us: if we measure nucleon 1 to be a proton, then with a probability of 1, nucleon 2 must be a neutron (and vice-versa). That is, this is an entangled state: by measuring just one of the nucleons, we know what the other nucleon must be by demanding that the total isospin of the two-nucleon state is conserved and 0.

Correspondingly, the individual properties of entangled particles are undetermined; all that is determined are the total values of some conserved quantity measured on the entire entangled state. Disentangling means collapsing the wavefunction to a state in which every particle has definite properties. For the example of the isospin singlet, the entangled state consists of a linear combination of all possible configurations of protons and neutrons that have a total of 0 isospin. By measuring nucleon 1 to be a proton, for example, we disentangle the two nucleons, collapsing to the state $|pn\rangle$ in which the nucleons are well defined.

Returning to the case of neutron decay, we can express the entangled state of the neutron decay products in the mass basis for the neutrinos. Because we just consider two generations of neutrinos, we can relate the flavor and mass bases with an orthogonal matrix:

$$\begin{pmatrix} |\nu_e\rangle \\ |\nu_\mu\rangle \end{pmatrix} = \begin{pmatrix} \cos\theta & \sin\theta \\ -\sin\theta & \cos\theta \end{pmatrix} \begin{pmatrix} |\nu_1\rangle \\ |\nu_2\rangle \end{pmatrix}, \tag{12.76}$$

where θ is the mixing angle. Of course, the general mixing matrix would be a 2×2 unitary matrix, but we can remove all phases in the two-generation case, as we did for the quarks earlier in this chapter. With this relationship between the bases established, we can equivalently express the neutron decay product state as

$$|pe^-\bar{\nu}_e\rangle = \cos\theta|pe^-\bar{\nu}_1\rangle + \sin\theta|pe^-\bar{\nu}_2\rangle. \tag{12.77}$$

On the right side of this expression, each entangled state implicitly conserves momentum and energy individually. Note that this is a bit non-trivial because anti-neutrinos $\bar{\nu}_1$ and $\bar{\nu}_2$ have different masses. However, the precise details of how this works aren't important for this discussion.

Written in this form, each entangled state on the right of Eq. 12.77 consists entirely of mass eigenstates. Therefore, they each individually propagate with a well-defined momentum. This is a necessary consideration because the point of neutron decay (the Reed reactor) and the detector (in eastern Oregon) are significantly separated in space, and so these states must propagate from one point to the other. Then, once the states are detected, we then reverse the decomposition of Eq. 12.77, expressing the mass eigenstates as flavor eigenstates, as it is through their interactions that they are detected.

However, this leads to a potential problem. If the decay products remain entangled throughout the propagation and detection, then you will, with probability 1, just observe an electron anti-neutrino in your detector. This is because the entangled state in Eq. 12.77 evolves with a well-defined momentum, and that momentum is constrained by the decay of the neutron. The anti-neutrino that you observe in your detector must be electron-type because all of the initial conservation laws from decay still hold. So, to observe neutrino oscillations, or to have a non-zero probability that you observe a muon anti-neutrino in your detector, the decay products must be disentangled so that the initial conservation laws do not apply. A way to accomplish this disentanglement is to observe all of the neutron decay products, and not just the anti-neutrino. Disentanglement happens because once you observe all decay products, the wavefunctions collapse, and the results of measurements are represented by classical probabilities and not quantum mechanical amplitudes.

We therefore assume that we have measured all decay products from the neutron, and not just the anti-neutrino, and so the states disentangle. As such, because the states have been disentangled, we can ignore the other decay products in our analysis, and exclusively focus on the properties of the neutrinos. This is what we do in the following.

Requirements for Oscillation

With the assumption that the neutron decay products are disentangled, we then only have to consider the dynamics of the anti-neutrinos (which is what we wanted to do in the first place!). At the time of neutron decay, the anti-neutrino is electron flavor and can be decomposed into mass eigenstates as

$$|\bar{\nu}_e\rangle = \cos\theta|\bar{\nu}_1\rangle + \sin\theta|\bar{\nu}_2\rangle. \tag{12.78}$$

As fermions, anti-neutrinos $\bar{\nu}_1$ and $\bar{\nu}_2$ satisfy the Dirac equation, and therefore propagate according to the Klein–Gordon equation. As they are mass eigenstates (eigenstates of the squared-momentum operator), their respective Klein–Gordon equations are

$$(\partial_\mu\partial^\mu + m_1^2)|\bar{\nu}_1\rangle = 0, \qquad\qquad (\partial_\mu\partial^\mu + m_2^2)|\bar{\nu}_2\rangle = 0. \tag{12.79}$$

These wavefunctions therefore propagate (have position and time dependence) as

$$|\bar{\nu}_1\rangle \propto e^{-ip_1\cdot x}, \qquad\qquad |\bar{\nu}_2\rangle \propto e^{-ip_2\cdot x}, \tag{12.80}$$

where p_1 and p_2 are the four-momenta of the mass eigenstates and $x = (t,\vec{x})$ is the spacetime position four-vector. Then, at a generic time t after neutron decay, the electron anti-neutrino is composed of the mass eigenstates as

$$|\bar{\nu}_e(t)\rangle = e^{-ip_1\cdot x}\cos\theta|\bar{\nu}_1\rangle + e^{-ip_2\cdot x}\sin\theta|\bar{\nu}_2\rangle. \tag{12.81}$$

While this is time evolution of the electron anti-neutrino, there are other things that need to be accounted for to ensure that oscillations exist. First, the time of neutron decay is not fixed: it is exponentially distributed according to its lifetime, τ_n. For the neutron, the lifetime is approximately $\tau_n = 900$ s. This lifetime sets the size of the wave-packet of the propagating anti-neutrinos $\bar{\nu}_1$ and $\bar{\nu}_2$. That is, the neutron has approximately uniform probability to decay anytime within 900 seconds. Anti-neutrinos with a velocity v from neutron decay will be spread in space over a characteristic distance $\sigma = \tau_n v$. Therefore, the wave-packets of anti-neutrinos $\bar{\nu}_1$ and $\bar{\nu}_2$ have characteristic spreads of $\sigma_1 = \tau_n v_1$ and $\sigma_2 = \tau_n v_2$, respectively.

Additionally, because the wave-packets for $\bar{\nu}_1$ and $\bar{\nu}_2$ travel at different velocities in general, they will separate from one another in space. Thus, after propagating for a time t, the two wave-packets will have traveled different distances from the point of decay. In particular, the distance d_1 that $\bar{\nu}_1$ travels in a time t is $d_1 = tv_1$, while the distance d_2 that $\bar{\nu}_2$ travels is $d_2 = tv_2$. By the way, the velocities v_1 or v_2 of the two anti-neutrinos are their **group velocity**, because we are considering eigenstates of momentum. The group velocity is the ratio of the relativistic momentum to the energy of the anti-neutrinos. Recall that for a particle of mass m, the magnitude of relativistic momentum $|\vec{p}|$ and relativistic energy E are (in natural units)

$$|\vec{p}| = \gamma mv, \qquad\qquad E = \gamma m, \tag{12.82}$$

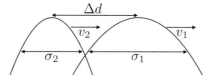

Fig. 12.2   Illustration of the anti-neutrino wave-packets at time t. In this figure, we assume that $v_1 > v_2$, and so the width of wave-packet 1 is larger than that of 2: $\sigma_1 > \sigma_2$. The separation of the centers of the wave-packets is denoted by Δd.

where γ is the boost factor. Then, the velocity v is

$$v = \frac{|\vec{p}|}{E}. \tag{12.83}$$

With this set-up, after the anti-neutrino wave-packets have propagated for a time t, we have a configuration that looks like that displayed in Fig. 12.2. For there to be neutrino oscillations observed at the detector, the wave-packets must have significant overlap; otherwise they cannot interfere quantum mechanically. Requiring that the wave-packets significantly overlap leads to constraints on the spread of the wave-packets relative to their separation. In particular, the separation of or distance between the two wave-packets Δd after propagating for a time t is just

$$\Delta d = |d_1 - d_2| = t|v_1 - v_2|. \tag{12.84}$$

Additionally, the combined spread or width of the two wave-packets σ_{12} is just the sum of their individual spreads:

$$\sigma_{12} = \sigma_1 + \sigma_2 = \tau_n(v_1 + v_2). \tag{12.85}$$

Therefore, for the wave-packets to have significant overlap at the detector, we need to require that their separation Δd be much smaller than their combined width σ_{12}:

$$\Delta d \ll \sigma_{12} \quad \Rightarrow \quad \left| \frac{v_1 - v_2}{v_1 + v_2} \right| \ll \frac{\tau_n}{t}. \tag{12.86}$$

To summarize, to observe neutrino oscillation (that is, to observe initial flavor eigenstate neutrinos turning into an orthogonal flavor eigenstate at the detector) requires two things to happen:

1 the neutron decay products must be disentangled to eliminate strict conservation laws of the quantum states, and
2 the neutrino wave-packets must overlap significantly (Eq. 12.86) so that they can interfere quantum mechanically.

This second constraint tells us how far away we can put our detector to observe neutrino oscillations. If the distance is too great, then the mass eigenstate wave-packets will have separated too much, will not interfere, and exhibit no oscillations. For neutrinos from neutron decay, then, how far can the detector be from the source to observe oscillations?

Example 12.2 Over what distance do neutrinos' wave-packets overlap, enabling oscillations?

Solution

To good approximation, the velocities of the neutrinos are both the speed of light $c = 1$ because their masses are so small, and so we approximate

$$v_1 + v_2 \simeq 2. \tag{12.87}$$

Additionally, the distance traveled from the source to the detector by the neutrinos d over time t is approximately

$$t = d, \tag{12.88}$$

in natural units. So, we have the constraint

$$|v_1 - v_2| \ll \frac{2\tau_n}{d}. \tag{12.89}$$

We now need to evaluate the velocity difference $|v_1 - v_2|$ in a useful form.

As discussed above, we can express the velocities of the anti-neutrinos as the ratio of relativistic momentum to energy. The relativistic velocity is

$$v = \frac{|\vec{p}|}{E} = \frac{1}{E}\sqrt{E^2 - m^2}. \tag{12.90}$$

For neutron decays, the energy of the anti-neutrinos will be on the order of MeV, while their masses are on the order of eV, so we can Taylor expand. The velocity is then approximately

$$v = \sqrt{1 - \frac{m^2}{E^2}} \simeq \left(1 - \frac{m^2}{2E^2}\right). \tag{12.91}$$

Therefore, assuming that the anti-neutrinos have similar energies but different masses, their velocity difference is

$$|v_1 - v_2| \simeq \frac{|m_1^2 - m_2^2|}{2E^2}. \tag{12.92}$$

Then, the requirement that the anti-neutrinos have significant wave-packet overlap is

$$\frac{|m_1^2 - m_2^2|}{2E^2} \ll \frac{2\tau_n}{d}, \tag{12.93}$$

or, that they stay overlapping as long as the distance d satisfies

$$d \ll 4\frac{E^2}{|m_1^2 - m_2^2|}\tau_n. \tag{12.94}$$

With the masses of the anti-neutrinos about 1 eV, the energy–mass ratio is approximately

$$\frac{E^2}{|m_1^2 - m_2^2|} \simeq \left(\frac{1 \text{ MeV}}{1 \text{ eV}}\right)^2 \simeq 10^{12}. \tag{12.95}$$

Additionally, the lifetime of the neutron in meters is approximately

$$c\tau_n \simeq (3 \times 10^8) \times (900) \text{ m} \simeq 3 \times 10^{11} \text{ m}. \tag{12.96}$$

Therefore, the two mass eigenstates of anti-neutrinos will remain coherent with significant wave-packet overlap over distances d such that

$$d \ll 4 \times 10^{12} \times 3 \times 10^{11} \text{ m} \simeq 10^{24} \text{ m} \simeq 10^8 \text{ ly}. \tag{12.97}$$

So neutrinos will exhibit oscillations over millions of light-years!

Oscillation Probability Calculation

We have now identified the relevant requirements for neutrino oscillation to occur: the neutron decay products must be disentangled, and the neutrino detector cannot be too far from the source. This latter requirement is quite trivial for any experiment we might imagine, from the result of Example 12.2.

With this set-up, we want to see how this mixing affects the probability for an electron anti-neutrino to be observed far from its source. A picture of the experiment, with artistic license regarding the non-physics parts, is:

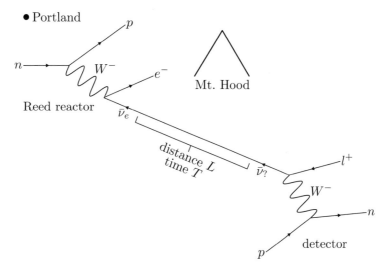

That is, the neutron decays in the reactor at Reed College, in Portland, Oregon, the anti-neutrino from the decay propagates from the source into eastern Oregon, and then is subsequently observed in our detector. From earlier, and following from the assumptions that we discussed above, the electron anti-neutrino at the source (time $t = 0$) can be decomposed into mass eigenstates as

$$|\bar{\nu}_e(t = 0)\rangle = \cos\theta|\bar{\nu}_1\rangle + \sin\theta|\bar{\nu}_2\rangle. \tag{12.98}$$

After time $t = T$, the anti-neutrino reaches the detector, and the mass eigenstates have picked up appropriate phases from their propagation. So, at time $t = T$ the electron anti-neutrino has evolved into

$$|\bar{\nu}_e(t = T)\rangle = e^{-ip_1 \cdot x} \cos\theta|\bar{\nu}_1\rangle + e^{-ip_2 \cdot x} \sin\theta|\bar{\nu}_2\rangle. \tag{12.99}$$

Here, the spacetime position four-vector is

$$x = (T, 0, 0, L), \tag{12.100}$$

where L is the distance from the source to the detector. If the mixing angle θ is 0 (that is, if the mass eigenbasis is identical to the flavor basis), then we always observe an electron anti-neutrino in eastern Oregon.

With the assumption that the anti-neutrino wave-packets have significant overlap at the detector, we can rewrite the spacetime position vector x in terms of the momentum and energy of the anti-neutrinos. The region of maximal overlap corresponds to the center of mass of the two wave-packets. The velocity of the center of mass of the wave-packets is just the average group velocity $\bar{v}$ of the anti-neutrinos. The average group velocity is just the ratio of the average momentum to the average energy:

$$\bar{v} = \frac{|\vec{p}_1 + \vec{p}_2|}{E_1 + E_2}. \tag{12.101}$$

To see that this corresponds to the center-of-mass velocity, it is useful to take the non-relativistic limit. In that limit the sum of the momenta is

$$|\vec{p}_1 + \vec{p}_2| \xrightarrow{\text{n-r}} |m_1 \vec{v}_1 + m_2 \vec{v}_2|, \tag{12.102}$$

while the sum of the energies is

$$E_1 + E_2 \xrightarrow{\text{n-r}} m_1 + m_2, \tag{12.103}$$

because the kinetic energies are negligible compared to the masses. Then, the average velocity is

$$\bar{v} \xrightarrow{\text{n-r}} \frac{|m_1 \vec{v}_1 + m_2 \vec{v}_2|}{m_1 + m_2}, \tag{12.104}$$

which is indeed the velocity of the center of mass. For particles traveling in the same direction, we can of course drop the vectors.

With this average velocity, we identify it with the ratio of the distance to the detector L over the time elapsed T:

$$\bar{v} = \frac{|\vec{p}_1 + \vec{p}_2|}{E_1 + E_2} = \frac{L}{T}. \tag{12.105}$$

Then, the spacetime position four-vector is

$$x = (T, 0, 0, L) = T\left(1, 0, 0, \frac{|\vec{p}_1 + \vec{p}_2|}{E_1 + E_2}\right) = \frac{T}{E_1 + E_2}(p_1 + p_2), \tag{12.106}$$

proportional to the sum of the anti-neutrinos' momentum four-vectors. Using this result, the time-evolved electron anti-neutrino at the detector is

$$|\bar{\nu}_e(t = T)\rangle = e^{-ip_1 \cdot x}\left[\cos\theta |\bar{\nu}_1\rangle + e^{i(p_1 - p_2)\cdot x} \sin\theta |\bar{\nu}_2\rangle\right] \tag{12.107}$$

$$= e^{-ip_1 \cdot x}\left[\cos\theta |\bar{\nu}_1\rangle + e^{i\frac{T}{E_1 + E_2}(p_1 - p_2)\cdot(p_1 + p_2)} \sin\theta |\bar{\nu}_2\rangle\right]$$

$$= e^{-ip_1 \cdot x}\left[\cos\theta |\bar{\nu}_1\rangle + \exp\left[i\frac{T(m_1^2 - m_2^2)}{E_1 + E_2}\right] \sin\theta |\bar{\nu}_2\rangle\right].$$

We can simplify this expression a bit. First, in the extreme relativistic limit, the energies of the two mass eigenstate anti-neutrinos from neutron decay will be very similar, so we will just set

$$E_1 + E_2 = 2E,\tag{12.108}$$

where E is a characteristic energy. Additionally, up to negligible corrections, we can replace the time of flight from the source to the detector T with the distance L. Also, the overall phase of a wavefunction has no measurable consequences, so we can safely ignore the factor outside the square brackets in Eq. 12.107. Then, the wavefunction of the electron anti-neutrino at time T can be expressed as

$$|\bar{\nu}_e(t=T)\rangle = \cos\theta|\bar{\nu}_1\rangle + \exp\left[i\frac{L(m_1^2 - m_2^2)}{2E}\right]\sin\theta|\bar{\nu}_2\rangle.\tag{12.109}$$

This is then the anti-neutrino that makes it to our detector. Our detector works by identifying the flavor of the charged lepton produced in the scattering of the anti-neutrino and a proton. While mass eigenstates have simple propagation properties, we need to re-express the anti-neutrino at time $t = T$ in terms of its flavor or weak eigenstates to determine how it interacts and what we observe.

From earlier, the mass eigenstate anti-neutrinos decompose into flavor eigenstates as

$$|\bar{\nu}_1\rangle = \cos\theta|\bar{\nu}_e\rangle - \sin\theta|\bar{\nu}_\mu\rangle,\tag{12.110}$$
$$|\bar{\nu}_2\rangle = \sin\theta|\bar{\nu}_e\rangle + \cos\theta|\bar{\nu}_\mu\rangle.$$

In terms of $\bar{\nu}_e$ and $\bar{\nu}_\mu$ states, then, the anti-neutrino at our detector is

$$|\bar{\nu}_e(t=T)\rangle = \cos^2\theta|\bar{\nu}_e\rangle - \cos\theta\sin\theta|\bar{\nu}_\mu\rangle + \exp\left[i\frac{L(m_1^2 - m_2^2)}{2E}\right]\sin^2\theta|\bar{\nu}_e\rangle\tag{12.111}$$

$$+ \exp\left[i\frac{L(m_1^2 - m_2^2)}{2E}\right]\cos\theta\sin\theta|\bar{\nu}_\mu\rangle$$

$$= \left(\cos^2\theta + \exp\left[i\frac{L(m_1^2 - m_2^2)}{2E}\right]\sin^2\theta\right)|\bar{\nu}_e\rangle$$

$$- \cos\theta\sin\theta\left(1 - \exp\left[i\frac{L(m_1^2 - m_2^2)}{2E}\right]\right)|\bar{\nu}_\mu\rangle.$$

Then, the amplitudes for an electron anti-neutrino to be produced at Reed and observed as an electron anti-neutrino or a muon anti-neutrino in our detector in Baker City are given by the wavefunction overlaps:

$$\mathcal{M}(\bar{\nu}_e \to \bar{\nu}_e) = \langle\bar{\nu}_e|\bar{\nu}_e(t=T)\rangle = \cos^2\theta + \exp\left[i\frac{L(m_1^2 - m_2^2)}{2E}\right]\sin^2\theta,\tag{12.112}$$

$$\mathcal{M}(\bar{\nu}_e \to \bar{\nu}_\mu) = \langle\bar{\nu}_\mu|\bar{\nu}_e(t=T)\rangle = -\cos\theta\sin\theta\left(1 - \exp\left[i\frac{L(m_1^2 - m_2^2)}{2E}\right]\right).$$

The corresponding detection probabilities are the absolute squares of these expressions:

$$P(\bar{\nu}_e \to \bar{\nu}_e) = |\mathcal{M}(\bar{\nu}_e \to \bar{\nu}_e)|^2 = 1 - \sin^2(2\theta)\sin^2\left(\frac{L}{4E}(m_1^2 - m_2^2)\right), \qquad (12.113)$$

$$P(\bar{\nu}_e \to \bar{\nu}_\mu) = |\mathcal{M}(\bar{\nu}_e \to \bar{\nu}_\mu)|^2 = \sin^2(2\theta)\sin^2\left(\frac{L}{4E}(m_1^2 - m_2^2)\right).$$

To write these expressions, double- and half-angle trigonometric identities were used liberally. Note, importantly, that the sum of probabilities is 1; that is, the electron anti-neutrino turns into something in its travel across the Cascades.

These expressions are exceptionally interesting. Equation 12.113 manifests why this is referred to as neutrino *oscillations*. The probability for detecting a particular flavor of neutrino oscillates with the distance L from the source to the detector. Note that the anti-neutrinos only mix if they are massive, which we already knew from our previous study of the quarks. However, even more interesting is the fact that they only mix if they have different masses! That is, if $\bar{\nu}_1$ and $\bar{\nu}_2$ had the same mass $m = m_1 = m_2$, we could diagonalize the flavor and mass operators simultaneously. (Can you convince yourself of this?) If that were the case, then there would be 0 probability to observe a muon anti-neutrino.

We have also put our detector somewhere in eastern Oregon to observe neutrino oscillations. Why is this reasonable? Or, rather, to what range of mass-squared differences does this make us sensitive? Plugging back in the cs and $\hbar$s, the argument of the $\sin^2$ is

$$\sin^2\left(\frac{c^3 L}{4\hbar E}(m_1^2 - m_2^2)\right) = \sin^2\left(1.27 \times 10^3 (m_1^2 - m_2^2)(\text{eV}^2)\frac{L(\text{km})}{E(\text{MeV})}\right). \qquad (12.114)$$

Here, the dimensions have been removed, and we evaluate the distance L in kilometers and the energy of the anti-neutrino in MeV, and measure the mass splitting in eV^2. That is, to measure eV neutrino masses, we should indeed have the detector many kilometers away from our reactor. The distance L between the source of neutrinos and the detector is referred to as the **baseline**.

Some other things to note: measuring neutrino oscillations is only sensitive to the squared-mass difference. It says nothing about the absolute mass scale; that is, what m is. So, all we can tell from neutrino oscillation is that at least one neutrino has mass that is not zero, and is different from the masses of the other neutrinos. Additionally, to exhibit oscillations, we needed to have the anti-neutrino propagate a long distance. This means that we make no measurements from the source to the final detection point. If instead we continually measured the momentum of the neutron decay products, then the mass eigenstate anti-neutrinos would never pick up the phase factor $\exp[-ip \cdot x]$. If that phase factor were not present, then there would be 0 probability to observe a muon anti-neutrino at any point. This is an example of the **quantum Zeno effect**.

This analysis contains most of the physics of neutrino oscillation, but can be extended to the general case of arbitrary numbers of neutrinos. In particular, the Standard Model has three neutrinos, which in general mix if they have different masses. As with the CKM matrix that defined the mixings of the quarks, there is a 3×3 matrix that describes the mixing of the neutrinos. It is called the **Pontecorvo–Maki–Nakagawa–Sakata (PMNS)**

matrix and is parametrized by three angles (the Euler angles) and one complex phase.[14] As with the quarks, the existence of a complex phase means that, generically, neutrino oscillations of three generations violate CP. However, unlike the case for quarks, the complex phase of the PMNS matrix has not yet been measured, so it is not known with certainty what the CP violation of the neutrino sector is.

12.3.2 Neutrino Oscillation Measurement

This analysis of neutrino oscillation makes a clear prediction: if the mass of at least one of the neutrinos is non-zero and at least two masses are distinct, then the probability of detecting a particular flavor of neutrino depends on the distance from the source and the energy of the neutrino, as defined through Eq. 12.113. With precise enough measurements, it would then be possible to measure mixing angles of the PMNS matrix and the squared-mass differences of the neutrinos. Like the unitarity triangle for the CKM matrix, a measurement of the neutrino mixing angles would then provide indirect evidence for the existence or non-existence of a fourth generation of leptons.

To perform this experiment, we need an extremely high flux of (anti-)neutrinos whose energies are well calibrated. As suggested earlier, such a source of anti-neutrinos is from a nuclear reactor, as electron anti-neutrinos are produced in processes related to neutron decay. One of the most prolific sources of electron anti-neutrinos from a nuclear reactor is at the Daya Bay nuclear power complex, located in Daya Bay, China.[15] There are three experimental halls which house anti-neutrino detectors, at distances ranging from a few hundred meters to nearly two kilometers from the nuclear power plants. These multiple detectors enable corroborating measurements of anti-neutrinos with various baseline distance to energy (L/E) ratios that determine the probability oscillation frequency. As the Daya Bay experiment uses nuclear reactors as the source of electron anti-neutrinos, it is particularly sensitive to the mixing of mass eigenstate $\bar{\nu}_1$ with the other anti-neutrinos. Measurements of other neutrino properties through numerous and varied experiments are collected in the PDG. Here, we just discuss the Daya Bay experiment and its observation of neutrino oscillations.

The easiest charged lepton to measure and observe is the electron (or positron), as the electron is a stable particle and does not decay. The Daya Bay experiment only observes positrons through **inverse β-decay** interactions between electron anti-neutrinos and protons in the atomic nuclei of the detectors. The inverse β-decay process is

$$\bar{\nu}_e + p \rightarrow n + e^+ , \tag{12.115}$$

which is related to neutron decay by crossing symmetry. Because of this, the Daya Bay experiment only measures the survival probability of electron anti-neutrinos: the probability that an anti-neutrino produced at the nuclear reactor survives to be observed

[14] Z. Maki, M. Nakagawa and S. Sakata, "Remarks on the unified model of elementary particles," Prog. Theor. Phys. **28**, 870 (1962); B. Pontecorvo, "Neutrino experiments and the problem of conservation of leptonic charge," Sov. Phys. JETP **26**, 984 (1968) [Zh. Eksp. Teor. Fiz. **53**, 1717 (1967)].

[15] X. Guo *et al.* [Daya Bay Collaboration], "A precision measurement of the neutrino mixing angle θ_{13} using reactor antineutrinos at Daya-Bay" [arXiv:hep-ex/0701029].

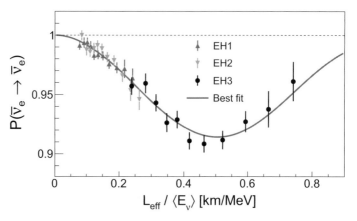

Fig. 12.3 Plot of the electron anti-neutrino survival probability $P(\bar{\nu}_e \to \bar{\nu}_e)$ as a function of the ratio of the baseline L to the anti-neutrino energy E_ν. Results from the three detectors (experimental halls, or EH) are shown and the errors only include statistical uncertainties. Reprinted figure with permission from F. P. An *et al.* [Daya Bay Collaboration], Phys. Rev. Lett. **115**, no. 11, 111802 (2015). Copyright 2015 by the American Physical Society.

by a detector. That is, the Daya Bay experiment measures the probability $P(\bar{\nu}_e \to \bar{\nu}_e)$ as a function of distance from the source and anti-neutrino energy.

A complete treatment of three-generation neutrino oscillation to which the Daya Bay experiment is sensitive requires a parametrization of the PMNS matrix and a more detailed analysis than we presented in the two-generation case in the previous section. However, the measured mass differences between the neutrinos significantly simplify the analysis. It has been observed that the mass difference between the first two mass eigenstates $\bar{\nu}_1$ and $\bar{\nu}_2$ is significantly less than the mass difference between the first and third $\bar{\nu}_3$ mass eigenstates:[16]

$$|m_2^2 - m_1^2| \ll |m_3^2 - m_1^2| . \tag{12.116}$$

Therefore, the rate of oscillation between the first two mass eigenstates is significantly less than between the first and third eigenstates. Therefore, to good approximation, we can just assume that there are two relevant generations of neutrinos. As such, the expression for the survival probability is just what we found in the previous section:

$$P(\bar{\nu}_e \to \bar{\nu}_e) \simeq 1 - \sin^2 2\theta_{13} \sin^2 \left(1.27 \times 10^3 (m_3^2 - m_1^2)(\text{eV}^2) \frac{L(\text{km})}{E(\text{MeV})} \right) . \tag{12.117}$$

We have also written explicitly the units of the quantities in this expression.

The results of the Daya Bay experiment are shown in Fig. 12.3. The survival probability $P(\bar{\nu}_e \to \bar{\nu}_e)$ is plotted versus the baseline in kilometers divided by the energy of the anti-neutrino in MeV, as this is the relevant ratio in the survival probability. The oscillation of the survival probability is unambiguous and provides a very precise measurement of the mixing angle θ_{13} and the mass difference $m_3^2 - m_1^2$. From these results, Daya Bay found

$$\sin^2 2\theta_{13} = 0.084 \pm 0.005 , \qquad |m_3^2 - m_1^2| = (2.37 \pm 0.14) \times 10^{-3} \text{ eV}^2 . \tag{12.118}$$

---

[16] T. Araki *et al.* [KamLAND Collaboration], "Measurement of neutrino oscillation with KamLAND: Evidence of spectral distortion," Phys. Rev. Lett. **94**, 081801 (2005) [arXiv:hep-ex/0406035].

These are among the most precise measurements of these mixing parameters and mass differences.

Example 12.3 Using Fig. 12.3, can we estimate these values for the mixing angle $\sin^2 2\theta_{13}$ and mass difference $|m_3^2 - m_1^2|$?

Solution

To do this, note that the mixing angle sets the amplitude of oscillation and the mass difference sets the frequency of oscillation. The first trough appears at approximately

$$\frac{L}{E} = 0.5 \frac{\text{km}}{\text{MeV}}, \tag{12.119}$$

and $\sin^2$ is maximized when its argument is equal to $\pi/2$. (Note that the maximum corresponds to the trough because it is subtracted from 1.) Setting the argument of $\sin^2$ in Eq. 12.117 to $\pi/2$, we have

$$1.27 \times 10^3 (m_3^2 - m_1^2)(\text{eV}^2) \frac{L(\text{km})}{E(\text{MeV})} \simeq 1.27 \times 10^3 (m_3^2 - m_1^2) \cdot 0.5 = \frac{\pi}{2}, \tag{12.120}$$

or that

$$|m_3^2 - m_1^2| \simeq 2.47 \times 10^{-3} \text{ eV}^2. \tag{12.121}$$

This is within uncertainties of the value that Daya Bay extracted.

For the mixing angle, we just look at the amplitude of the trough. At the trough, the amplitude is approximately 0.92, corresponding to

$$1 - \sin^2 2\theta_{13} \simeq 0.92, \tag{12.122}$$

or that

$$\sin^2 2\theta_{13} \simeq 0.08. \tag{12.123}$$

This is within uncertainties of the Daya Bay value.

12.3.3 Neutrino Astrophysics

Neutrino oscillations imply that neutrinos must be massive, which is in contrast to the Standard Model prediction. Because they interact so weakly with matter, they are really hard to measure directly, unless there is an exceedingly large flux of neutrinos. One way to measure neutrinos that has a sufficient flux and can be used to set bounds on the absolute scale of neutrino masses is through observation of a **supernova**. In the collapse of a star under its own gravity, unimaginable densities are created that fuel the largest explosions in the universe. In the explosion of a supernova, the light produced in the explosion must pass through the plasma of high-temperature matter. This plasma consists of ionized protons and electrons (i.e., charged particles), so the photons have a very small mean free path. They bounce around in this plasma like ping pong balls in a washing machine, interacting with the protons and electrons that impede them from propagating out. Photons get stuck in the

A composite image from the Hubble Space Telescope of the remnants of SN 1987a, identified as the double ring structure in the center of the image. Credit: NASA, ESA, R. Kirshner (Harvard-Smithsonian Center for Astrophysics and Gordon and Betty Moore Foundation), and M. Mutchler and R. Avila (STScI).

supernova for an extended period of time. Neutrinos, unlike photons, really do not like to interact with other particles. So, while the photons are stuck in the supernova, neutrinos pass right through, with essentially no impedance.

In February of 1987, a supernova explosion was directly observed on Earth.[17] It was observed optically and studied extensively by astronomers for several months, and was even visible to the naked eye. A composite image of SN 1987a taken over a number of years after the original explosion is displayed in Fig. 12.4. Unlike most astrophysical observations, the first detection of SN 1987a was actually via neutrinos. Because the explosion of a star or supernova is governed by nuclear processes, we should expect copious neutrino production. Because of the delay in the photons escaping the supernova, neutrinos from SN 1987a actually arrived at Earth before the photons. There was a good 2 hours between them!

Several neutrino detection sites around the globe measured a total of 25 neutrinos from SN 1987a.[18] This was enough, however, to place an upper bound on the mass of the neutrinos of about 16 eV. If the neutrinos were more massive, then they would have taken longer to get to Earth, and might not have beaten the light in getting here. Also, though only

[17] W. Kunkel and B. Madore, *IAU Circular No. 4316 (24 February 1987)*, International Astronomical Union (1987).

[18] R. M. Bionta *et al.*, "Observation of a neutrino burst in coincidence with supernova SN 1987a in the Large Magellanic Cloud," Phys. Rev. Lett. **58**, 1494 (1987); K. S. Hirata *et al.*, "Observation in the Kamiokande-II detector of the neutrino burst from supernova SN 1987a," Phys. Rev. D **38**, 448 (1988); E. N. Alekseev, L. N. Alekseeva, I. V. Krivosheina and V. I. Volchenko, "Detection of the neutrino signal from SN1987A in the LMC using the Inr Baksan underground scintillation telescope," Phys. Lett. B **205**, 209 (1988).

a whopping 25 neutrinos were observed on Earth, it is believed that 99% of the energy in a supernova explosion is contained in neutrinos. In SN 1987a, these neutrinos observed on Earth were consistent with models that predicted 10^{58} neutrinos produced for a total energy of 10^{46} Joules. The Sun would have to shine for about a trillion years to match the energy in neutrinos in SN 1987a!

Exercises

12.1 *Mass and Flavor Basis Commutator.* We started this chapter with a discussion of the non-commutativity of the mass and flavor bases for expressing the different generations of quarks. In this exercise, we study the commutator of the mass and flavor operators. For the up-type quarks, for example, the mass matrix from Eq. 12.40 is

$$\mathbb{M}_u = \begin{pmatrix} \frac{y_u v}{\sqrt{2}} & & \\ & \frac{y_c v}{\sqrt{2}} & \\ & & \frac{y_t v}{\sqrt{2}} \end{pmatrix} . \tag{12.124}$$

The CKM matrix was formed from the product of unitary matrices that related the up and down quark flavor and mass bases:

$$V_{\text{CKM}} = \mathbb{U}_u^\dagger \mathbb{U}_d . \tag{12.125}$$

We argued that the CKM matrix V_{CKM} could not be brought into diagonal form, but we are still free to choose the down quark mass and flavor bases to be identical. This sets the mixing matrix $\mathbb{U}_d$ to be the identity:

$$\mathbb{U}_d = \mathbb{I} . \tag{12.126}$$

We will work with in this basis in this exercise.

(a) In this basis, the up quark mixing matrix is then just the CKM matrix:

$$V_{\text{CKM}} = \mathbb{U}_u^\dagger . \tag{12.127}$$

A particularly useful parametrization of the CKM matrix is called the **Wolfenstein parametrization**.[19] In this form, we express the CKM matrix as

$$V_{\text{CKM}} \simeq \begin{pmatrix} 1 - \lambda^2/2 & \lambda & A\lambda^3(\rho - i\eta) \\ -\lambda & 1 - \lambda^2/2 & A\lambda^2 \\ A\lambda^3(1 - \rho - i\eta) & -A\lambda^2 & 1 \end{pmatrix} , \tag{12.128}$$

where A, λ, ρ, and η are real numbers. λ is assumed to be very small, $\lambda \ll 1$, which is consistent with the measured elements of the CKM matrix. Show that this matrix is unitary, up to corrections of order λ^4. That is,

$$V_{\text{CKM}} V_{\text{CKM}}^\dagger = \mathbb{I} + \mathcal{O}(\lambda^4) . \tag{12.129}$$

---

[19] L. Wolfenstein, "Parametrization of the Kobayashi–Maskawa matrix," Phys. Rev. Lett. **51**, 1945 (1983).

(b) As a unitary matrix, the CKM matrix can be expressed as an exponentiated Hermitian matrix:

$$V_{\text{CKM}} = \mathbb{U}_u^\dagger = \exp[-i\mathbb{F}_u], \qquad (12.130)$$

where $\mathbb{F}_u = \mathbb{F}_u^\dagger$ is Hermitian. With the CKM matrix in the Wolfenstein parametrization, determine $\mathbb{F}_u$ through $\mathcal{O}(\lambda^3)$.

Hint: To do this, express the CKM matrix as $V_{\text{CKM}} = \mathbb{I} + \mathbb{X}$, for a matrix $\mathbb{X}$. Then, Taylor expand both sides to cubic order in $\mathbb{X}$:

$$\mathbb{F}_u = i\log(V_{\text{CKM}}) = i\log(\mathbb{I} + \mathbb{X}) = i\mathbb{X} - i\frac{\mathbb{X}^2}{2} + i\frac{\mathbb{X}^3}{3} + \mathcal{O}(\lambda^4). \quad (12.131)$$

Only keep those terms up through order λ^3.

(c) The Hermitian matrix $\mathbb{F}_u$ is the flavor matrix for the up-type quarks. Evaluate the commutator of this flavor matrix and the mass matrix $\mathbb{M}_u$ defined in Eq. 12.124. That is, calculate

$$[\mathbb{M}_u, \mathbb{F}_u]. \qquad (12.132)$$

Is it non-zero? At what order in λ are the first non-zero terms?

(d) Can you provide a physical interpretation of the value of this commutator? For example, what experiment would you do to measure this commutator?

12.2 *Unitarity of the CKM Matrix.* Verify that the parametrizations of the CKM matrix in equations 12.62 and 12.63 are unitary matrices.

12.3 *Jarlskog Invariant.* As discussed in Section 12.2.4, the Jarlskog invariant is a measure of the amount of CP violation in the CKM matrix, assuming that the CKM matrix is unitary.

(a) From the unitary condition

$$V_{ud}V_{ub}^* + V_{cd}V_{cb}^* + V_{td}V_{tb}^* = 0 \qquad (12.133)$$

on the entries of the CKM matrix, calculate the Jarlskog invariant in terms of the CKM matrix entries. Recall that the expression above can be interpreted as a triangle in the complex plane, and the Jarlskog invariant is the area of that triangle.

Hint: It might help to draw a picture of this triangle. Remember, the area of a triangle is unchanged under rotation.

(b) The CP violation in the CKM matrix happens to be quite small (not enough to explain the matter–anti-matter asymmetry, for example). What is the largest possible value of the Jarlskog invariant J? How much larger is this than the current measured value of J from Eq. 12.72?

Hint: What type of triangle corresponds to the largest possible J? What are the lengths of its sides?

12.4 *Extra Quark Generations.* With more than three generations of quarks, the structure of the corresponding CKM matrix would be different than we studied here. We argued that the CKM quark mixing matrix, V_{CKM}, has four parameters because there are three generations of quarks: three angles and one complex phase. The existence of an irreducible complex phase means that particles and anti-particles couple to W bosons differently, leading to CP violation.

Consider a universe in which there are N generations of quarks. How many independent, irreducible parameters of the corresponding quark mixing matrix would there be? How many are angles? How many are complex phases?

Hint: If you are in N dimensions, how many distinct, orthogonal planes are there in which you can rotate? Don't forget to eliminate phases by redefining the quark fields.

12.5 *Measuring the Cabibbo Angle.* Each element of the CKM matrix can be determined by studying particular hadron decay modes or production processes that involve a W boson mixing quark flavors. An extensive presentation of all of the measurements that go into determining the CKM matrix can be found in the PDG. In this exercise, we will work to just extract the Cabibbo angle θ_C from a comparison of different decays of a D meson.

The D mesons are a class of hadron that consist of a charm quark and an up, down, or strange quark. The D^+ meson is a positively charged particle that is formed from the bound state of a charm quark and an anti-down quark: $D^+ = c\bar{d}$. There are numerous decay modes of this meson; the two we will study here are decays to a neutral kaon $\bar{K}^0$ and a neutral pion π^0:

$$D^+ \to \bar{K}^0 e^+ \nu_e \,, \qquad\qquad D^+ \to \pi^0 e^+ \nu_e \,. \qquad (12.134)$$

Feynman diagrams that are used to calculate the decay rates for these processes are

$$\mathcal{M}(D^+ \to \bar{K}^0 e^+ \nu_e) = D^+ \begin{pmatrix} \bar{d} \\ c \end{pmatrix} \qquad\qquad , \qquad (12.135)$$

$$\mathcal{M}(D^+ \to \pi^0 e^+ \nu_e) = D^+ \begin{pmatrix} \bar{d} \\ c \end{pmatrix} \qquad\qquad .$$

From the PDG, the fraction of the time that the D^+ meson decays to these final states (called the **branching fractions**) is

$$\text{Br}(D^+ \to \bar{K}^0 e^+ \nu_e) = (8.6 \pm 0.5) \times 10^{-2}, \tag{12.136}$$
$$\text{Br}(D^+ \to \pi^0 e^+ \nu_e) = (4.4 \pm 0.7) \times 10^{-3}.$$

Using these Feynman diagrams and the branching fractions, estimate the sine of the Cabibbo angle, $\sin \theta_C$.

Hint: How is a Feynman diagram related to a decay rate?

12.6 *Non-Relativistic Limit of Neutrino Oscillations.* Equation 12.113 was our final result for neutrino oscillation probabilities for two generations of neutrinos. In that derivation, we assumed that the neutrinos were relativistic; what does the non-relativistic limit look like? Restore factors of c and $\hbar$ in the probabilities of Eq. 12.113, and take the non-relativistic limit. Further, exchange the baseline distance L for the propagation time T. The result you find should look very similar to the oscillation rate between the energy eigenstates in a two-state system. For example, you should find the survival probability to be

$$P(\bar{\nu}_e \to \bar{\nu}_e) = 1 - \sin^2(2\theta) \sin^2 \left(\frac{T}{4\hbar} |m_1 c^2 - m_2 c^2| \right). \tag{12.137}$$

12.7 *Neutrinos for Nuclear Non-Proliferation.* A fascinating proposal for a practical use of neutrinos and neutrino detectors is as a detection tool for production of nuclear weapons. A vital ingredient in nuclear weapons is radioactive elements enriched with isotopes that can sustain fission reactions. An example of this would be enriched uranium in which the ^{235}U content is significantly increased with respect to the dominant isotope ^{238}U. Nuclear chain reactions with ^{235}U (controlled in a reactor or uncontrolled in a weapon) would release a huge number of neutrinos which could be detected a long distance away, and the yield of the reaction determined by the number of neutrinos detected.

One such neutrino detector that is proposed for potential non-proliferation applications is the WATCHMAN experiment.[20] The WATCHMAN experiment would consist of a 10.8 m high and 10.8 m diameter cylinder filled with gadolinum-doped water. Anti-neutrinos are detected through the inverse β-decay process

$$\bar{\nu}_e + p \to e^+ + n, \tag{12.138}$$

so it is vital to observe both the positron and the neutron in the final state. The purpose of the gadolinium (mostly ^{158}Gd) in the water (which only needs to be about 0.1% by weight) is to capture as many of the neutrons as possible, which can be up to about 85% of all neutrons produced.

(a) At 0.1% doping by weight, how many ^{158}Gd atoms would there be in the WATCHMAN detector?

---

[20] M. Askins *et al.* [WATCHMAN Collaboration], "The physics and nuclear nonproliferation goals of WATCHMAN: A WATer CHerenkov Monitor for ANtineutrinos" [arXiv:1502.01132 [physics.ins-det]]

(b) A first goal of WATCHMAN for non-proliferation applications is to detect a 10 megawatt (MW) nuclear reactor at a distance of about 25 km. The fission of one ^{235}U atom releases about 200 MeV of energy. In a 10 MW reactor, how many ^{235}U atoms undergo fission per second? What is the total mass of ^{235}U atoms that fission per second?

(c) On average, about two electron anti-neutrinos with MeV-scale energies are produced in the fission of ^{235}U. If the WATCHMAN experiment is 25 km from the 10 MW reactor, estimate the number of anti-neutrinos that pass through it per second.

(d) Assuming that the gadolinium doping ensures that 100% of the neutrons produced from inverse β-decay can be observed, how long would you have to wait to see just one reactor anti-neutrino interact with the water in the WATCHMAN experiment? You can assume that the cross section for inverse β-decay is about 10^{-19} b.

   Hint: Don't forget that water consists of two hydrogen atoms and an oxygen atom.

(e) 25 km isn't very far, and in many cases is unlikely to be a realistic distance from a nuclear reactor where a neutrino detector could be placed. Assuming that a neutrino detector could be realistically placed 200 km from a nuclear reactor, about how much water would be needed to detect electron anti-neutrinos at the same rate as calculated in part (d)? You can still assume that the nuclear reactor power output is 10 MW. Express your answer in kilograms of water.

12.8 *Neutrinos from SN 1987a.* While SN 1987a emitted 10^{58} neutrinos, only 25 were detected on Earth about 2 hours before light from the supernova was detected. In this exercise, we will use this observation to estimate neutrino properties.

(a) SN 1987a is about 168,000 light-years from Earth. If the Earth's radius is about 6000 km, estimate the number of neutrinos from SN 1987a that passed through Earth.

(b) The Kamiokande-II detector in Japan observed 11 neutrinos from SN 1987a. It was a cylindrical vat about 16 m high and 16 m in diameter filled with water. Neutrinos that passed through the water would occasionally hit a hydrogen or oxygen nucleus and produce a charged lepton that would emit Čerenkov radiation that was detected in photomultiplier tubes. Approximately how many neutrinos from SN 1987a passed through the Kamiokande-II detector?

(c) From the observed number of neutrinos and the total that passed through Kamiokande-II, estimate the cross section for neutrinos to interact with water molecules. You can safely assume that neutrinos are traveling at the speed of light and water molecules are at rest. The total time over which the neutrinos were observed was about 1 minute.

(d) The time difference between observing neutrinos and light from SN 1987a was used to place an upper bound on the mass of neutrinos of about 16 eV. To establish this bound requires modeling the supernova and estimating the mean free path of photons as they pass through the charged particles produced in the explosion. While this is quite detailed, we can still get a sense for how this mass bound was established.

If neutrinos were massless, how long would it take them to travel from SN 1987a to Earth? What if neutrinos saturated the mass bound of 16 eV? How does the time difference between the different neutrino mass assumptions compare to the observed 2 hour time difference between neutrinos and light arriving at Earth? You can assume that the energy of the neutrinos is 1 MeV.

12.9 *Solar Neutrino Problem.* In the process of nuclear fusion of hydrogen nuclei and other light elements in the Sun, a copious amount of neutrinos are produced. Because these nuclear fusion chains involve turning protons into neutrons, and vice-versa, exclusively electron neutrinos are produced. One can then observe electron neutrinos produced from the Sun (called "solar neutrinos") on Earth and compare to the number expected. Surprisingly, significantly fewer neutrinos have been observed on Earth than are expected to be produced by the Sun. This is referred to as the **solar neutrino problem**.

The resolution is that neutrinos oscillate from the initial electron-type to other neutrinos which are not observed. As a first estimate of the fraction of solar neutrinos observed on Earth, assume that the neutrinos are produced incoherently in the Sun and propagate to Earth. The relevant mixing angle for initial electron neutrinos is θ_{12}, for which

$$\sin^2 \theta_{12} = 0.307 \pm 0.013 \qquad (12.139)$$

from the PDG. What is the survival probability $P(\nu_e \to \nu_e)$ for electron neutrinos produced in the Sun and measured on Earth?

12.10 *Research Problem.* What is the absolute mass scale of neutrinos? Are any of the neutrinos massless?

13 The Higgs Boson

The weak interactions provided some significant surprises for the physics in the subatomic world in relation to our experiences of other forces, like electromagnetism or gravity. In addition to parity violation of nuclear decays that initiated our foray into the weak interactions, we have also identified the phenomena of spontaneous symmetry breaking, flavor violation, CP violation, and neutrino oscillations. Except for gluons, which interact exclusively through QCD, every other particle of the Standard Model is involved in the weirdness of the weak force.

In our discussion thus far, however, the Higgs boson has just been a consequence of Goldstone's theorem and the Higgs mechanism of the spontaneous symmetry breaking of the unified electroweak force. In a real sense, the Higgs boson is a lynchpin of essentially the entire Standard Model, and so its observation and existence is a huge validation of the theoretical structure that we have constructed.[1] So, if we are to claim victory in having a complete theoretical understanding and experimental verification of the Standard Model, we need to observe the Higgs boson.

Nevertheless, this is a very tall challenge. The Higgs boson was predicted in the early 1960s and wasn't observed through many generations of experiments. Its lack of observation stumped physicists over and over again. If it were to exist, we know all of its interactions by the structure of the weak force. However, we do not know its mass, as that is not constrained by any properties of the weak force. Scientists looked and looked, in new experiments and higher energies, but no Higgs was found. A now-classic book, *The Higgs Hunter's Guide*, was published in the late 1980s as a compendium of properties

---

[1] In popular culture, for this reason the Higgs boson is sometimes referred to as the "God particle," which induces pain in physicists the world over. The unfortunate term can be traced to Leon Lederman's book with that as the title (L. Lederman and D. Teresi, *The God Particle: If the Universe is the Answer, What is the Question?*, Houghton Mifflin (1993). 434 p.). In fact, in the book, Lederman explains how it came to be:

> This boson is so central to the state of physics today, so crucial to our final understanding of the structure of matter, yet so elusive, that I have given it a nickname: the God Particle. Why God Particle? Two reasons. One, the publisher wouldn't let us call it the Goddamn Particle, though that might be a more appropriate title, given its villainous nature and the expense it is causing. And two, there is a connection, of sorts, to another book, a *much* older one...

of the Higgs boson.[2] It wasn't until 2012 that these hunters caught their quarry, in the experiments of the Large Hadron Collider.[3]

In this chapter, we discuss efforts of searches for the Higgs boson and how it was eventually discovered at the LHC. This requires a detailed understanding of both the production and decay of the Higgs at colliders. We'll start with searches for the Higgs boson at LEP, and then searches at the Tevatron and early searches at the LHC, which provide robust lower and upper bounds that frame the eventual discovery.

13.1 Searching for the Higgs Boson at LEP

In searching for the Higgs boson at an electron–positron collider like LEP, we might first imagine discovering it in a similar way that the Z boson is observed. At our e^+e^- collider, we can continuously vary the center-of-mass collision energy. When the collision energy is near 91 GeV, we observe a significant increase in the likelihood of interaction, which we ascribe to the existence of the Z boson at that mass. Similarly, as we tune the collision energy, we might observe another energy at which the likelihood of interaction increases significantly, which we could associate with the Higgs boson. With this in mind, where should we look?

An important aspect of the Z boson is that we already knew where to look. As discussed in Section 11.3.4, by knowing the electric charge e, the Fermi constant G_F, and the mass of the W boson m_W, the electroweak theory uniquely predicts the mass of the Z boson, m_Z. So, if we didn't know where to look before, we can use the electroweak theory to guide the way to find the Z boson. Unfortunately, such a guide does not work with the Higgs boson. Within the Higgs mechanism, the Higgs boson acquires mass through its potential. Recall that the potential for the scalar field $\vec{\phi}$ is

$$V(\vec{\phi}) = \lambda \left(|\vec{\phi}|^2 - \frac{v^2}{2} \right)^2 \rightarrow \lambda \left(\left(\frac{v}{\sqrt{2}} + \frac{H(x)}{\sqrt{2}} \right)^2 - \frac{v^2}{2} \right)^2 \tag{13.1}$$

$$= \lambda v^2 H(x)^2 + \lambda v H(x)^3 + \frac{\lambda}{4} H(x)^4 \, .$$

In this expression, we have expanded the field $\vec{\phi}$ about its vev v and identified the Higgs boson $H(x)$ as the fluctuations about the vev. The mass m_H of the Higgs boson is the coefficient of the term quadratic in the Higgs field $H(x)$:

$$m_H = \sqrt{2\lambda} v \, . \tag{13.2}$$

We know the value of the vev v from other measurements, as mentioned above, but the parameter λ is thus far completely unconstrained. No properties of the W or Z boson

---

[2] J. F. Gunion, H. E. Haber, G. L. Kane and S. Dawson, *The Higgs Hunter's Guide*, Front. Phys. **80**, 1 (2000).

[3] G. Aad *et al.* [ATLAS Collaboration], "Observation of a new particle in the search for the Standard Model Higgs boson with the ATLAS detector at the LHC," Phys. Lett. B **716**, 1 (2012) [arXiv:1207.7214 [hep-ex]]; S. Chatrchyan *et al.* [CMS Collaboration], "Observation of a new boson at a mass of 125 GeV with the CMS experiment at the LHC," Phys. Lett. B **716**, 30 (2012) [arXiv:1207.7235 [hep-ex]].

provide insight into the value of λ, and so the electroweak theory does not tell us where to look for the Higgs. We have to look everywhere!

While this seems daunting (and indeed it is!), we'll be systematic and search for the Higgs boson where and how we can. As we saw with the Z boson, our first attempt at searching for the Higgs boson is to look for a of center-of-mass, s-channel resonance in e^+e^- collisions. In the case of the Z boson, we observed that as the center-of-mass collision energy was tuned to be around 91 GeV, the scattering cross section significantly increased. We can in principle do the exact same thing to find the Higgs boson: scan over a wide range of electron–positron collision energies and look for a bump, which would correspond to the Higgs boson as a resonance. This general procedure is colloquially called **bump hunting**. Unfortunately, this turns out to be essentially an impossible endeavor, unless your data set consists of an exceedingly large number of electron–positron collisions.

13.1.1 $e^+e^- \rightarrow Z$

To justify this claim, let's first estimate the cross section for producing a Z boson in e^+e^- collisions, $e^+e^- \rightarrow Z$. The Feynman diagram for this process is

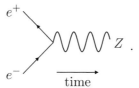

The Z boson in this diagram is a final-state particle, and as such, is something that we would detect in our experiment. We know, however, that the Z boson is an unstable particle and decays very quickly into less massive particles (like leptons or quarks) that we actually detect. Nevertheless, it is not unreasonable to assume that the Z boson is a final-state particle; or, equivalently, that it has a relatively long lifetime. The approximation that an unstable particle like the Z boson is taken to be on-shell and long-lived is referred to as the narrow width approximation. The narrow width approximation is justified for the Z boson because its decay width is about 2.5 GeV, which is much smaller than its mass. (See Exercise 1.6 of Chapter 1.) We use the narrow width approximation to analyze the production of both the Z boson and the Higgs boson here.

To calculate this Feynman diagram, we need to determine the strength with which electrons couple to the Z boson. This can be extracted from the covariant derivative of the electroweak theory. From the expression of the covariant derivative constructed in Section 11.3.3, electrons couple to the Z boson as

$$\mathcal{L} = e_L^\dagger i\bar{\sigma} \cdot D e_L + e_R^\dagger i\sigma \cdot D e_R \tag{13.3}$$

$$\supset \left(-\frac{1}{2} \frac{g_W^2}{\sqrt{g_W^2 + g_Y^2}} + \frac{1}{2} \frac{g_Y^2}{\sqrt{g_W^2 + g_Y^2}} \right) e_L^\dagger \bar{\sigma} \cdot Z e_L + \frac{g_Y^2}{\sqrt{g_W^2 + g_Y^2}} e_R^\dagger \sigma \cdot Z e_R \,.$$

To write this expression, we recall that the left-handed electron e_L is the lower component of an electroweak doublet (with the electron neutrino) with hypercharge $Y = -1$, while the right-handed electron e_R is an electroweak singlet with hypercharge $Y = -2$. The coefficients of the appropriate electron–Z coupling terms in the Lagrangian are the corresponding values of the vertices in the Feynman diagram calculation. So, with that, we can calculate these cross sections.

Starting with the left-handed electron coupling, the Feynman diagram is

$$(13.4)$$

$$= \left(-\frac{1}{2} \frac{g_W^2}{\sqrt{g_W^2 + g_Y^2}} + \frac{1}{2} \frac{g_Y^2}{\sqrt{g_W^2 + g_Y^2}} \right) v_R^\dagger(p_{e+}) \bar{\sigma}^\mu u_L(p_{e-}) \epsilon_\mu(p_Z) .$$

Because the Z boson is a spin-1 particle, its external wavefunction is a polarization four-vector, $\epsilon(p_Z)$, just like the photon or gluon we encountered earlier in Sections 2.2.3 and 7.3.2. The Feynman diagram with a right-handed electron is correspondingly

$$= \frac{g_Y^2}{\sqrt{g_W^2 + g_Y^2}} v_L^\dagger(p_{e+}) \sigma^\mu u_R(p_{e-}) \epsilon_\mu(p_Z) . \quad (13.5)$$

To calculate the squared matrix element, we need to individually absolute square these separate helicity configurations and then sum. Additionally, because the initial electrons and positrons are unpolarized in our experiment, we divide by a factor of 4 to account for the initial-state spin average. Then, we have

$$\frac{1}{4} \sum_{\text{spins}} |\mathcal{M}|^2 = \frac{1}{4} \left(-\frac{1}{2} \frac{g_W^2}{\sqrt{g_W^2 + g_Y^2}} + \frac{1}{2} \frac{g_Y^2}{\sqrt{g_W^2 + g_Y^2}} \right)^2 |v_R^\dagger(p_{e+}) \bar{\sigma} \cdot \epsilon(p_Z) u_L(p_{e-})|^2 \quad (13.6)$$

$$+ \frac{1}{4} \left(\frac{g_Y^2}{\sqrt{g_W^2 + g_Y^2}} \right)^2 |v_L^\dagger(p_{e+}) \sigma \cdot \epsilon(p_Z) u_R(p_{e-})|^2 .$$

To go further, we need to think about our experimental detector. We typically only measure the energy and momentum of final-state particles (and maybe their charge), and so we do not select different polarizations of the final-state Z boson. That is, from our detector's point of view, the Z boson is unpolarized. With this assumption, the Z boson cannot distinguish between the different initial electron and positron helicities. Therefore, the two different products of external particle wavefunctions in Eq. 13.6 are equal:

$$|v_R^\dagger(p_{e+}) \bar{\sigma} \cdot \epsilon(p_Z) u_L(p_{e-})|^2 = |v_L^\dagger(p_{e+}) \sigma \cdot \epsilon(p_Z) u_R(p_{e-})|^2 . \quad (13.7)$$

Because the Z is unpolarized, we need to sum over its spins, which we include implicitly here, but will address shortly. With this simplification, we still need to evaluate the product $|v_R^\dagger(p_{e+}) \bar{\sigma} \cdot \epsilon(p_Z) u_L(p_{e-})|^2$, for example. While we have discussed several ways to do this throughout this book, we'll introduce yet another way here. Doing this will be the first example of this chapter.

Example 13.1 What is the spinor product

$$\sum_{Z \text{ spins}} |v_R^\dagger(p_{e+}) \bar{\sigma} \cdot \epsilon(p_Z) u_L(p_{e-})|^2 \, ? \tag{13.8}$$

Solution

Let's deconstruct the product $|v_R^\dagger(p_{e+}) \bar{\sigma} \cdot \epsilon(p_Z) u_L(p_{e-})|^2$ carefully. As usual, we make the simplification that the electron is massless. We can expand out the absolute square:

$$|v_R^\dagger(p_{e+}) \bar{\sigma} \cdot \epsilon(p_Z) u_L(p_{e-})|^2 = v_R^\dagger(p_{e+}) \bar{\sigma} \cdot \epsilon(p_Z) u_L(p_{e-}) u_L^\dagger(p_{e-}) \bar{\sigma} \cdot \epsilon^*(p_Z) v_R(p_{e+}) . \tag{13.9}$$

First, note that $u_L(p_{e-}) u_L^\dagger(p_{e-})$ is the outer product of two spinors. In Exercise 6.2 of Chapter 6, you showed that this outer product is

$$u_L(p_{e-}) u_L^\dagger(p_{e-}) = \sigma \cdot p_{e-} . \tag{13.10}$$

So, we can make this replacement. Additionally, note that we could have equivalently written the absolute square as

$$|v_R^\dagger(p_{e+}) \bar{\sigma} \cdot \epsilon(p_Z) u_L(p_{e-})|^2 = u_L^\dagger(p_{e-}) \bar{\sigma} \cdot \epsilon^*(p_Z) v_R(p_{e+}) v_R^\dagger(p_{e+}) \bar{\sigma} \cdot \epsilon(p_Z) u_L(p_{e-}) . \tag{13.11}$$

In this expression, we can use the fact that the outer product of spinors is

$$v_R(p_{e+}) v_R^\dagger(p_{e+}) = \sigma \cdot p_{e+} . \tag{13.12}$$

To relate these two expressions for the absolute square, we note that the trace of the outer product of spinors is just the inner product:

$$\text{tr}[v_R(p_{e+}) v_R^\dagger(p_{e+})] = v_R^\dagger(p_{e+}) v_R(p_{e+}) = \text{tr}\, \sigma \cdot p_{e+} . \tag{13.13}$$

Therefore, we can express the absolute square as a trace of a product of Pauli matrices:

$$|v_R^\dagger(p_{e+}) \bar{\sigma} \cdot \epsilon(p_Z) u_L(p_{e-})|^2 = \text{tr}[\sigma \cdot p_{e+} \bar{\sigma} \cdot \epsilon(p_Z) \sigma \cdot p_{e-} \bar{\sigma} \cdot \epsilon^*(p_Z)] . \tag{13.14}$$

Now we're cooking. Let's work in the center-of-mass frame and align the three-momenta of the electron and positron along the $\hat{z}$-axis. Then, the four-vectors of the electron and positron are

$$p_{e-} = \frac{E_{\text{cm}}}{2}(1, 0, 0, 1), \qquad p_{e+} = \frac{E_{\text{cm}}}{2}(1, 0, 0, -1). \tag{13.15}$$

With these choices, the spinor outer product matrices are

$$\sigma \cdot p_{e-} = E_{\text{cm}} \begin{pmatrix} 0 & 0 \\ 0 & 1 \end{pmatrix}, \qquad \sigma \cdot p_{e+} = E_{\text{cm}} \begin{pmatrix} 1 & 0 \\ 0 & 0 \end{pmatrix}. \tag{13.16}$$

Also, we can express the Z boson polarization vector in terms of its explicit components as

$$\epsilon(Z) = (\epsilon_0, \epsilon_1, \epsilon_2, \epsilon_3), \tag{13.17}$$

for instance. Then, the matrix product with the polarization vector is

$$\bar{\sigma} \cdot \epsilon(Z) = \begin{pmatrix} \epsilon_0 + \epsilon_3 & \epsilon_1 - i\epsilon_2 \\ \epsilon_1 + i\epsilon_2 & \epsilon_0 - \epsilon_3 \end{pmatrix}. \tag{13.18}$$

Now that the matrices are all constructed, we can take their product and trace:

$$\text{tr}[\sigma \cdot p_{e^+} \bar{\sigma} \cdot \epsilon(p_Z) \sigma \cdot p_{e^-} \bar{\sigma} \cdot \epsilon^*(p_Z)] \tag{13.19}$$

$$= E_{\text{cm}}^2 \text{tr}\left[\begin{pmatrix} 1 & 0 \\ 0 & 0 \end{pmatrix} \begin{pmatrix} \epsilon_0 + \epsilon_3 & \epsilon_1 - i\epsilon_2 \\ \epsilon_1 + i\epsilon_2 & \epsilon_0 - \epsilon_3 \end{pmatrix} \begin{pmatrix} 0 & 0 \\ 0 & 1 \end{pmatrix} \begin{pmatrix} \epsilon_0^* + \epsilon_3^* & \epsilon_1^* - i\epsilon_2^* \\ \epsilon_1^* + i\epsilon_2^* & \epsilon_0^* - \epsilon_3^* \end{pmatrix}\right]$$

$$= E_{\text{cm}}^2 \text{tr}\left[\begin{pmatrix} \epsilon_0 + \epsilon_3 & \epsilon_1 - i\epsilon_2 \\ 0 & 0 \end{pmatrix} \begin{pmatrix} 0 & 0 \\ \epsilon_1^* + i\epsilon_2^* & \epsilon_0^* - \epsilon_3^* \end{pmatrix}\right]$$

$$= E_{\text{cm}}^2 \left(|\epsilon_1|^2 + |\epsilon_2|^2 + i(\epsilon_1 \epsilon_2^* - \epsilon_1^* \epsilon_2)\right).$$

As the Z boson is unpolarized in our experiment, we need to explicitly sum over its polarizations. So, we need to determine a complete basis of polarization vectors and sum their contributions. While we have often discussed circular polarization or helicity, the Z boson is a massive particle, and as such doesn't have a well-defined helicity. A useful polarization basis in this case is then **linear polarization**, in which the Z boson is polarized along one of the spatial axes. We can work in the frame in which the Z boson is at rest so that its momentum four-vector is

$$p_Z = m_Z(1, 0, 0, 0). \tag{13.20}$$

As we discussed in Chapter 3, the 0th component of a spin-1 boson does not propagate (regardless of its mass), and so the linear polarization vector ϵ has no 0th component either. A massive, spin-1 particle can then be polarized about the $\hat{x}$, $\hat{y}$, or $\hat{z}$ axes, and the corresponding polarization basis is

$$\epsilon^x = (0, 1, 0, 0), \qquad \epsilon^y = (0, 0, 1, 0), \qquad \epsilon^z = (0, 0, 0, 1). \tag{13.21}$$

The absolute squared spinor product, summed over Z boson polarizations, is then

$$\sum_{Z \text{ spins}} |v_R^\dagger(p_{e^+}) \bar{\sigma} \cdot \epsilon(p_Z) u_L(p_{e^-})|^2 \tag{13.22}$$

$$= E_{\text{cm}}^2 \sum_{Z \text{ pols. } i} \left(|\epsilon_1^i|^2 + |\epsilon_2^i|^2 + i(\epsilon_1^i \epsilon_2^{*i} - \epsilon_1^{*i} \epsilon_2^i)\right)$$

$$= 2 E_{\text{cm}}^2.$$

When the dust settles, we finally find that the squared matrix element, averaged over initial spins and summed over Z boson spins, is

$$\frac{1}{4} \sum_{Z \text{ spins}} |\mathcal{M}(e^+ e^- \to Z)|^2 \tag{13.23}$$

$$= \frac{E_{\text{cm}}^2}{2} \left(-\frac{1}{2} \frac{g_W^2}{\sqrt{g_W^2 + g_Y^2}} + \frac{1}{2} \frac{g_Y^2}{\sqrt{g_W^2 + g_Y^2}}\right)^2 + \frac{E_{\text{cm}}^2}{2} \left(\frac{g_Y^2}{\sqrt{g_W^2 + g_Y^2}}\right)^2.$$

To calculate the cross section, we then put this into Fermi's Golden Rule and integrate over the phase space of the final-state Z boson. From Fermi's Golden Rule (and skipping a few steps), we have

$$\sigma(e^+e^- \to Z) \tag{13.24}$$

$$= \frac{1}{2E_{cm}^2} \int \frac{d^4p_Z}{(2\pi)^4} 2\pi\delta(p_Z^2 - m_Z^2) \frac{1}{4} \sum_{\text{spins}} |\mathcal{M}(e^+e^- \to Z)|^2 (2\pi)^4 \delta^{(4)}(p_{e^-} + p_{e^+} - p_Z).$$

The integral over the Z boson momentum p_Z is one-body phase space, which is weird, but let's just do the integrals blindly, and think about the physical interpretation later. The momentum-conserving δ-function can be used to remove all four integrals, so we find

$$\sigma(e^+e^- \to Z) = \frac{\pi}{E_{cm}^2}\delta(E_{cm}^2 - m_Z^2) \frac{1}{4} \sum_{\text{spins}} |\mathcal{M}(e^+e^- \to Z)|^2 \tag{13.25}$$

$$= \frac{\pi}{2}\delta(E_{cm}^2 - m_Z^2) \left[\left(-\frac{1}{2}\frac{g_W^2}{\sqrt{g_W^2 + g_Y^2}} + \frac{1}{2}\frac{g_Y^2}{\sqrt{g_W^2 + g_Y^2}} \right)^2 + \left(\frac{g_Y^2}{\sqrt{g_W^2 + g_Y^2}} \right)^2 \right].$$

In this expression, note that momentum conservation enforces $p_Z^2 = m_Z^2 = E_{cm}^2$.

This cross section is interesting: its functional dependence on the center-of-mass energy E_{cm} is a δ-function. In the narrow width approximation, the Z boson is only produced in e^+e^- collisions if the center-of-mass energy is *precisely* m_Z. This is a bit unrealistic, and a finite decay width will smear this out a bit, but it's nevertheless a good approximation. We can also plug in values for the electroweak couplings $g_W = 0.642$ and $g_Y = 0.344$ that we extracted in Section 11.3.4, and find

$$\sigma(e^+e^- \to Z) = 0.0671 \frac{\pi}{2}\delta(E_{cm}^2 - m_Z^2). \tag{13.26}$$

This will be useful for comparing to the Higgs production cross section in the following.

13.1.2 $e^+e^- \to H$

Now, let's calculate the cross section for electron–positron scattering to directly produce a Higgs boson, $e^+e^- \to H$. The Feynman diagram for this process is

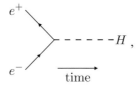

and to calculate it, we need to determine the strength with which the electron couples to the Higgs boson. The coupling of the Higgs boson to the electron follows from expanding the scalar field $\vec{\phi}$ about its vev. Following our work in Section 12.2.1, the mass and coupling of the electron to the Higgs boson H terms in the electroweak Lagrangian are

$$\mathcal{L} \supset -\frac{y_e v}{\sqrt{2}} e_L^\dagger e_R - \frac{y_e H}{\sqrt{2}} e_L^\dagger e_R + \text{h.c.} \tag{13.27}$$

The strength of coupling of electrons to Higgs bosons is thus proportional to the electron's Yukawa coupling, y_e. Note that, because the Higgs boson is a spin-0 particle, electrons and positrons of opposite spin or identical helicity interact with the Higgs boson. Thus, for example, the Feynman diagram for a left-handed electron and left-handed positron producing a Higgs boson is

$$\begin{array}{c} e_L^+ \\[6pt] \text{(diagram)} \quad \cdots\cdots\cdots H \\[6pt] e_L^- \quad \text{time} \end{array} \;=\; -\frac{y_e}{\sqrt{2}} v_L^\dagger(p_{e^+}) u_L(p_{e^-}). \tag{13.28}$$

The external wavefunction for the Higgs boson in the Feynman diagram is just 1 because it has no spin (i.e., it is always in the same spin state with probability 1).

The absolute square of this Feynman diagram is then

$$|\mathcal{M}(e_L^- e_R^+ \to H)|^2 = \frac{y_e^2}{2} |v_L^\dagger(p_{e^+}) u_L(p_{e^-})|^2. \tag{13.29}$$

To evaluate the spinor product, we can use the trace trick developed in Example 13.1. We have

$$|v_L^\dagger(p_{e^+}) u_L(p_{e^-})|^2 = v_L^\dagger(p_{e^+}) u_L(p_{e^-}) u_L^\dagger(p_{e^-}) v_L(p_{e^+}) \tag{13.30}$$

$$= \text{tr}[\sigma \cdot p_{e^-} \bar{\sigma} \cdot p_{e^+}]. \tag{13.31}$$

We have already evaluated the matrix $\sigma \cdot p_{e^-}$ in the center-of-mass collision frame, and the other matrix is actually identical:

$$\sigma \cdot p_{e^-} = \bar{\sigma} \cdot p_{e^+} = E_{\text{cm}} \begin{pmatrix} 0 & 0 \\ 0 & 1 \end{pmatrix}. \tag{13.32}$$

The spinor product then immediately follows:

$$|v_L^\dagger(p_{e^+}) u_L(p_{e^-})|^2 = E_{\text{cm}}^2. \tag{13.33}$$

Because the Higgs boson couples to all helicities of electrons identically, the matrix elements for left- and right-handed electron–positron scattering are identical:

$$|\mathcal{M}(e_L^- e_R^+ \to H)|^2 = |\mathcal{M}(e_R^- e_L^+ \to H)|^2 = \frac{y_e^2}{2} E_{\text{cm}}^2. \tag{13.34}$$

Then, by averaging over the spins of the electron and positron, we find

$$\frac{1}{4} \sum_{\text{spins}} |\mathcal{M}(e^+ e^- \to H)|^2 = \frac{y_e^2}{4} E_{\text{cm}}^2. \tag{13.35}$$

To calculate the cross section, we use Fermi's Golden Rule, and we now know how to interpret the one-body phase space integrals from the discussion of the previous section. The cross section is

$$\sigma(e^+e^- \to H) \tag{13.36}$$

$$= \frac{1}{2E_{cm}^2} \int \frac{d^4p_H}{(2\pi)^4} 2\pi\delta(p_H^2 - m_H^2) \frac{1}{4} \sum_{spins} |\mathcal{M}(e^+e^- \to H)|^2 (2\pi)^4\delta^{(4)}(p_{e^-} + p_{e^+} - p_H)$$

$$= \frac{y_e^2}{2} \frac{\pi}{2} \delta(E_{cm}^2 - m_H^2).$$

From Eq. 13.27 above, the electron mass m_e is

$$m_e = \frac{y_e v}{\sqrt{2}}. \tag{13.37}$$

With the vev $v = 246$ GeV and $m_e = 511$ keV, the electron Yukawa coupling y_e is

$$y_e = 2.94 \times 10^{-6}. \tag{13.38}$$

This is tiny! Plugging this into the expression for the cross section, we have

$$\sigma(e^+e^- \to H) = 4.31 \times 10^{-12} \frac{\pi}{2} \delta(E_{cm}^2 - m_H^2). \tag{13.39}$$

This is a factor of more than 10^{10} times smaller than the result for the production of the Z boson, Eq. 13.26! Another way to say this is that we would need to collect 10^{10} times more data than LEP to have measurements of the Higgs boson that have a comparable precision to the existing studies of the Z boson. LEP collected about 17 million Z bosons during its entire run, so for discovery and measurements of the Higgs boson through direct production with comparable precision to the Z boson, LEP would have had to collect about 10^{17} events. For a benchmark, the LHC collides protons every 25 nanoseconds, or 40 million times a second. The LHC would need to run continuously for about a century to possibly collect that many collision events.

13.1.3 $e^+e^- \to ZH$

Needless to say, direct production of the Higgs boson in electron–positron collisions is not a feasible method for discovery. This is because electrons have extremely low mass compared to the vev v. As such, they couple very, very weakly to the Higgs boson. For any hope of discovery of the Higgs boson in e^+e^- collisions, we need (1) the electrons and positrons to interact relatively strongly and (2) the Higgs boson to couple to a very massive particle. Thankfully, we know of one particle that can accomplish both tasks at the same time: the Z boson. The Z boson couples to electrons relatively strongly, and it has a large mass, and so couples to the Higgs boson relatively strongly. The cost of this, however, is that a Z boson is produced in the final state, along with the Higgs boson. We say that the Z boson is **produced in association** with the Higgs. The Feynman diagram for the $e^+e^- \to ZH$ process is

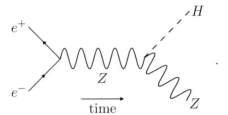

Because of this associative production of the Higgs, its presence is not manifest as a resonance as we scan over center-of-mass energies; instead, we observe the Higgs boson due to a threshold energy for the process $e^+e^- \to ZH$ to occur. But we're getting ahead of ourselves.

Let's estimate the cross section for this associative process and see this threshold explicitly. To evaluate the Feynman diagram above, we need to determine the coupling of the Higgs boson to the Z boson. Recall from Chapter 11 that in the electroweak theory we identified the mass of the Z boson. The Z boson mass term in the Lagrangian is

$$\mathcal{L}_{m^2} \supset \frac{1}{2}\frac{v^2}{4}(g_W^2 + g_Y^2)Z_\mu Z^\mu . \tag{13.40}$$

This was found by setting the scalar field $\vec{\phi}$ equal to its vev v, and ignoring all terms with the Higgs boson $H(x)$. To find the coupling to the Higgs boson, all we have to do is replace each instance of v with $H(x)$. Because this term in the Lagrangian is proportional to v^2, we pick up two terms that are each linear in the Higgs field $H(x)$. Then, the coupling of the Z boson to the Higgs boson is governed by the Lagrangian term

$$\mathcal{L}_{ZH} \supset \frac{v}{4}(g_W^2 + g_Y^2)HZ_\mu Z^\mu . \tag{13.41}$$

Therefore, the ZZH vertex in the Feynman diagram is proportional to the vev v and a combination of the weak isospin and hypercharge couplings:

$$\propto \frac{v}{4}(g_W^2 + g_Y^2) = \frac{m_Z}{2}\sqrt{g_W^2 + g_Y^2}. \tag{13.42}$$

On the right, we have exchanged the vev v with the mass of the Z boson m_Z, which justifies the claim that the Higgs boson couples proportionally to the Z boson's mass. With this coupling, we know everything else in this Feynman diagram, so we can evaluate it.

First, let's calculate the Feynman diagram with an initial left-handed electron and positron. This diagram is then

$$(13.43)$$

$$= \left(-\frac{1}{2} \frac{g_W^2}{\sqrt{g_W^2 + g_Y^2}} + \frac{1}{2} \frac{g_Y^2}{\sqrt{g_W^2 + g_Y^2}} \right) v_R^\dagger(p_{e^+}) \bar{\sigma}^\mu u_L(p_{e^-})$$

$$\times \frac{1}{(p_{e^-} + p_{e^+})^2 - m_Z^2} \frac{m_Z}{2} \sqrt{g_W^2 + g_Y^2} \epsilon_\mu(p_Z) \,.$$

In this expression, we have used the propagator for a massive particle, as developed in Section 11.3.4. Note also that the polarization vector of the Z boson couples directly to the electron and positron spinors. This is because the Higgs boson is spin-0, and so cannot affect the conservation of angular momentum. Simplifying the expression for the matrix element, we have

$$\mathcal{M}(e_R^+ e_L^- \to ZH) = \frac{m_Z}{4}(g_Y^2 - g_W^2)\, v_R^\dagger(p_{e^+}) \bar{\sigma} \cdot \epsilon(p_Z) u_L(p_{e^-}) \frac{1}{E_{\text{cm}}^2 - m_Z^2} \,. \qquad (13.44)$$

While we won't present the details here, the matrix element for the opposite electron–positron helicity configuration is

$$\mathcal{M}(e_L^+ e_R^- \to ZH) = \frac{m_Z}{4} g_Y^2\, v_L^\dagger(p_{e^+}) \sigma \cdot \epsilon(p_Z) u_R(p_{e^-}) \frac{1}{E_{\text{cm}}^2 - m_Z^2} \,. \qquad (13.45)$$

Squaring these matrix elements, averaging over initial spins, and summing over final spins, we then find

$$\frac{1}{4} \sum_{\text{spins}} |\mathcal{M}(e^+ e^- \to ZH)|^2 = \frac{m_Z^2}{64} \frac{1}{(E_{\text{cm}}^2 - m_Z^2)^2} \qquad (13.46)$$

$$\times \sum_{Z \text{ spins}} \left[(g_Y^2 - g_W^2)^2 |v_R^\dagger(p_{e^+}) \bar{\sigma} \cdot \epsilon(p_Z) u_L(p_{e^-})|^2 + g_Y^4 |v_L^\dagger(p_{e^+}) \sigma \cdot \epsilon(p_Z) u_R(p_{e^-})|^2 \right] \,.$$

To continue, we need to evaluate the spinor products explicitly. However, this is now not as easy as the case when the Z boson was the only final-state particle, and so we won't do it explicitly here. Nevertheless, we can extract essentially everything we need in order to estimate the cross section for $e^+ e^- \to ZH$, with a few observations. First, because the Z boson is unpolarized, the spinor products are the same, just as in the case of $e^+ e^- \to Z$:

$$\sum_{Z \text{ spins}} |v_R^\dagger(p_{e^+}) \bar{\sigma} \cdot \epsilon(p_Z) u_L(p_{e^-})|^2 = \sum_{Z \text{ spins}} |v_L^\dagger(p_{e^+}) \sigma \cdot \epsilon(p_Z) u_R(p_{e^-})|^2 \,. \qquad (13.47)$$

Additionally, the spinor product $|v_R^\dagger(p_{e+})\bar\sigma\cdot\epsilon(p_Z)u_L(p_{e-})|^2$ is just some function of Lorentz-invariant combinations of the momenta that compose it. What exactly this function is is not important; we just extract its mass dimension by multiplying and dividing by the center-of-mass energy E_{cm}:

$$\sum_{Z \text{ pols.}} |v_R^\dagger(p_{e+})\bar\sigma \cdot \epsilon(p_Z)u_L(p_{e-})|^2 \equiv E_{cm}^2 f(\theta), \tag{13.48}$$

where here $f(\theta)$ is a function of the scattering angle, θ. Then, we can express the squared matrix element in the compact form

$$\frac{1}{4}\sum_{\text{spins}} |\mathcal{M}(e^+e^- \to ZH)|^2 = \frac{(g_Y^2 - g_W^2)^2 + g_Y^4}{64} \frac{m_Z^2 E_{cm}^2}{(E_{cm}^2 - m_Z^2)^2} f(\theta). \tag{13.49}$$

Now, to calculate the cross section, we can put this expression into Fermi's Golden Rule. We have successfully extracted the scattering angle dependence of the matrix element, so we can use the compact form of Fermi's Golden Rule for two-body phase space in the center-of-mass frame established in Example 4.2 of Chapter 4. The cross section for $e^+e^- \to ZH$ scattering is thus

$$\sigma(e^+e^- \to ZH) = \frac{1}{2E_{cm}^2}\frac{2|\vec{p}|}{E_{cm}}\int \frac{d\cos\theta}{16\pi}\frac{(g_Y^2 - g_W^2)^2 + g_Y^4}{64}\frac{m_Z^2 E_{cm}^2}{(E_{cm}^2 - m_Z^2)^2}f(\theta) \tag{13.50}$$

$$= \frac{(g_Y^2 - g_W^2)^2 + g_Y^4}{64}\frac{m_Z^2}{(E_{cm}^2 - m_Z^2)^2}\frac{|\vec{p}|}{E_{cm}}\int \frac{d\cos\theta}{16\pi}f(\theta).$$

Here, $|\vec{p}|$ is the magnitude of three-momentum of the Z or Higgs boson in the final state. It can be found by constructing its momentum consistent with conservation laws. In particular, we can rotate to the frame in which we can write the Z and Higgs momentum four-vectors as

$$p_Z = \left(\sqrt{|\vec{p}|^2 + m_Z^2}, 0, 0, |\vec{p}|\right), \qquad p_H = \left(\sqrt{|\vec{p}|^2 + m_H^2}, 0, 0, -|\vec{p}|\right), \tag{13.51}$$

which are both on-shell and have zero net three-momentum. To determine $|\vec{p}|$, we use conservation of energy:

$$E_{cm} = \sqrt{|\vec{p}|^2 + m_Z^2} + \sqrt{|\vec{p}|^2 + m_H^2}. \tag{13.52}$$

One finds that $|\vec{p}|$ is

$$|\vec{p}| = \frac{E_{cm}}{2}\sqrt{1 - 2\frac{m_Z^2 + m_H^2}{E_{cm}^2} + \frac{(m_Z^2 - m_H^2)^2}{E_{cm}^4}}. \tag{13.53}$$

The integral over the scattering angle is just some number, so let's call it c_{int}:

$$\int \frac{d\cos\theta}{16\pi}f(\theta) \equiv c_{int}. \tag{13.54}$$

This number may have dependence on the Z and Higgs boson masses, but we ignore any residual dependence on them in c_{int}. Then, after all this effort, the cross section for the process $e^+e^- \to ZH$ is

$$\sigma(e^+e^- \to ZH) \tag{13.55}$$

$$= \frac{(g_Y^2 - g_W^2)^2 + g_Y^4}{128} \frac{m_Z^2}{(E_{\text{cm}}^2 - m_Z^2)^2} \sqrt{1 - 2\frac{m_Z^2 + m_H^2}{E_{\text{cm}}^2} + \frac{(m_Z^2 - m_H^2)^2}{E_{\text{cm}}^4}} c_{\text{int}}.$$

It may be unclear if all of this work was worth it, but let's evaluate the overall coupling dependence to see how relevant this term is. The dependence on the couplings g_W and g_Y is

$$\frac{(g_Y^2 - g_W^2)^2 + g_Y^4}{128} = 7.84 \times 10^{-4}, \tag{13.56}$$

where we use that $g_W = 0.642$ and $g_Y = 0.344$. While this is a small number, it is about eight orders of magnitude larger than the suppression of the direct production process $e^+e^- \to H$! It is therefore much more feasible to discover the Higgs boson in the associative process $e^+e^- \to ZH$.

Our procedure for discovering the Higgs boson and measuring its mass at a lepton collider would then be the following. We scan over center-of-mass energies for the electron–positron initial state E_{cm} and look for a Z boson in the final state. We of course observe the resonance at $E_{\text{cm}} = m_Z$, but the hope is that at some higher energy, we will observe a threshold for Higgs boson production. Note that the cross section calculated above is only non-zero if

$$E_{\text{cm}} > m_Z + m_H. \tag{13.57}$$

So, at the center-of-mass energy where we observe the threshold E_{th}, the Higgs boson mass is just m_Z less than that:

$$m_H = E_{\text{th}} - m_Z. \tag{13.58}$$

This was the search strategy of the four experiments at LEP (ALEPH, DELPHI, L3, and OPAL) through the end of its physics program. The maximum center-of-mass collision energy of electrons and positrons at LEP was (just over) 206 GeV, and so this would be the maximum threshold energy that could be observed. Therefore, the maximum mass of the Higgs boson that could just be detected at LEP is 91 GeV less than this, or about 115 GeV. The total integrated luminosity of data collected by LEP at the highest energies was 536 pb^{-1}, corresponding to a total of about 10,000 total electron–positron collision events. (From the PDG, the total e^+e^- collision cross section at $E_{\text{cm}} = 200$ GeV is a few tens of picobarns.) Though this is a relatively small number of events, it was sufficient to say with 95% confidence that no such energy threshold was observed at LEP.[4] Therefore, if the Higgs boson exists, then LEP determined that its mass was larger than about 115 GeV:

$$m_H > 115 \text{ GeV (LEP)}. \tag{13.59}$$

---

[4] R. Barate *et al.* [ALEPH and DELPHI and L3 and OPAL Collaborations and LEP Working Group for Higgs boson searches], "Search for the standard model Higgs boson at LEP," Phys. Lett. B **565**, 61 (2003) [arXiv:hep-ex/0306033].

13.2 Searching for the Higgs Boson
at Tevatron and LHC

The LEP collider and associated experiments ended data collection in 2000 and no Higgs boson was found. At the time of LEP's shutdown, the only other high-energy particle collision experiment was the Tevatron, which was located on the Illinois prairie at Fermi National Accelerator Laboratory (Fermilab). The Tevatron collided protons and anti-protons at a maximum center-of-mass energy of 1.96 TeV at two detectors: the CDF (Collider Detector at Fermilab) and D∅ (read: "dee-zero") experiments. Until the completion of the LHC, the Tevatron was the highest-energy and largest-particle collision experiment in the world, with a 4-mile (6.4-kilometer) circumference main accelerator ring. The greatest accomplishment of the Tevatron by the early 2000s was the discovery of the top quark in 1995.[5] An aerial photo of the Tevatron ring and surrounding area of Fermilab is presented in Fig. 13.1.

Tevatron had its sights on discovering the Higgs boson. With the lower bound of 115 GeV established by LEP, the search region of the Tevatron was narrowed, but there was still no firm upper bound on the Higgs boson mass. The hope was that the nearly 2 TeV center-of-mass collision energy of the Tevatron would be sufficient to produce the Higgs

Fig. 13.1   Aerial view of Fermilab. The Main Injector is the ring in the lower left of the photograph, the Tevatron ring is visible in the upper-right, Wilson Hall (the main building at Fermilab) is the Brutalist tower visible in the upper left, and the bison paddock is located at the top center of the image.

[5] F. Abe *et al.* [CDF Collaboration], "Observation of top quark production in $\bar{p}p$ collisions," Phys. Rev. Lett. **74**, 2626 (1995) [arXiv:hep-ex/9503002]; S. Abachi *et al.* [D0 Collaboration], "Observation of the top quark," Phys. Rev. Lett. **74**, 2632 (1995) [arXiv:hep-ex/9503003].

Historical Profile: Benjamin Lee

Benjamin Lee was a Korean-American theoretical physicist. He earned his Ph.D. in 1961 from the University of Pennsylvania, studying under Abraham Klein. Lee is perhaps best known for work on developing a robust upper bound on the mass of the Higgs boson of about 1 TeV, above which the weak force would interact very differently than observed.[6] Among a broad particle physics research program, Lee also predicted the mass of the charm quark[7] (along with Mary Gaillard) and provided constraints from cosmology on neutrino masses.[8] In 1970, Lee lectured on the weak force at the Cargèse Summer School located on the island of Corsica, France. A young Gerardus 't Hooft was a student at that school, and was inspired by Lee's lecture to work on the theory of the weak force. Lee moved to Fermilab in 1973, and became the head of its theory division shortly thereafter. While traveling to Aspen, Colorado, from Fermilab in 1977 for a meeting, Lee tragically died in a car accident.

boson and verify its existence with certainty. The production of the Higgs boson at a hadron collider like Tevatron or LHC is very different than that at LEP, which in turn implies that the search strategy of these experiments is different than that of LEP. Though the Tevatron collided protons on anti-protons while the LHC collides protons on protons, this difference is essentially irrelevant for the dominant production mechanism of the Higgs boson at these experiments. As such, in what follows we just discuss the production of the Higgs boson at a proton–proton collider like the LHC, but will justify the claim that the production at a proton–anti-proton collider would be the same.

13.2.1 $pp \rightarrow H$

As with any high-energy process at a proton collider, it is not the protons that are directly responsible for the process, but rather their constituent quarks and gluons. The simplest mechanism for the production of the Higgs boson at a proton collider like the LHC would then be the direct process through the interactions of quarks. A pseudo-Feynman diagram that represents such a production process would be something like

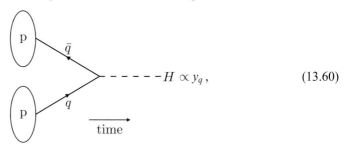

$$ \cdots\,H \propto y_q\,, \tag{13.60} $$

---

[6] B. W. Lee, C. Quigg and H. B. Thacker, "Weak interactions at very high energies: The role of the Higgs boson mass," Phys. Rev. D **16**, 1519 (1977).

[7] M. K. Gaillard and B. W. Lee, "Rare decay modes of the K-mesons in gauge theories," Phys. Rev. D **10**, 897 (1974).

[8] B. W. Lee and S. Weinberg, "Cosmological lower bound on heavy neutrino masses," Phys. Rev. Lett. **39**, 165 (1977).

where y_q is the Yukawa coupling of the quark pulled from the proton. The cross section for such a process is therefore proportional to y_q^2, which is problematic for the same reason as the process $e^+e^- \to H$ studied in the previous chapter: the Yukawa couplings of quarks in the proton are tiny! The heaviest quark that exists in the proton in reasonable amounts is the strange quark, which has a mass of $m_s \simeq 100$ MeV (from the PDG, of course). The strange-quark Yukawa coupling is therefore

$$y_s = \frac{\sqrt{2}m_s}{v} \simeq \frac{140 \text{ MeV}}{246 \text{ GeV}} \simeq 5.7 \times 10^{-4}. \tag{13.61}$$

When squared, this suppresses the cross section by a factor of about 10^{-7}. So, just as we found for the process $e^+e^- \to H$, the direct production of the Higgs boson at a proton collider is extremely rare and therefore very challenging to observe.

A possible resolution is to consider the process $pp \to ZH$, similar to the resolution for searching for the Higgs boson at LEP. This has a significantly larger cross section than $pp \to H$, but the discovery of the Higgs boson at a proton collider via this process is almost impossible. Recall what the search strategy was for discovering the Higgs boson in $e^+e^- \to ZH$. We are able to control the center-of-mass energy of the electron–positron collision very precisely, and scan over different energies looking for a threshold. The threshold energy then corresponds to the sum of the Z and Higgs boson masses. At a hadron collider we can use a similar technique, by scanning over a range of proton collision center-of-mass energies. Now, however, we don't have control over the actual collision energy of the partons within the colliding protons, so we don't know what the partonic center-of-mass energy was. The energy distribution of quarks and gluons in the proton is determined by the parton distribution functions. As a consequence, scanning for threshold energies is very subtle at a hadron collider, since these do not manifest in a clearly useful way as they did at LEP. Thus, searching for the Higgs as a resonance – or bump hunting – at a hadron collider is a much more fruitful approach.

Gluon–Gluon Fusion

That said, from the estimation above, we can't expect to produce a sufficient number of Higgs bosons from the partonic process $q\bar{q} \to H$, as the Yukawa couplings of quarks in the proton are extremely small. How can we produce the Higgs boson at a hadron collider in numbers that enable discovery? From the s-channel (resonant) production of the Higgs boson, let's work backward to see how this can be done. To produce a Higgs boson via a process with a large cross section, the Higgs should couple to as massive a particle as possible. The more massive the particle, the stronger the coupling. So, let's couple the Higgs to the most massive particle of the Standard Model, the top quark. The mass of the top quark is $m_t = 173.1$ GeV and so its Yukawa coupling is

$$y_t = \frac{\sqrt{2}m_t}{v} \simeq \sqrt{2}\frac{173.1 \text{ GeV}}{246 \text{ GeV}} \simeq 1.0. \tag{13.62}$$

Coupling a Higgs to top quarks is unsuppressed. So, to produce the Higgs boson with the largest possible cross section, we want to do it through top quarks as

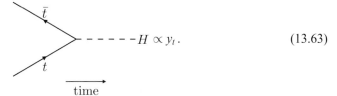

$$ H \propto y_t . \qquad (13.63) $$

At a hadron collider, then, how do we produce top quarks? With a mass that's about 175 times that of the proton, there's no way that the top quark can be a parton in the proton. Instead, we need a method to produce top quarks from honest constituents of the proton. Gluons, of course, couple to top quarks, and gluons exist in copious amounts in the proton. So, we can use gluons to produce top quarks, but this comes at a price. Just as electrons emit photons, top quarks can emit gluons, and so each gluon from the protons splits into a top quark and an anti-top quark. One possible diagram representing this process is

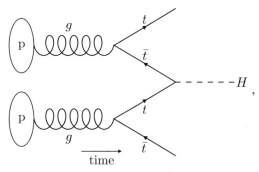

,

but this actually corresponds to the process $pp \rightarrow t\bar{t}H$, which is not what we want. We're very close, though; we just need to eliminate the top and anti-top quarks from the final state. We can accomplish this by connecting them into a loop of top quarks. That is, the diagram that just corresponds to $pp \rightarrow H$ through gluons is

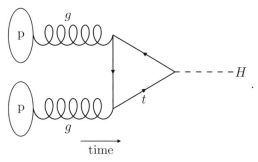

.

This diagram is special; topologically, it has a loop of top quarks, and so is referred to as a **one-loop diagram**. This loop of top quarks accomplishes a specific task. Gluons, which are massless and numerous in the proton, are coupled to the Higgs boson, whose strength of interaction with particles is proportional to their mass. In the Lagrangian of the Standard Model, there is no direct coupling of the Higgs boson to gluons, because gluons are massless. Another way to say this is that gluons do not appear in the classical equations of motion for the Higgs boson. However, particle physics is quantum mechanical, and the Heisenberg uncertainty principle allows for the production of virtual top quarks that connect gluons to the Higgs boson. For the production of the Higgs boson at a hadron collider, this quantum mechanical property is absolutely vital. If $\hbar = 0$, the value of this diagram would be zero. We'll see how this works in a second.

Let's estimate the matrix element of the process $gg \to H$ and the corresponding value of the cross section for the process $pp \to H$ through a top quark loop in the following example.

Example 13.2 What is the matrix element for the process $pp \to H$ through a top quark loop?

Solution

Let's first collect all of the coupling factors. As identified earlier, the coupling of the top quark to the Higgs boson is proportional to the Yukawa coupling of the top quark, $y_t \simeq 1$. The coupling of gluons to top quarks is proportional to the coupling constant of the strong force, $g = \sqrt{4\pi\alpha_s}$. So, the matrix element for the process $gg \to H$ is proportional to

$$\mathcal{M}(gg \to H) \propto g^2 y_t \propto 4\pi\alpha_s y_t . \tag{13.64}$$

So far, the factors we have identified in the matrix element are dimensionless, and we will need to include appropriate energy factors to get the dimensions correct. To determine the mass dimension of the matrix element, recall that the phase space integral in Fermi's Golden Rule has mass dimension 0:

$$\left[\int \frac{d^4 p_H}{(2\pi)^4} 2\pi\delta(p_H^2 - m_H^2) \, |\mathcal{M}(gg \to H)|^2 \, \delta^{(4)}(p_1 + p_2 - p_H)\right] = [\text{mass}]^0 . \tag{13.65}$$

Here, p_H is the four-momentum of the Higgs boson and p_1 and p_2 are the four-momenta of the initial gluons. For this to have the correct mass dimension, the matrix element must have mass dimension 1. (Recall that δ-functions have mass dimension equal to the inverse mass dimension of their argument.) So, we need to account for this mass dimension.

Let's consider what happens if the mass of the top quark $m_t \to \infty$. In this limit, the Compton wavelength of the top quark goes to 0 and the virtual top quarks in the loop exist for zero time. Thus, the matrix element must vanish, because there is zero probability that the infinitely massive virtual top quarks could mediate this interaction. Therefore, the matrix element is proportional to an inverse power of the top quark mass:

$$\mathcal{M}(gg \to H) \propto \frac{4\pi\alpha_s y_t}{m_t^a} , \tag{13.66}$$

where $a > 0$. Additionally, in the limit that $m_t \to \infty$, the only relevant mass scale of the weak force is the Higgs vev, v. The quark masses in the Standard Model are a consequence of the Higgs mechanism and are proportional to the Higgs vev. That is, the only mechanism in the Standard Model for $m_t \to \infty$ is if the vev also diverges: $v \to \infty$. If the top quark Yukawa coupling y_t were to diverge, the Higgs mechanism could not account for this phenomenon, as the Yukawa coupling is just a parameter in the Standard Model. Therefore, the power a must be such that the matrix element is only a function of the vev v. As

$$m_t = \frac{y_t v}{\sqrt{2}}, \tag{13.67}$$

we find that $a = 1$. As written, the matrix element doesn't yet have mass dimension 1. The only other relevant mass scale in this matrix element is the Higgs boson mass, and this must account for the difference in mass dimension. That is,

$$\mathcal{M}(gg \to H) \propto 4\pi\alpha_s y_t \frac{m_H^2}{m_t}, \tag{13.68}$$

which has the correct mass dimension.

With dimensional analysis, we have identified the dependence of the matrix element $\mathcal{M}(gg \to H)$ on the various couplings and masses relevant to the process. Our argument for the dependence on the top quark and Higgs boson masses relied on the limit of $m_t \to \infty$, or the limit in which $m_H \ll m_t$. Relaxing this assumption allows for dependence on an arbitrary function of the ratio m_H/m_t, but we will see that $m_H \ll m_t$ is justified experimentally. So, we won't consider any other mass dependence. However, as the diagram for the process $gg \to H$ contains a loop, there are additional pure number factors that should be included to accurately estimate the value of the matrix element. These factors can be systematically determined with a procedure called **naïve dimensional analysis**, or NDA.[9] For every loop in a Feynman diagram, NDA states that we should multiply by a factor of $1/(4\pi)^2$. Doing this, the final result for the matrix element is

$$\mathcal{M}(gg \to H) \simeq \frac{4\pi\alpha_s y_t}{(4\pi)^2} \frac{m_H^2}{m_t} = \frac{\alpha_s y_t}{4\pi} \frac{m_H^2}{m_t}. \tag{13.69}$$

While the rules of NDA can be derived directly from consideration of the mathematical expression for the Feynman diagram with a loop, we will provide some motivation of this $1/(4\pi)^2$ factor.

Unlike the diagrams that we studied and calculated before, the loop in this diagram has an unconstrained momentum flowing around it. Let's see how this works. With the gluons' momenta denoted by p_1 and p_2, the momentum of the Higgs boson is therefore $p_H = p_1 + p_2$. Enforcing momentum conservation at every vertex, we then find that the momentum in the loop is

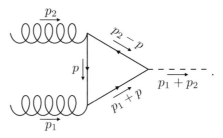

---

[9] A. Manohar and H. Georgi, "Chiral quarks and the nonrelativistic quark model," Nucl. Phys. B **234**, 189 (1984).

The momentum p is the unconstrained, arbitrary momentum. By the rules of quantum mechanics, we must sum over all possibilities consistent with measurement, which corresponds to an improper integral over all four-momentum components of p. As with the discussion of Fermi's Golden Rule and Lorentz-invariant phase space in Chapter 4, whenever we integrate over the four-momentum components, there is an associated factor of $1/(2\pi)^4$ in the integration measure. This factor comes from Fourier transforming from position space x to momentum space p, and ensures that inverse Fourier transforming back to position space has the proper normalization. The NDA factor of $1/(4\pi)^2$ comes from this Fourier-transform normalization.

With the matrix element established, we can now evaluate the cross section. Using Fermi's Golden Rule and the one-body phase space integral established in previous sections, the cross section for the process $gg \to H$ is

$$\sigma(gg \to H) \simeq \frac{1}{2m_H^2} \int \frac{d^4 p_H}{(2\pi)^4} 2\pi \delta(p_H^2 - m_H^2) \, |\mathcal{M}(gg \to H)|^2 \, (2\pi)^4 \delta^{(4)}(p_1 + p_2 - p_H)$$

$$\simeq \frac{\alpha_s^2 y_t^2}{16\pi} \frac{m_H^2}{m_t^2} \delta(\hat{s} - m_H^2). \tag{13.70}$$

The combination of overall couplings and numerical factors is approximately

$$\frac{\alpha_s^2 y_t^2}{16\pi} \simeq 2.8 \times 10^{-4}, \tag{13.71}$$

where we have used the value of the strong coupling, $\alpha_s = 0.118$, established by the PDG evaluated at the Z boson mass. This is several orders of magnitude larger than direct production of the Higgs boson from quarks considered earlier, and will be central for discovery of the Higgs.

To predict the cross section for the process $pp \to H$, we need to include the probability of pulling gluons out of protons. This is accomplished by the gluon parton distribution function $f_g(x)$. The squared partonic center-of-mass energy $\hat{s}$ is related to the squared proton center-of-mass energy s by

$$\hat{s} = x_1 x_2 s, \tag{13.72}$$

where x_1 and x_2 are the momentum fractions of the gluons with respect to the protons' momenta. Then, the cross section differential in both x_1 and x_2 is

$$\frac{d^2 \sigma(pp \to H)}{dx_1 \, dx_2} \simeq \frac{\alpha_s^2 y_t^2}{16\pi} \frac{m_H^2}{m_t^2} f_g(x_1) f_g(x_2) \delta(x_1 x_2 s - m_H^2). \tag{13.73}$$

Finally, the total cross section for Higgs boson production from proton–proton collisions through a top quark loop is

$$\sigma(pp \to H) \simeq \int_0^1 dx_1 \int_0^1 dx_2 \frac{\alpha_s^2 y_t^2}{16\pi} \frac{m_H^2}{m_t^2} f_g(x_1) f_g(x_2) \delta(x_1 x_2 s - m_H^2). \tag{13.74}$$

The integral can be massaged and simplified, but to completely evaluate it requires a functional form of the gluon parton distribution function, $f_g(x)$. This is the exact same value of the cross section of Higgs boson production as at the Tevatron, $p\bar{p} \to H$. Anti-protons are (obviously) the anti-particles of protons, and so we should replace one of the gluon parton distributions with the anti-gluon's parton distribution. However, the gluon is its own anti-particle, so the cross section is unchanged.

This process for producing a Higgs boson is called **gluon–gluon fusion**. The diagram that provides the first non-zero contribution to a cross section of interest with the fewest number of loops is called a **leading order diagram.** For many processes, like $e^+e^- \to \mu^+\mu^-$, leading order diagrams are also called **tree diagrams**, which have no loops and look topologically like trees. For Higgs production, however, the leading order diagram has one loop. As more loops are added, one builds up a better and better approximation to the desired result. One refers to diagrams with more loops as next-to-leading (one more loop), next-to-next-to-leading (one further loop, NNLO for short), etc. The state-of-the-art high-precision calculations are now being performed at next-to-next-to-next-to-leading order (N^3LO; read "N-three-ell-oh"), which adds three (!) loops to the leading order diagram. These are Herculean calculations that require teams to complete. The first N^3LO calculation was presented for Higgs production in 2015.[10]

13.2.2 $pp \to H \to W^+W^-$

Hadron colliders are often colloquially referred to as **discovery machines** because protons can be collided at enormous center-of-mass energies and the collision energy of the individual partons can range over many decades. With the production of the Higgs boson understood, we want to use this powerful lever to focus our search for the Higgs boson. From LEP, we were able to establish a general and robust lower bound on the Higgs mass; can we establish a similarly robust upper bound at a hadron collider?

The Higgs boson is an unstable particle and decays almost immediately after it is produced, so we can only identify it through its decay products. Our strategy for establishing an upper bound on the Higgs boson mass is to search for evidence of a Higgs boson decay to massive particles. With the lower bound from LEP of $m_H > 115\,\text{GeV}$, we want as low an upper bound as possible. The Higgs boson decays to pairs of massive particles, and so the upper bound we are looking for will be twice the mass of a decay product. The lightest particle of the Standard Model with twice its mass greater than 115 GeV is the W boson, $m_W \simeq 80$ GeV. So, looking for Higgs boson decays to W bosons enables us to establish an upper bound on the Higgs boson mass of about 160 GeV. Actually, we are able to do better than this: the W boson itself is an unstable particle, so we look for the decay products of W bosons at invariant masses a bit below the threshold for on-shell W^+W^- production.

What makes this search powerful is that W bosons decay to both quarks and leptons of the Standard Model. So, we have our choice of which decay products to measure and attempt to discover the Higgs boson. Searching for W bosons through their decays to quarks is especially challenging. While it is true that the W boson decays to quarks nearly 70%

[10] C. Anastasiou, C. Duhr, F. Dulat, F. Herzog and B. Mistlberger, "Higgs boson gluon-fusion production in QCD at three loops," Phys. Rev. Lett. **114**, 212001 (2015) [arXiv:1503.06056 [hep-ph]].

of the time, in an experiment, quarks manifest themselves as hadronic jets. So, if we look for the Higgs boson through hadronic decays of the W boson, the process that we would see in our detector would be $pp \to$ four jets, with the transverse momentum of each of the jets about half the W boson mass, 40 GeV. This is problematic because the rate for such a process to occur through standard QCD channels is enormous: it is very easy at the LHC to produce many, relatively low-energy, jets. Whatever signal we would be looking for would be overwhelmed by background.

Instead of hadronic W decays, we then focus on leptonic W decays. We do take a hit to the rate of leptonic decays, but this is more than made up for by the extremely clean and low-background final state. The W boson couples charged leptons (electrons, muons, taus) to their corresponding neutrinos, so the leptonic decay channel process is

$$pp \to H \to W^+ W^- \to l^+ \nu_l \, l'^- \bar{\nu}_{l'} \, ,$$

where l and l' are two flavors of leptons. The pseudo-Feynman diagram that represents this process is

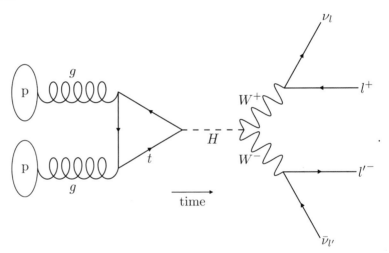

Electrons and muons are stable on time and distance scales measured at the LHC's detectors, so they can be identified by their charged track and a deposit in the electromagnetic calorimetry (for electrons) or as a track in the muon chamber (for muons). (Tau leptons decay quickly, and introduce additional experimental challenges and so are ignored for now.) The two final-state neutrinos cannot, of course, be directly measured in the detector. Indirectly, we would infer their existence by the measurement of missing transverse momentum (MET). A complication of this final state is that there are two neutrinos and therefore the MET is the vector sum of the momentum of the neutrinos, and not the momentum of a single neutrino. Nevertheless, this subtlety is worth the cost because it is so easy to cleanly identify the charged leptons, and therefore to significantly reduce the possible backgrounds.

So, our experimental task is clear: we look for the process

$$pp \to l^+ l'^- + \text{MET},$$

where l and l' are electrons or muons, in our detector and compare the rate for this process in the Standard Model with and without the assumption of the existence of the Higgs boson. The mass of the Higgs boson can be inferred from a measurement of the invariant mass of the leptons and missing transverse momentum of the final state. This is of course complicated by the fact that only two components of the missing transverse momentum are constrained by momentum conservation and the individual neutrinos' momenta is not resolved. Nevertheless, in Exercise 13.3, you will study this configuration and show that the mass of the Higgs boson can be identified from this final state.

The experiments at Tevatron and LHC employed this Higgs boson search strategy in their data collected through the end of 2011. Their corresponding analyses were all published by mid-2012, and no Higgs boson was found by searching for decays through W bosons.[11] A robust upper bound on the Higgs mass was established to be about 140 GeV at these hadron colliders. Combined with the lower bound from LEP, the window in which the Higgs boson could exist was getting extremely narrow by July 2012:

$$115 \text{ GeV (LEP)} < m_H < 140 \text{ GeV (Tevatron and LHC)}. \qquad (13.75)$$

13.2.3 The Golden Channels: $pp \to H \to \gamma\gamma$ and $pp \to H \to 4\ell$

Either the Standard Model Higgs boson has a mass in the window established by LEP, Tevatron, and LHC, or it does not exist. To unambiguously discover the Higgs boson, we want to clearly observe a resonance mass peak. This requires that all final-state particles from Higgs boson decay be visible in the detector, so there can't be any neutrinos produced. Additionally, we want small backgrounds and clear final-state particle identification, so we don't want any jets. These requirements eliminate the $H \to W^+W^-$ decay channel as a vehicle for discovery. (Observing this decay is vital for verifying that the Higgs boson is indeed the Higgs boson, however.) We also want the **decay mode** to have a reasonably large rate, otherwise it will be very hard to collect enough candidate events. This eliminates direct Higgs decays to leptons, as their small masses mean very small couplings to the Higgs. The Higgs can, however, decay to four charged leptons through Z bosons in the process

$$H \to ZZ \to l^+l^-\,l'^+l'^-\,.$$

Z bosons have a large coupling to the Higgs because of their large mass, and Z bosons decay to electrons or muons about 7% of the time. The cost of this decay is that at least one of the Z bosons must be off-shell, as the largest possible Higgs mass is less than two times the Z boson mass. The Feynman diagram for this decay of the Higgs boson is

[11] G. Aad *et al.* [ATLAS Collaboration], "Search for the Higgs boson in the $H \to \text{WW}(^*) \to \ell_\nu\ell_\nu$ decay channel in pp collisions at $\sqrt{s} = 7$ TeV with the ATLAS detector," Phys. Rev. Lett. **108**, 111802 (2012) [arXiv:1112.2577 [hep-ex]]; S. Chatrchyan *et al.* [CMS Collaboration], "Search for the standard model Higgs boson decaying to W^+W^- in the fully leptonic final state in pp collisions at $\sqrt{s} = 7$ TeV," Phys. Lett. B **710**, 91 (2012) [arXiv:1202.1489 [hep-ex]]; Tevatron New Physics Higgs Working Group [CDF and D0 Collaborations], "Updated combination of CDF and D0 searches for Standard Model Higgs boson production with up to 10.0 fb^{-1} of data" [arXiv:1207.0449 [hep-ex]].

For ease of detection, the leptons l and l' are electrons or muons.

A leptonic final state isn't the only configuration that satisfies the requirements for discovery established above. The Higgs boson can also decay to photons, $H \to \gamma\gamma$. Just like the production mechanism through gluons, there is no Lagrangian coupling of the Higgs boson to photons, because the photons are massless. However, the Higgs boson couples strongly to the top quark, and top quarks carry electric charge, and so can radiate photons. Photons have a very clean signature in the detector: as they are uncharged, they leave no track, but deposit all of their energy in the electromagnetic calorimeter. An additional feature of the $pp \to H \to \gamma\gamma$ process is that the background process $pp \to \gamma\gamma$ for direct production of two photons in the Standard Model has no intrinsic mass scales associated with it. Photons are massless and they can just be emitted from any charged particle, so the background distribution must be smooth with no resonances or thresholds. Therefore, a resonance from Higgs decay will clearly stick out above background. The Feynman diagram for Higgs decay to photons is

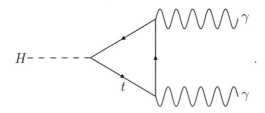

These two Higgs decay modes, $H \to$ four leptons and $H \to \gamma\gamma$, are colloquially referred to as the **golden channels** because they have a distinct experimental signature, small backgrounds, and an unambiguous resonance peak if the Higgs does indeed exist. The ATLAS and CMS experiments at the LHC extensively measured these golden channels in the established mass window in search for the Higgs boson. On July 4, 2012, in back-to-back presentations, the spokespersons for CMS and ATLAS, Joseph Incandela and Fabiola Gianotti, conclusively demonstrated that each of their experiments had discovered the Higgs boson with greater than 5σ significance. Rolf-Dieter Heuer, the CERN Director-

General at the time, remarked after the presentations, "As a layman, I would say, I think we have it!"

Evidence of Discovery

The primary evidence that Incandela and Gianotti presented on behalf of CMS and ATLAS for the discovery of the Higgs boson were the invariant mass distributions for the $H \to$ four leptons and $H \to \gamma\gamma$ decays. Starting with the $H \to$ four leptons channel, the measurement by the CMS collaboration is presented in Fig. 13.2. To ensure that the final state could have come from a Higgs decay, CMS required that the four leptons were electrons or muons, two were positively charged and two were negatively charged, and that pairs of opposite-signed charged leptons were the same flavor. The main plot shows the invariant mass of the four leptons $m_{4\ell}$ versus the number of events at that mass that were observed, ranging from about 70 GeV to 180 GeV. There is a clear resonance at about 90 GeV, which corresponds to the Z boson. The Z boson couples to itself and so can decay to four leptons through two off-shell Z bosons. At high masses, the number of events is seen to rise, corresponding to threshold production of a pair of Z bosons around 180 GeV ($\simeq 2m_Z$). The shaded regions correspond to an estimate of the background hypothesis for this distribution corresponding to no Higgs boson existing.

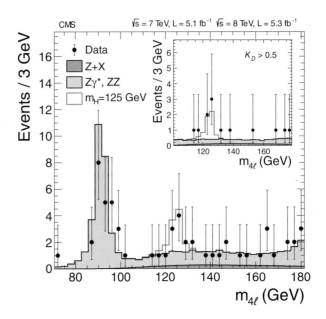

Fig. 13.2 Plot of the four-lepton invariant mass $m_{4\ell}$ versus the number of measured events by the CMS experiment. The data points come from a total of about 10 inverse femtobarns collected by CMS at center-of-mass collision energies of both 7 and 8 TeV. Reprinted from Phys. Lett. B **716**, S. Chatrchyan *et al.* [CMS Collaboration], "Observation of a new boson at a mass of 125 GeV with the CMS experiment at the LHC," 30 (2012), with permission from Elsevier.

At a mass of about 125 GeV, however, there is a clear excess of events above the expected shaded background. While this excess doesn't seem to be much, 10 events were observed in the range between 120 and 130 GeV, compared to an expected number of about 4 events. Assuming Poisson statistics, this is an excess of 3σ. Importantly, however note, that this is the local significance, not the global significance. The location of the excess was not known beforehand, and could have in principle been anywhere in the allowed range of 115 to 140 GeV. This is referred to as the look-elsewhere effect and has the consequence of reducing the global significance versus the local significance. (See Exercise 5.6 of Chapter 5 for details.) The plot in the inset of Fig. 13.2 shows the same data over a narrower mass range with the restriction that those events have additional requirements on them that force them to look more like they came from the Standard Model Higgs boson. This has the effect of reducing the number of events in the 120–130 GeV window to five, but the expected number of events (the integral of the shaded region) is reduced to about one over that same range. The local significance (not including the look-elsewhere effect) is now about 4σ!

A plot of the invariant mass of two photons versus the number of events from the search for the $H \to \gamma\gamma$ decay by the ATLAS collaboration is presented in Fig. 13.3. The range of di-photon masses $m_{\gamma\gamma}$ in this plot extends from 100 to 160 GeV, and the background distribution corresponding to direct photon production (modeled by the dashed curve) is smooth with no structure, as expected. On top of this background, there is a clear excess of events at around 125 GeV, the same mass where there was an excess in the $m_{4\ell}$ distribution.

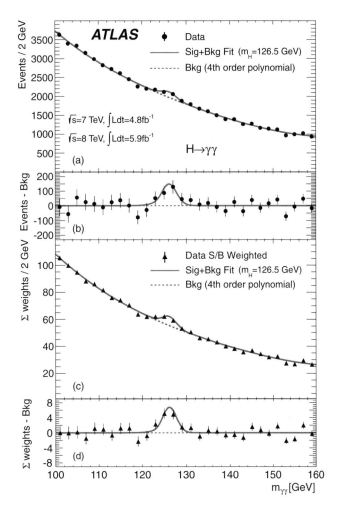

Fig. 13.3   Plots of the di-photon invariant mass $m_{\gamma\gamma}$ versus the number of measured events by the ATLAS experiment. The data points come from a total of about 10 inverse femtobarns collected by ATLAS at center-of-mass collision energies of both 7 and 8 TeV. The upper plot shows the raw number of events as a function of $m_{\gamma\gamma}$ while the bottom plot enhances the excess near 125 GeV by weighting the events. Reprinted from Phys. Lett. B **716**, G. Aad *et al.* [ATLAS Collaboration], "Observation of a new particle in the search for the Standard Model Higgs boson with the ATLAS detector at the LHC," 1 (2012), with permission from Elsevier.

Focusing on the upper plot, in the mass range of 123–129 GeV about 8400 events were observed, while only about 8000 events were expected, assuming that there is no Standard Model Higgs boson. The statistical standard deviation of the expected number of events is therefore about 90 events, so this excess has a significance of nearly 5σ alone! Note that there is no look-elsewhere effect in these data. From the $H \rightarrow$ four leptons search, we had already determined that masses near 125 GeV were interesting, so only if the di-photon masses exhibited an excess in this region would such an excess provide additional evidence for the Higgs boson.

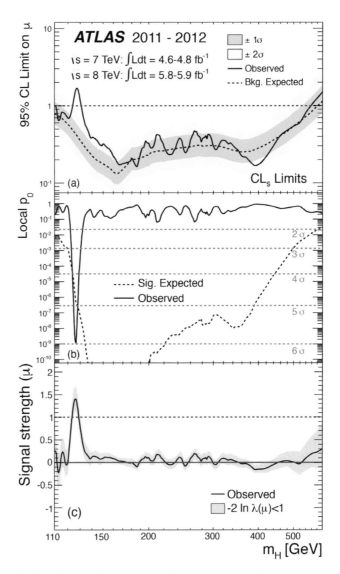

Fig. 13.4   Summary plots of ATLAS's combination of searches in the $H \rightarrow$ four leptons and $H \rightarrow \gamma\gamma$ channels. The top two plots demonstrate a nearly 6σ excess over background. Reprinted from Phys. Lett. B **716**, G. Aad *et al.* [ATLAS Collaboration], "Observation of a new particle in the search for the Standard Model Higgs boson with the ATLAS detector at the LHC," 1 (2012), with permission from Elsevier.

These two results from $H \rightarrow$ four leptons and $H \rightarrow \gamma\gamma$ searches were then combined by both experiments separately to establish their evidence for existence of the Higgs boson. The combined summary plots from ATLAS are shown in Fig. 13.4. The top two plots are the most relevant for our discussion here; the bottom plot presents the interpretation of these results. First, the top plot. Such a plot is colloquially referred to as a **Brazil plot** because of its (admittedly weak) similarity to the colors (green and yellow in the color version) of the Brazilian flag. This plot shows the 95% confidence limit (2σ) on the signal strength μ versus the corresponding mass of the Higgs boson, m_H. The signal strength

μ is 0 if the Higgs boson does not exist and 1 if the Higgs is exactly as assumed in the Standard Model. A different value of μ corresponds to a Higgs boson with properties different than that of the Standard Model. The dashed line is the expected limit on μ as established by simulating events at the LHC assuming the absence of the existence of a Higgs boson. The shaded bands around this are the 1σ and 2σ uncertainties in the expected limit. The solid curve is the established limit on μ from data, and over most of the mass range, this measured limit lies comfortably within the 1σ uncertainties. However, there is a huge excess around a mass of 125 GeV, implying that such a limit on μ can't be established. That is, at a mass of around 125 GeV, the data are not consistent with the null hypothesis of there being no Higgs boson.

This is quantified a bit more in the middle plot. This plot shows the local p-value of the data from the null hypothesis (i.e., no Higgs boson) versus mass, m_H. The p-value is the probability that the null hypothesis could produce an excess at least as large as the observed excess. Over the vast majority of the mass range this p-value is large (at the level of tens of percent), meaning that the probability for the null hypothesis to explain the data is high. However, at a mass of about 125 GeV, the p-value takes a nosedive, and gets down to about 10^{-9}. That is, there is a probability of about 1 in 1 billion that the excesses observed at $m_H \simeq 125$ GeV could be explained by the null hypothesis. This corresponds to a local significance of about 6σ. The look-elsewhere effect will reduce this significance slightly, but cannot remove it. Thus these results establish, with greater than 5σ confidence at both ATLAS and CMS, that the Higgs boson exists. Combining measurements from ATLAS and CMS yields the Higgs boson mass m_H to be

$$m_H = 125.09 \pm 0.21 \text{ (stat)} \pm 0.11 \text{ (sys) GeV}. \tag{13.76}$$

The first uncertainty is statistical while the second is an estimate of systematic uncertainty in the measurement.

Example 13.3 Can we estimate how ATLAS did this combination of searches for the Higgs boson in the two golden channels? Specifically, let's consider two measurements A and B in which there are $N_{\exp}^A$ events expected from the null hypothesis for A and N_{meas}^A events actually measured for A (and similarly for B). What is the σ deviation from the null hypothesis for the combination of measurements A and B? Assume that the measurements are uncorrelated and that $N_{\exp}^A$ and $N_{\exp}^B$ are very large.

Solution

This problem asks for the statistical significance or probability that two outcomes occurred. This is an "and" statement in probability, meaning that the probability for both to occur is just the product of each individually happening. Importantly, this joint probability is simply the product of probabilities because we assume that measurements A and B are uncorrelated. This means that knowledge about one measurement tells you nothing about the other measurement. If there were correlation between measurements, this would complicate the analysis we present here, and in general decrease the effect that we find.

With those caveats out of the way, let's determine the probability for each of these measurements. As always, we assume Poisson statistics, so that the standard deviation on a measurement of N events is $\sqrt{N}$. Thus, the number of standard deviations of measurements A and B from their null hypotheses σ_A and σ_B is

$$\sigma_A = \frac{N^A_{\text{meas}} - N^A_{\text{exp}}}{\sqrt{N^A_{\text{exp}}}}, \qquad\qquad \sigma_B = \frac{N^B_{\text{meas}} - N^B_{\text{exp}}}{\sqrt{N^B_{\text{exp}}}} . \qquad (13.77)$$

The subscripts meas and exp represent the measured and expected number of events, respectively. Because we assume that the expected number of events is large, we can approximate the distribution as Gaussian. The probability that there was a fluctuation in the null hypotheses at least as large as σ_A and σ_B is the product of their p-values:

$$p_{A\&B} = p_A p_B = \left[\int_{\sigma_A}^{\infty} dx \, \frac{e^{-\frac{x^2}{2}}}{\sqrt{2\pi}} \right] \left[\int_{\sigma_B}^{\infty} dy \, \frac{e^{-\frac{y^2}{2}}}{\sqrt{2\pi}} \right] \qquad (13.78)$$

$$= \left[\frac{1}{2} - \frac{\text{erf}\left(\frac{\sigma_A}{\sqrt{2}}\right)}{2} \right] \left[\frac{1}{2} - \frac{\text{erf}\left(\frac{\sigma_B}{\sqrt{2}}\right)}{2} \right] .$$

We could end here, but it is common practice (though somewhat misleading) to reinterpret this p-value itself in σ deviation from the null hypothesis. There's really no elegant way to do this, but we want to solve for the deviation $\sigma_{A\&B}$ such that

$$\int_{\sigma_{A\&B}}^{\infty} dz \, \frac{e^{-\frac{z^2}{2}}}{\sqrt{2\pi}} = \left[\frac{1}{2} - \frac{\text{erf}\left(\frac{\sigma_A}{\sqrt{2}}\right)}{2} \right] \left[\frac{1}{2} - \frac{\text{erf}\left(\frac{\sigma_B}{\sqrt{2}}\right)}{2} \right] . \qquad (13.79)$$

We can solve for $\sigma_{A\&B}$ by evaluating the right side of this equation and then adjusting the value of $\sigma_{A\&B}$ in the integral on the left to equal it.

13.3 Properties of the Higgs Boson

The discovery of the Higgs boson was based simply on observation of two of its decay modes. To verify that this particle is indeed *the* Higgs boson of the Standard Model, we need to observe many more decay modes and properties. While we survey a few of these validation measurements here, there are still numerous experimental questions to be answered about the Higgs boson. Nevertheless, every measurement performed so far is consistent with the Standard Model Higgs boson.

13.3.1 Scalar Potential Coupling λ

The form of the Higgs boson potential $V(H)$ is what is ultimately responsible for the spontaneous symmetry breaking of the electroweak theory. Recall that the potential is

$$V(H) = \lambda v^2 H^2 + \lambda v H^3 + \frac{\lambda}{4} H^4 , \qquad (13.80)$$

where λ is the quartic Higgs coupling. The vev $v = 246$ GeV sets the mass scale of the electroweak bosons; the mass of the Higgs from this potential is

$$m_H = \sqrt{2\lambda}v. \tag{13.81}$$

With a measured value for the Higgs boson mass of $m_H \simeq 125$ GeV, we then extract the value of λ to be

$$\lambda \simeq 0.13. \tag{13.82}$$

It is important to emphasize that this is an extraction of λ, and not a measurement. The value of λ also determines the strength of coupling of three and four Higgs bosons to one another. A measurement of λ, and therefore a validation of the Higgs potential of the Standard Model, Eq. 13.80, requires measuring the coupling of multiple Higgs bosons. So, to measure λ, we not only need to observe the production of one Higgs boson to measure its mass, but also of two and three Higgs bosons to ensure that the rate for those processes is controlled by λ. No such observation of multiple Higgs production exists, so ATLAS and CMS are only able to place (weak) limits on the value of λ from these searches. We will discuss more about the searches for multi-Higgs production and the Higgs potential in the following chapter.

13.3.2 Coupling Strength Proportional to Mass

A robust prediction of the Standard Model Higgs boson is that its couplings to fermions and electroweak bosons are proportional to the masses of those particles. That is, in the Standard Model Lagrangian, the Higgs couples to particles via

$$\mathcal{L}_H \supset -\frac{m_f}{v}H\bar{f}f + 2g_W m_W HW_\mu^+ W^{-\,\mu} + \frac{\sqrt{g_W^2 + g_Y^2}}{2}m_Z HZ_\mu Z^\mu, \tag{13.83}$$

where f is any fermion of the Standard Model (quark or lepton). By measuring the rates at which the Higgs boson decays through different particles, we are able to experimentally determine the coupling strengths to fermions and electroweak bosons. This approach is limited, however, by the fact that fermion masses can be very, very small compared to the vev, and so decay of Higgs boson to light fermions is extremely rare. Nevertheless, for fermions with masses of hundreds of MeV and above, the LHC should collect enough data to measure their Yukawa couplings.

As of now, ATLAS and CMS have been able to set bounds on Higgs boson couplings to fermions, and a combination of their results is presented in Fig. 13.5. This plot shows the coupling strength of the particle to the Higgs (for fermions, the Yukawa coupling) versus the mass of the particle. The Standard Model prediction is shown by the dashed line, corresponding to a 1-to-1 relationship. For the muon, tau lepton, and bottom quark, the Yukawa couplings are determined by direct searches for Higgs decays to these particles. For the W and Z bosons, the coupling to the Higgs bosons is determined by searches for four-lepton final states, as we have discussed earlier in this chapter. Determining the Yukawa coupling of the top quark to the Higgs boson is much more challenging. This is accomplished by multiple fits to various Higgs production processes in which the top

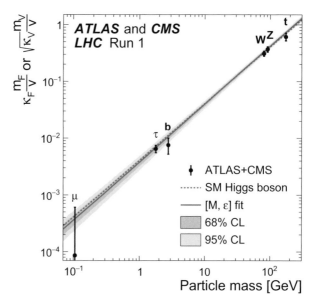

Fig. 13.5   Plot of the coupling of the Higgs boson to particles of the Standard Model as a function of particle mass. The Standard Model prediction is the dashed line, corresponding to a 1-to-1 relationship. From G. Aad *et al.* [ATLAS and CMS Collaborations], "Measurements of the Higgs boson production and decay rates and constraints on its couplings from a combined ATLAS and CMS analysis of the LHC pp collision data at $\sqrt{s} = 7$ and 8 TeV," J.High Energy Phys. **1608**, 045 (2016) [arXiv:1606.02266 [hep-ex]].

quark is essential, such as gluon–gluon fusion through a loop of top quarks. There is excellent agreement between the Standard Model prediction and measured values, but note that Fig. 13.5 is a log–log plot. The experimental results still have substantially large uncertainties, which are suppressed on this display of the plot. So, there is still a possibility of deviations from Standard Model predictions that may hint at new physical processes, but, for now, everything is nicely consistent with expectation.

13.3.3 Spin-0

A very generic prediction of spontaneous breaking of electroweak symmetry is that the Higgs is a spin-0 particle. (See Exercise 11.2 of Chapter 11 for an argument why.) However, just the observation of decays to four leptons or two photons only provides limited constraints on the spin of this newly discovered particle. First, we directly observe decays to particles of the same spin (the two photons, for example), and so the new particle must be a boson, with integer intrinsic spin. So, it could have spin 0, spin 1, or spin 2. Higher spins than 2 are also possible, but typically not considered because there is no consistent way to couple a high-spin point particle to other particles in a Lorentz-invariant quantum field theory.[12] Further, the Higgs boson cannot be a spin-1 particle because of a

---

[12] A quick argument for why this is true is the following. In a Lorentz-invariant quantum field theory, the only conserved quantities are vector currents (such as electromagnetic currents) and the stress–energy tensor $T_{\mu\nu}$.

result known as the Landau–Yang theorem.[13] The Landau–Yang theorem states that there is no way for a massive spin-1 particle to decay to two identical massless spin-1 particles. As we have observed this new particle decay to two photons (identical, massless, spin-1 particles), the Landau–Yang theorem prohibits it from being spin-1. You will prove the Landau–Yang theorem in Exercise 13.6. So, just from the observation that this new particle decays to two photons, we have argued that its spin can only be 0 or 2!

To determine whether this new particle is spin-0 or spin-2, there are a few things we can do. A higher-spin particle has more degrees of freedom as it has more spin states. As such, the production cross section of a high-spin particle will be larger than a low-spin particle. So, a measurement of the cross section can provide constraints on the spin of the particle. This is a bit challenging, however, because one can also change the values of couplings to affect the production cross section. A more direct method for measuring the spin is to study angular distributions of the final state. We are familiar with angular distributions containing information about spin. Because muons are spin-1/2 particles, we observed a characteristic $1 + \cos^2 \theta$ distribution of events at an angle θ with respect to the electron–positron beam. This was used to argue that quarks, through jet production, are also spin-1/2 particles. We can do a similar thing for the Higgs boson at a hadron collider, though we have to be a bit careful.

We need to define relevant angles for the Higgs production and decay process. Because all final-state particles from Higgs decay are measured, we can boost to the frame in which the Higgs is at rest. We assume that this can be done by boosting exclusively along the proton beam axis, which is a good approximation as the transverse momentum of the Higgs will almost always be small compared to its mass. From this set-up, Fig. 13.6 shows the relevant angles of the process, specifically for $pp \to H \to$ four leptons. The angles that we consider here are θ^*, the angle between the two photons or Z bosons from Higgs decay to the proton beam, and Φ, the angle between the planes defined by the pairs of leptons from the subsequent Z boson decays. Let's first analyze the difference in dependence of this process on θ^* if the Higgs is spin-0 versus spin-2.

For the process $gg \to H \to \gamma\gamma$, a configuration of gluon and photons spins that corresponds to a spin-0 Higgs boson is the following:

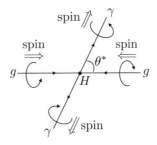

Vector currents couple to spin-1 particles (like the photon) to produce Lorentz-invariant terms in a Lagrangian. $T_{\mu\nu}$ is a symmetric two-index tensor and so can couple in a Lorentz-invariant way to spin-2 particles. There are no conserved quantities of higher rank (= more Lorentz indices) in a quantum field theory, and so higher-spin point particles cannot couple to anything else.

[13] L. D. Landau, "On the angular momentum of a system of two photons," Dokl. Akad. Nauk Ser. Fiz. **60**, no. 2, 207 (1948); C. N. Yang, "Selection rules for the dematerialization of a particle into two photons," Phys. Rev. **77**, 242 (1950).

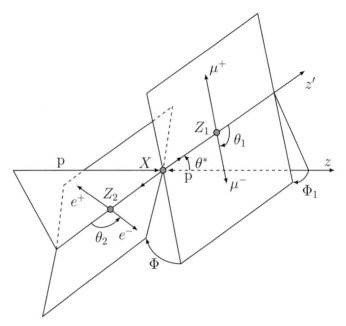

Fig. 13.6 Figure illustrating the relative angles between pairs of final-state leptons from a $H \rightarrow$ four leptons decay. In the frame where the Higgs boson is at rest, θ^* is the angle between the proton beam and the direction of momentum of one pair of leptons and Φ is the relative angle between the planes defined by the momenta of pairs of leptons and their net momentum. From G. Aad *et al.* [ATLAS Collaboration], "Study of the spin and parity of the Higgs boson in diboson decays with the ATLAS detector," Eur. Phys. J. C **75**, no. 10, 476 (2015), Erratum: [Eur. Phys. J. C **76**, no. 3, 152 (2016)] [arXiv:1506.05669 [hep-ex]].

Because the Higgs has spin 0, the spins of the gluons must be anti-aligned, so their sum is, of course, 0. Similarly, the spins of the photons must also be anti-aligned. This correlation between spins is an example of entanglement: if one of the photons is measured to have right-handed helicity (as in the figure), then the helicity of the other photon must also be right-handed. Also, note that the spins of the photons are completely uncorrelated with the spins of the gluons (other than both of their net spins being 0). Regardless of the angle θ^*, the projection of the photon spins onto the gluons' axis always sums to 0. Another way to say this is that, in the evaluation of the matrix element for the process $gg \rightarrow H \rightarrow \gamma\gamma$ with a spin-0 Higgs, the dependence on the gluons' and photons' polarization vectors is

$$\mathcal{M}(gg \rightarrow H_{\text{spin-0}} \rightarrow \gamma\gamma) \propto (\epsilon_{g1}^* \cdot \epsilon_{g2}^*)(\epsilon_{\gamma 1} \cdot \epsilon_{\gamma 2}), \quad (13.84)$$

where $g1$ and $g2$ are the two gluons, for example, and their polarization vectors are complex conjugated because they are in the initial state. This matrix element is therefore independent of θ^* because, by momentum conservation, the two gluons' and two photons' momenta are back-to-back.

The case if the Higgs boson is spin-2 is quite different. Here is an example of spin configurations that can exist if the Higgs boson is spin-2:

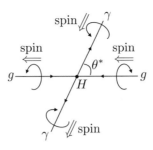

Now, the gluons' and photons' spins are aligned. Because the initial state has net spin 2 pointing to the left, only the component of the photons' spins with the same net spin contributes to the matrix element, by angular momentum conservation. If $\theta^* = 0$, the photon spins are perfectly aligned with the gluon spins, so that configuration has high probability for occurring. By contrast, if $\theta^* = \pi$, the photon spins are anti-aligned with the gluon spins, and there is zero component of the photon spins in the direction of initial angular momentum. Therefore, this configuration has zero probability of occurring because angular momentum cannot be conserved. This suggests that the matrix element for this configuration of spins for a spin-2 Higgs boson is

$$\mathcal{M}(gg \to H_{\text{spin-2}} \to \gamma\gamma) \propto (\epsilon_{g1}^* \cdot \epsilon_{\gamma1})(\epsilon_{g2}^* \cdot \epsilon_{\gamma2}) + (\epsilon_{g1}^* \cdot \epsilon_{\gamma2})(\epsilon_{g2}^* \cdot \epsilon_{\gamma1}). \qquad (13.85)$$

Note that this matrix element is symmetric in $1 \leftrightarrow 2$, which it must be because gluons and photons are both bosons. Because this matrix element consists of mixed dot products of polarization vectors of gluons and photons, there will be non-trivial dependence on the angle θ^*. You will determine the functional dependence on θ^* for this collection of spins in Exercise 13.8. This matrix element is therefore different than the spin-0 hypothesis, and the distribution of θ^* can be measured to distinguish the two hypotheses.

While we won't discuss it in detail, one can also measure the angle Φ between the planes defined by the pairs of leptons from $H \to$ four leptons decays. The pairs of leptons that should be correlated to define a plane are identified in experiment by studying the process $H \to e^+ e^- \mu^+ \mu^-$. The angle between the electron plane and the muon plane contains information about the spin of the Higgs boson, and the spin-0 and spin-2 hypotheses will have different distributions. Using these and other angular distributions, the ATLAS and CMS experiments have tested spin-0 and spin-2 hypotheses and find agreement with the Standard Model prediction of a spin-0 Higgs boson at 99.9% confidence (3σ). Nevertheless, this is a very subtle procedure that requires specific assumptions about modeling the spin-2 hypothetical Higgs boson. Because of this challenge, while all current evidence is strongly in favor of a spin-0 Higgs, the PDG has not yet established an official determination of the spin of the Higgs.

Exercises

13.1 *W Boson Decays.* The W boson decays to hadrons about 70% of the time, while it decays to leptons the remaining 30% of the time. In searching for the Higgs boson through $H \to W^+W^-$, how often is the final state composed exclusively of leptons? How often is it composed exclusively of hadrons? The **semileptonic decay** of the Higgs corresponds to one W boson decaying to leptons and the other to hadrons. How often does this semileptonic decay occur?

Hint: Do the probabilities add up to 1?

13.2 $pp \to W^+W^-$ *Backgrounds.* Estimate the cross section at the 13 TeV LHC in picobarns for the process $pp \to W^+W^- \to l^+\nu_l l'^- \bar{\nu}_{l'}$ in the Standard Model. Use the Stairway to Heaven plot from Exercise 4.7 of Chapter 4 and the fact that the W boson decays to leptons about 30% of the time.

13.3 *Searching for $H \to W^+W^-$.* To set a robust experimental upper bound on the mass of the Higgs boson, we look for its decays through pairs of W bosons, which subsequently decay to leptons. This is a bit tricky because decays of W bosons produce neutrinos, which aren't directly measured. Nevertheless, the charged leptons from the W decays still contain a significant amount of information about the Higgs boson, which we will explore in this exercise.

(a) Consider the following possible decay of the Higgs:

$$H \to W^+W^- \to e^+\nu_e\mu^-\bar{\nu}_\mu . \tag{13.86}$$

Assume that the Higgs is sufficiently massive such that the two W bosons are both on-shell. What is the largest possible value for the positron–muon invariant mass $m_{e\mu}^2 = (p_{e^+} + p_{\mu^-})^2$, in terms of the Higgs mass m_H and the W boson mass m_W? Assume that all leptons are massless.

Hint: It's simplest to work in the rest frame of the Higgs boson, and you can align the momentum of the W bosons along the $\hat{z}$-axis.

(b) What is the minimum possible value of the positron–muon invariant mass $m_{e\mu}^2$?

(c) What is the largest possible value of missing transverse momentum from the neutrinos? You can again assume that the Higgs boson is at rest, but you can't, in general, assume that the W bosons' momenta lie along the $\hat{z}$-axis. Express the missing transverse momentum in terms of the Higgs and W boson masses, m_H and m_W.

Hint: Recall that the missing transverse momentum is the magnitude of the sum of the transverse momenta of the two neutrinos in the final state.

13.4 $H \to \gamma\gamma$ *Rate.* In this exercise, we will estimate the decay rate for $H \to \gamma\gamma$.

(a) Using the approximation that $m_H \ll m_t$ and naïve dimensional analysis, estimate the Feynman diagram that represents the matrix element $\mathcal{M}(H \to \gamma\gamma)$ presented in Section 13.2.3.

(b) Using this result, estimate the rate of Higgs decay to photons, $\Gamma_{H \to \gamma\gamma}$.

Hint: You'll need Fermi's Golden Rule for decays, which we worked out for muon decays in Section 10.4.1.

(c) The dominant decay of the Higgs boson is through bottom quarks: $H \to b\bar{b}$. As it is a fermion, the coupling of the bottom quark to the Higgs boson is controlled by its Yukawa coupling, y_b. With the mass of the bottom quark $m_b = 4.18$ GeV, determine the value of the Yukawa coupling y_b.

(d) Estimate the matrix element for Higgs decay to bottom quarks, $\mathcal{M}(H \to b\bar{b})$. You can safely assume that $m_b \ll m_H$ in evaluating the spinor product in the matrix element. How much larger is the decay rate of the Higgs to bottom quarks than to photons?

13.5 *Higgs Production Rate.* In Eq. 13.74, we put together the Higgs production cross section for gluon–gluon fusion. We stopped a bit short there, and in this exercise, we'll simplify the cross section further. For simplicity, we reprint that expression here:

$$\sigma(pp \to H) \simeq \int_0^1 dx_1 \int_0^1 dx_2 \, \frac{\alpha_s^2 y_t^2}{16\pi} \frac{m_H^2}{m_t^2} f_g(x_1) f_g(x_2) \delta(x_1 x_2 s - m_H^2) . \qquad (13.87)$$

(a) Use the δ-function to integrate over the x_2 gluon momentum fraction. Don't forget about the Jacobian factor.

(b) Now, in the remaining integral over gluon momentum fraction x_1, change variables from x_1 to the rapidity y of the Higgs boson. Recall that the rapidity is defined as

$$y = \frac{1}{2} \log \frac{E + p_z}{E - p_z} . \qquad (13.88)$$

(c) At the 13 TeV LHC, what is the largest possible value of the Higgs boson's rapidity, y?

(d) In Fig. 13.7, we show the measurement of the Higgs boson rapidity distribution as measured at the ATLAS detector. ATLAS determined this distribution from combining results from the $H \to \gamma\gamma$ and $H \to$ four leptons decay channels. From the expression you derived in part (b), can you describe a procedure for determining the gluon parton distribution function $f_g(x)$ from these data?

13.6 *Landau–Yang Theorem.* In this exercise, we will work through a proof of the Landau–Yang theorem, which states that a massive, spin-1 particle cannot decay to two massless, identical spin-1 particles. To do this proof, we'll refer to the massive, spin-1 particle as H, and will take massless particles to be photons γ, which we denote as 1 and 2. As spin-1 particles, they all have polarization three-vectors $\vec{\epsilon}_H$, $\vec{\epsilon}_1$, and $\vec{\epsilon}_2$. We will attempt to calculate the matrix element for the decay $\mathcal{M}(H \to \gamma\gamma)$ and find that it must be 0.

(a) Go to the frame in which H is at rest, so its three-momentum vector is $\vec{p}_H = 0$. Trivially, we then have that $\vec{p}_H \cdot \vec{\epsilon}_H = 0$. For the two photons' momenta, $\vec{p}_1$ and $\vec{p}_2$, argue that

$$\vec{p}_1 \cdot \vec{\epsilon}_1 = \vec{p}_1 \cdot \vec{\epsilon}_2 = \vec{p}_2 \cdot \vec{\epsilon}_2 = \vec{p}_2 \cdot \vec{\epsilon}_1 = 0 . \qquad (13.89)$$

(b) Using this result and that photons are identical bosons, argue that the matrix element must be of the form

$$\mathcal{M}(H \to \gamma\gamma) = m_H \kappa_{ijk} \epsilon_H^i (\epsilon_1^j \epsilon_2^k + \epsilon_1^k \epsilon_2^j) , \qquad (13.90)$$

for some constant, three-index object κ_{ijk} and H mass m_H.

Hint: Why is the factor of m_H there? Why is there no dependence on any of the momentum vectors?

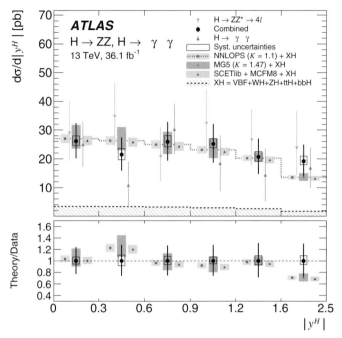

Fig. 13.7 Plot of the absolute value of the rapidity of the Higgs boson in 13 TeV proton collisions measured at the ATLAS experiment. The combined result from $H \to \gamma\gamma$ and $H \to$ four leptons decays is shown by the black dots. From M. Aaboud *et al.* [ATLAS Collaboration], "Combined measurement of differential and total cross sections in the $H \to \gamma\gamma$ and the $H \to ZZ^* \to 4\ell$ decay channels at $\sqrt{s} = 13$ TeV with the ATLAS detector," Phys. Lett. B **786**, (2018), doi:10.1016/j.physletb.2018.09.019.

(c) Now, perform a rotation on the polarization vectors. The matrix element is Lorentz invariant, so it is necessarily rotationally invariant. That is, for a matrix $\mathbb{M} \in SO(3)$ with matrix elements M_{ij}, the matrix element transforms as

$$\mathcal{M}(H \to \gamma\gamma) \to m_H \kappa_{ijk}(M_{il}\epsilon_H^l)[(M_{jm}\epsilon_1^m)(M_{kn}\epsilon_2^n) + (M_{km}\epsilon_1^m)(M_{jn}\epsilon_2^n)]$$
$$= m_H(\kappa_{ijk}M_{il}M_{jm}M_{kn} + \kappa_{ijk}M_{il}M_{km}M_{jn})\epsilon_H^l\epsilon_1^m\epsilon_2^n$$
$$= \mathcal{M}(H \to \gamma\gamma). \tag{13.91}$$

Therefore,

$$\kappa_{ijk}M_{il}M_{jm}M_{kn} = \kappa_{lmn}. \tag{13.92}$$

That is, the object κ is also an invariant of rotations, just like the identity matrix, $\mathbb{I}$.

By contracting pairs of indices l, m, or n, argue that κ_{ijk} is only non-zero if i, j, and k are all different. Recall that $\mathbb{M}^\intercal\mathbb{M} = \mathbb{I}$.

(d) As an element of $SO(3)$, the determinant of $\mathbb{M}$ is 1. Show that the determinant can be written as

$$\det \mathbb{M} = \epsilon_{ijk}M_{1i}M_{2j}M_{3k} = 1, \tag{13.93}$$

where ϵ_{ijk} is the completely anti-symmetric symbol defined by

$$\epsilon_{ijk} = \epsilon_{jki} = \epsilon_{kij} = -\epsilon_{jik}, \qquad \epsilon_{123} = 1, \qquad (13.94)$$

and $\epsilon_{iik} = 0$. Use this to argue that $\kappa_{ijk} = \kappa\epsilon_{ijk}$, for some constant κ.

Hint: What is ϵ_{ijk} after a rotation?

(e) Then, the matrix element is

$$\mathcal{M}(H \to \gamma\gamma) = \kappa m_H \epsilon_{ijk} \epsilon_H^i (\epsilon_1^j \epsilon_2^k + \epsilon_1^k \epsilon_2^j). \qquad (13.95)$$

Argue that this implies that $\mathcal{M}(H \to \gamma\gamma) = 0$, which proves the Landau–Yang theorem.

13.7 *Combining Uncorrelated Measurements.* In Example 13.3, we discussed how to combine and interpret the statistical significance of uncorrelated measurements. In this exercise, we will apply this technique to discover the Higgs boson. In Fig. 13.8, we show ATLAS's measurement of the four-lepton invariant mass in the search for the Higgs boson. This plot, in addition to Fig. 13.3, was used as evidence by ATLAS to claim discovery of the Higgs boson. From these two plots, estimate the deviation in σ from the null hypothesis that there is no Higgs boson. Assume Poisson statistics and that the measurements are uncorrelated. Don't worry about subtleties with the look-elsewhere effect; just focus on the significant deviations around a mass of 125 GeV. How does the estimated significance you find compare to ATLAS's official combination from Fig. 13.4?

These measurements are strictly not uncorrelated and the ATLAS combination requires careful accounting for uncertainties in the measurements, so this will only be a rough approximation. Nevertheless, you should find a deviation that is reasonably close to the 6σ from Fig. 13.4.

13.8 *Testing the Spin-2 Higgs Boson Hypothesis.* We argued on very general grounds in Section 13.3.3 that the matrix element for the spin-0 Higgs boson is independent of the scattering angle θ^*. The result of this analysis was the matrix element in Eq. 13.84. In this exercise, we will evaluate the matrix element for the hypothesized spin-2 Higgs boson and demonstrate that there is explicit dependence on θ^*. We will use the assumed form of the matrix element from Eq. 13.85.

The set-up of this scattering is the following. Working in the frame in which the Higgs is at rest, we will take the momentum four-vectors of the gluons and photons in this process to be

$$p_{g1} = \frac{m_H}{2}(1,0,0,1), \qquad\qquad p_{g2} = \frac{m_H}{2}(1,0,0,-1), \quad (13.96)$$

$$p_{\gamma1} = \frac{m_H}{2}(1,0,\sin\theta^*,\cos\theta^*), \quad p_{\gamma2} = \frac{m_H}{2}(1,0,-\sin\theta^*,-\cos\theta^*).$$

(a) For the matrix element $\mathcal{M}(g_{1L}g_{2R} \to H_{\text{spin-2}} \to \gamma_{1L}\gamma_{2R})$, write down the polarization four-vectors of the gluons, ϵ_{g1} and ϵ_{g2}.

(b) In this same process, with the momentum four-vectors identified as above, calculate the polarization four-vectors for the photons, $\epsilon_{\gamma1}$ and $\epsilon_{\gamma2}$.

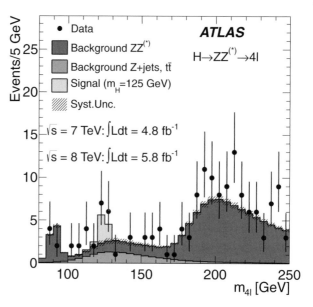

Fig. 13.8   Plot of the invariant mass of four leptons in the search for the Higgs boson at the ATLAS experiment. Reprinted from Phys. Lett. B **716**, G. Aad *et al.* [ATLAS Collaboration], "Observation of a new particle in the search for the Standard Model Higgs boson with the ATLAS detector at the LHC," 1 (2012), with permission from Elsevier.

> *Hint*: To find these polarization vectors, can you just rotate the polarization vectors for the gluons?

(c) Now, with the polarization vectors in hand, evaluate the dot products that compose the matrix element:

$$\mathcal{M}(gg \to H_{\text{spin-2}} \to \gamma\gamma) \propto (\epsilon_{g1}^* \cdot \epsilon_{\gamma1})(\epsilon_{g2}^* \cdot \epsilon_{\gamma2}) + (\epsilon_{g1}^* \cdot \epsilon_{\gamma2})(\epsilon_{g2}^* \cdot \epsilon_{\gamma1}) . \quad (13.97)$$

For what value of θ^* is this matrix element maximized? For what angle θ^* is it 0? Does that agree with our conclusions in Section 13.3.3?

13.9 *Research Problem.* Is the particle first observed on July 4, 2012 **the** Higgs boson of the Standard Model? What measurements are required for you be convinced either way?

Particle Physics at the Frontier

What next? The discovery of the Higgs boson is the culmination of the Standard Model. We know the properties of the quarks, the leptons, and the gauge bosons, and their interactions are constrained by requirements of invariance under the various gauge symmetries of the Standard Model. However, along the way, it seems like every discovery only raises more questions. Why are the quark and lepton Yukawa couplings what they are? Where does mass come from ultimately? Why are there three generations of fermions? How is electroweak symmetry broken? Why are the gauge couplings of the Standard Model the values that they are? Is this the final story? Is there any deeper structure to Nature? It is these open questions that motivate every particle physicist, and in this final chapter we'll survey a few of them. I hope that you can contribute to their solutions!

14.1 Neutrino Masses

The observation that different flavors of neutrinos oscillate into one another means that some of the neutrinos must have distinct and non-zero masses. This is not strictly allowed by the structure of the Standard Model, which requires neutrinos to be exclusively left-handed fermions. With only left-handed neutrinos in the game, it doesn't seem possible to include a mass for them in the Standard Model.

This isn't to say that neutrino masses invalidate the Standard Model; at typical energies, neutrinos can be thought of as massless, to extremely good approximation. Adding a mass for neutrinos in the Standard Model is actually not that hard, either. Let's just consider the case with one neutrino, ν. Most naïvely, we can just add a neutrino mass m_ν to the Lagrangian as

$$\mathcal{L}_m \supset -m_\nu \nu_L^\dagger \nu_R - m_\nu \nu_R^\dagger \nu_L . \tag{14.1}$$

This requires the introduction of a right-handed neutrino, ν_R. Now, unlike all other fermions of the Standard Model, neutrino masses can be completely unrelated to the Higgs mechanism. One can posit a new source for neutrino masses. Further, the weak interactions still only couple to left-handed neutrinos, and this is the only way that we know how to create neutrinos. In fact, the right-handed neutrino must be completely inert and not carry charge under any of the Standard Model gauge groups. The right-handed neutrino is therefore referred to as a **sterile neutrino**. If we included such a mass term in the Standard Model, then the neutrino and anti-neutrino would be distinct particles. Such a mass term is called a **Dirac mass**.

However, it is perfectly consistent with all symmetries to postulate that a neutrino is its own anti-particle. A neutrino is electrically neutral, and so its anti-particle is also electrically neutral. In the Standard Model, examples of particles that are their own anti-particles are the photon and the Z boson. By contrast, quarks and charged leptons cannot be their own anti-particles because the electric charges of their anti-particles would be different. For example, the electron and positron have electric charge $-e$ and $+e$, respectively. For neutrinos, there is no such restriction and one can write down a mass for the neutrino as its own anti-particle. Such a mass term is called a **Majorana mass** after Italian physicist Ettore Majorana.[1] Majorana disappeared at age 32 under exceptionally suspicious circumstances, traveling by boat from Palermo to Naples in 1938.

If the neutrino is its own anti-particle, we say that it is a **Majorana fermion**. The action of charge conjugation $\mathbb{C}$ turns a particle into its anti-particle, and so a Majorana fermion is an eigenstate of charge conjugation. Concretely, acting $\mathbb{C}$ on a left-handed spinor ν_L yields

$$\mathbb{C}\nu_L = -\epsilon \nu_R^* \,, \tag{14.2}$$

where * denotes complex conjugation and ϵ is the anti-symmetric symbol, which is represented in matrix form as

$$\epsilon = \begin{pmatrix} 0 & 1 \\ -1 & 0 \end{pmatrix} \,. \tag{14.3}$$

A left-handed Majorana spinor is one for which charge conjugation just returns itself:

$$\mathbb{C}\nu_L = \nu_L \,. \tag{14.4}$$

So, a Majorana fermion is just right- or left-handed, and not both. If the neutrino is a Majorana fermion, its Lagrangian can be exclusively expressed in terms of a left-handed neutrino ν_L. The mass term for such a neutrino is

$$\mathcal{L}_m \supset -\frac{m_\nu}{2} \left(\nu_L^\mathsf{T} \epsilon \nu_L - \nu_L^\dagger \epsilon \nu_L^* \right) \,. \tag{14.5}$$

Note that the relative signs between the two terms ensure that this mass is Hermitian: the transpose-conjugate of ϵ is $-\epsilon$. This mass is a bit weird, and might seem like it is just identically 0. A Majorana mass is only non-zero if the components of the spinor ν_L are themselves anti-commuting quantities, called **Grassmann numbers**. We have seen the necessity of anti-commuting quantities to describe spin 1/2 before, and this is connected to the relationship between fermions and anti-symmetry under identical particle exchange (called the **spin-statistics theorem**).

There are experiments currently running that are attempting to determine if the neutrino is Dirac or Majorana type (or some admixture). The largest such experiment, called EXO (<u>E</u>nriched <u>X</u>enon <u>O</u>bservatory) consists of a vat of 200 kg of enriched liquid ^{136}Xe.[2] An image of the EXO detector is presented in Fig. 14.1. The vat is observed continuously for evidence of a double β-decay of xenon to barium:

$$^{136}\text{Xe} \;\rightarrow\; ^{136}\text{Ba} + e^+ + e^- + \nu_e + \bar{\nu}_e.$$

[1] E. Majorana, "Teoria simmetrica dell'elettrone e del positrone," Nuovo Cim. **14**, 171 (1937).
[2] M. Auger *et al.*, "The EXO-200 detector, part I: Detector design and construction," J. Instrum. **7**, P05010 (2012) [arXiv:1202.2192 [physics.ins-det]].

Fig. 14.1 The EXO detector (the cylinder to the right) as it enters the cryostat to the right. The detector houses the liquid xenon and is approximately 70 liters in volume. From M. Auger *et al.*, "The EXO-200 detector, part I: Detector design and construction," J. Instrum. **7**, P05010 (2012) [arXiv:1202.2192 [physics.ins-det]].

This decay has been observed and has a lifetime of about 10^{21} years.[3] If the neutrino is its own anti-particle, then it is possible for the neutrino and anti-neutrino to annihilate one another, and so the final state would be

$$^{136}\text{Xe} \ \rightarrow \ ^{136}\text{Ba} + e^+ + e^-.$$

This decay is called "neutrinoless double β-decay," and is a smoking-gun signature of a Majorana fermion. The distinction between neutrinoless double β-decay and standard β-decay is that the emitted electron and positron should have equal and opposite momenta, as there are no final-state neutrinos. EXO is looking for this but hasn't found anything yet,[4] setting a lower limit on the half-life of this decay of about 10^{25} years. A follow-up experiment called nEXO consisting of a 5 ton vat of liquid xenon is currently being proposed.[5] Stay tuned. . .

[3] N. Ackerman *et al.* [EXO-200 Collaboration], "Observation of two-neutrino double-beta decay in ^{136}Xe with EXO-200," Phys. Rev. Lett. **107**, 212501 (2011) [arXiv:1108.4193 [nucl-ex]].
[4] J. B. Albert *et al.* [EXO-200 Collaboration], "Search for Majorana neutrinos with the first two years of EXO-200 data," Nature **510**, 229 (2014) [arXiv:1402.6956 [nucl-ex]].
[5] S. A. Kharusi *et al.* [nEXO Collaboration], *nEXO Pre-conceptual Design Report*, [arXiv:1805.11142 [physics.ins-det]].

14.2 Dark Matter

While the Standard Model seems to be a complete theory for three fundamental forces and particle interactions, and has been and still is rigorously verified, there is strong evidence that the 17 particles we identified in Chapter 1 are only 20% of the matter of the universe. There does not seem to be enough normal matter (that of the Standard Model) to explain observed astrophysical phenomena. Perhaps the canonical example of evidence that more matter is needed in the universe is the velocity distribution of stars within galaxies. The velocity of a star orbiting a galaxy is dependent on the amount of total mass within its orbital radius to the center of the galaxy. For a circular orbit of radius R, Newton's law of gravitation and Newton's second law imply that the orbital velocity v of the star is

$$v^2 = \frac{G_N M_{\text{tot}}}{R}, \tag{14.6}$$

where M_{tot} is the total mass within the orbital radius. The mass within the orbital radius can be determined by counting luminous stars, and stars dominantly exist near the center of galaxies. Therefore, at large orbital radii, the mass within the orbital radius is independent of the radius. Therefore, we would predict that $v \propto R^{-1/2}$. However, it has been found that this relationship of the orbital velocity is nearly constant at large orbital radii,[6] suggesting that there is extra mass responsible for maintaining star velocities far from the center of the galaxy.

This necessary extra mass is not visible and does not seem to interact electromagnetically, so it is referred to as **dark matter**. Extensive evidence for dark matter now exists, from many galactic rotation curves, to the distribution of mass after galactic collisions,[7] to detailed measurements of the cosmic microwave background.[8] As of this writing, all of the evidence for dark matter is exclusively through its gravitational interactions and astrophysical observations. As such, all that is conclusively known regarding dark matter is that it interacts gravitationally; that is, it has a non-zero energy density.

Many experiments have collected data, are currently running, or are proposed to search for a particle origin of dark matter. Throughout this book, we have seen a few such experiments, for example the LUX experiment in the exercises in Chapters 2 and 4. LUX, in particular, is sensitive to a possible type of dark matter called a WIMP, or weakly interacting massive particle. "Weakly interacting" in this case refers to its features of coupling with low strength with other particles and of interacting through the weak force. Another possible candidate as the particle of dark matter is the axion, which we introduced as a solution to the strong CP problem within QCD in the Historical Profile in Box 12.1.

---

[6] V. C. Rubin and W. K. Ford, Jr., "Rotation of the Andromeda nebula from a spectroscopic survey of emission regions," Astrophys. J. **159**, 379 (1970).

[7] M. Markevitch *et al.*, "Direct constraints on the dark matter self-interaction cross-section from the merging galaxy cluster 1E0657-56," Astrophys. J. **606**, 819 (2004) [arXiv:astro-ph/0309303].

[8] D. N. Spergel *et al.* [WMAP Collaboration], "First year Wilkinson Microwave Anisotropy Probe (WMAP) observations: Determination of cosmological parameters," Astrophys. J. Suppl. **148**, 175 (2003) [arXiv:astro-ph/0302209].

There are terrestrial experiments searching for the axion, as well as astrophysical searches, but no conclusive evidence has been identified yet. Several other possibilities for dark matter have been proposed, from micro-black holes to modifications of Newtonian gravity at extragalactic distances. As mentioned earlier, the only evidence that we have for dark matter is through gravity, so we do not even know if dark matter interacts through the forces of the Standard Model at all. Searching for the particle nature of dark matter is one of the greatest outstanding problems in particle physics and will remain a major research direction in the future.

14.3 Higgs Self-Coupling

In our understanding of the Standard Model, we set up the potential of the Higgs boson so that the origin, where $\vec{\phi} = 0$, is an unstable equilibrium point. The Higgs then "rolls" down the potential and settles in the minimum. Because $\langle |\vec{\phi}| \rangle \neq 0$, this spontaneously breaks the electroweak gauge symmetry, giving masses to the W and Z bosons. We argued that this unified electroweak theory had a lot of constraints. What was not constrained, however, was the precise shape of the Higgs potential. Can we nail down this mysterious piece of the Standard Model?

Let's remind ourselves what the scalar potential of the Higgs boson was in the Standard Model. The potential is

$$V(\vec{\phi}) = \lambda \left(|\vec{\phi}|^2 - \frac{v^2}{2} \right)^2, \tag{14.7}$$

where λ is the quartic coupling of the Higgs and v is the vev. Expanding about v as

$$\vec{\phi} = \begin{pmatrix} 0 \\ \frac{v}{\sqrt{2}} + \frac{H(x)}{\sqrt{2}} \end{pmatrix}, \tag{14.8}$$

the spontaneously broken potential is

$$V(H) = \frac{\lambda}{4}(H^2 + 2vH)^2 = \lambda v^2 H^2 + \lambda v H^3 + \frac{\lambda}{4} H^4. \tag{14.9}$$

The mass of the Higgs boson is $m_H^2 = 2\lambda v^2$ and so the quartic coupling is

$$\lambda = \frac{m_H^2}{2v^2} = \frac{(125 \text{ GeV})^2}{2(246 \text{ GeV})^2} = 0.13. \tag{14.10}$$

To evaluate λ, the mass of the Higgs boson is $m_H = 125$ GeV and the value of the vev is $v = 246$ GeV.

Currently, as discussed in Chapter 13, all we know about the Higgs boson is its mass, spin, and (some of) its decay modes. This is not enough information to determine the shape of the potential, and therefore the mechanism for the breaking of electroweak symmetry. Effectively, we have Taylor expanded the Higgs potential $V(\vec{\phi})$ about its minimum, where the Higgs field H is the expansion parameter. By measuring the mass of the Higgs boson, we have measured the curvature of the potential's minimum. High curvature or small radius of curvature means a large mass, while low curvature or large radius of curvature means small mass. Compare this to energy levels of states in a potential as determined by the Schrödinger equation. As a second-order differential equation, the Schrödinger equation sets energy eigenvalues proportional to the curvature of a potential. To measure the full Higgs potential, however, means determining the cubic and quartic terms in the Taylor expansion in H.

These higher-order terms in the potential correspond to Higgs boson self-interaction terms. The cubic term, $\lambda v H^3$, describes the interaction of three Higgs bosons. Just measuring the mass is not sufficient to determine the coefficient of the cubic term; one needs to consider a process in which three Higgs bosons interact. Perhaps the simplest process to study the cubic Higgs interaction at the LHC is the production of two Higgs bosons in the final state: $pp \rightarrow HH$. One contribution to the cross section for this process is from a diagram that is proportional to the coupling λ directly:

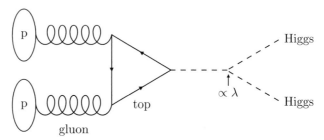

Similarly, the production of the three Higgs bosons in the final state has a contribution that is also sensitive to λ directly. Therefore, the measurement of the cross section for multi-Higgs production is sensitive to the shape of the potential.

Multi-Higgs production has not yet been observed at the LHC,[9] and it is likely that it will only be measured at a future, yet-to-be-constructed facility. Nevertheless, this will be a hugely exciting test for the Standard Model. Additionally, when and/or if observed, it will be the first direct measurement of a fundamental particle's self-interaction.

Another interesting question is how the potential got to be formed in the first place. The standard picture is that in the early universe, when it was very hot and dense, the vev of the Higgs was 0: $\langle |\vec{\phi}| \rangle = 0$. As the temperature cooled, there was a phase transition (like water into ice) that modified the potential and gave the Higgs a non-zero vev: $\langle |\vec{\phi}| \rangle = v \neq 0$. The exact dynamics of this phase transition depends on properties of the early universe. So, if we are able to measure Higgs self-interactions, this could provide information about what was happening right after the Big Bang! Cool!

[9] A. M. Sirunyan *et al.* [CMS Collaboration], "Search for Higgs boson pair production in events with two bottom quarks and two tau leptons in proton–proton collisions at $\sqrt{s} = 13$ TeV," Phys. Lett. B **778**, 101 (2018) [arXiv:1707.02909 [hep-ex]]; M. Aaboud *et al.* [ATLAS Collaboration], "Search for pair production of Higgs bosons in the $b\bar{b}b\bar{b}$ final state using proton–proton collisions at $\sqrt{s} = 13$ TeV with the ATLAS detector" [arXiv:1804.06174 [hep-ex]].

14.4 End of Feynman Diagrams?

Throughout this book, we have used Feynman diagrams as the language in which we expressed processes in particle physics. Feynman diagrams have a nice physical picture and a precise mathematical formulation, and are systematically improvable as a perturbation theory. That is, we can calculate more complicated Feynman diagrams in a systematic way to obtain a better and more precise answer for a particular cross section. Behind this beautiful façade lies an ugly truth: the perturbation theory of Feynman diagrams does not converge. In fact, the radius of convergence of Feynman diagram perturbation theory is precisely zero. This would seem to suggest that Feynman diagrams are totally useless, which doesn't jibe with our experience of comparing their results to data. So what's going on?

14.4.1 Failure of Convergence of Feynman Diagrams

The claim that Feynman diagram perturbation theory doesn't converge and has zero radius of convergence is a big one, but easy to prove. The importance of Feynman diagrams contributing to a particular process is ordered by the number of loops they have. Each loop in a Feynman diagram adds another factor of the coupling constant, like α in QED or α_s in QCD. To see this, compare the tree-level diagram to a one-loop diagram for the process $e^+e^- \to \mu^+\mu^-$:

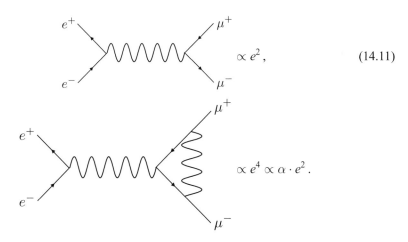

$$\propto e^2, \qquad (14.11)$$

$$\propto e^4 \propto \alpha \cdot e^2.$$

Each vertex with two fermions and a photon is proportional to the electric charge e, and $4\pi\alpha = e^2$. Therefore, the one-loop diagram is formally suppressed with respect to the tree-level diagram by a factor of $\alpha \simeq 1/137$. A two-loop diagram would be suppressed by α^2, and a diagram with ℓ loops would be suppressed by a factor of α^ℓ. Therefore, Feynman diagram perturbation theory for the process $e^+e^- \to \mu^+\mu^-$ is an expansion in the fine structure constant about $\alpha = 0$.

We can use the ratio test to determine the radius of convergence of this series expansion. Let's express the matrix element for $e^+e^- \to \mu^+\mu^-$ scattering as a series in α:

$$\mathcal{M}(e^+e^- \to \mu^+\mu^-) = \mathcal{M}^{(0)}(e^+e^- \to \mu^+\mu^-) \sum_{\ell=0}^{\infty} c_\ell \, \alpha^\ell, \qquad (14.12)$$

where $\mathcal{M}^{(0)}(e^+e^- \to \mu^+\mu^-)$ is the tree-level diagram, c_ℓ are some constant coefficients, and ℓ counts the numbers of loops. Note that $c_0 = 1$. Then, the ratio test tells us that the radius of convergence R in α of this series is

$$R = \lim_{\ell \to \infty} \frac{c_\ell}{c_{\ell+1}}. \qquad (14.13)$$

The claim that the radius of convergence $R = 0$ for Feynman diagrams implies that coefficients c_ℓ get larger as ℓ increases. This is really weird and very unfamiliar from studying series like the exponential function, for example.

The argument for zero radius of convergence for Feynman diagrams is due to Freeman Dyson.[10] A simplified version of the argument goes as follows. For $\alpha > 0$, it costs energy for a photon to split into an electron–positron pair. However, if $\alpha < 0$ (that is, electric charge e is imaginary), then the system can lose energy by photons splitting to electron–positron pairs. So, the system can shed electron–positron pairs and keep losing energy. This can in principle continue ad infinitum and so the system with $\alpha < 0$ does not have a lower bound on the ground state energy. If Feynman diagrams are an expansion about $\alpha = 0$, and for any $\alpha < 0$ there is no ground state, then the radius of convergence of the Feynman diagram perturbation theory is 0. This means that, as you calculate more and more Feynman diagrams with more and more loops, the result you find does not converge. Such a series with a zero radius of convergence is called an **asymptotic series**.

This might suggest that Feynman diagrams are exceptionally useless. If there is no hope of convergence of the perturbation theory, then what do we do? It turns out that asymptotic series are actually exceptionally useful, often more useful than convergent series. Asymptotic series have some super crazy properties. For many asymptotic series, the result you get after a finite order in the perturbation theory is arbitrarily close to the exact result. This is not what happens with convergent series. Additionally, the precise way that asymptotic series diverge as you include more terms contains a huge amount of information for properties of the exact result. There is an effort in the theoretical physics community to understand the behavior of asymptotic series in quantum field theory, a program called **resurgence**.[11]

Why, then, do Feynman diagrams seem to give such a good description of processes in particle physics? It is because only calculating low orders of the expansion effectively

[10] F. J. Dyson, "Divergence of perturbation theory in quantum electrodynamics," Phys. Rev. **85**, 631 (1952).

[11] Recent reviews of this field are D. Dorigoni, "An Introduction to resurgence, trans-series and alien calculus" [arXiv:1411.3585 [hep-th]]; G. V. Dunne and M. Ünsal, "What is QFT?: Resurgent trans-series, Lefschetz thimbles, and new exact saddles," in *Proceedings of the 33rd International Symposium on Lattice Field Theory*, PoS(LATTICE2015)010 (2016) [arXiv:1511.05977 [hep-lat]].

exhibits convergence properties. To see this feature, let's model the Feynman diagram expansion as perhaps the simplest asymptotic series:

$$\mathcal{M}(e^+e^- \to \mu^+\mu^-) \simeq \mathcal{M}^{(0)}(e^+e^- \to \mu^+\mu^-) \sum_{\ell=0}^{\infty} \ell! \, \alpha^\ell . \qquad (14.14)$$

It's easy to show with the ratio test that this has 0 radius of convergence. Let's now go back to the ratio test, but let's not take the limit $\ell \to \infty$. The ratio L_ℓ between terms at order $\ell + 1$ and ℓ is then

$$L_\ell = \frac{c_{\ell+1}}{c_\ell}\alpha = (\ell+1)\alpha . \qquad (14.15)$$

The ratio test says that if, as $\ell \to \infty$, this is less than 1, then the series converges. But we can also interpret it differently. We know the value of $\alpha \simeq 1/137$ and so we can determine the order ℓ at which the series starts to diverge. Demanding that $(\ell+1)\alpha < 1$, the order ℓ up to which it *appears* that the Feynman diagram series converges is then

$$\ell \lesssim 1/\alpha \simeq 137 . \qquad (14.16)$$

So, one needs to calculate diagrams with over 100 loops to start seeing divergence! Some modern calculations in QED now include up to five loops, so we are a very, very long way from being sensitive to the asymptotic nature of the Feynman diagram expansion.

Example 14.1 A powerful technique for making sense of asymptotic series is the method of Borel summation. What is the Borel sum of the series

$$\sum_{\ell=0}^{\infty} \ell! \, \alpha^\ell ? \qquad (14.17)$$

Solution

To Borel sum, we first multiply and divide each term in the series by the factorial:

$$\sum_{\ell=0}^{\infty} \ell! \, \alpha^\ell = \sum_{\ell=0}^{\infty} \ell! \, \alpha^\ell \times \frac{\ell!}{\ell!} = \sum_{\ell=0}^{\infty} \alpha^\ell \int_0^\infty dt \, t^\ell e^{-t} . \qquad (14.18)$$

In the second equation, we have used the fact that the integral evaluates to $\ell!$. Now, we exchange the sum and integral. This is not typically mathematically allowed, but we can just define the Borel sum by this procedure. Exchanging the sum and integral, we then find

$$\int_0^\infty dt \, e^{-t} \sum_{\ell=0}^{\infty} \alpha^\ell t^\ell = \int_0^\infty dt \, \frac{e^{-t}}{1 - \alpha t} , \qquad (14.19)$$

where we have used the geometric series

$$\sum_{\ell=0}^{\infty} \alpha^\ell t^\ell = \frac{1}{1 - \alpha t} . \qquad (14.20)$$

Remarkably, the integral that remains is finite for $\alpha \le 0$! When the Borel sum converges, it can be used to define the value of a divergent series.

14.4.2 More Efficient Calculational Techniques

In addition to corresponding to the expansion of an asymptotic series, Feynman diagrams also are less than optimal for efficiency of calculation. As we have discussed in this book, Feynman diagrams encode momentum, energy, angular momentum, and charge conservation at every vertex, and interactions are mediated by force-carrier bosons. Because of these properties, Feynman diagrams have a beautiful physical interpretation, but this can also come with a huge amount of baggage. To illustrate this, let's consider the calculation of the tree-level diagram for five-gluon scattering. We label the gluons 1, 2, 3, 4, 5, where each has a definite helicity. The Feynman diagram calculation of this process includes diagrams like

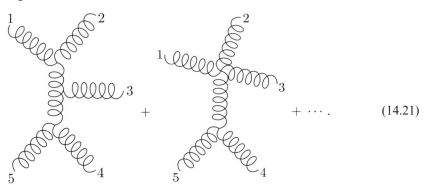

$$+ \cdots . \tag{14.21}$$

However, to calculate the corresponding matrix element requires calculating 25 Feynman diagrams with a total of 3600 terms. This requires a computer to evaluate.

If, after evaluating the Feynman diagrams and summing them, you are able to simplify the result, the final answer is exceedingly compact. If gluons 1 and 2 have left-handed helicity and 3, 4, and 5 have right-handed helicity, one finds that the matrix element is

$$\mathcal{M}(g_{1L}, g_{2L}, g_{3R}, g_{4R}, g_{5R}) = -ig^3 \frac{\langle 12 \rangle^4}{\langle 12 \rangle \langle 23 \rangle \langle 34 \rangle \langle 45 \rangle \langle 51 \rangle}. \tag{14.22}$$

This ridiculously simple result was first found by Stephen Parke and Tomasz Taylor in 1986.[12] Here, $\langle ij \rangle$ represents the spinor-helicity notation, and is just a shorthand for the spinor product

$$\langle ij \rangle = u_R^\dagger(p_i) u_L(p_j). \tag{14.23}$$

Why are there so many Feynman diagrams and what are Feynman diagrams hiding to make them so complicated?

The answer is related to gauge invariance. Though Feynman diagrams manifest momentum conservation, individual Feynman diagrams are not themselves gauge invariant, in general. Only after summing over all possible Feynman diagrams do you find that the result is gauge invariant. We have argued that gauge invariance is a powerful tool for constructing Lagrangians. We just had to posit the existence of a symmetry under

---

[12] S. J. Parke and T. R. Taylor, "An amplitude for n gluon scattering," Phys. Rev. Lett. **56**, 2459 (1986).

complex linear combinations of three "colors," and QCD was the unique result. However, gauge invariance has no direct physical consequence. Unlike a true symmetry which transforms physical states (like a physical rotation), a gauge symmetry leaves physical states unchanged. We can never "see" a gauge symmetry in the laboratory. It seems like clinging to gauge invariance as a guiding principle is responsible for the complexity of Feynman diagrams.

Can we then develop a new technique for calculating scattering amplitudes that doesn't reference gauge invariance at all? Let's determine what pieces of information we would need. The scattering amplitude encodes the helicity information of external particles. Written in spinor helicity notation, it is easy to identify the helicity of particles; all we have to do is count the effective number of right- or left-handed spinors. For example, in Eq. 14.22, consider gluon 1. Gluon 1 appears in the numerator four times and in the denominator two times. From the numerator, there are four left-handed spinors (recall that $u_R^\dagger$ is left-handed), which corresponds to a total left-handed spin of 2 ($= 4 \times 1/2$). The denominator therefore has a total left-handed spin of 1, and so the net spin of gluon 1 is just one unit, left-handed. The spins of all gluons are similarly accounted for. One has an additional constraint on the matrix element because the mass dimension is also known.

The most powerful constraints on the matrix element come from imposing unitarity, though in a way with which you may not be familiar. A propagator that goes on-shell corresponds to the existence of a real particle. Real particles live "forever" (because the uncertainty on their energy is 0), so whenever a propagator goes on-shell the amplitude must decompose into the product of simpler amplitudes. In gluon scattering, there are two physical configurations in which propagators go on-shell: either two gluons become collinear or one gluon has much lower energy than all the others. To see this concretely, let's take the amplitude from Eq. 14.22 in the limit where gluon 3 has a much lower energy than the other gluons. In that case, the amplitude simplifies:

$$\mathcal{M}(g_{1L}, g_{2L}, g_{3R}, g_{4R}, g_{5R}) \xrightarrow[p_3 \to 0]{} -ig \frac{\langle 24 \rangle}{\langle 23 \rangle \langle 34 \rangle} \cdot g^2 \frac{\langle 12 \rangle^4}{\langle 12 \rangle \langle 24 \rangle \langle 45 \rangle \langle 51 \rangle} . \tag{14.24}$$

The four-point amplitude is

$$\mathcal{M}(g_{1L}, g_{2L}, g_{4R}, g_{5R}) = g^2 \frac{\langle 12 \rangle^4}{\langle 12 \rangle \langle 24 \rangle \langle 45 \rangle \langle 51 \rangle} , \tag{14.25}$$

and the factor that is singular as $p_3 \to 0$ is called the **soft amplitude**. The general feature of the amplitude breaking up into simpler pieces when a propagator goes on-shell is called **factorization**, and it provides extremely strong constraints on the functional form of the amplitude.

For the past 20 years or so, there has been a large effort in the theoretical community to develop new methods for calculation that are much more efficient than Feynman diagrams. This includes using techniques like factorization and helicity management as discussed above, but many other properties of the amplitude as well. This effort goes by the name of S-matrix or **amplitudes program**.[13] Importantly, gauge invariance is nowhere to be

---

[13] A good modern book that reviews developments in this field is H. Elvang and Y. t. Huang, *Scattering Amplitudes in Gauge Theory and Gravity*, Cambridge University Press (2015) [arXiv:1308.1697 [hep-th]].

seen in the amplitudes program. With this approach, deep connections between Feynman diagrams, function theory, the structure of transcendental numbers, algebraic geometry, and other fascinating mathematics have been established. There's still a long way to go before Feynman diagrams are totally outmoded, but this seems to be an extremely promising approach.

14.5 The Future of Collider Physics

After a few minor hiccoughs, the Large Hadron Collider has been collecting proton collision data since 2009, and is planning to continue taking data well into the 2030s. Its performance has exceeded expectations and even exceeded design specifications in some ways, so a long and successful run is anticipated. Nevertheless, it is not too early to think about the next collider physics experiment, especially because the LHC (or any proton collider more generally) is not necessarily the right tool for some questions in particle physics. We end this chapter with a discussion of a few of the collider experiment projects that are being considered, following the LHC.

14.5.1 International Linear Collider (ILC)

With the discovery of the Higgs boson, we want to learn as much about it as possible. The LHC has already taught and will continue to teach us a significant amount: we know the mass of the Higgs boson, have observed some of its decays, and have strong evidence for its spin. The proton collision environment of the LHC is not ideal, however, for detailed and dedicated studies of the Higgs boson. The parton collision center-of-mass energy is not known, and so the collision frame is unknown. At the very least, this makes searches for the Higgs boson challenging, and at worst, it means that a huge number of Higgs bosons that are produced are never identified as such.

An electron–positron collider is therefore ideal to precisely produce and identify Higgs bosons, as the collision energy is directly controllable and final states are very clean. One such electron–positron collider that has been in the works for about 20 years is the International Linear Collider, or ILC.[14] The ILC is proposed to collide electrons and positrons at energies up to 500 GeV (and potentially up to 1 TeV), which would more than double the collision energy achieved at LEP. The 500 GeV benchmark enables study of a number of processes that involve the Higgs boson:

- $e^+e^- \to ZH$. The threshold for this process is about 215 GeV and this is the dominant production process of the Higgs boson at an electron–positron collider. This process enables a measurement of the total decay width of the Higgs boson because you can isolate this process by identification of the Z boson and total momentum conservation.

---

[14] J. Brau *et al.* [ILC Collaboration], "ILC reference design report, Volume 1: Executive summary" [arXiv:0712.1950 [physics.acc-ph]]; A. Djouadi *et al.* [ILC Collaboration], *International Linear Collider Reference Design Report Volume 2: Physics at the ILC* [arXiv:0709.1893 [hep-ph]].

- $e^+e^- \rightarrow HH$. The threshold for this process is about 250 GeV. Because multiple Higgs bosons are produced in the final state, this is sensitive to the self-coupling of the Higgs boson, and therefore to the Higgs potential parameter λ.
- $e^+e^- \rightarrow t\bar{t}H$. The threshold for this process is about 475 GeV, and this is the most sensitive way to directly determine the top quark's Yukawa coupling; i.e., its coupling to the Higgs boson.

The ILC would be hosted in Japan, and likely constructed in northern Honshu within the next decade. There is international support for the ILC and scientists in the United States, in particular, emphasize participation in the ILC as imperative for a strong future in particle physics. The Japanese are very excited about this project as well; high schools have made promotional videos to support the building of the machine in or near their town. Even Hello Kitty has endorsed the science of the ILC!

14.5.2 Circular Electron Positron Collider (CEPC)

In addition to the relatively near-term schedule of the ILC, there are a couple of collider physics projects being planned now that may come online in the mid-twenty-first century. One is an electron–positron collider proposed by scientists in China, called the Circular Electron Positron Collider, or CEPC.[15] The CEPC, as the name suggests, would be a circular collider, like LEP. However, the collision energies would be much higher, starting at about 250 GeV and possibly extending higher. Because of the higher energies, the CEPC accelerator ring would be much larger than LEP or LHC, with a circumference of about 80 km. The CEPC is thus a complementary collider to the ILC that can shine more light on properties of the Higgs boson.

Locations for the CEPC within China are being determined now, with the most likely location being Qinhuangdao city, situated east of Beijing on the Bohai Sea. There is a lot of support from the local government to host the CEPC, and that area of China is a popular tourist destination because it is near the eastern end of the Great Wall. The CEPC is just the first step in China's collider proposal. Much like LEP and LHC, the CEPC site would be repurposed for a proton collider at the end of its data taking. The proposed proton collider, called the Super Proton–Proton Collider, or SPPC, would collide protons at energies exceeding 50 TeV, many times the LHC's energy. The China collider proposal is moving forward very rapidly, and with an extended program consisting of both an electron–positron and a proton collider, China could be the center of particle physics later this century.

14.5.3 Future Circular Collider (FCC)

After LHC, CERN is also looking to continue particle collision experiments into the foreseeable future. The CERN proposal, called the Future Circular Collider, or FCC,

---

[15] CEPC-SPPC Study Group, *CEPC-SPPC Preliminary Conceptual Design Report. 1: Physics and Detector* IHEP-CEPC-DR-2015-01, IHEP-TH-2015-01, IHEP-EP-2015-01; CEPC-SPPC Study Group, *CEPC-SPPC Preliminary Conceptual Design Report. 2: Accelerator*, IHEP-CEPC-DR-2015-01, IHEP-AC-2015-01.

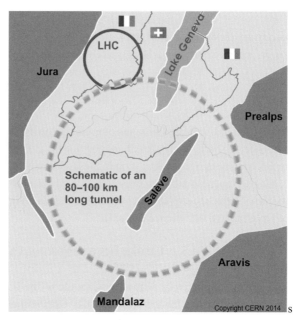

Fig. 14.2 Schematic aerial view of the proposed FCC accelerator ring in the area near Geneva. The LHC ring is also shown for scale. Credit: CERN © CERN.

would include a proton collider achieving collision energies of up to 100 TeV.[16] The FCC would be located in the Geneva area; a schematic map of the proposed FCC accelerator ring is presented in Fig. 14.2. To achieve these exceedingly high collision energies, the circumference of the FCC would be 100 km, with 16 T bending magnets. The LHC itself would be used as a pre-accelerator ring that would inject protons into the 100 km ring.

The physics program of the FCC would be exceptionally broad. The Higgs boson is of course central, and the very high energies enable observations of multiple Higgs boson production, which would provide direct sensitivity to the Higgs potential. Correspondingly, a determination of the Higgs potential informs the process by which electroweak symmetry is broken. At such high energies, neutrino interactions with other particles start becoming non-negligible. There is a significant probability that a neutrino with multi-TeV energy can radiate Z bosons, just like an electron can radiate photons, which will enable neutrinos to be "seen" by the detectors. Jets produced at the FCC will have a dynamic range that can extend over four decades in energy, enabling a more detailed probe of their properties than ever before. The FCC will also enable a detailed study of the energy dependence of the running gauge couplings of the Standard Model. While this energy dependence is determined in the Standard Model, deviations from the Standard Model prediction would be a clear indication of an energy scale for new physics.

---

[16] M. L. Mangano, *Physics at the FCC-hh, a 100 TeV pp Collider*, CERN Yellow Report 2017-003-M, CERN (2017) [arXiv:1710.06353 [hep-ph]].

These proposed experiments, the ILC, the CEPC and SPPC, and the FCC, will be at the next frontier of probing the universe at ever-decreasing distances, in order to understand a bit more about how we got here.

Example 14.2 The Superconducting Super Collider (SSC) was a proposed proton–proton collider that was to be built in Waxahachie, Texas, in the 1990s. It would have accelerated protons to 20 TeV in an 87 km ring and therefore would have probed energies several times that of the LHC about a decade before the LHC. The SSC project was terminated in 1993, however, after public and political support dwindled.

How strong were the SSC's bending magnets supposed to be, and how do they compare to the LHC and the proposed FCC magnets?

Solution

To keep charged particles moving in a circle, the strength of the magnetic field B is, from Chapter 5,

$$B = \frac{vE}{eR}, \tag{14.26}$$

where v is the particle's velocity, E is the particle energy, e is the fundamental charge, and R is the radius of the circle. To extremely good approximation, we can set $v = c$ and the SSC's radius is

$$R = \frac{87 \text{ km}}{2\pi} \simeq 14 \text{ km}. \tag{14.27}$$

An energy of 20 TeV corresponds to an energy in joules of

$$20 \text{ TeV} \times 1.6 \times 10^{-19} \simeq 3 \times 10^{-6} \text{ J}. \tag{14.28}$$

The strength of the SSC's bending magnets is therefore

$$B_{\text{SSC}} \simeq 4.8 \text{ T}. \tag{14.29}$$

The magnets at the LHC have a maximum strength of 8.3 T, so are a bit stronger than those for the SSC. By contrast, the FCC would accelerate would protons to 50 TeV energy and have a radius of 100 km, so the strength of the bending magnets would need to be

$$B_{\text{FCC}} \simeq 10.5 \text{ T}. \tag{14.30}$$

Exercises

14.1 *Neutrinoless Double-β Decay at EXO.* How many ^{136}Xe atoms are there in the EXO-200 vat? How long does it need to be watched to observe just one neutrinoless double β-decay if the half-life is 10^{25} years?

14.2 *Double-Higgs Production.* In Section 14.3, we discussed the process $pp \to HH$, which is sensitive to the Higgs self coupling, λ. Using the NDA method developed in Example 13.2, estimate the matrix element $\mathcal{M}(gg \to HH)$ represented by the Feynman diagram in Section 14.3. Unlike the application of NDA to the matrix element $\mathcal{M}(gg \to H)$, the momentum flowing through the loop of top quarks is now about $2m_H < m_t$. Nevertheless, to get the mass dimension of the diagram correct you can still assume that $m_H \ll m_t$.

Hint: A $2 \to 2$ matrix element is dimensionless.

14.3 *Borel Summation of a Convergent Series.* Borel summation is a powerful technique for taming divergent series. In this exercise, we will see how it reproduces the correct result when the series converges.

Perform the Borel summation technique for the convergent series

$$\frac{1}{1-x} = \sum_{n=0}^{\infty} x^n, \tag{14.31}$$

for $|x| < 1$. Do you find the same function of x for the Borel-summed result? For what values of x does the Borel sum converge?

14.4 *The International Linear Collider.* The ILC is a linear collider, in contrast to LEP, for example, which was a circular electron–positron collider. What are some advantages of a linear electron–positron collider over a circular collider? Can you think of some disadvantages?

14.5 *The Largest Possible Collider.* The FCC pushes the limits of how large we think a particle physics experiment can be. However, the largest possible terrestrial circular collider would completely circle the Earth. While this is a bit silly, it is a good benchmark to keep in mind when asking what particle physics questions a collider can answer. In this exercise, we will estimate the maximum collision energy of such a collider. Assume that the number of protons in a bunch in the collider is the same as at the LHC and that the bunches are collided every 25 nanoseconds.

(a) What is the highest possible energy to which this collider could accelerate protons? You'll need the expression for the power emitted by synchrotron radiation from Chapter 5. Assume that the synchrotron losses are the same as at the LHC. The radius of the Earth is approximately 6000 km.

(b) Do the same exercise, but now assume that all of the power generated on Earth is used to counter losses from synchrotron radiation. The power consumption of humans on Earth is approximately 10^{13} W.

(c) Getting really silly, what is the highest possible energy of protons in a circular collider that extended along Earth's orbit around the Sun? Use all of Earth's resources as before and the radius of Earth's orbit, approximately 150 million km.

14.6 *Research Question.* Why is there something rather than nothing?

Appendix A Useful Identities

A.1 Complex Numbers

The imaginary number, i, is the square-root of -1:

$$i = \sqrt{-1}. \tag{A.1}$$

A general complex number z is a linear combination of a real number and a purely imaginary number and can be expressed as

$$z = a + ib, \tag{A.2}$$

for two real numbers $a, b \in \mathbb{R}$. a is referred to as the real part of z and b is referred to as the imaginary part of z. The complex conjugate of a complex number z is identical, except that all imaginary numbers i are negated. The complex conjugate of z is denoted in various contexts by z^*, $\bar{z}$, or sometimes $z^\dagger$ and is

$$z^* = \bar{z} = z^\dagger = a - ib. \tag{A.3}$$

Real numbers and purely imaginary numbers are distinct and so can be naturally thought of as orthogonal to one another. As such, the magnitude or length of a complex number follows from the Pythagorean theorem in two dimensions, corresponding to its real and imaginary parts. For the complex number z, its magnitude is denoted by $|z|$ and is

$$|z| = \sqrt{a^2 + b^2}. \tag{A.4}$$

Note that the sum of the squares of the real and imaginary parts of the complex number z is also equal to the product of itself times its complex conjugate:

$$|z|^2 = zz^* = (a + ib)(a - ib) = a^2 + b^2. \tag{A.5}$$

An equivalent representation of a complex number is through its magnitude and phase angle. The real and imaginary parts of a complex number can be defined as the projections along the real and imaginary axes, respectively. We can define

$$a = |z| \cos \phi, \qquad\qquad b = |z| \sin \phi, \tag{A.6}$$

for a phase angle $\phi \in [0, 2\pi)$. Then, the complex number z is

$$z = a + ib = |z| \cos \phi + i|z| \sin \phi = |z|e^{i\phi}. \tag{A.7}$$

The rightmost equality is known as **Euler's formula**.

A.2 δ-Function

The Dirac δ-function $\delta(x)$ is properly a distribution or generalized function whose value is 0 if its argument x is non-zero and infinite if $x = 0$. Though infinite, the δ-function is still integrable:

$$\int_{-\epsilon}^{\epsilon} dx\, \delta(x) = 1 \,, \tag{A.8}$$

for any $\epsilon > 0$. When integrated against any function $f(x)$, the δ-function just fixes the argument of the function to be 0:

$$\int dx\, f(x)\, \delta(x) = f(0) \,. \tag{A.9}$$

The argument of the δ-function can be displaced by some constant a. Then, its integral on the domain $x \in [0, 1]$ is

$$\int_{0}^{1} dx\, \delta(x - a) = \Theta(a)\Theta(1 - a) \,. \tag{A.10}$$

Here, $\Theta(x)$ is the Heaviside Θ-function, which is 1 if its argument x is greater than 0, and 0 if $x < 0$. That is, the way to read Eq. A.10 is that the integral over the δ-function $\delta(x - a)$ is only 1 if $a > 0$ and $a < 1$. Otherwise, the δ-function is zero over the whole integration domain, and so the integral is 0.

If the δ-function has a non-linear argument, then a change of variables must be performed before the integral can be done. Consider the function $f(x)$ and the integral over the δ-function whose argument is $f(x)$:

$$\int dx\, \delta\left(y - f(x)\right) \,, \tag{A.11}$$

where y is some number. Changing variables to $t = f(x)$, the integral becomes

$$\int dx\, \delta\left(y - f(x)\right) = \int dt \frac{1}{\left|\frac{df(x)}{dx}\right|_{x=f^{-1}(t)}} \delta(y - t) \,. \tag{A.12}$$

The inverse derivative is just the Jacobian of the change of variables. In this form, one can then integrate the δ-function.

A.3 Fourier Transforms

The Fourier transform of a function of one argument $f(x)$ is defined to be

$$\tilde{f}(p) = \int_{-\infty}^{\infty} dx\, f(x) e^{ipx} \,, \tag{A.13}$$

where p is a real number. We say that x and p are Fourier conjugate variables to one another, and the Fourier transform can be inverted as

$$f(x) = \int_{-\infty}^{\infty} \frac{dp}{2\pi} \tilde{f}(p) e^{-ipx} \,. \tag{A.14}$$

For the Fourier transform and the inverse Fourier transform to be consistent, we require that the Fourier transform of the function $f(x) = 1$ is the δ-function:

$$f(x) = \int_{-\infty}^{\infty} \frac{dp}{2\pi} \tilde{f}(p) e^{-ipx} = \int_{-\infty}^{\infty} \frac{dp}{2\pi} \left[\int_{-\infty}^{\infty} dx' f(x') e^{ipx'} \right] e^{-ipx} \tag{A.15}$$

$$= \int_{-\infty}^{\infty} dx' f(x') \int_{-\infty}^{\infty} \frac{dp}{2\pi} e^{ip(x'-x)} = f(x) \,.$$

That is,

$$\delta(x' - x) = \int_{-\infty}^{\infty} \frac{dp}{2\pi} e^{ip(x'-x)} \,. \tag{A.16}$$

In multiple dimensions, the Fourier transform generalizes naturally. Consider the n-dimensional vector $\vec{x}$ and a function $f(\vec{x})$ of this vector. Its Fourier transform is

$$\tilde{f}(\vec{p}) = \int d^n x f(\vec{x}) e^{i\vec{p} \cdot \vec{x}} \,, \tag{A.17}$$

where $\vec{p}$ is an n-dimensional vector of real numbers. The integral extends over all possible values of each component of the vector $\vec{x}$. The inverse Fourier transform in multiple dimensions is correspondingly

$$f(\vec{x}) = \int \frac{d^n p}{(2\pi)^n} \tilde{f}(\vec{p}) e^{-i\vec{p} \cdot \vec{x}} \,. \tag{A.18}$$

For a function $f(x)$ of the spacetime four-vector x, its Fourier transform is

$$\tilde{f}(p) = \int d^4 x f(x) e^{-ip \cdot x} \,, \tag{A.19}$$

where p is the momentum four-vector. The inverse Fourier transform is then

$$f(x) = \int \frac{d^4 p}{(2\pi)^4} \tilde{f}(p) e^{ip \cdot x} \,. \tag{A.20}$$

A.4 Spin-0

The Klein–Gordon equation that describes a massive, free, spin-0 field $\phi(x)$ is

$$\left(\partial^2 + m^2 \right) \phi(x) = 0 \,, \tag{A.21}$$

where m is the mass of the field. A general solution to the Klein–Gordon equation can be expressed as

$$\phi(x) = \tilde{\phi}(p) e^{-ip \cdot x} \,, \tag{A.22}$$

where p is the momentum four-vector that is on-shell: $p^2 = m^2$. $\tilde{\phi}(p)$ is the Fourier-transformed field of $\phi(x)$ that exclusively depends on momentum p.

A.5 Spin-1/2

The Dirac equation that describes a massive, spin-1/2 field $\phi(x)$ is

$$(i\gamma \cdot \partial - m)\psi(x) = 0. \tag{A.23}$$

In this book, we typically just considered massless solutions to the Dirac equation. In the massless case, the Dirac equation separates into two equations for two-component spinors $\psi_R(x)$ and $\psi_L(x)$, called the Weyl equations:

$$i\sigma \cdot \partial\psi_R(x) = 0, \qquad\qquad i\bar{\sigma} \cdot \partial\psi_L(x) = 0. \tag{A.24}$$

σ_μ and $\bar{\sigma}_\mu$ are four-component objects of the Pauli spin matrices:

$$\sigma_\mu = (\mathbb{I}, \sigma_1, \sigma_2, \sigma_3)_\mu, \qquad\qquad \bar{\sigma}_\mu = (\mathbb{I}, -\sigma_1, -\sigma_2, -\sigma_3)_\mu. \tag{A.25}$$

$\mathbb{I}$ is the 2×2 identity matrix, and the σ matrices are

$$\sigma_1 = \begin{pmatrix} 0 & 1 \\ 1 & 0 \end{pmatrix}, \qquad \sigma_2 = \begin{pmatrix} 0 & -i \\ i & 0 \end{pmatrix}, \qquad \sigma_3 = \begin{pmatrix} 1 & 0 \\ 0 & -1 \end{pmatrix}. \tag{A.26}$$

The positive-energy solutions to the Weyl equations, $p_0 > 0$, describe particles and can be expressed as

$$\psi_R^+(x) = u_R(p)e^{-ip\cdot x} = \sqrt{2E}\begin{pmatrix} e^{-i\phi/2}\cos\frac{\theta}{2} \\ e^{i\phi/2}\sin\frac{\theta}{2} \end{pmatrix}e^{-ip\cdot x}, \tag{A.27}$$

$$\psi_L^+(x) = u_L(p)e^{-ip\cdot x} = \sqrt{2E}\begin{pmatrix} e^{-i\phi/2}\sin\frac{\theta}{2} \\ -e^{i\phi/2}\cos\frac{\theta}{2} \end{pmatrix}e^{-ip\cdot x}. \tag{A.28}$$

In these expressions, the momentum four-vector is expressed in spherical coordinates as

$$p = E(1, \cos\phi\sin\theta, \sin\phi\sin\theta, \cos\theta), \tag{A.29}$$

where θ and ϕ are the polar and azimuthal angles, respectively. The negative-energy solutions, $p_0 < 0$, describe anti-particles and can be expressed as

$$\psi_R^-(x) = v_R(p)e^{-ip\cdot x} = \sqrt{2E}\begin{pmatrix} e^{-i\phi/2}\sin\frac{\theta}{2} \\ -e^{i\phi/2}\cos\frac{\theta}{2} \end{pmatrix}e^{-ip\cdot x}, \tag{A.30}$$

$$\psi_L^-(x) = v_L(p)e^{-ip\cdot x} = \sqrt{2E}\begin{pmatrix} e^{-i\phi/2}\cos\frac{\theta}{2} \\ e^{i\phi/2}\sin\frac{\theta}{2} \end{pmatrix}e^{-ip\cdot x}. \tag{A.31}$$

A.6 Spin-1

The equations of motion for a source-free, massless, spin-1 field $A_\mu(x)$, such as a photon, is

$$\left(\eta^{\mu\nu}\partial^2 - \partial^\nu\partial^\mu\right) A_\mu(x) = 0. \tag{A.32}$$

These equations enjoy a gauge invariance in which the field can transform inhomogeneously without affecting the equations of motion. Under a transformation of

$$A_\mu(x) \to A_\mu(x) + \partial_\mu\lambda(x), \tag{A.33}$$

where $\lambda(x)$ is any scalar function of the spacetime coordinates x, the equations of motion are unchanged. In this book, we introduced the Lorenz gauge in which we enforce

$$\partial \cdot A(x) = 0. \tag{A.34}$$

In Lorenz gauge, the equations of motion are

$$\partial^2 A_\mu(x) = 0, \tag{A.35}$$

which is just the Klein–Gordon equation.

The solution to the spin-1 field equations of motion is

$$A_\mu(x) = \epsilon_\mu(p)e^{-ip\cdot x}, \tag{A.36}$$

for massless, on-shell momentum p, where $p^2 = 0$. $\epsilon(p)$ is called the polarization vector of the field and satisfies

$$\epsilon(p) \cdot p = 0. \tag{A.37}$$

There are two physical polarization states which can naturally be expressed as right-handed and left-handed polarization. For momentum aligned along the $\hat{z}$-axis,

$$p = E(1,0,0,1), \tag{A.38}$$

where E is the energy, the right- and left-handed polarization vectors are

$$\epsilon_R(p) = \frac{1}{\sqrt{2}}(0,1,i,0), \qquad\qquad \epsilon_L(p) = \frac{1}{\sqrt{2}}(0,1,-i,0). \tag{A.39}$$

Appendix B **Review of Quantum Mechanics**

In this appendix, we present a review of quantum mechanics from a perspective that complements the approach to particle physics exhibited in this book. Nothing in this appendix is necessarily required for understanding the results in the body of the textbook, but it will hopefully provide some insight, if you only have a minimal introduction to the subject. We start from some fundamental axioms, and from there derive, among many other things, the Schrödinger equation. The two axioms that we will take are

Axiom 1: The wavefunction $\psi(x,t)$ is a complex probability amplitude for the quantum system to be at point x at time t.

Axiom 2: The total probability is unity and it does not change with time:

$$\int_{-\infty}^{\infty} \psi^*(x,t)\psi(x,t)\, dx = 1 ,$$

(B.1)

for all t.

B.1 Unitary Operators

Starting from these axioms, what can we do? One thing we can do is perform a transformation of $\psi(x,t)$ in a way that preserves probability. We implement a transformation by an operator $\hat{U}$ whose goal in life is to manipulate $\psi(x,t)$. That is, we consider transforming $\psi(x,t)$ as

$$\psi(x,t) \to \hat{U}\psi(x,t) .$$

(B.2)

What are the properties of $\hat{U}$? Performing this transformation within the calculation of the probability, we have

$$\int_{-\infty}^{\infty} \psi^*(x,t)\psi(x,t)\, dx \to \int_{-\infty}^{\infty} \left(\hat{U}\psi(x,t)\right)^* \left(\hat{U}\psi(x,t)\right)\, dx$$

(B.3)

$$= \int_{-\infty}^{\infty} \psi^*(x,t)\hat{U}^\dagger\hat{U}\psi(x,t)\, dx .$$

Here, $\hat{U}^\dagger$ is the complex conjugate operator of U; formally, this is the Hermitian conjugate, in which we complex conjugate and transpose (when thought of as a matrix). We do the transpose because $\hat{U}$ acts to the right, while $\hat{U}^\dagger$ acts to the left.

To ensure that probability is conserved, we must enforce that

$$\hat{U}^{\dagger}\hat{U} = 1 \,. \tag{B.4}$$

If the $\hat{U}$ operator were a matrix, this would be the condition on unitary matrices. We then say that $\hat{U}$ is a unitary operator (i.e., it preserves unit probability). This result is known as Wigner's theorem.[1] Actually, Wigner's theorem also allows for the possibility that $\hat{U}$ is anti-unitary. An anti-unitary operator still satisfies Eq. B.4, but additionally complex conjugates the wavefunction:

$$\hat{U}_{\text{anti-unitary}}\psi(x,t) = \hat{U}_{\text{unitary}}\psi^{*}(x,t) \,. \tag{B.5}$$

This clearly also preserves probability, but is related to time reversal, so we won't study it further here.

With that assumption $\hat{U}$ is unitary, can we leverage that property to express it in a nice way? Indeed we can. Note that if $\hat{U}$ were just a complex number, we could write $\hat{U}$ as

$$\hat{U} = e^{-i\phi} \,, \tag{B.6}$$

for some real number ϕ. Then, $\hat{U}^{\dagger}\hat{U} = e^{i\phi}e^{-i\phi} = 1$, which is the condition of unitarity. With this motivation, let's then express $\hat{U}$ as

$$\hat{U} = e^{-i\hat{T}} \,, \tag{B.7}$$

for some other operator $\hat{T}$. If these were matrices, $e^{-i\hat{T}}$ looks scary; however, it's just shorthand for the Taylor series:

$$\hat{U} = e^{-i\hat{T}} = 1 - i\hat{T} - \frac{\hat{T}^2}{2} + i\frac{\hat{T}^3}{6} + \cdots = \sum_{j=0}^{\infty} \frac{\left(-i\hat{T}\right)^j}{j!} \,. \tag{B.8}$$

What properties does $\hat{T}$ have? From the unitarity constraint, we have

$$\hat{U}^{\dagger}\hat{U} = e^{i\hat{T}^{\dagger}}e^{-i\hat{T}} = e^{-i(\hat{T}-\hat{T}^{\dagger})} = 1 \,. \tag{B.9}$$

This multiplication, found by just adding exponents, can be justified with the Taylor series. Then, for this to hold true, we must enforce

$$\hat{U}^{\dagger}\hat{U} = 1 \qquad \Rightarrow \qquad \hat{T} = \hat{T}^{\dagger} \,, \tag{B.10}$$

or that $\hat{T}$ is a Hermitian operator. Hermitian operators or matrices are the generalization of real numbers, as their eigenvalues are strictly real. We say that Hermitian operators "generate" unitary transformations of the wavefunction. The verb "generate" will make more sense shortly.

This has been quite abstract; let's bring it back to reality and explicitly construct a unitary operator. Let's consider the transformation that translates the wavefunction to the right by Δx:

$$\psi(x,t) \to \psi(x - \Delta x, t) = \hat{U}\psi(x,t) \,. \tag{B.11}$$

[1] E. P. Wigner, *Gruppentheorie und ihre Anwendung auf die Quantenmechanik der Atomspektren*, Vieweg (1931).

Note the minus sign for moving right; that is, if I move left, then it looks like the wavefunction has moved right. Note that this transformation indeed preserves probability:

$$\int_{-\infty}^{\infty} \psi^*(x,t)\psi(x,t)\,dx \rightarrow \int_{-\infty}^{\infty} \psi^*(x-\Delta x,t)\psi(x-\Delta x,t)\,dx = 1. \tag{B.12}$$

This is true because we can make the change of variables $y = x - \Delta x$ in the integral.

So what is $\hat{U}$? Now we see the power of expressing $\hat{U}$ as an exponential. We have

$$\hat{U}\psi(x,t) = e^{-i\hat{T}}\psi(x,t) = \psi(x-\Delta x,t). \tag{B.13}$$

To continue, let's Taylor expand both sides to linear order. This then yields

$$\left(1 - i\hat{T} + \cdots\right)\psi(x,t) = \left(1 - \Delta x \frac{\partial}{\partial x} + \cdots\right)\psi(x,t). \tag{B.14}$$

We then immediately find that

$$\hat{T} = -i\Delta x \frac{\partial}{\partial x}. \tag{B.15}$$

It then follows that

$$\hat{U} = e^{-\Delta x \frac{\partial}{\partial x}}, \tag{B.16}$$

which is a pretty crazy operator! However, let's see what it does to $\psi(x,t)$:

$$\hat{U}\psi(x,t) = e^{-\Delta x \frac{\partial}{\partial x}}\psi(x,t) = \sum_{j=0}^{\infty} \frac{1}{j!}\left(\Delta x \frac{\partial}{\partial x}\right)^j \psi(x,t) \tag{B.17}$$

$$= \psi(x,t) - \Delta x \frac{\partial}{\partial x}\psi(x,t) + \frac{\Delta x^2}{2}\frac{\partial^2}{\partial x^2}\psi(x,t) + \cdots = \psi(x-\Delta x,t).$$

This is just the Taylor expansion of the wavefunction about the point x.

Let's study this Hermitian operator in some more detail. Let's suggestively write

$$\hat{T} = \frac{\Delta x}{\hbar}\hat{p}, \tag{B.18}$$

where

$$\hat{p} = -i\hbar \frac{\partial}{\partial x} \tag{B.19}$$

and $\hbar$ is some constant. So far, this is a tautology. However, we can determine what this operator $\hat{p}$ is and give it a name. As shown above, the derivative operator displaces the wavefunction from an initial position x. Let's imagine that we have a collection of point masses $\{m_i\}$ located at positions $\{x_i\}$. Then, the center of mass of the system of masses c is the mass-weighted average of positions:

$$c = \frac{\sum_i m_i x_i}{\sum_i m_i}. \tag{B.20}$$

If we act on the center of mass with the derivative operator, it moves it by an amount Δx:

$$c' = \left(1 - \Delta x \frac{\partial}{\partial x}\right)c = \frac{\sum_i m_i(x_i - \Delta x)}{\sum_i m_i}. \tag{B.21}$$

That is, the derivative moves the center of mass. If the center of mass moves, then the system of masses must have momentum. Therefore, we identify the operator $\hat{p}$ with momentum.

Then, if $\hat{p}$ has dimensions of momentum, it follows that $\hbar$ has dimensions of energy $\times$ time:

$$[\hbar] = [\text{energy}][\text{time}]. \tag{B.22}$$

Let's see what that can get us.

B.2 Time Translation and the Schrödinger Equation

Let's do the same exercise but for time translation. That is, let's consider the transformation that moves the wavefunction in time from t to $t + \Delta t$:

$$\psi(x, t) \to \psi(x, t + \Delta t) = \hat{U}\psi(x, t). \tag{B.23}$$

Note that moving forward in time means that we shift t by $+\Delta t$; this is the opposite of the case with position x. By our axioms, this preserves the total probability:

$$\int_{-\infty}^{\infty} \psi^*(x, t + \Delta t)\psi(x, t + \Delta t)\, dx = \int_{-\infty}^{\infty} \psi^*(x, t)\psi(x, t)\, dx = 1, \tag{B.24}$$

as we assume that the total probability is independent of time. Let's write this time translation operator $\hat{U}$ in the suggestive exponential form

$$\hat{U}\psi(x, t) = e^{-i\frac{\Delta t \hat{H}}{\hbar}}\psi(x, t) = \psi(x, t + \Delta t), \tag{B.25}$$

for some Hermitian operator $\hat{H}$. Taylor expanding both sides of this equation to linear order in Δt, we find

$$\left(1 - i\frac{\Delta t \hat{H}}{\hbar}\right)\psi(x, t) = \left(1 + \Delta t\frac{\partial}{\partial t}\right)\psi(x, t), \tag{B.26}$$

or that

$$\hat{H} = i\hbar\frac{\partial}{\partial t}, \tag{B.27}$$

the time translation operator. Now, we can use our earlier result for the units of $\hbar$. Because $[\hbar] = [\text{energy}][\text{time}]$, it follows that the units of $\hat{H}$ are

$$[\hat{H}] = [\text{energy}]. \tag{B.28}$$

Therefore, $\hat{H}$ is some measure of the energy of the state represented by $\psi(x, t)$. The operator $\hat{H}$ is called the **Hamiltonian** and its eigenvalues correspond to possible energy states of the system.

This isn't a derivation, per se, but at least a plausibility argument that the energy (Hamiltonian) operator generates time translations.

From here, we can then derive the differential equation that governs time evolution of the wavefunction. From earlier, we had identified

$$- i\frac{\Delta t \hat{H}}{\hbar}\psi(x,t) = \Delta t\frac{\partial}{\partial t}\psi(x,t)\,, \tag{B.29}$$

or that

$$i\hbar\frac{\partial}{\partial t}\psi = \hat{H}\psi. \tag{B.30}$$

This is known as the **Schrödinger equation**. Given the Hamiltonian of a system (defining its kinetic and potential energies appropriately), we can determine the wavefunction at any later time.

In fact, we can explicitly solve the Schrödinger equation, simply by integrating over time. Integrating both sides over time, we find

$$\psi(x,t) = e^{-\frac{i}{\hbar}\int_{t_0}^{t}\hat{H}(t')\,dt'}\psi(x,t_0)\,, \tag{B.31}$$

where the wavefunction $\psi(x,t_0)$ is defined at some initial time, t_0. It is easy to verify that this expression satisfies the Schrödinger equation, using the fundamental theorem of calculus. In the exponent, we have also explicitly written the time t' at which the Hamiltonian is evaluated.

The Taylor expansion of the exponential factor is called the **Dyson series**. We can write it as

$$e^{-\frac{i}{\hbar}\int_{t_0}^{t}\hat{H}(t')\,dt'} = 1 - \frac{i}{\hbar}\int_{t_0}^{t}\hat{H}(t')\,dt' + \left(-\frac{i}{\hbar}\right)^2\frac{1}{2}\left(\int_{t_0}^{t}\hat{H}(t')\,dt'\right)^2 + \cdots \tag{B.32}$$

$$= 1 - \frac{i}{\hbar}\int_{t_0}^{t}\hat{H}(t_1)\,dt_1 + \left(-\frac{i}{\hbar}\right)^2\int_{t_0}^{t}dt_1\int_{t_0}^{t_1}dt_2\,\hat{H}(t_1)\hat{H}(t_2) + \cdots \tag{B.33}$$

$$= 1 + \sum_{n=1}^{\infty}\left(-\frac{i}{\hbar}\right)^n\int_{t_0}^{t}dt_1\int_{t_0}^{t_1}dt_2\cdots\int_{t_0}^{t_{n-1}}dt_n\,\hat{H}(t_1)\hat{H}(t_2)\cdots\hat{H}(t_n)\,.$$

At quadratic and higher order in the Hamiltonian, we can rewrite the integrals as nested integrals that are time-ordered. Note that, in general, Hamiltonians at different times are not identical, and so

$$\hat{H}(t_1)\hat{H}(t_2) \neq \hat{H}(t_2)\hat{H}(t_1)\,, \tag{B.34}$$

in general. Thus, we must be careful with the order of integrals. In quantum mechanics, this is often accomplished by just explicitly stating that the operators are time-ordered. When they are time-ordered, their series corresponds to a unitary operator that can be expressed as an exponential. The Dyson series is the starting point for Feynman diagrams and their mathematical justification.

B.3 Heisenberg Equations of Motion

There is another aspect of quantum mechanics that we can define. The Schrödinger equation governs the time evolution of the wavefunction $\psi(x, t)$; what governs the time evolution of an operator $\hat{O}$? There are two answers to this question. The first is that operators are time-independent, and time dependence is carried by the states, governed by the Schrödinger equation. This is called the **Schrödinger picture** of quantum mechanics. The second answer is that states are time-independent, while the operators carry time dependence. This is called the **Heisenberg picture** of quantum mechanics and produces identical physical consequences as the Schrödinger picture. We would like to derive the time evolution equation for operators in the Heisenberg picture.[2]

Starting again from the definition of probability, let's calculate the expectation value of an operator $\hat{O}$:

$$\langle \hat{O}(t) \rangle \equiv \int_{-\infty}^{\infty} \psi^*(x, t) \hat{O} \psi(x, t)\, dx. \tag{B.35}$$

We will assume that $\hat{O}$ is Hermitian in what follows. Evaluating the expectation value at a later time $t + \Delta t$, we have

$$\langle \hat{O}(t + \Delta t) \rangle = \int_{-\infty}^{\infty} \psi^*(x, t + \Delta t) \hat{O} \psi(x, t + \Delta t)\, dx \tag{B.36}$$

$$= \int_{-\infty}^{\infty} \psi^*(x, t) \hat{U}^\dagger(\Delta t) \hat{O} \hat{U}(\Delta t) \psi(x, t)\, dx.$$

Here, $\hat{U}(\Delta t)$ is the time translation operator,

$$\hat{U}(\Delta t) = e^{-\frac{i}{\hbar} \Delta t \hat{H}} = e^{\Delta t \frac{\partial}{\partial t}}. \tag{B.37}$$

Note that the Hermitian conjugate operator $\hat{U}^\dagger(\Delta t)$ is therefore

$$\hat{U}^\dagger(\Delta t) = e^{\frac{i}{\hbar} \Delta t \hat{H}} = e^{-\Delta t \frac{\partial}{\partial t}}. \tag{B.38}$$

Now, let's Taylor expand and equate the two expression for this expectation value. First, with the time derivative and expanding to linear order in Δt, we have

$$\psi^*(x, t) \left(1 - \Delta t \frac{\partial}{\partial t} + \cdots \right) \hat{O} \left(1 + \Delta t \frac{\partial}{\partial t} + \cdots \right) \psi(x, t) \tag{B.39}$$

$$= \psi^*(x, t) \psi(x, t) - \Delta t \psi^*(x, t) \frac{\partial}{\partial t} \left(\hat{O} \psi(x, t) \right) + \Delta t \psi^*(x, t) \hat{O} \frac{\partial}{\partial t} \psi(x, t) + \cdots$$

$$= \psi^*(x, t) \psi(x, t) - \Delta t \psi^*(x, t) \left(\frac{\partial}{\partial t} \hat{O} \right) \psi(x, t) + \cdots.$$

Here, the $\cdots$ denotes terms at order Δt^2 and higher. To get the second equality, we used the Leibniz product rule.

---

[2] There is a third picture in which to formulate quantum mechanics, called the **interaction picture**. The interaction picture is a hybrid of the Schrödinger and Heisenberg pictures and is typically the starting point for quantization of relativistic fields.

Using the expression for the time translation operator with the Hamiltonian, we have

$$\psi^*(x,t)\left(1 + \frac{i}{\hbar}\Delta t\hat{H} + \cdots\right)\hat{\mathcal{O}}\left(1 - \frac{i}{\hbar}\Delta t\hat{H} + \cdots\right)\psi(x,t) \tag{B.40}$$

$$= \psi^*(x,t)\psi(x,t) + \frac{i}{\hbar}\Delta t\psi^*(x,t)\hat{H}\hat{\mathcal{O}}\psi(x,t) - \frac{i}{\hbar}\Delta t\psi^*(x,t)\hat{\mathcal{O}}\hat{H}\psi(x,t) + \cdots$$

$$= \psi^*(x,t)\psi(x,t) + \frac{i}{\hbar}\Delta t\psi^*(x,t)[\hat{H},\hat{\mathcal{O}}]\psi(x,t) + \cdots.$$

Here, $[\hat{H},\hat{\mathcal{O}}]$ is the commutator

$$[\hat{H},\hat{\mathcal{O}}] = \hat{H}\hat{\mathcal{O}} - \hat{\mathcal{O}}\hat{H}. \tag{B.41}$$

Setting terms at each order in Δt equal to one another, we then find

$$-\Delta t\psi^*(x,t)\left(\frac{\partial}{\partial t}\hat{\mathcal{O}}\right)\psi(x,t) = \frac{i}{\hbar}\Delta t\psi^*(x,t)[\hat{H},\hat{\mathcal{O}}]\psi(x,t). \tag{B.42}$$

If this is to hold for any wavefunction $\psi(x,t)$, then we must have that

$$i\hbar\frac{\partial}{\partial t}\hat{\mathcal{O}} = [\hat{H},\hat{\mathcal{O}}]. \tag{B.43}$$

This is called the **Heisenberg equation of motion**.

Note in particular that if we take $\hat{\mathcal{O}} = \hat{H}$, then energy is conserved if the Hamiltonian does not depend explicitly on time. If the Hamiltonian does have time dependence, then the commutator involves the Hamiltonian evaluated at different times. As argued earlier, the Hamiltonian does not necessarily commute with itself evaluated at a different time. Another way to interpret the Heisenberg equation of motion is that the quantity associated with $\hat{\mathcal{O}}$ is conserved if $\hat{\mathcal{O}}$ commutes with the Hamiltonian. That is, $\hat{\mathcal{O}}$ generates a symmetry of the quantum system, by Noether's theorem.

Particle physics is infamous for using jargon, much of which is initially colloquial phrases that become de facto technical terms. A number of these terms are defined and used in the main text, but we collect them here for ease of identification.

3σ: the size of deviation from the null hypothesis that is considered evidence of new physics; corresponds to a probability of approximately 1 in 1000 to be described by the null hypothesis

4π **hermetic detector**: a particle collider detector that, to the greatest extent possible, captures particles produced from a collision point throughout the 4π steradians of the sphere

5σ: the size of deviation from the null hypothesis that is considered discovery of new physics; corresponds to a probability of approximately 1 in 3.5 million to be described by the null hypothesis

associative production: a process in which a particle of interest is produced in collisions, but also requiring the existence of another particle or other particles in the final state; e.g., to search for the Higgs boson at LEP, one searches for the Higgs in the final state and a Z boson produced in association with the Higgs

barrel: the main cylindrical component of a particle collider detector like ATLAS or CMS at the LHC

baseline: the distance between the point of particle production and the location of detector experiments; typically refers to the distance between the location of neutrino production (at a reactor, for example) and a neutrino detector

bin: one element in a histogram defined by its upper and lower limits; a bin is "filled" with events that satisfy the range of the bin

Bjorken scaling: the observation that, at high energies, the dynamics of quarks inside the proton is approximately independent of energy or wavelength used to probe the proton; this is a consequence of the approximate scale invariance of QCD at high energies

branching fraction: the fraction or probability of the time that a particle decays to a particular set of other particles; the sum of all branching fractions is 1

Brazil plot: a plot used to quantify deviations from expected sensitivity to new physics; its etymology derives from the bright green and yellow bands used to denote 1σ and 2σ deviations, respectively

bump hunting: a colloquial name for the experimental search process of measuring distributions of invariant masses of collections of particles produced in collisions and searching for "bumps" which would indicate the presence of a new particle

bunch: a collection of numerous particles accelerated together and set to collide with another such collection; typically, this refers to the collection of about 10^{11} protons accelerated and collided at the LHC

decay mode: the transformation of a particle to a particular collection of other particles; each possible set of particles from a decay is a "mode"

detector-stable: refers to a particle whose lifetime is large compared to the size of a particle physics detector so that it can be directly measured

dijet: literally "two jet"; a scattering process in which two jets are produced in the final state

dimensional transmutation: the process by which the energy dependence of a dimensionless quantity (like a running coupling) introduces a dimensionful energy scale

discovery machine: a term used to describe hadron colliders because their center-of-mass collision energies can be enormous while the individual parton collision energies can range over several decades of energy; this feature enables searches for new particles and physics without having to scan over collision energies

doublet: an object that transforms in the two-dimensional representation of SU(2); examples of doublets include a spin-1/2 field (for SU(2) rotations), the nucleon doublet (for SU(2) isospin), and the electron and electron neutrino (for SU(2)$_W$)

duality: two distinct yet physically equivalent descriptions of a phenomenon; an example is QCD at low energies, which can be described either through strongly interacting quarks and gluons or through weakly interacting hadrons

exclusive cross section: a cross section for a process in which there exists one or more constraints on the phase space of the final state

external particle: a particle in a Feynman diagram that is in either the initial state or the final state and whose representation on the diagram consists of one free end; external particles are always on-shell

flavor-changing neutral current: a process in which a lepton or quark turns into a different flavor lepton or quark, without changing its electric charge; such processes are forbidden or highly suppressed in the Standard Model

flavor-diagonal: the property of electromagnetism and QCD that their interactions with fermions preserve the type or flavor of fermion; for example, on emitting a photon, an electron is still an electron

gluon–gluon fusion: the dominant collision process for producing a Higgs boson at the LHC or any hadron collider; refers to the collision of gluon partons in protons to "fuse" into a Higgs boson through a loop of virtual top quarks

golden channels: decay modes of the Higgs boson that have especially distinct, simple, and clean experimental signatures; this typically refers to the processes $H \rightarrow l^+ l^- \, l'^+ l'^-$, where l and l' are electrons or muons, and $H \rightarrow \gamma\gamma$

hit: an observation in an experiment that lies in a region of interest

inclusive cross section: a cross section for a process in which the final state consists of a chosen selection of particles and anything else

internal symmetry: a symmetry whose transformations do not affect spacetime properties such as direction of spin, momentum, and energy of a particle

IRC safety: IRC (or "infrared and collinear") safety is the property of an observable that ensures that its distribution can be calculated from Feynman diagrams and Fermi's Golden Rule; an IRC-safe observable is one for which divergences from infrared (low energy) or collinear particle emission are localized at isolated points on phase space

Jacobian zero: a zero of a distribution that arises from a change of variables and not because of the impossibility of a physical configuration

leading order (LO): the collection of Feynman diagrams that have the fewest number of loops that contribute to a process of interest; Feynman diagrams with one additional loop are called "next-to-leading order" (NLO), those with two additional loops "next-to-next-to-leading order" (NNLO), etc.

Lego plot: a plot of the transverse momentum deposited in calorimeter cells versus azimuthal angle and pseudorapidity; its etymology derives from the similarity of towers of transverse momenta to towers of stacked Lego toys

look-elsewhere effect: a statistical effect which reduces the significance of an excess over the null hypothesis in a particular region because there was no a-priori reason to prefer that region over any other

luminosity: the number of pairs of particles that could interact in a particle collision experiment per unit area per unit time

manifest symmetry: a symmetry is said to be manifest if it is easily seen to be a symmetry of the Lagrangian of a system; more precisely, a manifest symmetry is one for which the action of the symmetry on the Lagrangian and on states is the same

matrix element: a synonym for scattering amplitude that represents the probability amplitude for a process; its name derives from the scattering matrix or S-matrix which encodes the probabilities of all initial–to–final-state transitions in a quantum system

Matthew effect: a principle due to social scientist Robert Merton which states that a more famous scientist will typically get credit over a less famous scientist if their work was similar or even less important; the term is derived from the parable of the talents in the Synoptic Gospels

maximally violated: a symmetry is said to be maximally violated if, under the action of the transformation, a physical state turns into a state that has zero probability to exist

Mercedes-Benz configuration: the configuration of three jets produced with approximately equal energy and approximately $120°$ from one another; the term derives from the similarity to the emblem of the car manufacturer

Mexican hat potential: the colloquial, yet standard, term for describing the shape of the potential energy well of the Higgs boson; the term derives from its similar appearance to a sombrero

missing transverse momentum or energy (MET): the transverse momentum necessary to add to the visible final-state particles to ensure that the total transverse momentum at a collider is zero; typically, missing transverse momentum is associated with the presence of neutrinos in the final state

naïve: a somewhat common word in particle physics that effectively means the application of a rule or equation without consideration of conservation laws, symmetries, or other constraints

off-shell: a particle for which the square of its four-momentum does not equal the square of its mass; particles that are off-shell are also called "virtual"

on-shell: a particle for which the square of its four-momentum equals the square of its mass; particles that are on-shell are also called "real"

particle: an object that is localized in space and whose intrinsic properties are unchanged under the action of any symmetry group; particles are defined by the irreducible representations of symmetry groups under which they transform

parton: a constituent or "part" of the proton at high energies; partons include the low-mass quarks and gluons

phenomenology: the study of a particular physical phenomenon; theoretical physicists who work closely with experiment are often called "phenomenologists"

propagator: an internal line on a Feynman diagram whose two ends are connected at vertices; propagators express the momentum dependence of the potential of the field that they represent

resummation: the procedure for systematically including the effects of multiple mass or energy scales in an exclusive cross section; "resummation" refers to the necessity for summing an infinite tower of terms to all orders in the coupling of the theory of interest that are enhanced by the presence of large logarithms of dimensionless energy ratios

robust: a result that is unaffected, or only weakly affected, by changes in assumptions; a related term is "model-independent," which refers to the consequences of a result being independent of the model (or theoretical assumptions) under which it is interpreted

rule out: to statistically eliminate the possibility of a deviation from the null hypothesis, typically with 95% confidence

running coupling: a coupling whose value depends on the energy scale at which it is being probed; its value is said to "run" with energy scale

scattering angle: the angle between the line along which the momenta of the two final-state particles lie and the line along which the momenta of the two initial-state particles lie in a $2 \to 2$ collision process in the center-of-mass frame

semileptonic decay: a process in which a massive particle and its anti-particle are produced, one of which decays to leptons and the other of which decays to quarks; examples of final states that can decay semileptonically are $W^+ W^-$, ZZ, and $t\bar{t}$

singlet: an object that transforms in the one-dimensional representation of a group; such an object transforms to itself up to an overall phase under the action of the group

soft: synonym for "low-energy"; typically refers to massless force-carrying bosons such as photons and gluons that have low energy compared to a fiducial energy scale

spontaneous symmetry breaking: when a system whose physical description enjoys a symmetry but whose ground state does not exhibit that symmetry; an example includes the double-well potential which is symmetric for $x \to -x$, but the ground state is localized in only one of the two minima

Stairway to Heaven plot: a summary plot of cross sections, measured at the LHC, of fundamental processes in the Standard Model; its name derives from the "stairway" of cross sections distributed over decades of picobarns

Stigler's law of eponymy: the principle, due to statistician Stephen Stigler, which states that no scientific discovery is named after its original discoverer; Stigler himself credits Robert Merton with this principle

threshold energy: the minimum energy above which a process is kinematically allowed

tower: an individual element or cell of a particle detector calorimeter that measures the energy deposited in a finely segmented region of pseudorapidity η and azimuthal angle ϕ

track: typically used as a synonym for charged particle; references a collection of ionization hits in the tracking system of a particle detector that is due to a single charged particle

tree diagram: a Feynman diagram that contains no loops, and topologically similar to a tree; an example of a tree diagram is the leading-order diagram for the process $e^+e^- \to \mu^+\mu^-$

trigger: a minimal requirement on the final state that a collision event must pass for it to be recorded for further analysis

triplet: an object that transforms in the three-dimensional representation of a group; examples include the three pions π^+, π^-, π^0 (for SU(2) isospin) or a quark (for SU(3) color)

trivial: a transformation is trivial if its action is equivalent to the identity

vev: acronym for "vacuum expectation value," the magnitude of the scalar field $|\vec{\phi}|$ in vacuum; fluctuations about this value correspond to the Higgs boson

Bibliography

Books

Undergraduate

Thomson, Mark. *Modern Particle Physics*, Cambridge University Press, 2013. A modern particle physics textbook for undergraduates published after the discovery of the Higgs boson.

Griffiths, David. *Introduction to Elementary Particles*, John Wiley & Sons, 2008. An elementary introduction that provides a historical introduction to the subject.

Henley, Ernest M. and Garcia, Alejandro. *Subatomic Physics*, World Scientific, 2007. Includes an extensive introduction to nuclear physics, particle detectors, and accelerators.

Haywood, Stephen. *Symmetries and Conservation Laws in Particle Physics: An Introduction to Group Theory for Particle Physicists*, World Scientific, 2011. In-depth introduction to group theory in particle physics, focused around the groups U(1), SU(2), and SU(3) that are central to the Standard Model.

Graduate

Coleman, Sidney. *Aspects of Symmetry*, Cambridge University Press, 1985. Notes from the Erice summer school from Sidney Coleman. Coleman's quantum field theory class at Harvard was legendary and lectures from 1975 were recorded and are available to view online at: `www.physics.harvard.edu/events/videos/Phys253`. Notes from that class were typed up by students and can be found at `https:arxiv.org/pdf/1110.5013.pdf`.

Cheng, Ta-Pei and Li, Ling-Fong. *Gauge Theory of Elementary Particle Physics*, Oxford University Press, 1984. Graduate-level textbook surveying many fundamental topics in particle physics.

Schwartz, Matthew. *Quantum Field Theory and the Standard Model*, Cambridge University Press, 2014. A modern graduate-level book on quantum field theory published after the discovery of the Higgs boson.

Zee, Anthony. *Quantum Field Theory in a Nutshell*, Princeton University Press, 2003. More of a monograph than a textbook, an approachable, readable introduction to ideas in quantum field theory.

Srednicki, Mark. *Quantum Field Theory*, Cambridge University Press, 2007. A modern textbook that is somewhat unique in its organization of introducing quantum fields ordered by their spin.

Peskin, Michael E. and Schroeder, Daniel V. *An Introduction to Quantum Field Theory*, Addison-Wesley, 1995. A standard for a long time, quickly gets to tools and techniques for doing calculations in quantum field theory.

Weinberg, Steven. *The Quantum Theory of Fields*, 3 vols, Cambridge University Press, 2005. The encyclopedic reference for quantum field theory and its mathematical construction.

Popular

Polchinski, J. *Memories of a Theoretical Physicist,* [arXiv:1708.09093 [physics.hist-ph]]. Polchinski's memoir of his career in physics, from childhood to his diagnosis with brain cancer in 2015. Polchinski died in 2018.

Taubes, Gary. *Nobel Dreams: Power, Deceit and the Ultimate Experiment*, Microsoft Press, 1988. A story mostly focused around Carlo Rubbia and his quest to stop at nothing to win another Nobel Prize for discovering supersymmetry at CERN in the 1980s.

Traweek, Sharon. *Beamtimes and Lifetimes: The World of High Energy Physicists*, Harvard University Press, 1992. A social scientist observes the interactions of particle physicists.

Butterworth, Jon. *Smashing Physics*, Headline, 2014. Butterworth is an experimentalist on the ATLAS experiment and provides an insider's perspective of the discovery of the Higgs boson at the LHC.

Wilczek, Frank. *The Lightness of Being: Mass, Ether, and the Unification of Forces*, Basic Books, 2010. Wilczek provides his perspective on what we are made of and how the universe came to be.

Quinn, Helen R. and Nir, Yossi. *The Mystery of the Missing Antimatter*, Princeton University Press, 2008. Quinn and Nir probe the question of why we are made out of matter, instead of anti-matter, and the cosmology necessary for this to happen.

Lederman, Leon M. and Hill, Christopher T. *Beyond the God Particle*, Prometheus Books, 2013. The sequel to Lederman's *The God Particle,* this begins with the discovery of the Higgs boson in 2012 and where to go next.

Gaillard, Mary K. *A Singularly Unfeminine Profession*, World Scientific, 2015. Gaillard's memoir of her career in physics and especially the obstacles she faced as a woman.

Films

Documentaries

"The Atom Smashers," www.pbs.org/independentlens/atomsmashers A documentary about the search for the Higgs boson at Fermilab, several years before its discovery at the LHC.

"Big Bang Machine," PBS Nova, `www.pbs.org/wgbh/nova/physics/big-bang-machine.html` Another documentary on the discovery of the Higgs boson.

"CERN People," `www.youtube.com/user/CERNPeople` A collection of short documentaries by filmmaker Liz Mermin. She has interviewed numerous scientists and provides a rather intimate and very human perspective of the people that work at CERN.

"Particle Fever," `http://particlefever.com` A documentary on the discovery of the Higgs boson that includes interviews with scientists at the forefront. A unique aspect is that the narrator, David E. Kaplan, is a physicist at Johns Hopkins University and a producer of the film.

Entertainment

"Large Hadron Rap," `www.youtube.com/watch?v=j50ZssEojtM` Science journalist Kate McAlpine raps about the physics of the LHC and the things it may discover.

"Daily Show at CERN," `www.cc.com/video-clips/hzqmb9/the-daily-show-with-jon-stewart-large-hadron-collider` Daily Show correspondent John Oliver travels to CERN to determine if the LHC will destroy the world.

"Rolling in the Higgs," `www.youtube.com/watch?v=VtItBX1l1VY` YouTube user A Capella Science performs an ode to the Higgs boson, inspired by Adele's song "Rolling in the Deep."

"DECAY," `www.decayfilm.com` In 2012, a group of CERN graduate students wrote, directed, and starred in their own zombie flick that is set at CERN. It isn't a great movie by any means, but in some sense is one of the most realistic depictions of particle physics graduate students in film.

"ILC Promotion Video," `www.youtube.com/watch?v=jf2WlQcVXIM` A high school in Japan produced a video to promote the construction of the International Linear Collider in their prefecture. It is remarkably well produced, and tells a story of two students representing an electron and a positron that want to be together.

"All About That Higgs," `www.youtube.com/watch?v=mx64FHIOcKE` Participants at a workshop in Les Houches, high in the French alps, sing an ode to the Higgs boson, inspired by Meghan Trainor's song "All About That Bass."

"Collide," `https://home.cern/news/news/cern/musician-howie-day-records-love-song-physics` Graduate students made an LHC-themed parody of Howie Day's song "Collide." After seeing the video on YouTube, Howie Day contacted CERN for a visit, and made a music video of the parody at CERN.

In-Text References

Collected below are the in-text references presented in alphabetical order. This is formally superfluous, as all are provided as footnotes on the page where they are referenced. However, collecting them in one place can make finding a particular reference easier.

Aaboud, M. *et al.* [ATLAS Collaboration], "Search for resonances in diphoton events at $\sqrt{s}$=13 TeV with the ATLAS detector," J. High Energy Phys. **1609**, 001 (2016) [arXiv:1606.03833 [hep-ex]].

Aaboud, M. *et al.* [ATLAS Collaboration], "Measurement of the *W*-boson mass in pp collisions at $\sqrt{s} = 7$ TeV with the ATLAS detector," Eur. Phys. J. C **78**, no. 2, 110 (2018) [arXiv:1701.07240 [hep-ex]].

Aaboud, M. *et al.* [ATLAS Collaboration], "Search for pair production of Higgs bosons in the $b\bar{b}b\bar{b}$ final state using proton–proton collisions at $\sqrt{s} = 13$ TeV with the ATLAS detector" [arXiv:1804.06174 [hep-ex]].

Aaboud, M. *et al.* [ATLAS Collaboration], "Combined measurement of differential and total cross sections in the $H \rightarrow \gamma\gamma$ and the $H \rightarrow ZZ^* \rightarrow 4\ell$ decay channels at $\sqrt{s} = 13$ TeV with the ATLAS detector," Phys. Lett. B **786**, 114 (2018) [arXiv:1805.10197 [hep-ex]].

Aad, G. *et al.* [ATLAS Collaboration], "The ATLAS experiment at the CERN Large Hadron Collider," J. Instrum. **3**, S08003 (2008).

Aad, G. *et al.* [ATLAS Collaboration], "Search for the Higgs boson in the $H \rightarrow \mathrm{WW}(*)$ $\rightarrow \ell_\nu \ell_\nu$ decay channel in *pp* collisions at $\sqrt{s} = 7$ TeV with the ATLAS detector," Phys. Rev. Lett. **108**, 111802 (2012) [arXiv:1112.2577 [hep-ex]].

Aad, G. *et al.* [ATLAS Collaboration], "Jet mass and substructure of inclusive jets in $\sqrt{s} = 7$ TeV *pp* collisions with the ATLAS experiment," J. High Energy Phys. **1205**, 128 (2012) [arXiv:1203.4606 [hep-ex]].

Aad, G. *et al.* [ATLAS Collaboration], "Observation of a new particle in the search for the Standard Model Higgs boson with the ATLAS detector at the LHC," Phys. Lett. B **716**, 1 (2012) [arXiv:1207.7214 [hep-ex]].

Aad, G. *et al.* [ATLAS Collaboration], "Measurements of the pseudorapidity dependence of the total transverse energy in proton-proton collisions at $\sqrt{s} = 7$ TeV with ATLAS," J. High Energy Phys. **1211**, 033 (2012) [arXiv:1208.6256 [hep-ex]].

Aad, G. *et al.* [ATLAS Collaboration], "Study of the spin and parity of the Higgs boson in diboson decays with the ATLAS detector," Eur. Phys. J. C **75**, no. 10, 476 (2015), Erratum: [Eur. Phys. J. C **76**, no. 3, 152 (2016)] [arXiv:1506.05669 [hep-ex]].

Aad, G. *et al.* [ATLAS and CMS Collaborations], "Measurements of the Higgs boson production and decay rates and constraints on its couplings from a combined ATLAS and CMS analysis of the LHC pp collision data at $\sqrt{s} = 7$ and 8 TeV," J. High Energy Phys. **1608**, 045 (2016) [arXiv:1606.02266 [hep-ex]].

Aartsen, M. G. *et al.* [IceCube Collaboration], "Observation of high-energy astrophysical neutrinos in three years of IceCube data," Phys. Rev. Lett. **113**, 101101 (2014) [arXiv:1405.5303 [astro-ph.HE]].

Abachi, S. *et al.* [D0 Collaboration], "Observation of the top quark," Phys. Rev. Lett. **74**, 2632 (1995) [hep-ex/9503003].

Abazov, V. M. *et al.* [D0 Collaboration], "Measurement of the shape of the boson rapidity distribution for $p\bar{p} \rightarrow Z/gamma^* \rightarrow e^+e^- + X$ events produced at $\sqrt{s}$ of 1.96-TeV," Phys. Rev. D **76**, 012003 (2007) [arXiv:hep-ex/0702025 [HEP-EX]].

Abbiendi, G. *et al.* [OPAL Collaboration], "Measurement of event shape distributions and moments in $e^+e^- \rightarrow$ hadrons at 91 GeV – 209 GeV and a determination of α_s," Eur. Phys. J. C **40**, 287 (2005) [hep-ex/0503051].

Abe, F. *et al.* [CDF Collaboration], "Observation of top quark production in $\bar{p}p$ collisions," Phys. Rev. Lett. **74**, 2626 (1995) [arXiv:hep-ex/9503002].

Abe, K. *et al.* [SLD Collaboration], "Measurement of $\alpha_s(m_Z^2)$ from hadronic event observables at the Z0 resonance," Phys. Rev. D **51**, 962 (1995) [arXiv:hep-ex/9501003].

Abe, K. *et al.* [SLD Collaboration], "A study of the orientation and energy partition of three jet events in hadronic Z0 decays," Phys. Rev. D **55**, 2533 (1997) [arXiv:hep-ex/9608016].

Abreu, P. *et al.* [DELPHI Collaboration], "Determination of α_s using the next-to-leading log approximation of QCD," Z. Phys. C **59**, 21 (1993).

Abreu, P. *et al.* [DELPHI Collaboration], "Measurement of the triple gluon vertex from four-jet events at LEP," Z. Phys. C **59**, 357 (1993).

Abreu, P. *et al.* [DELPHI Collaboration], "Improved measurements of cross-sections and asymmetries at the Z0 resonance," Nucl. Phys. B **418**, 403 (1994).

Abreu, P. *et al.* [DELPHI Collaboration], "Tuning and test of fragmentation models based on identified particles and precision event shape data," Z. Phys. C **73**, 11 (1996).

Achard, P. *et al.* [L3 Collaboration], "Measurement of the running of the electromagnetic coupling at large momentum-transfer at LEP," Phys. Lett. B **623**, 26 (2005) [arXiv:hep-ex/0507078].

Ackerman, N. *et al.* [EXO-200 Collaboration], "Observation of two-neutrino double-beta decay in ^{136}Xe with EXO-200," Phys. Rev. Lett. **107**, 212501 (2011) [arXiv:1108.4193 [nucl-ex]].

Adloff, C. *et al.* [H1 Collaboration], "Measurement of neutral and charged current cross-sections in electron–proton collisions at high Q^2," Eur. Phys. J. C **19**, 269 (2001) [arXiv:hep-ex/0012052].

Akerib, D. S. *et al.* [LUX Collaboration], "Results from a search for dark matter in the complete LUX exposure," Phys. Rev. Lett. **118**, no. 2, 021303 (2017) [arXiv:1608.07648 [astro-ph.CO]].

Akers, R. *et al.* [OPAL Collaboration], "Measurement of single photon production in e^+e^- collisions near the Z0 resonance," Z. Phys. C **65**, 47 (1995).

Albert, J. B. *et al.* [EXO-200 Collaboration], "Search for Majorana neutrinos with the first two years of EXO-200 data," Nature **510**, 229 (2014) [arXiv:1402.6956 [nucl-ex]].

Alekseev, E. N., Alekseeva, L. N., Krivosheina, I. V. and Volchenko, V. I. "Detection of the neutrino signal from SN1987A in the LMC using the Inr Baksan underground scintillation telescope," Phys. Lett. B **205**, 209 (1988).

Altarelli, G. and Parisi, G. "Asymptotic freedom in parton language," Nucl. Phys. B **126**, 298 (1977).

Altarelli, G., Cabibbo, N., Corbo, G., Maiani, L. and Martinelli, G. "Leptonic decay of heavy flavors: A theoretical update," Nucl. Phys. B **208**, 365 (1982).

Altarelli, G. and Feruglio, F. "Discrete flavor symmetries and models of neutrino mixing," Rev. Mod. Phys. **82**, 2701 (2010) [arXiv:1002.0211 [hep-ph]].

Altarelli, G. "The QCD running coupling and its measurement," in *Proceedings of the Corfu Summer Institute 2012*, PoS(Corfu2012)002 (2013) [arXiv:1303.6065 [hep-ph]].

An, F. P. *et al.* [Daya Bay Collaboration], "New measurement of antineutrino oscillation with the full detector configuration at Daya Bay," Phys. Rev. Lett. **115**, no. 11, 111802 (2015) [arXiv:1505.03456 [hep-ex]].

Anastasiou, C., Duhr, C., Dulat, F., Herzog, F. and Mistlberger, B. "Higgs boson gluon-fusion production in QCD at three loops," Phys. Rev. Lett. **114**, 212001 (2015) [arXiv:1503.06056 [hep-ph]].

Anderson, C. D. "The positive electron," Phys. Rev. **43**, 491 (1933).

Anderson, P. W. "Plasmons, gauge Invariance, and mass," Phys. Rev. **130**, 439 (1963).

Andersson, B., Gustafson, G., Lonnblad, L. and Pettersson, U. "Coherence effects in deep inelastic scattering," Z. Phys. C **43**, 625 (1989).

Antchev, G. *et al.* [TOTEM Collaboration], "Luminosity-independent measurement of the proton–proton total cross section at $\sqrt{s} = 8$ TeV," Phys. Rev. Lett. **111**, no. 1, 012001 (2013).

Araki, T. *et al.* [KamLAND Collaboration], "Measurement of neutrino oscillation with KamLAND: Evidence of spectral distortion," Phys. Rev. Lett. **94**, 081801 (2005) [arXiv:hep-ex/0406035].

Askins, M. *et al.* [WATCHMAN Collaboration], "The physics and nuclear nonproliferation goals of WATCHMAN: A WATer CHerenkov Monitor for ANtineutrinos" [arXiv:1502.01132 [physics.ins-det]].

Aubert, J. J. *et al.* [E598 Collaboration], "Experimental observation of a heavy particle J," Phys. Rev. Lett. **33**, 1404 (1974).

Auger, M. *et al.*, "The EXO-200 detector, part I: Detector design and construction," J. Instrum. **7**, P05010 (2012) [arXiv:1202.2192 [physics.ins-det]].

Augustin, J. E. *et al.* [SLAC-SP-017 Collaboration], "Discovery of a narrow resonance in e^+e^- annihilation," Phys. Rev. Lett. **33**, 1406 (1974).

Barate, R. *et al.* [ALEPH Collaboration], "Study of the muon pair production at center-of-mass energies from 20 GeV to 136 GeV with the ALEPH detector," Phys. Lett. B **399**, 329 (1997).

Barate, R. *et al.* [ALEPH and DELPHI and L3 and OPAL Collaborations and LEP Working Group for Higgs boson searches], "Search for the standard model Higgs boson at LEP," Phys. Lett. B **565**, 61 (2003) [arXiv:hep-ex/0306033].

Barber, D. P. *et al.*, "Discovery of three jet events and a test of quantum chromodynamics at PETRA energies," Phys. Rev. Lett. **43**, 830 (1979).

Bardeen, J., Cooper, L. N. and Schrieffer, J. R. "Theory of superconductivity," Phys. Rev. **108**, 1175 (1957).

Barnes, V. E. *et al.*, "Observation of a hyperon with strangeness -3," Phys. Rev. Lett. **12**, 204 (1964).

Bartel, W. *et al.* [JADE Collaboration], "Observation of planar three jet events in e^+e^- annihilation and evidence for gluon bremsstrahlung," Phys. Lett. **91B**, 142 (1980).

Bell, J. S. "Time reversal in field theory," Proc. Roy. Soc. Lond. A **231**, 479 (1955).

Benvenuti, A. C. *et al.* [BCDMS Collaboration], "A high statistics measurement of the proton structure functions $F_2(x, Q^2)$ and R from deep inelastic muon scattering at high Q^2," Phys. Lett. B **223**, 485 (1989).

Berger, C. *et al.* [PLUTO Collaboration], "Evidence for gluon bremsstrahlung in e^+e^- annihilations at high energies," Phys. Lett. **86B**, 418 (1979).

Bethe, H. "Theory of the passage of fast corpuscular rays through matter," Annalen Phys. **397**, 325 (1930).

Bhabha, H. J. "The scattering of positrons by electrons with exchange on Dirac's theory of the positron," Proc. Roy. Soc. Lond. A **154**, 195 (1936).

Bionta, R. M. *et al.*, "Observation of a neutrino burst in coincidence with supernova SN 1987a in the Large Magellanic Cloud," Phys. Rev. Lett. **58**, 1494 (1987).

Bjorken, J. D. "Current algebra at small distances," in *Selected Topics in Particle Physics: Proceedings of the International School of Physics "Enrico Fermi,"* Course XLI, Academic Press (1968), pp. 55–81.

Bjorken, J. D. "Asymptotic sum rules at infinite momentum," Phys. Rev. **179**, 1547 (1969).

Bjorken, J. D. and Brodsky, S. J. "Statistical model for electron–positron annihilation into hadrons," Phys. Rev. D **1**, 1416 (1970).

Bloch, F. and Nordsieck, A. "Note on the radiation field of the electron," Phys. Rev. **52**, 54 (1937).

Brandelik, R. *et al.* [TASSO Collaboration], "Evidence for planar events in e^+e^- annihilation at high energies," Phys. Lett. **86B**, 243 (1979).

Brau, J. *et al.* [ILC Collaboration], "ILC reference design report, volume 1: Executive summary" [arXiv:0712.1950 [physics.acc-ph]].

Buras, A. J., Ellis, J. R., Gaillard, M. K. and Nanopoulos, D. V. "Aspects of the grand unification of strong, weak and electromagnetic interactions," Nucl. Phys. B **135**, 66 (1978).

Cabibbo, N. "Unitary symmetry and leptonic decays," Phys. Rev. Lett. **10**, 531 (1963).

Callan, C. G. Jr., "Broken scale invariance in scalar field theory," Phys. Rev. D **2**, 1541 (1970).

Catani, S., Turnock, G. and Webber, B. R. "Jet broadening measures in e^+e^- annihilation," Phys. Lett. B **295**, 269 (1992).

Catani, S., Dokshitzer, Y. L., Seymour, M. H. and Webber, B. R. "Longitudinally invariant K_t clustering algorithms for hadron hadron collisions," Nucl. Phys. B **406**, 187 (1993).

CEPC-SPPC Study Group, *CEPC-SPPC Preliminary Conceptual Design Report. 1: Physics and Detector*, IHEP-CEPC-DR-2015-01, IHEP-TH-2015-01, IHEP-EP-2015-01.

CEPC-SPPC Study Group, *CEPC-SPPC Preliminary Conceptual Design Report. 2: Accelerator*, IHEP-CEPC-DR-2015-01, IHEP-AC-2015-01.

Chadwick, J. "Possible existence of a neutron," Nature **129**, 312 (1932).

Chao, C-Y. "The absorption coefficient of hard γ-rays," Proc. Natl. Acad. Sci. U. S. A. **16**, no. 6, 431 (1930).

Chatrchyan, S. *et al.* [CMS Collaboration], "The CMS experiment at the CERN LHC," J. Instrum. **3**, S08004 (2008).

Chatrchyan, S. *et al.* [CMS Collaboration], "Search for the standard model Higgs boson decaying to W^+W^- in the fully leptonic final state in pp collisions at $\sqrt{s} = 7$ TeV," Phys. Lett. B **710**, 91 (2012) [arXiv:1202.1489 [hep-ex]].

Chatrchyan, S. *et al.* [CMS Collaboration], "Observation of a new boson at a mass of 125 GeV with the CMS experiment at the LHC," Phys. Lett. B **716**, 30 (2012) [arXiv:1207.7235 [hep-ex]].

Chau, L. L. and Keung, W. Y. "Comments on the parametrization of the Kobayashi–Maskawa matrix," Phys. Rev. Lett. **53**, 1802 (1984).

Chibisov, G. V. "Astrophysical upper limits on the photon rest mass," Sov. Phys. Usp. **19**, 624 (1976) [Usp. Fiz. Nauk **119**, 551 (1976)].

Christenson, J. H., Cronin, J. W., Fitch, V. L. and Turlay, R. "Evidence for the 2π decay of the K_2^0 meson," Phys. Rev. Lett. **13**, 138 (1964).

Cohen, A. G., Glashow, S. L. and Ligeti, Z. "Disentangling neutrino oscillations," Phys. Lett. B **678**, 191 (2009) [arXiv:0810.4602 [hep-ph]].

Coleman, S. R. and Mandula, J. "All possible symmetries of the S matrix," Phys. Rev. **159**, 1251 (1967).

Collins, J. C., Soper, D. E. and Sterman, G. F. "Transverse momentum distribution in Drell-Yan pair and W and Z boson production," Nucl. Phys. B **250**, 199 (1985).

Compton, A. H. "A quantum theory of the scattering of X-rays by light elements," Phys. Rev. **21**, 483 (1923).

DeGrand, T. and Detar, C. E. *Lattice Methods for Quantum Chromodynamics*, World Scientific (2006).

Derrick, M. *et al.* [HRS Collaboration], "New results on the reaction $e^+e^- \to \mu^+\mu^-$ at $\sqrt{s} = 29$ GeV," Phys. Rev. D **31**, 2352 (1985).

Dick, A. *Emmy Noether 1882–1935*, Springer Nature (1981).

Dirac, P. A. M. "Quantum theory of emission and absorption of radiation," Proc. Roy. Soc. Lond. A **114**, 243 (1927).

Dirac, P. A. M. "The quantum theory of the electron," Proc. Roy. Soc. Lond. A **117**, 610 (1928).

Dixon, L. J. "Calculating scattering amplitudes efficiently," in *QCD and Beyond: Proceedings, Theoretical Advanced Study Institute in Elementary Particle Physics, TASI-95, Boulder, CO, June 4–30, 1995*, World Scientific (1996), pp. 539–582 [arXiv:hep-ph/9601359].

Djouadi, A. *et al.* [ILC Collaboration], *International Linear Collider Reference Design Report Volume 2: Physics at the ILC* [arXiv:0709.1893 [hep-ph]].

Dokshitzer, Y. L. "Calculation of the structure functions for deep inelastic scattering and e^+e^- annihilation by perturbation theory in quantum chromodynamics," Sov. Phys. JETP **46**, 641 (1977) [Zh. Eksp. Teor. Fiz. **73**, 1216 (1977)].

Dokshitzer, Y. L., Leder, G. D., Moretti, S. and Webber, B. R. "Better jet clustering algorithms," J. High Energy Phys. **9708**, 001 (1997) [arXiv:hep-ph/9707323].

Donoghue, J. F., Low, F. E. and Pi, S. Y. "Tensor analysis of hadronic jets in quantum chromodynamics," Phys. Rev. D **20**, 2759 (1979).

Dorigoni, D. "An Introduction to resurgence, trans-series and alien calculus" [arXiv:1411.3585 [hep-th]].

Drell, S. D. and Yan, T. M. "Massive lepton pair production in hadron–hadron collisions at high energies," Phys. Rev. Lett. **25**, 316 (1970), Erratum: [Phys. Rev. Lett. **25**, 902 (1970)].

Dunne, G. V. and Ünsal, M. "What is QFT?: Resurgent trans-series, Lefschetz thimbles, and new exact saddles," in *Proceedings of the 33rd International Symposium on Lattice Field Theory*, PoS(LATTICE2015)010 (2016) [arXiv:1511.05977 [hep-lat]].

Durr, S. *et al.*, "Ab-initio determination of light hadron masses," Science **322**, 1224 (2008) [arXiv:0906.3599 [hep-lat]].

Dyson, F. J. "The radiation theories of Tomonaga, Schwinger, and Feynman," Phys. Rev. **75**, 486 (1949).

Dyson, F. J. "The S matrix in quantum electrodynamics," Phys. Rev. **75**, 1736 (1949).

Dyson, F. J. "Divergence of perturbation theory in quantum electrodynamics," Phys. Rev. **85**, 631 (1952).

Ellis, J. "The discovery of the gluon," Int. J. Mod. Phys. A **29**, no. 31, 1430072 (2014) [arXiv:1409.4232 [hep-ph]].

Ellis, J. R., Gaillard, M. K. and Ross, G. G. "Search for gluons in e^+e^- annihilation," Nucl. Phys. B **111**, 253 (1976), Erratum: [Nucl. Phys. B **130**, 516 (1977)].

Ellis, J. R., Gaillard, M. K., Nanopoulos, D. V. and Rudaz, S. "The phenomenology of the next left-handed quarks," Nucl. Phys. B **131**, 285 (1977), Erratum: [Nucl. Phys. B **132**, 541 (1978)].

Ellis, R. K., Ross, D. A. and Terrano, A. E. "The perturbative calculation of jet structure in e^+e^- annihilation," Nucl. Phys. B **178**, 421 (1981).

Ellis, R. K. and Webber, B. R. "QCD jet broadening in hadron hadron collisions," in *Physics of the Superconducting Supercollider: Proceedings of the 1986 Summer Study Meeting, June 23 to July 11, 1986, Snowmass, CO*, American Physical Society (1986), pp. 74–76.

Ellis, S. D. and Soper, D. E. "Successive combination jet algorithm for hadron collisions," Phys. Rev. D **48**, 3160 (1993) [arXiv:hep-ph/9305266].

Elvang, H. and Huang, Y. T. *Scattering Amplitudes in Gauge Theory and Gravity*, Cambridge University Press (2015) [arXiv:1308.1697 [hep-th]].

Englert, F. and Brout, R. "Broken symmetry and the mass of gauge vector mesons," Phys. Rev. Lett. **13**, 321 (1964).

Farhi, E. "A QCD test for jets," Phys. Rev. Lett. **39**, 1587 (1977).

Fermi, E. "Tentativo di una teoria dell'emissione dei raggi beta," Ric. Sci. **4**, 491 (1933).

Fermi, E. "An attempt of a theory of beta radiation. 1," Z. Phys. **88**, 161 (1934).

Fermi, E. *A Course Given by Enrico Fermi at The University of Chicago, 1949*, University of Chicago Press, revised edn. (1950).

Feynman, R. P. "Space-time approach to quantum electrodynamics," Phys. Rev. **76**, 769 (1949).

Feynman, R. P. and Gell-Mann, M. "Theory of Fermi interaction," Phys. Rev. **109**, 193 (1958).

Feynman, R. P. *The Character of Physical Law*, MIT Press (1967).

Feynman, R. P. "The behavior of hadron collisions at extreme energies," in *Proceedings of the 3rd International Conference on High Energy Collisions, Stony Brook, NY*, Gordon and Breach (1969), pp. 237–258.

Feynman, R. P. "Very high-energy collisions of hadrons," Phys. Rev. Lett. **23**, 1415 (1969).

Fock, V. "On the invariant form of the wave equation and the equations of motion for a charged point mass," Z. Phys. **39**, 226 (1926) [Surveys High Energ. Phys. **5**, 245 (1986)].

Ford, W. T. *et al.*, "Measurement of α_s from hadron jets in e^+e^- annihilation at $s^{1/2}$ of 29 GeV," Phys. Rev. D **40** (1989) 1385.

Friedman, J. I. and Kendall, H. W. "Deep inelastic electron scattering," Ann. Rev. Nucl. Part. Sci. **22**, 203 (1972).

Furry, W. H. "A symmetry theorem in the positron theory," Phys. Rev. **51**, 125 (1937).

Gaillard, M. K. and Lee, B. W. "Rare decay modes of the K-mesons in gauge theories," Phys. Rev. D **10**, 897 (1974).

Garwin, R. L., Lederman, L. M. and Weinrich, M. "Observations of the failure of conservation of parity and charge conjugation in meson decays: The magnetic moment of the free muon," Phys. Rev. **105**, 1415 (1957).

Gattringer, C. and Lang, C. B. "Quantum chromodynamics on the lattice," Lect. Notes Phys. **788**, 1 (2010).

Gell-Mann, M. "The Eightfold Way: A theory of strong interaction symmetry," CTSL-20, TID-12608.

Gell-Mann, M. "A schematic model of baryons and mesons," Phys. Lett. **8**, 214 (1964).

Georgi, H., Quinn, H. R. and Weinberg, S. "Hierarchy of interactions in unified gauge theories," Phys. Rev. Lett. **33**, 451 (1974).

Gerlach, W. and Stern, O. "Experimental test of the applicability of the quantum theory to the magnetic field," Z. Phys. **9**, 349 (1922).

Glashow, S. L. "Partial symmetries of weak interactions," Nucl. Phys. **22**, 579 (1961).

Glashow, S. L., Iliopoulos, J. and Maiani, L. "Weak interactions with lepton–hadron symmetry," Phys. Rev. D **2**, 1285 (1970).

Goldstone, J. "Field theories with superconductor solutions," Nuovo Cim. **19**, 154 (1961).

Goodman, M. W. and Witten, E. "Detectability of certain dark matter candidates," Phys. Rev. D **31**, 3059 (1985).

Gordon, W. "Der Comptoneffekt nach der Schrödingerschen theorie," Z. Phys. **40**, 117 (1926).

Greisen, K. "End to the cosmic ray spectrum?," Phys. Rev. Lett. **16**, 748 (1966).

Gribov, V. N. and Lipatov, L. N. "Deep inelastic e p scattering in perturbation theory," Sov. J. Nucl. Phys. **15**, 438 (1972) [Yad. Fiz. **15**, 781 (1972)].

Gross, D. J. and Wilczek, F. "Ultraviolet behavior of nonabelian gauge theories," Phys. Rev. Lett. **30**, 1343 (1973).

Grossheim, A. *et al.* [TWIST Collaboration], "Decay of negative muons bound in Al-27," Phys. Rev. D **80**, 052012 (2009) [arXiv:0908.4270 [hep-ex]].

Gunion, J. F., Haber, H. E., Kane, G. L. and Dawson, S. *The Higgs Hunter's Guide*, Front. Phys. **80**, 1 (2000).

Guo, X. *et al.* [Daya Bay Collaboration], "A precision measurement of the neutrino mixing angle θ_{13} using reactor antineutrinos at Daya-Bay" [arXiv:hep-ex/0701029].

Guralnik, G. S., Hagen, C. R. and Kibble, T. W. B. "Global conservation laws and massless particles," Phys. Rev. Lett. **13**, 585 (1964).

Haag, R., Lopuszanski, J. T. and Sohnius, M. "All possible generators of supersymmetries of the S-matrix," Nucl. Phys. B **88**, 257 (1975).

Hanson, G. *et al.*, "Evidence for jet structure in hadron production by e^+e^- annihilation," Phys. Rev. Lett. **35**, 1609 (1975).

Heisenberg, W. "On the structure of atomic nuclei," Z. Phys. **77**, 1 (1932).

Heister, A. *et al.* [ALEPH Collaboration], "Studies of QCD at e^+e^- centre-of-mass energies between 91 GeV and 209 GeV," Eur. Phys. J. C **35**, 457 (2004).

Higgs, P. W. "Broken symmetries and the masses of gauge bosons," Phys. Rev. Lett. **13**, 508 (1964).

Hirata, K. S. *et al.*, "Observation in the Kamiokande-II detector of the neutrino burst from supernova SN 1987a," Phys. Rev. D **38**, 448 (1988).

Holloway, M. G. and Baker, C. P. "How the barn was born," Phys. Today **25**, no. 7, 9 (1972).

Hooft, G. 't. "Renormalization of massless Yang–Mills fields," Nucl. Phys. B **33**, 173 (1971).

Hooft, G. 't. "Renormalizable Lagrangians for massive Yang–Mills fields," Nucl. Phys. B **35**, 167 (1971).

Hooft, G. 't. and Veltman, M. J. G. "Regularization and renormalization of gauge fields," Nucl. Phys. B **44**, 189 (1972).

Hooft, G. 't. "A planar diagram theory for strong interactions," Nucl. Phys. B **72**, 461 (1974).

Hooft, G. 't. and Veltman, M. J. G. "One loop divergencies in the theory of gravitation," Ann. Inst. H. Poincare Phys. Theor. A **20**, 69 (1974).

Hooft, G. 't. "Computation of the quantum effects due to a four-dimensional pseudoparticle," Phys. Rev. D **14**, 3432 (1976), Erratum: [Phys. Rev. D **18**, 2199 (1978)].

Hooft, G. 't. "When was asymptotic freedom discovered? Or the rehabilitation of quantum field theory," Nucl. Phys. Proc. Suppl. **74**, 413 (1999) [arXiv:hep-th/9808154].

Hudson, R. P. "Reversal of the parity conservation law in nuclear physics," in Lide, D. R. *A Century of Excellence in Measurements, Standards, and Technology*, CRC Press (2001), pp. 111–115.

Jarlskog, C. "Commutator of the quark mass matrices in the Standard Electroweak Model and a measure of maximal CP violation," Phys. Rev. Lett. **55**, 1039 (1985).

Jost, R. "A remark on the C.T.P. theorem," Helv. Phys. Acta **30**, 409 (1957).

Kadler, M. *et al.*, "Coincidence of a high-fluence blazar outburst with a PeV-energy neutrino event," Nature Phys. **12**, no. 8, 807 (2016) [arXiv:1602.02012 [astro-ph.HE]].

Khachatryan, V. *et al.* [CMS Collaboration], "Search for resonant production of high-mass photon pairs in proton–proton collisions at $\sqrt{s} =$8 and 13 TeV," Phys. Rev. Lett. **117**, no. 5, 051802 (2016) [arXiv:1606.04093 [hep-ex]].

Khachatryan, V. *et al.* [CMS Collaboration], "Measurement and QCD analysis of double-differential inclusive jet cross sections in pp collisions at $\sqrt{s} = 8$ TeV and cross section ratios to 2.76 and 7 TeV," J. High Energy Phys. **1703**, 156 (2017) [arXiv:1609.05331 [hep-ex]].

Kharusi, S. A. *et al.* [nEXO Collaboration], *nEXO Pre-conceptual Design Report* [arXiv:1805.11142 [physics.ins-det]].

Kinoshita, T. "Mass singularities of Feynman amplitudes," J. Math. Phys. **3**, 650 (1962).

Klein, O. "Elektrodynamik und wellenmechanik vom standpunkt des korresponden-zprinzips," Z. Phys. **41**, 407 (1927).

Kobayashi, M. and Maskawa, T. "CP violation in the renormalizable theory of weak interaction," Prog. Theor. Phys. **49**, 652 (1973).

Kunkel, W. and Madore, B. *IAU Circular No. 4316 (24 February 1987)*, International Astronomical Union (1987).

Landau, L. D. "On the angular momentum of a system of two photons," Dokl. Akad. Nauk Ser. Fiz. **60**, no. 2, 207 (1948).

Landau, L. D. "On the quantum theory of fields," in *Niels Bohr and the Development of Physics*, Pauli, W. (ed.), Pergamon Press (1955), pp. 52–69.

Landau, L. D. "On the conservation laws for weak interactions," Nucl. Phys. **3**, 127 (1957).

Lederman, L. and Teresi, D. *The God Particle: If the Universe is the Answer, What is the Question?*, Houghton Mifflin (1993).

Lee, B. W., Quigg, C. and Thacker, H. B. "Weak interactions at very high energies: The role of the Higgs boson mass," Phys. Rev. D **16**, 1519 (1977).

Lee, B. W. and Weinberg, S. "Cosmological lower bound on heavy neutrino masses," Phys. Rev. Lett. **39**, 165 (1977).

Lee, T. D. and Yang, C. N. "Question of parity conservation in weak interactions," Phys. Rev. **104**, 254 (1956).

Lee, T. D. and Nauenberg, M. "Degenerate systems and mass singularities," Phys. Rev. **133**, B1549 (1964).

Lellouch, L. *et al.* (eds), *Modern Perspectives in Lattice QCD: Quantum Field Theory and High Performance Computing*, Lecture notes of the Les Houches Summer School, August 2009, Vol. 93, Oxford University Press (2011).

Lüders, G. "On the equivalence of invariance under time reversal and under particle–antiparticle conjugation for relativistic field theories," Kong. Dan. Vid. Sel. Mat. Fys. Med. **28**, no. 5, 1 (1954).

Majorana, E. "Teoria simmetrica dell'elettrone e del positrone," Nuovo Cim. **14**, 171 (1937).

Maki, Z., Nakagawa, M. and Sakata, S. "Remarks on the unified model of elementary particles," Prog. Theor. Phys. **28**, 870 (1962).

Mandelbrot, B. B. "How long is the coast of Britain?," Science **156**, no. 3775, (1967), 636.

Mandelstam, S. "Determination of the pion–nucleon scattering amplitude from dispersion relations and unitarity: General theory," Phys. Rev. **112**, 1344 (1958).

Mandelstam, S. "Light cone superspace and the ultraviolet finiteness of the N=4 model," Nucl. Phys. B **213**, 149 (1983).

Mangano, M. L. and Parke, S. J. "Multiparton amplitudes in gauge theories," Phys. Rept. **200**, 301 (1991) [arXiv:hep-th/0509223].

Mangano, M. L. *Physics at the FCC-hh, a 100 TeV pp Collider*, CERN Yellow Report 2017-003-M, CERN (2017) [arXiv:1710.06353 [hep-ph]].

Manohar, A. and Georgi, H. "Chiral quarks and the nonrelativistic quark model," Nucl. Phys. B **234**, 189 (1984).

Markevitch, M. *et al.*, "Direct constraints on the dark matter self-interaction cross-section from the merging galaxy cluster 1E0657-56," Astrophys. J. **606**, 819 (2004) [arXiv:astro-ph/0309303].

Migdal, A. A. and Polyakov, A. M. "Spontaneous breakdown of strong interaction symmetry and the absence of massless particles," Sov. Phys. JETP **24**, 91 (1967) [Zh. Eksp. Teor. Fiz. **51**, 135 (1966)].

Merton, R. K. "The Matthew effect in science," Science **159**, no. 3810, 56 (1968).

Møller, C., "Zurtheorie des durchgangs schneller elektronen durch materie," Annalen Phys. **406**, no. 5, 531 (1932).

Nadolsky, P. *et al.*, "Progress in CTEQ-TEA PDF Analysis," in *XX International Workshop on Deep-Inelastic Scattering and Related Subjects, March 26–30, 2012, Bonn*, Desy (2012), pp. 417–420 [arXiv:1206.3321 [hep-ph]].

Nakahara, M. *Geometry, Topology and Physics*, Taylor & Francis (2003).

Nambu, Y. "Quasiparticles and gauge invariance in the theory of superconductivity," Phys. Rev. **117**, 648 (1960).

Nambu, Y. and Jona-Lasinio, G. "Dynamical model of elementary particles based on an analogy with superconductivity, 1," Phys. Rev. **122**, 345 (1961).

Ne'eman, Y. "Derivation of strong interactions from a gauge invariance," Nucl. Phys. **26**, 222 (1961).

Noether, E. "Invariant variation problems," Gott. Nachr. **1918**, 235 (1918) [Transp. Theory Statist. Phys. **1**, 186 (1971)] [arXiv:physics/0503066].

Parisi, G. "Super inclusive cross-sections," Phys. Lett. **74B**, 65 (1978).

Parke, S. J. and Taylor, T. R. "An amplitude for n gluon scattering," Phys. Rev. Lett. **56**, 2459 (1986).

Pauli, W. "Dear radioactive ladies and gentlemen," from a letter to Lise Meither, dated Dec. 1930 [reprinted in Phys. Today **31**, no. 9, 27 (1978)].

Pauli, W., Rosenfeld, L. and Weisskopf, V. (eds) *Niels Bohr and the Development of Physics*, Pergamon Press (1955).

Peccei, R. D. and Quinn, H. R. "CP conservation in the presence of instantons," Phys. Rev. Lett. **38**, 1440 (1977).

Peccei, R. D. and Quinn, H. R. "Constraints imposed by CP conservation in the presence of instantons," Phys. Rev. D **16**, 1791 (1977).

Peebles, P. J. E. "Recombination of the primeval plasma," Astrophys. J. **153**, 1 (1968).

Penzias, A. A. and Wilson, R. W. "A measurement of excess antenna temperature at 4080 Mc/s," Astrophys. J. **142**, 419 (1965).

Peon, B. "Is Hinchliffe's rule true?," submitted to Annals Gnosis.

Planck, M. "Über irreversible Strahlungsvorgänge," *Sitzungsberichte der Königlich Preußischen Akademie der Wissenschaften Zu Berlin* **5**, 440 (1899), and Ann. Phys. **306**, no. 1, 69 (1900).

Politzer, H. D. "Reliable perturbative results for strong interactions?," Phys. Rev. Lett. **30**, 1346 (1973).

Pontecorvo, B. "Neutrino experiments and the problem of conservation of leptonic charge," Sov. Phys. JETP **26**, 984 (1968) [Zh. Eksp. Teor. Fiz. **53**, 1717 (1967)].

Rakow, P. E. L. and Webber, B. R. "Transverse momentum moments of hadron distributions in QCD jets," Nucl. Phys. B **191**, 63 (1981).

Riordan, E. M. "The discovery of quarks," Science **256**, 1287 (1992).

Rochester, G. D. and Butler, C. C. "The new unstable cosmic-ray particles," Rep. Prog. Phys. **16**, no. 1, 364 (1953).

Rubin, V. C. and Ford, Jr., W. K. "Rotation of the Andromeda nebula from a spectroscopic survey of emission regions," Astrophys. J. **159**, 379 (1970).

De Rujula, A., Georgi, H., Glashow, S. L. and Quinn, H. R. "Fact and fancy in neutrino physics," Rev. Mod. Phys. **46**, 391 (1974).

Rutherford, E. "The scattering of alpha and beta particles by matter and the structure of the atom," Phil. Mag. Ser. 6 **21**, 669 (1911).

Sakharov, A. D. "Violation of CP invariance, C asymmetry, and baryon asymmetry of the universe," Pisma Zh. Eksp. Teor. Fiz. **5**, 32 (1967) [JETP Lett. **5**, 24 (1967)] [Sov. Phys. Usp. **34**, 392 (1991)] [Usp. Fiz. Nauk **161**, 61 (1991)].

Sakurai, J. J., "Mass reversal and weak interactions," Nuovo cim. **7**, 649 (1958).

Salam, A. "On parity conservation and neutrino mass," Nuovo Cim. **5**, 299 (1957).

Salam, A. "Weak and electromagnetic interactions," in *Elementary Particle Theory: Proceedings of the 8th Nobel Symposium, Lerum, Sweden, 1968*, John Wiley & Sons and Almqvist and Wiksell (1968), pp. 367–377.

Salam, G. P. "The strong coupling: A theoretical perspective" [arXiv:1712.05165 [hep-ph]], in Forte, S., Levy, A. and Ridolfi, G. (eds), *From My Vast Repertoire: The Legacy of Guido Altarelli*, World Scientific (2018), pp. 101–121.

Schael, S. *et al.* [ALEPH and DELPHI and L3 and OPAL and LEP Electroweak Collaborations], "Electroweak measurements in electron–positron collisions at W-boson-pair energies at LEP," Phys. Rept. **532**, 119 (2013) [arXiv:1302.3415 [hep-ex]].

Schwinger, J. S. "The theory of quantized fields. 1," Phys. Rev. **82**, 914 (1951).

Schwinger, J. S. "Gauge invariance and mass," Phys. Rev. **125**, 397 (1962).

Schwinger, J. S. "Gauge invariance and mass. 2," Phys. Rev. **128**, 2425 (1962).

Shifman, M. A. "Foreword," in Shifman, M. (ed.), *ITEP Lectures in Particle Physics and Field Theory*, Vol. 1. World Scientific (1999), pp. v-xi [arXiv:hep-ph/9510397].

Sirunyan, A. M. *et al.* [CMS Collaboration], "Measurement of the $t\bar{t}$ production cross section using events with one lepton and at least one jet in pp collisions at $\sqrt{s} = 13$ TeV," J. High Energy Phys. **1709**, 051 (2017) [arXiv:1701.06228 [hep-ex]].

Sirunyan, A. M. *et al.* [CMS Collaboration], "Search for Higgs boson pair production in events with two bottom quarks and two tau leptons in proton–proton collisions at $\sqrt{s} = 13$ TeV," Phys. Lett. B **778**, 101 (2018) [arXiv:1707.02909 [hep-ex]].

Snyder, A. E. and Quinn, H. R. "Measuring CP asymmetry in $B \to \rho\pi$ decays without ambiguities," Phys. Rev. D **48**, 2139 (1993).

Spergel, D. N. *et al.* [WMAP Collaboration], "First year Wilkinson Microwave Anisotropy Probe (WMAP) observations: Determination of cosmological parameters," Astrophys. J. Suppl. **148**, 175 (2003) [arXiv:astro-ph/0302209].

Stigler, S. M. "Stigler's law of eponymy," Trans. N. Y. Acad. Sci. **39**, 147 (1980).

Streater, R. F. and Wightman, A. S. *PCT, Spin and Statistics, and All That*, Princeton University Press (2000).

Stueckelberg, E. C. G. "Interaction forces in electrodynamics and in the field theory of nuclear forces," Helv. Phys. Acta **11**, 299 (1938).

Sudakov, V. V. "Vertex parts at very high energies in quantum electrodynamics," Sov. Phys. JETP **3**, 65 (1956) [Zh. Eksp. Teor. Fiz. **30**, 87 (1956)].

Sudarshan, E. C. G. and Marshak, R. E. "Chirality invariance and the universal Fermi interaction," Phys. Rev. **109**, 1860 (1958).

Symanzik, K. "Small distance behavior in field theory and power counting," Commun. Math. Phys. **18**, 227 (1970).

Symanzik, K. "Small distance behavior analysis and Wilson expansion," Commun. Math. Phys. **23**, 49 (1971).

Tanabashi, M. *et al.* [Particle Data Group], "Review of particle physics," Phys. Rev. D **98**, 030001 (2018).

Tevatron New Physics Higgs Working Group [CDF and D0 Collaborations], "Updated combination of CDF and D0 searches for Standard Model Higgs boson production with up to 10.0 fb^{-1} of data" [arXiv:1207.0449 [hep-ex]].

Thomson, J. J., "Cathode rays," Phil. Mag. Ser. 5 **44**, 293 (1897).

Vainshtein, A. I., Zakharov, V. I. and Shifman, M. A. "A possible mechanism for the $\Delta T = 1/2$ rule in nonleptonic decays of strange particles," JETP Lett. **22**, 55 (1975) [Pisma Zh. Eksp. Teor. Fiz. **22**, 123 (1975)].

Weinberg, S. "A model of leptons," Phys. Rev. Lett. **19**, 1264 (1967).

Wigner, E. P. *Gruppentheorie und ihre Anwendung auf die Quantenmechanik der Atomspektren*, Vieweg (1931).

Wigner, E. P. "On unitary representations of the inhomogeneous Lorentz group," Annals Math. **40**, 149 (1939) [Nucl. Phys. Proc. Suppl. **6**, 9 (1989)].

Wigner, E. P. "The unreasonable effectiveness of mathematics in the natural sciences," Comm. Pure Appl. Math. **13**, no. 1, 1 (1960).

Wilson, K. G. "Confinement of quarks," Phys. Rev. D **10**, 2445 (1974).

Wilson, K. G. "The renormalization group: Critical phenomena and the Kondo problem," Rev. Mod. Phys. **47**, 773 (1975).

Wobisch, M. and Wengler, T. "Hadronization corrections to jet cross-sections in deep inelastic scattering," in, Doyle, A. T., Grindhammer, G., Ingleman, G. and Jung, H. (eds), *Monte Carlo Generators for HERA Physics*, Desy (1999), pp. 270–279 [arXiv:hep-ph/9907280].

Wolfenstein, L. "Parametrization of the Kobayashi–Maskawa matrix," Phys. Rev. Lett. **51**, 1945 (1983).

Wu, C. S., Ambler, E., Hayward, R. W., Hoppes, D. D., and Hudson, R. P. "Experimental test of parity conservation in beta decay," Phys. Rev. **105**, 1413 (1957).

Wu, S. L. and Zobernig, G. "A method of three jet analysis in e^+e^- annihilation," Z. Phys. C **2**, 107 (1979).

Yang, C. N. "Selection rules for the dematerialization of a particle into two photons," Phys. Rev. **77**, 242 (1950).

Yang, C. N. and Mills, R. L. "Conservation of isotopic spin and isotopic gauge invariance," Phys. Rev. **96**, 191 (1954).

Yukawa, H. "On the interaction of elementary particles I," Proc. Phys. Math. Soc. Jap. **17**, 48 (1935) [Prog. Theor. Phys. Suppl. **1**, 1].

Zatsepin, G. T. and Kuzmin, V. A. "Upper limit of the spectrum of cosmic rays," JETP Lett. **4**, 78 (1966) [Pisma Zh. Eksp. Teor. Fiz. **4**, 114 (1966)].

Zeldovich, Y. B., Kurt, V. G. and Sunyaev, R. A. "Recombination of hydrogen in the hot model of the universe," Sov. Phys. JETP **28**, 146 (1969) [Zh. Eksp. Teor. Fiz. **55**, 278 (1968)].

Zweig, G. *An SU(3) Model for Strong Interaction Symmetry and Its Breaking: Version 1*, CERN-TH-401, CERN(1964).

Zweig, G. "An SU(3) model for strong interaction symmetry and its breaking: Version 2," in Lichtenberg, D. and Rosen, S. (eds), *Developments in the Quark Theory of Hadrons*, Vol. 1, Hadronic Press (1980), pp. 22–101.

Index